THE THEORY OF GROUPS AND QUANTUM MECHANICS

BY

HERMANN WEYL

PROFESSOR OF MATHEMATICS IN THE UNIVERSITY OF GÖTTINGEN

TRANSLATED FROM THE SECOND (REVISED) GERMAN EDITION BY

H. P. ROBERTSON

ASSOCIATE PROFESSOR OF MATHEMATICAL PHYSICS IN PRINCETON UNIVERSITY

WITH 3 DIAGRAMS

DOVER PUBLICATIONS, INC.

This Dover edition, first published in 1950, is an unabridged and unaltered republication of the English translation of *Gruppentheorie und Quantenmechanik* originally published in 1931.

The work is reprinted by special arrangement with Methuen and Company, Ltd., the original publisher of this translation.

Library of Congress Catalog Card Number: 49-11960

International Standard Book Number

ISBN-13: 978-0-486-60269-1
ISBN-10: 0-486-60269-9

Printed in Canada
60269929 2024
www.doverpublications.com

TO MY FRIEND

WALTER DÄLLENBACH

FROM THE AUTHOR'S PREFACE TO THE FIRST GERMAN EDITION

THE importance of the standpoint afforded by the theory of groups for the discovery of the general laws of quantum theory has of late become more and more apparent. Since I have for some years been deeply concerned with the theory of the representation of continuous groups, it has seemed to me appropriate and important to give an account of the knowledge won by mathematicians working in this field in a form suitable to the requirements of quantum physics. An additional impetus is to be found in the fact that, from the purely mathematical standpoint, it is no longer justifiable to draw such sharp distinctions between finite and continuous groups in discussing the theory of their representations as has been done in the existing texts on the subject. My desire to show how the concepts arising in the theory of groups find their application in physics by discussing certain of the more important examples has necessitated the inclusion of a short account of the foundations of quantum physics, for at the time the manuscript was written there existed no treatment of the subject to which I could refer the reader. In brief this book, if it fulfills its purpose, should enable the reader to learn the essentials of the theory of groups and of quantum mechanics as well as the relationships existing between these two subjects ; the mathematical portions have been written with the physicist in mind, and vice versa. I have particularly emphasized the "reciprocity" between the representations of the symmetric permutation group and those of the complete linear group ; this reciprocity has as yet been unduly neglected in the physical literature, in spite of the fact that it follows most naturally from the conceptual structure of quantum mechanics.

There exists, in my opinion, a plainly discernible parallelism between the more recent developments of mathematics and physics. Occidental mathematics has in past centuries broken away from the Greek view and followed a course which seems to have originated in India and which has been transmitted, with additions, to us by the Arabs; in it the concept of number appears as logically prior to the concepts of geometry. The result of this has been that we have applied this systematically developed number concept to all branches, irrespective of whether it is most appropriate for these particular applications. But the present trend in mathematics is clearly in the direction of a return to the Greek standpoint; we now look upon each branch of mathematics as determining its own characteristic domain of quantities. The algebraist of the present day considers the continuum of real or complex numbers as merely one "field" among many; the recent axiomatic foundation of projective geometry may be considered as the geometric counterpart of this view. This newer mathematics, including the modern theory of groups and "abstract algebra," is clearly motivated by a spirit different from that of "classical mathematics," which found its highest expression in the theory of functions of a complex variable. The continuum of real numbers has retained its ancient prerogative in physics for the expression of physical measurements, but it can justly be maintained that the essence of the new Heisenberg-Schrödinger-Dirac quantum mechanics is to be found in the fact that there is associated with each physical system a set of quantities, constituting a non-commutative algebra in the technical mathematical sense, the elements of which are the physical quantities themselves.

ZURICH, *August, 1928*

AUTHOR'S PREFACE TO THE SECOND GERMAN EDITION

DURING the academic year 1928-29 I held a professorship in mathematical physics in Princeton University. The lectures which I gave there and in other American institutions afforded me a much desired opportunity to present anew, and from an improved pedagogical standpoint, the connection between groups and quanta. The experience thus obtained has found its expression in this new edition, in which the subject has been treated from a more thoroughly *elementary* standpoint. Transcendental methods, which are in group theory based on the calculus of *group characteristics*, have the advantage of offering a rapid view of the subject as a whole, but true understanding of the relationships is to be obtained only by following an explicit elementary development. I may mention in this connection the derivation of the *Clebsch-Gordan* series, which is of fundamental importance for the whole of spectroscopy and for the applications of quantum theory to chemistry, the section on the *Jordan-Hölder* theorem and its analogues, and above all the careful investigation of the connection between the algebra of symmetric transformations and the symmetric permutation group. The reciprocity laws expressing this connection, which were proved by transcendental methods in the first edition, as well as the group-theoretic problem arising from the existence of spin have also been treated from the elementary standpoint. Indeed, the whole of Chapter V—which was, in the opinion of many readers, much too condensed and more difficult to understand than the rest of the book—has been entirely re-written. The algebraic standpoint has been emphasized, in harmony with the recent development of " abstract algebra," which has proved so useful in simplifying and unifying general concepts. It seemed impossible to avoid presenting the principal part of the theory of representations twice; first in Chapter III, where the representations are taken as given and their properties examined, and again in Chapter V, where the method of constructing the representations of a given group and of deducing their properties is developed. But I believe the reader will find this two-fold treatment an advantage rather than a hindrance.

To come to the changes in the more physical portions, in Chapter IV the rôle of the group of virtual rotations of space is more clearly presented. But above all several sections have been added which deal with the energy-momentum theorem of quantum physics and with the quantization of the wave equation in accordance with the recent work of *Heisenberg and Pauli*. This extension already leads so far away from the fundamental purpose of the book that I felt forced to omit the formulation of the quantum laws in accordance with the general theory of relativity, as developed by *V. Fock* and myself, in spite of its desirability for the deduction of the energy-momentum tensor. The fundamental problem of the proton and the electron has been discussed in its relation to the symmetry properties of the quantum laws with respect to the interchange of right and left, past and future, and positive and negative electricity. At present no solution of the problem seems in sight; I fear that the clouds hanging over this part of the subject will roll together to form a new crisis in quantum physics. I have intentionally presented the more difficult portions of these problems of spin and second quantization in considerable detail, as they have been for the most part either entirely ignored or but hastily indicated in the large number of texts which have now appeared on quantum mechanics.

It has been rumoured that the "group pest" is gradually being cut out of quantum physics. This is certainly not true in so far as the rotation and Lorentz groups are concerned; as for the permutation group, it does indeed seem possible to avoid it with the aid of the *Pauli* exclusion principle. Nevertheless the theory must retain the representations of the permutation group as a natural tool in obtaining an understanding of the relationships due to the introduction of spin, so long as its specific dynamic effect is neglected. I have here followed the trend of the times, as far as justifiable, in presenting the group-theoretic portions in as elementary a form as possible. The calculations of perturbation theory are widely separated from these general considerations ; I have therefore restricted myself to indicating the method of attack without either going into details or mentioning the many applications which have been based on the ingenious papers of *Hartree*, *Slater*, *Dirac* and others.

The constants c and h, the velocity of light and the quantum of action, have caused some trouble. The insight into the significance of these constants, obtained by the theory of relativity on the one hand and quantum theory on the other, is most forcibly expressed by the fact that they do not occur in the laws of Nature in a thoroughly systematic development of these theories. But physicists prefer to retain the usual c.g.s. units—principally because they are of the order of magnitude of the physical quantities with which we deal in everyday life. Only a wavering compromise is possible between these practical considerations and the ideal of the systematic theorist ; I initially adopt, with some regret, the current physical usage, but in the course of Chapter IV the theorist gains the upper hand.

An attempt has been made to increase the clarity of the exposition by numbering the formulæ in accordance with the sections to which they belong, by emphasizing the more important concepts by the use of boldface type on introducing them, and by lists of operational symbols and of letters having a fixed significance.

H. WEYL.

GÖTTINGEN, *November, 1930*

TRANSLATOR'S PREFACE

THIS translation was first planned, and in part completed, during the academic year 1928-29, when the translator was acting as assistant to Professor Weyl in Princeton. Unforeseen delays prevented the completion of the manuscript at that time, and as Professor Weyl decided shortly afterward to undertake the revision outlined in the preface above it seemed desirable to follow the revised edition. In the preparation of this manuscript the German has been followed as closely as possible, in the conviction that any alterations would but detract from the elegant and logical treatment which characterizes Professor Weyl's works. While an attempt has been made to follow the more usual English terminology in general, this programme is limited by the fact that the fusion of branches of knowledge which have in the past been so widely separated as the theory of groups and quantum theory can be accomplished only by adapting the existing terminology of each to that of the other ; a minor difficulty of a similar nature is to be found in the fact that the development of "fields" and "algebras" in Chapter V is accomplished in a manner which makes it appear desirable to deviate from the accepted English terminology.

It is a pleasure to express my indebtedness to Professor Weyl for general encouragement and assistance, to Professor R. E. Winger of Union College for the assistance he has rendered in correcting proof and in preparing the index, and to the publishers for their coöperation in adhering as closely as possible to the original typography.

H. P. ROBERTSON

PRINCETON, *September, 1931*

CONTENTS

C. Explicit Algebraic Construction

INTRODUCTION

THE *quantum theory of atomic processes* was proposed by NIELS BOHR in the year 1913, and was based on the atomic model proposed earlier by RUTHERFORD. The deduction of the *Balmer* series for the line spectrum of hydrogen and of the *Rydberg* number from universal atomic constants constituted its first convincing confirmation. This theory gave us the key to the understanding of the regularities observed in optical and X-ray spectra, and led to a deeper insight into the structure of the periodic system of chemical elements. The issue of *Naturwissenschaften*, dedicated to BOHR and entitled "Die ersten zehn Jahre der Theorie von NIELS Bohr über den Bau der Atome" (Vol. **11,** p. 535 (1923)), gives a short account of the successes of the theory at its peak. But about this time it began to become more and more apparent that the BOHR theory was a compromise between the old "classical" physics and a new quantum physics which has been in the process of development since Planck's introduction of energy quanta in 1900. BOHR described the situation in an address on "Atomic Theory and Mechanics" (appearing in *Nature*, **116,** p. 845 (1925)) in the words: "From these results it seems to follow that, in the general problem of the quantum theory, one is faced not with a modification of the mechanical and electrodynamical theories describable in terms of the usual physical concepts, but with an essential failure of the pictures in space and time on which the description of natural phenomena has hitherto been based." The rupture which led to a new stage of the theory was made by HEISENBERG, who replaced Bohr's negative prophecy by a positive guiding principle.

The foundations of the new quantum physics, or at least its more important theoretical aspects, are to be treated in this book. For supplementary references on the physical side, which are urgently required, I name above all the fourth edition of SOMMERFELD's well-known "Atombau und Spektrallinien" (Braunschweig, 1924), or the English translation "Atomic Structure and Spectral Lines" (London, 1923) of the third edition, together with the recent (1929) "Wellenmechanischer Ergänzungsband" or its English translation "Wave Mechanics" (1930). An equivalent original English book is that of RUARK AND UREY, "Atoms, Molecules and Quanta" (New York, 1930), which appears in the "International Series in Physics," edited by RICHTMEYER. I should also recommend GERLACH's short but valuable survey "Experimentelle Grundlagen der Quantentheorie" (Braunschweig, 1921). The spectroscopic data, presented in accordance with the new quantum theory, together with complete references to the literature, are given in the following three volumes of the series "Struktur der Materie," edited by BORN AND FRANCK:—

F. HUND, "Linienspektren und periodisches System der Elemente" (1927);

E. BACK AND A. LANDÉ, "Zeemaneffekt und Multiplettstruktur der Spektrallinien" (1925);

W. GROTRIAN, "Graphische Darstellung der Spektren von Atomen und Ionen mit ein, zwei und drei Valenzelektronen" (1928).

The spectroscopic aspects of the subject are also discussed in PAULING AND GOUDSMIT's recent "The Structure of Line Spectra" (1930), which also appears in the "International Series in Physics."

The development of quantum theory has only been made possible by the enormous *refinement of experimental technique*, which has given us an almost direct insight into atomic processes. If in the following little is said concerning the experimental facts, it should not be attributed to the mathematical haughtiness of the author; to report on these things lies outside his field. Allow me to express now, once and for all, my deep respect for the work of the experimenter and for his fight to wring significant facts from an inflexible Nature, who says so distinctly "No" and so indistinctly "Yes" to our theories.

Our generation is witness to a development of physical knowledge such as has not been seen since the days of KEPLER, GALILEO AND NEWTON, and mathematics has scarcely ever experienced such a stormy epoch. Mathematical thought removes the spirit from its worldly haunts to solitude and renounces the unveiling of the secrets of Nature. But as recompense, mathematics is less bound to the course of worldly events than physics. While the quantum theory can be traced back only as far as 1900, the origin of the *theory of groups* is lost in a past scarcely accessible to history; the earliest works of art show that the symmetry groups of plane figures were even then already known, although the theory of these was only given definite form in the latter part of the eighteenth and in the nineteenth centuries. F. KLEIN considered the group concept as most characteristic of nineteenth century mathematics. Until the present, its most important application to natural science lay in the description of the symmetry of crystals, but it has recently been recognized that group theory is of fundamental importance for quantum physics; it here reveals the essential features which are not contingent on a special form of the dynamical laws nor on special assumptions concerning the forces involved. We may well expect that it is just this part of quantum physics which is most certain of a lasting place. Two groups, *the group of rotations in 3-dimensional space* and *the permutation group*, play here the principal rôle, for the laws governing the possible electronic configurations grouped about the stationary nucleus of an atom or an ion are spherically symmetric with respect to the nucleus, and since the various electrons of which the atom or ion is composed are identical, these possible configurations are invariant under a permutation of the individual electrons. The investigation of groups first becomes a connected and complete theory in *the theory of the representation of groups by linear transformations*, and it is exactly this mathematically most important part which is necessary for an adequate description of the quantum mechanical relations. *All quantum numbers, with the exception of the so-called principal quantum number, are indices characterizing representations of groups.*

This book, which is to set forth the *connection between groups and quanta*, consists of five chapters. The first of these is concerned with *unitary geometry.* It is somewhat distressing that the theory of linear algebras must again and again be developed from the beginning, for the fundamental concepts of this branch of mathematics crop up everywhere in mathematics and physics, and a knowledge of them should be as widely disseminated as the elements of differential calculus. In this chapter many details will be introduced with an eye to future use in the applications; it is to be hoped that in spite of this the simple thread of the argument has remained plainly visible. Chapter II is devoted to preparation on the physical side; only that has been given which seemed to me indispensable for an understanding of the meaning and methods of *quantum theory.* A multitude of physical phenomena, which have already been dealt with by quantum theory, have been omitted. Chapter III develops the elementary portions of *the theory of representations of groups* and Chapter IV *applies them to quantum physics.* Thus mathematics and physics alternate in the first four chapters, but in Chapter V the two are fused together, showing how completely the mathematical theory is adapted to the requirements of quantum physics. In this last chapter *the permutation group and its representations*, together with the groups of linear transformations in an affine or unitary space of an arbitary number of dimensions, will be subjected to a thorough going study.

THE THEORY OF GROUPS AND QUANTUM MECHANICS

CHAPTER I

UNITARY GEOMETRY

§ 1. The n-dimensional Vector Space

THE mathematical field of operation of quantum mechanics, as well as of the theory of the representations of groups, is the multi-dimensional affine or unitary space. The axiomatic method of developing the geometry of such a space is no doubt the most appropriate, but for the sake of clearness I shall at first proceed along purely algebraic lines. I begin with the explanation that a **vector** $\mathfrak{x}$ in the n-dimensional linear space $\mathfrak{R} = \mathfrak{R}_n$ is a set of n ordered numbers $(x_1, x_2 \cdots, x_n)$; vector analysis is the calculus of such ordered sets. The two fundamental operations of the vector calculus are *the multiplication of a vector* $\mathfrak{x}$ *by a number* a and the *addition of two vectors* $\mathfrak{x}$ *and* $\mathfrak{y}$. On introducing the notation

$$\mathfrak{x} = (x_1, x_2, \cdots, x_n), \quad \mathfrak{y} = (y_1, y_2, \cdots, y_n)$$

these operations are defined by the equations

$$a\mathfrak{x} = (ax_1, ax_2, \cdots, ax_n), \quad \mathfrak{x} + \mathfrak{y} = (x_1 + y_1, x_2 + y_2, \cdots, x_n + y_n).$$

The fundamental rules governing these operations of multiplication by a number and addition are given in the following table of axioms, in which small German letters denote arbitrary vectors and small Latin letters arbitrary numbers:

(α) *Addition.*

1. $\mathfrak{a} + \mathfrak{b} = \mathfrak{b} + \mathfrak{a}$ (*commutative law*).
2. $(\mathfrak{a} + \mathfrak{b}) + \mathfrak{c} = \mathfrak{a} + (\mathfrak{b} + \mathfrak{c})$ (*associative law*).
3. $\mathfrak{a}$ *and* $\mathfrak{c}$ *being any two vectors, there exists one and only one vector* $\mathfrak{x}$ *for which* $\mathfrak{a} + \mathfrak{x} = \mathfrak{c}$. *It is called the difference* $\mathfrak{c} - \mathfrak{a}$ *of* $\mathfrak{c}$ *and* $\mathfrak{a}$ (*possibility of subtraction*).

(β) *Multiplication.*

1. $(a + b)\mathfrak{x} = (a\mathfrak{x}) + (b\mathfrak{x})$ (*first distributive law*).
2. $a(b\mathfrak{x}) = (ab)\mathfrak{x}$ (*associative law*).
3. $1\mathfrak{x} = \mathfrak{x}$.
4. $a(\mathfrak{x} + \mathfrak{y}) = (a\mathfrak{x}) + (a\mathfrak{y})$ (*second distributive law*).

The existence of a vector $\mathbf{0} = (0, 0, \cdots, 0)$ with the property

$$\mathfrak{x} + \mathbf{0} = \mathbf{0} + \mathfrak{x} = \mathfrak{x}$$

need not be postulated separately as it follows from the axioms.

Affine vector geometry concerns itself entirely with concepts which are defined in terms of the two fundamental operations with which the axioms (α) and (β) are concerned; we mention a few of the most important. A number of vectors $\mathfrak{a}_1, \mathfrak{a}_2, \cdots, \mathfrak{a}_h$ are said to be *linearly independent* if there exists between them no homogeneous linear relation

$$c_1\mathfrak{a}_1 + c_2\mathfrak{a}_2 + \cdots + c_h\mathfrak{a}_h = 0$$

except the trivial one with coefficients

$$c_1 = 0, \quad c_2 = 0, \quad \cdots, \quad c_h = 0.$$

h such vectors are said to *span* an *h-dimensional* (*linear*) ***sub-space*** $\mathfrak{R}'$ consisting of all vectors of the form

$$\mathfrak{x} = \xi_1\mathfrak{a}_1 + \xi_2\mathfrak{a}_2 + \cdots + \xi_h\mathfrak{a}_h \tag{1.1}$$

where the ξ's are arbitrary numbers. It follows from the fundamental theorem on homogeneous linear equations that there exists a non-trivial homogeneous relation between any $h + 1$ vectors of $\mathfrak{R}'$. The dimensionality h of $\mathfrak{R}'$ can therefore be characterized independently of the basis: every $h + 1$ vectors in $\mathfrak{R}'$ are linearly dependent, but there exist in it h linearly independent vectors. Any such system of h independent vectors $\mathfrak{a}_1, \mathfrak{a}_2, \cdots, \mathfrak{a}_h$ in $\mathfrak{R}'$ can be used as a ***co-ordinate system*** or *basis* in $\mathfrak{R}'$; the coefficients $\xi_1, \xi_2, \cdots, \xi_h$ in the representation (1.1) are then said to be the *components* of $\mathfrak{x}$ in the co-ordinate system $(\mathfrak{a}_1, \mathfrak{a}_2, \cdots, \mathfrak{a}_h)$.

The entire space $\mathfrak{R}$ is n-dimensional, and the vectors

$$\left.\begin{array}{l} \mathfrak{e}_1 = (1, 0, 0, \cdots, 0), \\ \mathfrak{e}_2 = (0, 1, 0, \cdots, 0), \\ \cdot \quad \cdot \quad \cdot \quad \cdot \quad \cdot \quad \cdot \quad \cdot \quad \cdot \\ \mathfrak{e}_n = (0, 0, 0, \cdots, 1) \end{array}\right\} \tag{1.2}$$

define a co-ordinate system in it in which the components of a vector

$$\mathfrak{x} = (x_1, x_2, \cdots, x_n)$$

agree with the " absolute components " x_i :

$$\mathfrak{x} = x_1\mathfrak{e}_1 + x_2\mathfrak{e}_2 + \cdots + x_n\mathfrak{e}_n.$$

From the standpoint of affine geometry, however, the " absolute co-ordinate system " (1.2) has no preference over any other which consists of n independent vectors of $\mathfrak{R}$. We now add to the previous axioms, which did not concern themselves with the dimensionality n, the following *dimensionality axiom :*

(γ) *The maximum number of linearly independent vectors in* $\mathfrak{R}$ *is* n.

These axioms (α), (β), and (γ) suffice for a complete formulation of vector calculus, for if $\mathfrak{e}_1, \mathfrak{e}_2, \cdots, \mathfrak{e}_n$ are any n independent vectors and $\mathfrak{x}$ is any other vector there must necessarily exist a linear dependence

$$a\mathfrak{x} + a_1\mathfrak{e}_1 + a_2\mathfrak{e}_2 + \cdots + a_n\mathfrak{e}_n = 0$$

between them. Since not all the coefficients may vanish we must in particular have $a \neq 0$, and consequently any vector $\mathfrak{x}$ can be expressed as a linear combination

$$\mathfrak{x} = x_1\mathfrak{e}_1 + x_2\mathfrak{e}_2 + \cdots + x_n\mathfrak{e}_n \tag{1.3}$$

of the " fundamental vectors " $\mathfrak{e}_1, \mathfrak{e}_2, \cdots, \mathfrak{e}_n$. We specify $\mathfrak{x}$ by the set $(x_1, x_2, \cdots, x_n)$ of components in this co-ordinate system. In accordance with axioms (α) and (β) for addition and multiplication we then have for any two vectors (1.3) and $\mathfrak{y}$

$$a\mathfrak{x}=(ax_1)\mathfrak{e}_1+\cdots+(ax_n)\mathfrak{e}_n,\ \mathfrak{x}+\mathfrak{y}=(x_1+y_1)\mathfrak{e}_1+\cdots+(x_n+y_n)\mathfrak{e}_n,$$

and we arrive at the definitions from which we started. The only—but important—difference between the arithmetic and the axiomatic treatment is that in the former the absolute co-ordinate system (1.2) is given the preference over any other, whereas in the latter treatment no such distinction is made.

Given any system of vectors, all vectors $\mathfrak{x}$ which are obtained, as (1.1), by linear combinations of a finite number of vectors $\mathfrak{a}_1, \mathfrak{a}_2, \cdots, \mathfrak{a}_h$ of the system constitute a (linear) sub-space—the sub-space " *spanned* " by the vectors $\mathfrak{a}$.

$\mathfrak{R}$ is said to be *decomposed* or *reduced* into two linear sub-spaces $\mathfrak{R}', \mathfrak{R}''$ ($\mathfrak{R} = \mathfrak{R}' + \mathfrak{R}''$) if an arbitrary vector $\mathfrak{x}$ can be expressed uniquely as the sum of a vector $\mathfrak{x}'$ of $\mathfrak{R}'$ and a vector $\mathfrak{x}''$ of $\mathfrak{R}''$. A co-ordinate system in $\mathfrak{R}'$ and a co-ordinate system in $\mathfrak{R}''$ constitute together a co-ordinate system for the entire space $\mathfrak{R}$; this co-ordinate system in $\mathfrak{R}$ is " *adapted* " to the decomposition $\mathfrak{R}' + \mathfrak{R}''$. The sum $n' + n''$ of the dimensionalities of $\mathfrak{R}'$ and $\mathfrak{R}''$ is equal to n, the dimensionality of $\mathfrak{R}$.

Conversely, if the sub-spaces $\mathfrak{R}'$, $\mathfrak{R}''$ have no vector except 0 in common, and if the sum of their dimensionalities is n, then $\mathfrak{R} = \mathfrak{R}' + \mathfrak{R}''$.

$\mathfrak{R}'$ being an n-dimensional sub-space, two vectors $\mathfrak{x}$ and $\mathfrak{y}$ are said to be *congruent modulo* $\mathfrak{R}'$:

$$\mathfrak{x} \equiv \mathfrak{y} \ (\text{mod. } \mathfrak{R}'),$$

if their difference lies in $\mathfrak{R}'$. Congruence satisfies the axioms postulated of any relation of equality : every vector is congruent to itself ; if $\mathfrak{x} \equiv \mathfrak{y}$ (mod. $\mathfrak{R}'$) then $\mathfrak{y} \equiv \mathfrak{x}$ (mod. $\mathfrak{R}'$) ; if $\mathfrak{x} \equiv \mathfrak{y}$ (mod. $\mathfrak{R}'$) and $\mathfrak{y} \equiv \mathfrak{z}$ (mod. $\mathfrak{R}'$), then $\mathfrak{x} \equiv \mathfrak{z}$ (mod. $\mathfrak{R}'$). It is therefore permissible to consider vectors which are congruent mod. $\mathfrak{R}'$ as differing in no wise from one another ; by this abstraction, which we call ***projection*** *with respect to* $\mathfrak{R}'$, the n-dimensional space $\mathfrak{R}$ gives rise to an $(n - n')$-dimensional space $\boldsymbol{\mathfrak{R}}$. $\boldsymbol{\mathfrak{R}}$ is also a vector space, for from

$$\mathfrak{x}_1 \equiv \mathfrak{x}_2, \quad \mathfrak{y}_1 \equiv \mathfrak{y}_2 \ (\text{mod. } \mathfrak{R}')$$

follow the relations

$$a\mathfrak{x}_1 \equiv a\mathfrak{x}_2, \quad \mathfrak{x}_1 + \mathfrak{y}_1 \equiv \mathfrak{x}_2 + \mathfrak{y}_2 \ (\text{mod. } \mathfrak{R}').$$

The operations of multiplication by a number and addition can therefore be considered ones which operate directly on the vectors $\boldsymbol{\mathfrak{x}}$ of $\boldsymbol{\mathfrak{R}}$. All vectors $\mathfrak{x}$ of $\mathfrak{R}$ which are congruent mod. $\mathfrak{R}'$ give rise to the same vector $\boldsymbol{\mathfrak{x}}$ of $\boldsymbol{\mathfrak{R}}$. If $\mathfrak{R}'$ is one-dimensional and is spanned by $\mathfrak{e}$ the above process is the familiar one of parallel projection in the direction of $\mathfrak{e}$; it is not necessary to give an $(n - 1)$-dimensional sub-space of $\mathfrak{R}$ *on to* which the projection is made.

If $\mathfrak{a}$ is a non-null vector, all vectors $\mathfrak{x}$ which arise by multiplying $\mathfrak{a}$ by a number are said to lie on the same ***ray*** as $\mathfrak{a}$. Two non-null vectors determine the same ray when, and only when, one is a multiple of the other. In a given co-ordinate system the *vector* $\mathfrak{a}$ is characterized by its components $a_1, a_2, \cdots, a_n$ whereas the *ray* $\mathfrak{a}$ is characterized by their ratios $a_1 . a_2 : \cdots : a_n$; these ratios have meaning only when the components of $\mathfrak{a}$ do not all vanish, i.e. only when $\mathfrak{a} \neq 0$.

The transition from one co-ordinate system $\mathfrak{e}_i$ to another $\mathfrak{e}_i'$ is accomplished by expressing the new co-ordinate vectors $\mathfrak{e}_i'$ in terms of the old :

$$\mathfrak{e}_k' = \sum_{i=1}^{n} a_{ik}\, \mathfrak{e}_i.$$

If x_i, x_i' are the components of an arbitrary vector $\mathfrak{x}$ in the old and in the new co-ordinate systems, respectively, then

$$\mathfrak{x} = \sum_i x_i \mathfrak{e}_i = \sum_k x_k' \mathfrak{e}_k',$$

from which the law of transformation

$$x_i = \sum_{k=1}^{n} a_{ik} x_k' \tag{1.4}$$

follows. The requirement that the co-ordinate vectors $\mathfrak{e}_k'$ also be linearly independent is expressed arithmetically by the non-vanishing of the determinant of the coefficients a_{ik}. The components of vectors $\mathfrak{x}, \mathfrak{y}, \cdots$ in $\mathfrak{R}$ undergo the same transformation on transition to the new co-ordinate system $\mathfrak{e}_i'$ and are said *to transform cogrediently.*

§ 2. Linear Correspondences. Matrix Calculus

The formula (1.4) can, however, be otherwise interpreted; it is the expression of a ***linear*** or ***affine correspondence*** or ***mapping*** of the space $\mathfrak{R}$ on itself. But for this purpose it will be found more convenient to interchange the rôles of the accented and the unaccented co-ordinates. On employing a definite co-ordinate system $\mathfrak{e}_i$, the equation

$$x_i' = \sum_{k=1}^{n} a_{ik} x_k \tag{2.1}$$

associates with an arbitrary vector $\mathfrak{x}$ with components x_i a vector $\mathfrak{x}'$ with components x_i'. This correspondence $A : \mathfrak{x} \rightarrow \mathfrak{x}'$ of $\mathfrak{R}$ on itself can be characterized as *linear* by the two assertions: if $\mathfrak{x}$, $\mathfrak{y}$ go over into $\mathfrak{x}'$, $\mathfrak{y}'$, then $a\mathfrak{x}$ goes over into $a\mathfrak{x}'$ and $\mathfrak{x} + \mathfrak{y}$ into $\mathfrak{x}' + \mathfrak{y}'$. Linear correspondences therefore leave all affine relations unaltered; hence their prominence in the theory of affine geometry. In order to show that these two conditions fully determine the linear correspondence (2.1), consider the following: if a correspondence A which satisfies these conditions sends the fundamental vector $\mathfrak{e}_k$ over into

$$\mathfrak{e}_k' = \sum_i a_{ik} \mathfrak{e}_i \tag{2.2}$$

then, in consequence of the above requirements,

$$\mathfrak{x} = x_1 \mathfrak{e}_1 + \cdots + x_n \mathfrak{e}_n$$

goes over into

$$\mathfrak{x}' = x_1 \mathfrak{e}_1' + \cdots + x_n \mathfrak{e}_n'.$$

On substituting (2.2) in this equation we see that the new vector $\mathfrak{x}'$ has in the co-ordinate system $\mathfrak{e}_i$ the components x_i' obtained from the components x_i of $\mathfrak{x}$ by means of (2.1). It has become customary in quantum physics to call the linear correspondences of a vector space $\mathfrak{R}$ ***operators*** which operate on the arbitrary vector $\mathfrak{x}$ of $\mathfrak{R}$.

Let A, B be two linear correspondences, the first of which sends the arbitrary vector $\mathfrak{x}$ over into $\mathfrak{x}' = A\mathfrak{x}$, while the second sends $\mathfrak{x}'$ into $\mathfrak{x}'' = B\mathfrak{x}' = B(A\mathfrak{x})$. The resultant correspondence C, which carries $\mathfrak{x}$ directly into $\mathfrak{x}''$, is also linear and is denoted by (BA) (*to be read from right to left!*) :

$$(BA)\mathfrak{x} = B(A\mathfrak{x}).$$

This "*multiplication*" satisfies laws which are similar to those of multiplication of ordinary numbers; in particular, the associative law

$$C(BA) = (CB)A$$

is here valid, but the commutative law is not—in general $AB \neq BA$. The "1" in this domain, which we here denote by $\mathbf{1}$, is the identity, i.e. that correspondence which associates every vector $\mathfrak{x}$ with itself: $\mathfrak{x} \to \mathfrak{x}$. Hence

$$A\mathbf{1} = \mathbf{1}A = A.$$

The correspondence A is then and only then reversible in case it is non-degenerate, i.e. if it carries no non-vanishing vector into the vector 0, or if distinct vectors are always carried over into distinct ones. The algebraic condition for this is the non-vanishing of the determinant $|a_{ik}| = \det A$; there then exists the inverse correspondence A^{-1}:

$$AA^{-1} = A^{-1}A = \mathbf{1}.$$

The multiplication theorem for determinants states that

$$\det (BA) = \det B \cdot \det A.$$

Not only can we "multiply" two correspondences, we can also "add" them. This concept of *addition* arises quite naturally: if the arbitrary vector $\mathfrak{x}$ is sent over into $\mathfrak{x}_1'$ by A and into $\mathfrak{x}_2'$ by B, then that correspondence which sends $\mathfrak{x}$ into $\mathfrak{x}_1' + \mathfrak{x}_2'$ is also linear and is denoted by $A + B$:

$$(A + B)\mathfrak{x} = A\mathfrak{x} + B\mathfrak{x}.$$

We may also introduce multiplication by an arbitrary number a: aA is that correspondence which sends $\mathfrak{x}$ into $a(A\mathfrak{x})$. Addition and multiplication by a number obey the same laws as the analogous operations on vectors. Addition is commutative, and has as its inverse subtraction. The rôle of 0 is played by the correspondence $\mathbf{0}$ which transforms every vector $\mathfrak{x}$ into the vector 0. Addition obeys the distributive law with respect to multiplication:

$$(A+B)C = AC + BC, \quad C(A+B) = CA + CB,$$
$$(aA)C = a(AC), \quad C(aA) = a(CA).$$

Before proceeding to the arithmetical expression of these operations in a given co-ordinate system, we consider another natural generalization. We can map an m-dimensional vector space $\mathfrak{R}$ linearly on an n-dimensional space $\mathfrak{S}$; this is accomplished when with each vector $\mathfrak{x}$ of $\mathfrak{R}$ a vector $\mathfrak{y}$ of $\mathfrak{S}$ is associated in such a way $\mathfrak{x} \rightarrow \mathfrak{y}$ that from $\mathfrak{x}_1 \rightarrow \mathfrak{y}_1$, $\mathfrak{x}_2 \rightarrow \mathfrak{y}_2$ it follows that

$$a\mathfrak{x}_1 \rightarrow a\mathfrak{y}_1, \quad \mathfrak{x}_1 + \mathfrak{x}_2 \rightarrow \mathfrak{y}_1 + \mathfrak{y}_2.$$

Such a correspondence $A : \mathfrak{x} \rightarrow \mathfrak{y}$ is expressed by equations of the form

$$y_k = \sum_{i=1}^{m} a_{ki} x_i \quad (k = 1, 2, \cdots, n) \tag{2.3}$$

where $x_1, \cdots, x_m$ are the components of $\mathfrak{x}$ in a given co-ordinate system in the space $\mathfrak{R}$ and $y_1, \cdots, y_n$ have the corresponding interpretation in $\mathfrak{S}$. With this correspondence A there is associated the matrix

$$\begin{Vmatrix} a_{11} & a_{12} & \cdots & a_{1m} \\ a_{21} & a_{22} & \cdots & a_{2m} \\ \cdot & \cdot & \cdot & \cdot \\ a_{n1} & a_{n2} & \cdots & a_{nm} \end{Vmatrix}$$

with n rows and m columns, and which we also denote by the same letter A. The first index indicates the row and the second the column to which a_{ki} belongs. We can also add correspondences of the *same* space $\mathfrak{R}$ on the *same* space $\mathfrak{S}$. Addition and multiplication by a number is accomplished on matrices by subjecting their $n \cdot m$ components to these operations: if

$$A = \| a_{ki} \| \quad \text{and} \quad B = \| b_{ki} \|$$

then

$$aA = \| a \cdot a_{ki} \|, \quad A + B = \| a_{ki} + b_{ki} \|.$$

If we have a third (p-dimensional) vector space $\mathfrak{T}$, the consecutive application of the correspondences $A : \mathfrak{x} \rightarrow \mathfrak{y}$ of $\mathfrak{R}$ on $\mathfrak{S}$ and $B : \mathfrak{y} \rightarrow \mathfrak{z}$ of $\mathfrak{S}$ on $\mathfrak{T}$ gives rise to the correspondence $C = BA : \mathfrak{x} \rightarrow \mathfrak{z}$ of $\mathfrak{R}$ on $\mathfrak{T}$. This composition is expressed in terms of matrix components by the law

$$\boxed{c_{li} = \sum_{k=1}^{n} b_{lk} a_{ki}} \quad \begin{pmatrix} l = 1, 2, \cdots, p \\ i = 1, 2, \cdots, m \end{pmatrix}. \tag{2.4}$$

B has p rows and n columns and A n rows and m columns; the composition of matrices is possible when the first factor B has the same number of columns as the second factor A has rows. The component or element c_{li}, which is found at the intersection of the lth row and the ith column, is formed in accordance with (2.4) from the components in the lth row of B and the ith column of A. An important special case is that in which $\mathfrak{T}$ is the same space as $\mathfrak{R}$; A is then a correspondence of $\mathfrak{R}$ on $\mathfrak{S}$, B of $\mathfrak{S}$ on $\mathfrak{R}$. Already here concepts of the theory of groups play an important

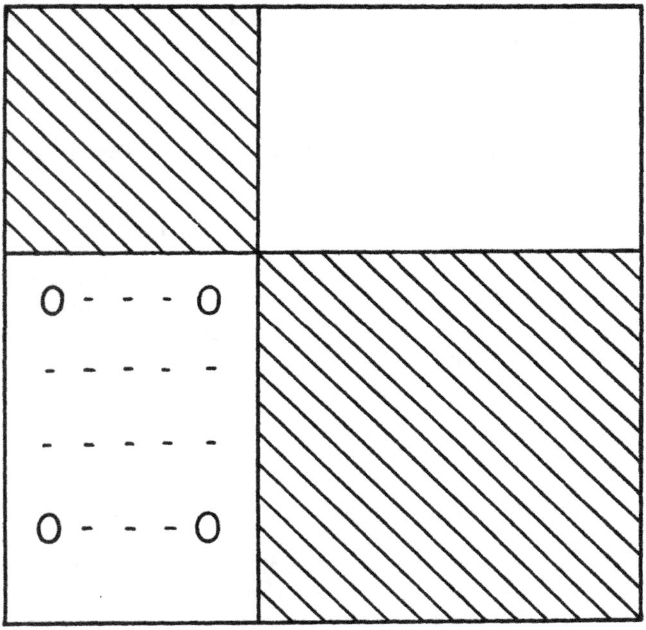

FIG. 1.

rôle; on beginning Chapter III, which deals with the theory of groups, the reader should return to the matter here discussed as an illustration.

The matrix calculus allows us to express the formulæ for a linear correspondence, such as (2.3), in an abbreviated form. We do this by denoting by x that matrix whose only column consists of the vector components $x_1, x_2, \cdots, x_m$; similarly for y. In accordance with the rule (2.4) for the composition of matrices, equations (2.3) can be written

$$y = Ax. \tag{2.5}$$

This form is particularly useful in examining the effect on the matrix A of a linear correspondence of a space $\mathfrak{R}$ on a space $\mathfrak{S}$ when the original co-ordinate systems are replaced by new ones. If this change of co-ordinates is effected by the transformations

$$x_i = \sum_j s_{ij} x_j' \quad \text{or} \quad x = Sx' \text{ in } \mathfrak{R},$$

$$y_k = \sum_h t_{kh} y_h' \quad \text{or} \quad y = Ty' \text{ in } \mathfrak{S},$$

then from (2.5)

$$Ty' = ASx' \qquad \text{or} \quad y' = (T^{-1}AS)x'.$$

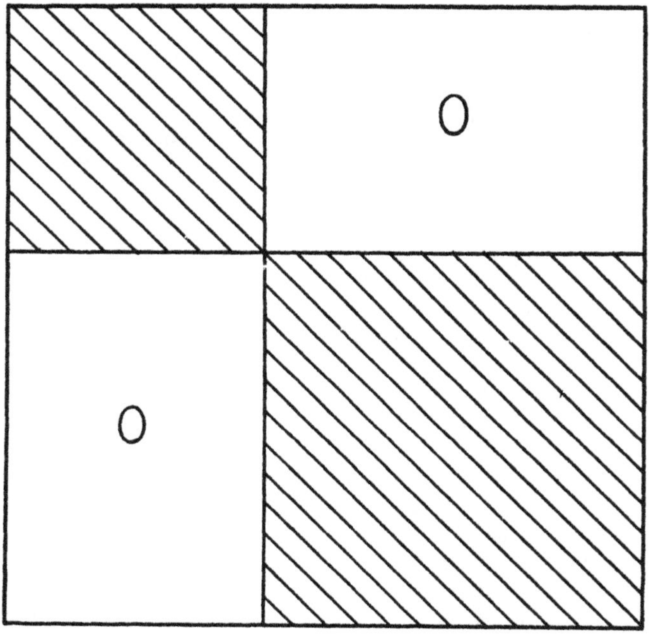

FIG. 2.

The same correspondence in the new co-ordinates is therefore expressed by the matrix

$$A' = T^{-1}AS. \tag{2.6}$$

Let us now return to the linear correspondence A of a space $\mathfrak{R}$ on to itself. If $\mathfrak{R}'$ is a linear n'-dimensional *sub-space* of $\mathfrak{R}'$ we say that *A leaves $\mathfrak{R}'$* ***invariant*** if it carries any vector of $\mathfrak{R}'$, over into a vector of $\mathfrak{R}'$. If the co-ordinate system is so chosen that the first n' fundamental vectors lie in $\mathfrak{R}'$, the matrix of a correspondence which leaves $\mathfrak{R}'$ invariant will assume the form given by Fig. 1. All elements in the rectangle of n' columns and $n - n'$ rows denoted by zeros in Fig. 1, vanish. A contains a correspondence of $\mathfrak{R}'$ on to itself and at the same time a correspondence of the space $\mathfrak{R}$, arising by projecting $\mathfrak{R}$ with respect to $\mathfrak{R}'$, on to itself. The matrices of these correspondences consist in the shaded squares. If $\mathfrak{R}$ is decomposed into $\mathfrak{R}_1 + \mathfrak{R}_2$ ($n_1 + n_2 = n$), and if the correspondence A leaves both sub-spaces $\mathfrak{R}_1$ and $\mathfrak{R}_2$ invariant, then A is ***completely reduced*** into a correspondence of $\mathfrak{R}_1$ on itself and a correspondence of $\mathfrak{R}_2$ on to itself. If the co-ordinate system is adapted to the decomposition $\mathfrak{R}_1 + \mathfrak{R}_2$, the matrix A is completely reduced into two square matrices arranged along the principal diagonal as in Fig. 2. The unshaded rectangles are empty—the elements situated in these portions are all zero.

Let the n-dimensional linear space $\mathfrak{R}$ be decomposed into sub-spaces $\mathfrak{R}_1 + \mathfrak{R}_2 + \cdots$, $\mathfrak{R}_\alpha$ having the dimensionality n_α; n is then equal to the sum $n_1 + n_2 + \cdots$. Any vector $\mathfrak{x}$ can then be written uniquely as the sum of components $\mathfrak{x}_1 + \mathfrak{x}_2 + \cdots$ which lie in the sub-spaces $\mathfrak{R}_1, \mathfrak{R}_2, \cdots$. The association $\mathfrak{x} \to \mathfrak{x}_\alpha$ is a linear correspondence E_α of $\mathfrak{R}$ on to $\mathfrak{R}_\alpha$. Given a correspondence $A : \mathfrak{x} \to \mathfrak{x}'$ of $\mathfrak{R}$ on to itself, we consider that linear correspondence $[A]_{\alpha\beta}$ which carries an arbitrary vector $\mathfrak{x}$ of $\mathfrak{R}_\beta$ over into the component $\mathfrak{x}_\alpha'$ in $\mathfrak{R}_\alpha$ of $\mathfrak{x}'$. We call $[A]_{\alpha\beta}$ *the portion of A in which $\mathfrak{R}_\alpha$ intersects $\mathfrak{R}_\beta$*. This terminology arises from the matrix representation of A; on adapting the co-ordinate system to the decomposition $\mathfrak{R}_1 + \mathfrak{R}_2 + \cdots$ the set of variables x_i, or rather their indices i which number the rows and columns of the matrix, is broken up into segments of lengths n_α ($\alpha = 1, 2, \cdots$). The matrix A is thereby divided into the single rectangles $[A]_{\alpha\beta}$ in which the αth set of rows intersects the βth set of columns, and which consist of $n_\alpha \cdot n_\beta$ elements.

If A is the matrix of a correspondence of $\mathfrak{R}$ on to itself in a given co-ordinate system, and A' its matrix in a co-ordinate system obtained from the first by means of the reversible transformation S, then in accordance with (2.6)

$$A' = S^{-1}AS. \tag{2.7}$$

The search for an invariantive characterization of correspondences may be formulated algebraically: to find expressions which are so formed from the components of an arbitrary matrix that they assume the same value for equivalent matrices, i.e. for matrices A, A' between which a relation (2.7) exists. The way in which this can be accomplished is indicated by the related problem of finding a vector $\mathfrak{x} \neq 0$ which is transformed into a multiple $\lambda\mathfrak{x}$ of itself under the influence of A. The column x of the components of $\mathfrak{x}$ must then satisfy the equation

$$\lambda x = Ax, \quad \text{or} \quad (\lambda\mathbf{1} - A)x = 0.$$

But n linear homogeneous equations in n unknowns have a non-vanishing solution only if their determinant vanishes; the multiplier λ is therefore necessarily a root of the "*characteristic polynomial*"

$$f(\lambda) = \det(\lambda\mathbf{1} - A) \tag{2.8}$$

of A. *This polynomial is an invariant* in the above sense, for from (2.7) or $SA' = AS$ it follows that

$$S(\lambda\mathbf{1} - A') = (\lambda\mathbf{1} - A)S,$$

whence by the theorem concerning the multiplication of determinants

$$\det S \cdot \det(\lambda\mathbf{1} - A') = \det(\lambda\mathbf{1} - A) \cdot \det S.$$

Since the determinant of the reversible transformation S cannot vanish, we can divide by it and obtain the required identity

$$|\lambda\mathbf{1} - A'| \equiv |\lambda\mathbf{1} - A|.$$

The characteristic polynomial is of degree n in λ:

$$f(\lambda) = \lambda^n - s_1\lambda^{n-1} + \cdots \pm s_n$$

whose coefficients, certain integral functions of the elements a_{ik}, are invariants of the correspondence A. The "*norm*" s_n is merely the determinant of A. The first coefficient s_1, the ***trace***

$$s_1 = a_{11} + a_{22} + \cdots + a_{nn} = \text{tr}A \tag{2.9}$$

is of more importance, as it depends linearly on the a_{ik}:

$$\text{tr}(A_1 + A_2) = \text{tr}A_1 + \text{tr}A_2.$$

If A is a linear correspondence of the m-dimensional vector space $\mathfrak{R}$ on the n-dimensional space $\mathfrak{S}$, and B is conversely a linear correspondence of $\mathfrak{S}$ on $\mathfrak{R}$, then we can build the correspondences BA of $\mathfrak{R}$ on to itself and AB of $\mathfrak{S}$ on to itself. These two correspondences have the same trace

$$\text{tr}(BA) = \text{tr}(AB) \tag{2.10}$$

for, in accordance with the rule of composition (2.4) and the definition (2.9) we have

$$\text{tr}(BA) = \sum_{i,k} b_{ik}\, a_{ki}, \quad \text{tr}(AB) = \sum_{i,k} a_{ki}\, b_{ik}$$

where i runs from 1 to m and k from 1 to n. The special case in which A and B are both correspondences of $\mathfrak{R}$ on to itself naturally deserves particular consideration.

§ 3. The Dual Vector Space

A function $L(\mathfrak{x})$ of the arbitrary vector $\mathfrak{x}$ of the form

$$\alpha_1 x_1 + \alpha_2 x_2 + \cdots + \alpha_n x_n \tag{3.1}$$

is called a *linear form*. This concept is invariant in the sense of affine geometry: it can be defined by means of the functional properties

$$L(\alpha\mathfrak{x}) = \alpha \cdot L(\mathfrak{x}), \quad L(\mathfrak{x} + \mathfrak{y}) = L(\mathfrak{x}) + L(\mathfrak{y}).$$

It is obvious that the expression (3.1) has these properties, and conversely, on introducing a co-ordinate system $\mathfrak{e}_i$ and setting $\mathfrak{x} = \Sigma x_i \mathfrak{e}_i$, it follows that

$$L(\mathfrak{x}) = \sum_i x_i L(\mathfrak{e}_i) = \sum_i \alpha_i x_i; \quad \alpha_i = L(\mathfrak{e}_i).$$

On going over to another co-ordinate system such that the components x_i of an arbitrary vector $\mathfrak{x}$ undergo the transformation (1.4), the linear form becomes

$$\Sigma \alpha_i x_i = \Sigma \alpha_i' x_i'$$

the coefficients α_i' of which are related to the original α_i by the equations

$$\alpha_k' = \sum_i a_{ik} \cdot \alpha_i.$$

The coefficients α_i of a linear form are said to transform ***contragrediently*** *to the variables x_i.*

It is, however, not necessary to consider the α_i as constants and the x_i as variables. When the α_i do not all vanish the equation $L(\mathfrak{x}) = 0$ defines a "plane," i.e. an $(n - 1)$-dimensional sub-space; a vector $\mathfrak{x}$ lies in the plane if its components satisfy this equation. But on the other hand we can ask for the equation of all planes which pass through a given non-vanishing vector $\mathfrak{x}^0$; the $x_i = x_i^0$ are then constants and the α_i variables. It is therefore most appropriate to consider the two sets $(x_1, x_2, \cdots, x_n)$, $(\alpha_1, \alpha_2, \cdots, \alpha_n)$ in parallel.

We therefore introduce in addition to the space $\mathfrak{R}$ a second n-dimensional vector space, the ***dual*** space $\mathbf{P}$. From the components $(\xi_1, \xi_2, \cdots, \xi_n)$ of a vector ξ of $\mathbf{P}$ and a vector $(x_1, x_2, \cdots, x_n)$ of $\mathfrak{R}$ we can construct the *inner* or *scalar product*

$$\xi_1 x_1 + \xi_2 x_2 + \cdots + \xi_n x_n \tag{3.2}$$

This product has, by definition, an invariantive significance, for when $\mathfrak{R}$ is referred to a new co-ordinate system by means of a transformation of the x_i the variables ξ_i of the dual space P undergo the contragredient transformation. This dual space is in fact introduced in order to enable us to associate a contragredient transformation with each one-to-one transformation. To repeat, two linear reversible transformations

$$x = Ax', \quad \xi = \mathsf{A}\xi' \tag{3.3}$$

are contragredient with respect to each other if they leave (3.2) unaltered:

$$\xi_1 x_1 + \xi_2 x_2 + \cdots + \xi_n x_n = \xi_1' x_1' + \xi_2' x_2' + \cdots + \xi_n' x_n'. \tag{3.4}$$

A vector $\mathfrak{x}$ of $\mathfrak{R}$ and a vector ξ of P are said to be *in involution* when their product (3.2) vanishes. A *ray* in $\mathfrak{R}$ determines a *plane* in P, i.e. the plane consisting of the vectors which are in involution with the given ray, and conversely. Duality is a reciprocal relationship.†

The ***dual*** or ***transposed*** matrix A^* of a matrix $A = \|a_{ki}\|$ is obtained by interchanging the rows and columns of A. $A^* = \|a^*_{ik}\|$ is therefore defined by $a^*_{ik} = a_{ki}$, and has m rows and n columns. We shall always employ the asterisk to indicate this process. And what is its geometrical interpretation? Let $\mathfrak{R}$ be an m-dimensional, $\mathfrak{S}$ an n-dimensional, vector space; $A : \mathfrak{x} \to \mathfrak{y}$ a linear correspondence of $\mathfrak{R}$ on $\mathfrak{S}$, specified in terms of given co-ordinate systems in $\mathfrak{R}$ and $\mathfrak{S}$ by the matrix A:

$$y_k = \sum_i a_{ki} x_i,$$

and let P, Σ be the dual spaces. The product

$$\sum_k \eta_k y_k = \sum_{k,i} a_{ki} \eta_k x_i \left(= \sum_i \xi_i x_i\right),$$

where η is an arbitrary vector of Σ with components η_k, has then an invariantive significance. A bilinear form which depends linearly on a vector η of Σ and a vector $\mathfrak{x}$ of $\mathfrak{R}$ is therefore invariantively associated with a linear correspondence of $\mathfrak{R}$ on $\mathfrak{S}$, and conversely. This gives rise, as the expression of the bilinear form given in parentheses shows, to a correspondence

$$\eta \to \xi : \quad \xi_i = \sum_k a_{ki} \eta_k$$

of Σ on P, i.e. the dual A^* of A. The reciprocal relation existing between the correspondence A and its dual A^* may be expressed

† In the theory of relativity it is usual to call vectors in $\mathfrak{R}$ and P *contravariant* and *covariant vectors*, respectively.

as follows: if $\mathfrak{x}$ is an arbitrary vector in $\mathfrak{R}$ and η is an arbitrary vector in Σ, then the product of the vectors $A\mathfrak{x}$ and η is equal to the product of $\mathfrak{x}$ and $A^*\eta$. The dual correspondences obey the linear laws

$$(A_1 + A_2)^* = A_1^* + A_2^*, \ (aA)^* = a \cdot A^*.$$

If A is a correspondence of $\mathfrak{R}$ on $\mathfrak{S}$ and B a correspondence of $\mathfrak{S}$ on $\mathfrak{T}$, then since

$$(BA)^* = A^*B^* \tag{3.5}$$

BA maps $\mathfrak{R}$ linearly on $\mathfrak{T}$, and A^*B^* maps the dual space T of $\mathfrak{T}$ on the dual P of $\mathfrak{R}$.

We have agreed once and for all to consider the set $x_1, x_2, \cdots, x_n$ of components of a vector $\mathfrak{x}$ as a column; the inner product of the vector $\mathfrak{x}$ in $\mathfrak{R}$ with the vector ξ in P can therefore be written in matrix notation as ξ^*x or $x^*\xi$. The transformations (3.3), from the first of which it follows that $x^* = x'^*A^*$, are consequently contragredient to one another if

$$A^*\mathsf{A} = 1 \quad \text{or} \quad \mathsf{A} = (A^*)^{-1}, \tag{3.6}$$

and we have arrived at an explicit expression for the contragredient transformation.

Let $\mathfrak{R}'$ be an n'-dimensional sub-space of $\mathfrak{R} = \mathfrak{R}_n$. All vectors of P which are in involution with the totality of vectors of $\mathfrak{R}'$ obviously constitute, in consequence of the simplest theorems on linear homogeneous equations, an $(n - n')$-dimensional sub-space P' of P. And from this we are led immediately to the result that *if a correspondence A of $\mathfrak{R}$ on itself leaves the sub-space $\mathfrak{R}'$ invariant, then the dual correspondence A^* of P on itself leaves the associated sub-space P' invariant.*

Let $\mathfrak{R}$ be decomposed into two or more sub-spaces $\mathfrak{R}_1 + \mathfrak{R}_2 + \cdots$ of dimensionalities $n_1, n_2, \cdots$, and let the sub-space of P which consists of all vectors in involution with all vectors of $\mathfrak{R}_2 + \mathfrak{R}_3 + \cdots$ be denoted by P_1, the dimensionality of which is also n_1. Defining P_2, P_3 analogously, we arrive at the decomposition $\mathsf{P} = \mathsf{P}_1 + \mathsf{P}_2 + \cdots$, for the sum of a vector of P_1 a vector of P_2, etc., can only be zero when each of the individual summands vanishes. In order to prove this latter statement, we note that if the sum is 0 then the first summand belongs to P_1 as well as to $\mathsf{P}_2 + \mathsf{P}_3 + \cdots$, i.e. it is in involution with all the vectors of $\mathfrak{R}_2 + \mathfrak{R}_3 + \cdots$ as well as with all those of $\mathfrak{R}_1$, and is therefore in involution with all the vectors of $\mathfrak{R}$. But this is only possible if this first, and therefore any, summand is zero. P_1 can be considered as the space dual to $\mathfrak{R}_1$, for if $\mathfrak{x}$ is an arbitrary vector in $\mathfrak{R}_1$ and η a vector in P

with components $\eta^{(\alpha)}$ in the various P_α, then the product of $\mathfrak{x}$ and η is equal to the product of $\mathfrak{x}$ and $\eta^{(1)}$.

If a correspondence A of $\mathfrak{R}$ on itself leaves the n'-dimensional sub-space $\mathfrak{R}'$ invariant, then the $(n - n')$-dimensional sub-space P' is invariant under the dual correspondence A^* of P on itself. If $\mathfrak{R}$ is decomposed into $\mathfrak{R}_1 + \mathfrak{R}_2 + \cdots$ and if A leaves each of the sub-spaces $\mathfrak{R}_\alpha$ invariant, then A^* leaves each of the sub-spaces P_α invariant. If A is any correspondence in $\mathfrak{R}$ and $[A]_{\alpha\beta}$ that portion in which $\mathfrak{R}_\alpha$ intersects $\mathfrak{R}_\beta$, then the portion $[A^*]_{\beta\alpha}$ of A^* in which P_β intersects P_α is dual to $[A]_{\alpha\beta}$:

$$[A^*]_{\beta\alpha} = [A]^*{}_{\alpha\beta}. \tag{3.7}$$

$[A]_{\alpha\beta}$ maps $\mathfrak{R}_\beta$ on $\mathfrak{R}_\alpha$ and $[A^*]_{\beta\alpha}$ maps the dual space P_α on P_β.

All these results are conceptually evident, but can be seen even more readily directly from the matrices on adapting the co-ordinate system to the decomposition $\mathfrak{R}_1 + \mathfrak{R}_2 + \cdots$.

§ 4. Unitary Geometry and Hermitian Forms

The *metric* is introduced into affine geometry by means of a new fundamental concept: *the absolute magnitude of a vector.* In Euclidean geometry the sum of the squares

$$\mathfrak{x}^2 = x_1^2 + x_2^2 + \cdots + x_n^2 \tag{4.1}$$

of the components of a vector $\mathfrak{x} = (x_1, x_2, \cdots, x_n)$ is taken as the square of its absolute value. The only co-ordinate systems which are then equally permissible are the Cartesian systems, in which the square of the absolute value of $\mathfrak{x}$ is given by (4.1) in terms of the components x_i; the range of values which the components may here assume is taken as the continuum of all real numbers. But the content of the preceding paragraphs is not bound to this choice; the only requirement is, in fact, that the range of permissible values constitute a "field" in which the four fundamental operations (excluding division by zero) can be performed. We shall hereafter consider the continuum of all *complex* numbers as the range of values which our components may assume. The expression (4.1) loses its definite character in this domain; the sum of the squares can vanish without implying that each term is zero. It is therefore desirable to replace the quadratic form (4.1) by the "unit Hermitian form"

$$\bar{x}_1 x_1 + \bar{x}_2 x_2 + \cdots + \bar{x}_n x_n \tag{4.2}$$

where $\bar{x}$ denotes the complex conjugate of a number x. The value $\mathfrak{x}^2$ of (4.2) will be taken as the *square of the* ***absolute***

magnitude *of the vector* $\mathfrak{x} = (x_1, x_2, \cdots, x_n)$ and the corresponding bilinear form

$$(\mathfrak{x}\mathfrak{y}) = x_1y_1 + x_2y_2 + \cdots + x_ny_n$$

as the ***scalar product*** $(\mathfrak{x}\mathfrak{y})$ of the two vectors $\mathfrak{x}$ and $\mathfrak{y} = (y_1, y_2, \cdots, y_n)$. A co-ordinate system is said to be ***normal*** when the square of the absolute magnitude of a vector $\mathfrak{x}$ is expressed in terms of its components x_i in this co-ordinate system by (4.2). In a normal co-ordinate system $\mathfrak{e}_i$ these components are the scalar products

$$x_i = (\mathfrak{e}_i\mathfrak{x}). \tag{4.3}$$

The transformations which lead from one normal co-ordinate system to another such, which therefore leave the form (4.2) invariant, are called ***unitary transformations***.†

The conditions which characterize unitary transformations are entirely analogous to those for orthogonal transformations, with which we are familiar from the elements of analytic geometry. Let $x = Sx'$ be such a transformation; under the influence of S the fundamental metric form (4.2) goes over into $\bar{x}'^*\bar{S}^*Sx'$. S is therefore unitary if and only if $\bar{S}^*S = \mathbf{1}$; the fact that det $S \neq 0$ follows immediately from this. Indeed, since a matrix S and its transposed S^* have the same determinant, it follows that *the determinant of a unitary transformation has the absolute value* 1 : $|\det S|^2 = 1$. These conditions may be expressed by the assertion that $\bar{S}^*$ is the matrix S^{-1} reciprocal to S, and therefore not only $\bar{S}^*S = \mathbf{1}$ but also $S\bar{S}^* = \mathbf{1}$. The first of these equations states that the sum of the squares of the absolute values of the elements of a column is 1 and that the sum of the mixed products $\sum_r \bar{s}_{ri}s_{rk}$ of two different columns $(i \neq k)$ is 0; the second equation contains the same assertion for the elements of the rows.

We carry over the terminology usual in Euclidean geometry. In particular, the vector $\mathfrak{y}$ is said to be *perpendicular* to $\mathfrak{x}$ if the scalar product $(\mathfrak{x}\mathfrak{y})$ vanishes. In virtue of the symmetry law

$$(\mathfrak{y}\mathfrak{x}) = \overline{(\mathfrak{x}\mathfrak{y})}$$

perpendicularity is a reciprocal relationship. There exists no vector $\mathfrak{a}$, except $\mathfrak{a} = 0$, to which all vectors are perpendicular; in fact, $\mathfrak{a} = 0$ is the only vector which is perpendicular to itself. Normal co-ordinate systems can be characterized by the fact

† The name "orthogonal" has been used in the physical literature to denote these transformations, but in mathematics it is necessary to have different names for these two different concepts.

that for them the scalar products of the fundamental vectors $\mathfrak{e}_i$ among themselves are

$$(\mathfrak{e}_i \mathfrak{e}_k) = \delta_{ik} = \begin{cases} 1 & (i = k) \\ 0 & (i \neq k). \end{cases}$$

On comparing the fundamental metric form (4.2) with (3.2) it is seen that the unitary space $\mathfrak{R}$ can be characterized by the fact that its conjugate complex $\bar{\mathfrak{R}}$ coincides with its dual P, or more precisely, that the conjugate complex $\bar{\mathfrak{x}}$ of a vector $\mathfrak{x}$ can at the same time be considered as its dual. We found that with a correspondence A of an m-dimensional unitary space $\mathfrak{R}$ on an n-dimensional $\mathfrak{S}$ is associated in an invariant manner the correspondence A^* of the dual space Σ on the dual P. As a consequence of the equation $\mathsf{P} = \bar{\mathfrak{R}}$ for unitary spaces

$$\bar{A}^* = \tilde{A}$$

is a correspondence of $\mathfrak{S}$ on $\mathfrak{R}$; we call it the "***Hermitian conjugate*** *of* A." $\tilde{A}A$ is a correspondence of $\mathfrak{R}$ on itself, $A\tilde{A}$ of $\mathfrak{S}$ on itself. A correspondence $\mathfrak{S}$ which carries the general vector $\mathfrak{x}$ over into $\mathfrak{x}' = S\mathfrak{x}$ is unitary if it leaves the absolute magnitude of $\mathfrak{x}$ unaltered: $\mathfrak{x}'^2 = \mathfrak{x}^2$. Two configurations consisting of vectors, either of which can be obtained from the other by a unitary transformation, are congruent in unitary geometry; i.e. unitary geometry is the theory of those relationships which are invariant under an arbitrary unitary transformation. The characteristic property of such transformations is expressed in terms of the matrix calculus by either of the two equations

$$\tilde{S}S = 1, \quad S\tilde{S} = 1.$$

Let $\mathfrak{R}'$ be an m-dimensional linear sub-space spanned by the linearly independent vectors $\mathfrak{a}_1, \mathfrak{a}_2, \cdots, \mathfrak{a}_m$. We consider a vector $\mathfrak{x}$ as belonging to the sub-space $\mathfrak{R}''$ if and only if it is perpendicular to $\mathfrak{R}'$, i.e. to all the vectors of $\mathfrak{R}'$; such a vector must therefore satisfy the equations

$$(\mathfrak{a}_1 \mathfrak{x}) = 0, \quad (\mathfrak{a}_2 \mathfrak{x}) = 0, \quad \cdots, \quad (\mathfrak{a}_m \mathfrak{x}) = 0.$$

From these it follows that $\mathfrak{R}''$ is $(n - m)$-dimensional. The relation between $\mathfrak{R}'$ and $\mathfrak{R}''$ is a reciprocal one: every vector of $\mathfrak{R}''$ is perpendicular to every vector of $\mathfrak{R}'$ and conversely. *We then have* $\mathfrak{R} = \mathfrak{R}' + \mathfrak{R}''$, for if the sum $\mathfrak{x}' + \mathfrak{x}''$ of a vector $\mathfrak{x}'$ in $\mathfrak{R}'$ and a vector $\mathfrak{x}''$ in $\mathfrak{R}''$ vanishes then $\mathfrak{x}' = -\mathfrak{x}''$ is a vector which belongs to both sub-spaces and is consequently perpendicular to itself, and this can only occur if $\mathfrak{x}' = 0$. A unitary correspondence which leaves $\mathfrak{R}'$ invariant will also leave $\mathfrak{R}''$ invariant since the relation of perpendicularity will not be destroyed by such a transformation. *In dealing with unitary correspondences or transformations it is therefore always possible to find an invariant sub-space $\mathfrak{R}''$ associated with a given invariant sub-space $\mathfrak{R}'$, such that $\mathfrak{R} = \mathfrak{R}' + \mathfrak{R}''$.* The previous remarks about projection suggest that here in the unitary geometry we identify the space generated by projecting $\mathfrak{R}$ with respect to $\mathfrak{R}'$ with the sub-space $\mathfrak{R}''$: we project *on to* the space $\mathfrak{R}''$ perpendicular to $\mathfrak{R}'$. To this end we remark that among all vectors $\mathfrak{a}$ in $\mathfrak{R}$ which are congruent mod. $\mathfrak{R}'$ there is one $(\mathfrak{a})$ which lies in $\mathfrak{R}''$; we then have

$$(a \cdot \mathfrak{a}) = a(\mathfrak{a}), \quad (\mathfrak{a} + \mathfrak{b}) = (\mathfrak{a}) + (\mathfrak{b}).$$

With an arbitrary linear correspondence A

$$\mathfrak{y} \to \mathfrak{y}' = A\,\mathfrak{y}: \quad y_i' = \sum_k a_{ik} y_k \tag{4.4}$$

of $\mathfrak{R}$ on itself is, as we have seen, associated a bilinear form

$$\sum_{ik} a_{ik} \xi_i y_k$$

which depends linearly on a vector ξ in **P** and a vector $\mathfrak{y}$ in $\mathfrak{R}$. In unitary space we can therefore associate the form

$$A(\mathfrak{x}, \mathfrak{y}) = \sum_{ik} a_{ik} \bar{x}_i y_k,$$

depending linearly on $\mathfrak{y} = (y_i)$ and $\bar{\mathfrak{x}} = (\bar{x}_i)$, with the correspondence (4.4). It is in fact the scalar product of $\mathfrak{x}$ and $A\mathfrak{y}$. The special case in which

$$A = \tilde{A} \quad \text{or} \quad \overline{A(\mathfrak{y}, \mathfrak{x})} = A(\mathfrak{x}, \mathfrak{y}) \quad \text{or} \quad a_{ki} = \bar{a}_{ik} \tag{4.5}$$

bears the name of the French mathematician *Hermite*. The correspondence (4.4) is consequently ***Hermitian*** if the scalar product of $\mathfrak{x}$ with $A\,\mathfrak{y}$ is the conjugate complex of the scalar product of $\mathfrak{y}$ with $A\,\mathfrak{x}$. On identifying $\mathfrak{y}$ with $\mathfrak{x}$ we obtain the "*Hermitian form*"

$$A(\mathfrak{x}) = A(\mathfrak{x}, \mathfrak{x}) = \Sigma a_{ik} \bar{x}_i x_k, \tag{4.6}$$

i.e. the scalar product of $\mathfrak{x}$ and $A\mathfrak{x}$; in consequence of (4.5) its value is real. An Hermitian form or correspondence A is said to be *non-degenerate* if there exists no vector $\mathfrak{x}$, except $\mathfrak{x} = 0$, whose transform $A\mathfrak{x}$ vanishes It is *positive definite* if the value of the form $A(\mathfrak{x}) > 0$ for all vectors $\mathfrak{x} \neq 0$; a positive definite form is non-degenerate.

The fundamental metric form (4.2) is one such positive

definite Hermitian form, the "unit form," the coefficients of which consist of the numbers

$$\delta_{ik} = \begin{cases} 1 \ (i = k) \\ 0 \ (i \neq k) \end{cases}.$$

On introducing an arbitrary co-ordinate system $\mathfrak{a}_i$ $(i = 1, 2, \cdots, n)$ into the n-dimensional space, the absolute magnitude of an arbitrary vector

$$\mathfrak{x} = x_1 \mathfrak{a}_1 + x_2 \mathfrak{a}_2 + \cdots + x_n \mathfrak{a}_n$$

is given by

$$\mathfrak{x}^2 = \Sigma g_{ik} \bar{x}_i x_k, \quad g_{ik} = (\mathfrak{a}_i \mathfrak{a}_k).$$

The expression for $\mathfrak{x}^2$ is accordingly always a definite Hermitian form; conversely, any positive definite Hermitian form $G(\mathfrak{x})$ could be taken as the fundamental metric form. To show this we employ the associated Hermitian bilinear form $G(\mathfrak{x}, \mathfrak{y})$ to carry through the following procedure, which is patterned after the step-by-step construction of a Cartesian co-ordinate system. Choose any non-vanishing vector $\mathfrak{e}_1$; since $G(\mathfrak{e}_1) > 0$ we may, on multiplying $\mathfrak{e}_1$ by an appropriate numerical factor, normalize it in accordance with the equation $G(\mathfrak{e}_1) = 1$. When the process of constructing a system of unitary-orthogonal vectors $\mathfrak{e}_1$

$$G(\mathfrak{e}_i, \mathfrak{e}_k) = \delta_{ik}$$

has been carried through m steps, $i = 1, 2, \cdots, m$, the next step is accomplished by choosing a solution $\mathfrak{x} = \mathfrak{e}_{m+1}$ of the $m < n$ homogeneous linear equations $G(\mathfrak{e}_i, \mathfrak{x}) = 0$ for the n unknown components of the vector $\mathfrak{x} \neq 0$ and normalizing it in accordance with the equation $G(\mathfrak{e}_{m+1}) = 1$. The procedure comes to an end after n steps; we then have n vectors $\mathfrak{e}_1, \mathfrak{e}_2, \cdots, \mathfrak{e}_n$ of such a kind that

$$G(\mathfrak{x}, \mathfrak{x}) = \bar{x}_1 x_1 + \bar{x}_2 x_2 + \cdots + \bar{x}_n x_n$$

where

$$\mathfrak{x} = x_1 \mathfrak{e}_1 + x_2 \mathfrak{e}_2 + \cdots + x_n \mathfrak{e}_n.$$

It follows from the equations themselves that $\mathfrak{x}$ can only vanish when all of its components x_i vanish, and consequently the $\mathfrak{e}_i$ are linearly independent and constitute a co-ordinate system in $\mathfrak{R}$.

The transition from affine to metric geometry can accordingly be accomplished by the introduction of the axiom:

(δ) *The square of the absolute magnitude of a vector $\mathfrak{x}$ is a real number $\mathfrak{x}^2$ which is a positive definite Hermitian form in the components of $\mathfrak{x}$.*

These last considerations are useful in another connection. If $\mathfrak{R}'$ is a linear sub-space of $\mathfrak{R}$ we can employ the construction used above to find m vectors $\mathfrak{e}_1, \mathfrak{e}_2, \cdots, \mathfrak{e}_m$ in $\mathfrak{R}'$ which span $\mathfrak{R}'$ and are mutually unitary-orthogonal in the sense of the equations $(\mathfrak{e}_i\mathfrak{e}_k) = \delta_{ik}$. By continuing the construction we can supplement these m fundamental vectors by $n - m$ additional ones $\mathfrak{e}_{m+1}, \cdots, \mathfrak{e}_n$ so that the two sets together form a co-ordinate system for the entire space $\mathfrak{R}$. We can therefore adapt our normal co-ordinate system to the separation of $\mathfrak{R}'$ out of $\mathfrak{R}$ or to the decomposition of $\mathfrak{R} = \mathfrak{R}' + \mathfrak{R}''$ into two perpendicular sub-spaces.

Since the correspondence A of $\mathfrak{R}$ on to itself is invariantively connected with the Hermitian form A in $\mathfrak{R}$, we may speak of the product BA of two Hermitian forms A, B in $\mathfrak{R}$, but this product is not in general Hermitian as

$$\widetilde{BA} = \tilde{A}\tilde{B} = AB.$$

The trace of an Hermitian form or correspondence A is real. The positive definite expression

$$\operatorname{tr}(A\tilde{A}) = \sum_{i,k} |a_{ik}|^2 \tag{4.7}$$

is of particular importance. When $\mathfrak{R}$ is decomposed into mutually perpendicular sub-spaces $\mathfrak{R}_\alpha$ $(\alpha = 1, 2, \cdots)$ the section $A_{\alpha\beta}$ of the correspondence or form A in which $\mathfrak{R}_\alpha$ intersects $\mathfrak{R}_\beta$ is uniquely determined; it is a correspondence of $\mathfrak{R}_\beta$ on $\mathfrak{R}_\alpha$, and $\tilde{A}_{\beta\alpha}$, the $\beta\alpha$-section of $\tilde{A}$, is a correspondence of $\mathfrak{R}_\alpha$ on $\mathfrak{R}_\beta$. When the co-ordinate system is adapted to the decomposition of $\mathfrak{R}$ we have

$$\operatorname{tr}(A_{\alpha\beta}\tilde{A}_{\beta\alpha}) = \operatorname{tr}(\tilde{A}_{\beta\alpha}A_{\alpha\beta}) = \sum |a_{ik}|^2 \tag{4.8}$$

where in the sum i runs through the αth, k through the βth set of indices.

Any non-vanishing vector $\mathfrak{a}$ determines a *ray* $\mathfrak{a}$ which consists of all vectors of the form $\lambda\mathfrak{a}$, λ being an arbitrary complex number. The generating vector $\mathfrak{a}$ can be so normalized that its absolute value $|\mathfrak{a}| = 1$; this does not, however, determine $\mathfrak{a}$ to within a change of sign, as in the real domain, as the normalization is unaltered on multiplying $\mathfrak{a}$ by an arbitrary (complex) number ε of modulus 1. We shall call the totality of vectors of $\mathfrak{R}$ the ***vector field*** $\mathfrak{R}$ and the totality of rays the ***ray field*** $\mathfrak{R}$. Any non-degenerate linear correspondence A of the vector field $\mathfrak{R}$ on itself is at the same time a correspondence of the ray field $\mathfrak{R}$ on itself, but this latter correspondence is unaltered by multiplication with any non-vanishing number. A unitary correspondence or transformation of the ray field on itself will be briefly referred to as a *rotation*. By the symbol $S' \simeq S$ we shall mean that the two transformations S, S' of the vector field on itself differ only by a numerical factor ε of modulus 1: $S' = \varepsilon S$, whence they both give rise to the same rotation of the ray field.

§ 5. Transformation to Principal Axes

The fundamental theorem on Hermitian forms is that concerning the *transformation to principal axes.* We are here concerned with the analogue of the familiar problem of finding the principal axes of an ellipse or ellipsoid in the ordinary geometry of two or three dimensions. *We wish to find a normal co-ordinate system* $\mathfrak{e}_i$ *associated with a Hermitian form* $A(\mathfrak{x})$ *such that in addition to*

$$\mathfrak{x} = x_1\mathfrak{e}_1 + x_2\mathfrak{e}_2 + \cdots + x_n\mathfrak{e}_n$$
$$\mathfrak{x}^2 = \bar{x}_1x_1 + \bar{x}_2x_2 + \cdots + \bar{x}_nx_n \tag{5.1}$$

we also have

$$A(\mathfrak{x}) = \alpha_1\bar{x}_1x_1 + \alpha_2\bar{x}_2x_2 + \cdots + \alpha_n\bar{x}_nx_n\,; \tag{5.2}$$

that is, A shall be brought into the normal form (5.2) by means of a unitary transformation. The real numbers $\alpha_1, \alpha_2, \cdots, \alpha_n$ are called the ***characteristic numbers*** of the form A, and $\mathfrak{e}_1, \mathfrak{e}_2, \cdots, \mathfrak{e}_n$ the corresponding ***characteristic vectors.***

To this end we first consider the correspondence $\mathfrak{x} \to \mathfrak{x}' = A\mathfrak{x}$ and seek those vectors $\mathfrak{x} \neq 0$ which are transformed into multiples $\mathfrak{x}' = \lambda\mathfrak{x}$ of themselves by A. We then obtain the "*secular equation*"

$$f(\lambda) \equiv \det(\lambda 1 - A) = 0$$

for the multipliers λ. According to the *fundamental theorem of algebra* this equation certainly has a root $\lambda = \alpha_1$; corresponding to it a non-vanishing vector $\mathfrak{x} = \mathfrak{e}_1$ can be found which satisfies the equation $A\mathfrak{e}_1 = \alpha_1\mathfrak{e}_1$, and on multiplying this vector by an appropriate numerical factor we may take it such that its modulus is unity. $\mathfrak{e}_1$ can then be supplemented by $n - 1$ further vectors $\mathfrak{e}_2, \cdots, \mathfrak{e}_n$ in such a way that these n vectors constitute a normal co-ordinate system. In these co-ordinates the formulæ

$$\mathfrak{e}_i' = A\mathfrak{e}_i = \sum_k a_{ki}\mathfrak{e}_k$$

for the correspondence A require, in accordance with the definition of $\mathfrak{e}_1$, that the coefficients $a_{21}, a_{31}, \cdots, a_{n1}$ vanish and that $a_{11} = \alpha_1$. Because of the symmetry conditions $a_{ki} = \bar{a}_{ik}$,

$a_{12}, a_{13}, \cdots, a_{1n}$ must also vanish. Hence in the new co-ordinates the matrix A assumes the form

$$\left\| \begin{matrix} \alpha_1 & 0 & 0 & \cdots & 0 \\ 0 & a_{22} & a_{23} & \cdots & a_{2n} \\ 0 & a_{32} & a_{33} & \cdots & a_{3n} \\ \cdot & \cdot & \cdot & \cdot & \cdot \\ 0 & a_{n2} & a_{n3} & \cdots & a_{nn} \end{matrix} \right\|, \qquad (5.3)$$

and the Hermitian form becomes

$$A(\mathfrak{x}) = \alpha_1 \bar{x}_1 x_1 + A'(\mathfrak{x}) \qquad (5.3)$$

where A' is an Hermitian form containing only the $n-1$ variables $x_2, x_3, \cdots, x_n$. Repeating this process, or calling on the method of mathematical induction, we establish the validity of the fundamental theorem stated above.

The characteristic polynomial of (5.2) is

$$\det(\lambda \mathbf{1} - A) = (\lambda - \alpha_1)(\lambda - \alpha_2) \cdots (\lambda - \alpha_n).$$

From this it follows that the characteristic numbers $\alpha_1, \alpha_2, \cdots, \alpha_n$, including their multiplicity, are uniquely determined by the Hermitian form A; their sum is the trace of A. What can we say concerning the characteristic vectors? Let α be a given real number; the vectors $\mathfrak{x}$ which satisfy the equation $A\mathfrak{x} = \alpha\mathfrak{x}$ constitute a linear sub-space $\mathfrak{R}(\alpha)$ of $\mathfrak{R}$, the ***characteristic space*** belonging to α. When the normal co-ordinate system $\mathfrak{e}_i$ is so chosen that A is in the normal form, the equation $A\mathfrak{x} = \alpha\mathfrak{x}$ is, in terms of its components,

$$\alpha_i x_i = \alpha x_i$$

from which it follows that $\mathfrak{R}(\alpha)$ is spanned by those vectors $\mathfrak{e}_i$ for which $\alpha_i = \alpha$. If, for example, the three roots $\alpha_1, \alpha_2, \alpha_3 = \alpha$ while all the others are different from α, the characteristic space $\mathfrak{R}(\alpha)$ is 3-dimensional. If none of the characteristic numbers α_i is equal to α, $\mathfrak{R}(\alpha)$ consists only of the vector 0. This again characterizes the characteristic numbers, including their multiplicity, in a way which is independent of the particular co-ordinate system chosen, and in addition it characterizes the corresponding sub-spaces $\mathfrak{R}(\alpha)$. $\mathfrak{R}$ is thus decomposed into the characteristic spaces $\mathfrak{R}(\alpha)$: $\mathfrak{R} = \sum_\alpha \mathfrak{R}(\alpha)$; only a finite number of terms occurs in this sum, i.e. those for which α is a characteristic number of A. A complete co-ordinate system $\mathfrak{e}_1, \mathfrak{e}_2, \cdots, \mathfrak{e}_n$ for the entire space $\mathfrak{R}$ can be obtained by choosing a normal co-ordinate system in each non-null sub-space $\mathfrak{R}(\alpha)$. The normal form (5.2) is undisturbed on subjecting the variables

x_i associated with the same characteristic number $\alpha_i = \alpha$ to an arbitrary unitary transformation.

If, for example, α is a triple characteristic number

$$\alpha_1 = \alpha_2 = \alpha_3 = \alpha$$

while the remaining $\alpha_i \neq \alpha$, then $x_1\mathfrak{e}_1 + x_2\mathfrak{e}_2 + x_3\mathfrak{e}_3$ is the normal projection $\mathfrak{x}_\alpha$ of the vector $\mathfrak{x}$ on $\mathfrak{R}(\alpha)$ and

$$E_\alpha(\mathfrak{x}) = \bar{x}_1x_1 + \bar{x}_2x_2 + \bar{x}_3x_3$$

is the scalar product of $\mathfrak{x}_\alpha$ with itself. The equations (5.1), (5.2) may then be written in the invariant form

$$\mathfrak{x}^2 = \sum_\alpha E_\alpha(\mathfrak{x}), \quad A(\mathfrak{x}) = \sum_\alpha \alpha \cdot E_\alpha(\mathfrak{x}). \tag{5.4}$$

$\mathfrak{R}'$ being a sub-space of $\mathfrak{R}$, any vector $\mathfrak{x}$ can be uniquely broken up into $\mathfrak{x}' + \mathfrak{x}_0$ where $\mathfrak{x}'$ lies in $\mathfrak{R}'$ and $\mathfrak{x}_0$ is perpendicular to $\mathfrak{R}'$. The "orthogonal projection" $\mathfrak{x} \to \mathfrak{x}' = E'\mathfrak{x}$ is a linear correspondence which obviously has the property

$$E'E' = E', \tag{5.5}$$

for the projection of $\mathfrak{x}'$ on $\mathfrak{R}'$ is simply $\mathfrak{x}'$ itself. Furthermore, the operator E' is *Hermitian*, for the scalar product of $\mathfrak{y}$ into $\mathfrak{x}'$ is equal to the scalar product of $\mathfrak{y}'$ into $\mathfrak{x}'$, where $\mathfrak{y}'$ is the projection of $\mathfrak{y}$ on $\mathfrak{R}'$. (The Hermitian form $E'(\mathfrak{x})$ is accordingly the square of the absolute value of $\mathfrak{x}'$.) We shall call Hermitian forms which satisfy equation (5.5) *idempotent.*

When the sub-spaces $\mathfrak{R}'$, $\mathfrak{R}''$ are orthogonal, the two corresponding projection operators E', E'' satisfy the equations

$$E'E'' = 0, \quad E''E' = 0, \tag{5.6}$$

for E' $(E''\mathfrak{x})$ is the component of $E''\mathfrak{x}$ lying in the space $\mathfrak{R}'$ perpendicular to $E''\mathfrak{x}$. Idempotent operators which satisfy these equations are said to be *independent.* The second equation is, moreover, a consequence of the first, as may be seen on going over to the Hermitian conjugate: $\tilde{E}''\tilde{E}' = 0$. If $\mathfrak{R}$ is decomposed into several mutually orthogonal sub-spaces $\mathfrak{R}'+\mathfrak{R}''+\cdots$, then

$$\mathfrak{x} = E'\mathfrak{x} + E''\mathfrak{x} + \cdots. \tag{5.7}$$

It is easily shown that the converses of all these assertions are also valid. If E' is an idempotent operator and $E'' = \mathbf{1} - E'$, all vectors of the form $E'\mathfrak{x}$ constitute a linear sub-space $\mathfrak{R}'$ and all vectors of the form $E''\mathfrak{x}$ a sub-space $\mathfrak{R}''$. The equation

$$\tilde{E}'E'' = E'E'' = E'(\mathbf{1} - E') = 0$$

shows that the scalar product of a vector $E'\mathfrak{x}$ in $\mathfrak{R}'$ and a vector $E''\mathfrak{y}$ in $\mathfrak{R}''$ is zero: $\bar{x}\bar{\tilde{E}}'E''y = 0$. The decomposition of a vector $\mathfrak{x}$ into a component lying in $\mathfrak{R}'$ and one perpendicular to $\mathfrak{R}'$ is accordingly expressed by

$$\mathfrak{x} = E'\mathfrak{x} + (\mathbf{1} - E')\mathfrak{x}.$$

If the two idempotent forms E', E'' satisfy the equation (5.6) then, as we have just seen, the two corresponding characteristic spaces $\mathfrak{R}'$, $\mathfrak{R}''$ are mutually perpendicular. If the sum (5.7) consists of independent idempotent forms, then by the above the corresponding mutually perpendicular sub-spaces $\mathfrak{R}'$, $\mathfrak{R}''$ exhaust the entire space $\mathfrak{R}$.

The theorem on transformation to principal axes can accordingly be stated: *An Hermitian form A associates with the real numbers α mutually independent idempotent Hermitian forms E_α such that*

$$\mathbf{1} = \sum_\alpha E_\alpha, \quad A = \sum_\alpha a \cdot E_\alpha; \tag{5.8}$$

E_α is non-vanishing for only a finite number of values α.

A correspondence A can be reiterated:

$$AA = A^2, \quad A^2A = A^3, \cdots$$

and we can accordingly obtain polynomials

$$f(A) = c_0\mathbf{1} + c_1A + c_2A^2 + \cdots + c_hA^h$$

in A with numerical coefficients c. On reiterating (5.8) $h - 1$ times

$$A^h = \sum_\alpha \alpha^h E_\alpha$$

whence for the general polynomial f

$$f(A) = \sum_\alpha f(\alpha)E_\alpha. \tag{5.9}$$

The characteristic numbers of $f(A)$ are therefore the values of the polynomial $f(\alpha)$ for the characteristic numbers α of A. This suggests defining the Hermitian form $f(A)$, where $f(\alpha)$ is *any* real function of the real variable α, by means of the equation (5.9)

Given *two Hermitian forms A, B*, under what conditions can they be brought *simultaneously into diagonal form*, i.e. when is it possible to find a normal co-ordinate system in which

$$\begin{aligned} A(\mathfrak{x}) &= \alpha_1\bar{x}_1x_1 + \alpha_2\bar{x}_2x_2 + \cdots + \alpha_n\bar{x}_nx_n \\ B(\mathfrak{x}) &= \beta_1\bar{x}_1x_1 + \beta_2\bar{x}_2x_2 + \cdots + \beta_n\bar{x}_nx_n \, ? \end{aligned} \tag{5.10}$$

A necessary condition is that they commute: $BA = AB$, for if A and B are in the normal form (5.10) BA as well as AB is the diagonal matrix with elements $\beta_i\alpha_i = \alpha_i\beta_i$. *This condition is also sufficient;* to prove this, choose a normal co-ordinate system in which A is already in normal form. The equation $BA = AB$ requires that the matrix $B = ||b_{ik}||$ satisfy

$$b_{ik}\alpha_k = \alpha_i b_{ik} \quad \text{or} \quad (\alpha_i - \alpha_k)b_{ik} = 0. \tag{5.11}$$

We divide the indices i, the fundamental vectors $\mathfrak{e}_i$ and the variables x_i into classes by considering i and k to be of the same class if $\alpha_i = \alpha_k$. Equation (5.11) states that $b_{ik} = 0$ when i and k belong to different classes. B is consequently decomposed into smaller matrices B', B'' aligned along the principal diagonal, corresponding to the way in which the α_i are distributed in classes α', α'', $\cdots$; the correspondence B consequently leaves each of the characteristic spaces $\mathfrak{R}(\alpha')$, $\mathfrak{R}(\alpha'')$, $\cdots$ of A invariant. But we can then choose a normal co-ordinate system in each of these characteristic sub-spaces $\mathfrak{R}(\alpha)$ in such a way that the Hermitian correspondences B', B'' in them are referred to principal axes; the normal form of A is undisturbed by this procedure.

This process can immediately be applied to any number of Hermitian forms: *Any number of Hermitian forms can be brought simultaneously into normal form if and only if they commute with one another.* By a slight modification we can further extend this theorem to *an arbitrary finite or infinite system Σ of Hermitian forms.* This will be briefly discussed here, although in general the consideration of systems of forms or correspondence is postponed until Chap. III. Let the space $\mathfrak{R}$ be decomposed into mutually perpendicular sub-spaces $\mathfrak{R}'$, $\mathfrak{R}''$, $\cdots$ in such a way that each correspondence of the system Σ takes place in these sub-spaces; on adapting the co-ordinate system to this decomposition each Hermitian matrix A of Σ consists of sub-matrices A', A'', $\cdots$ aligned along the principal diagonal. If all the A' are already multiples of the unit matrix 1 in $\mathfrak{R}'$ and similarly for all A'', $\cdots$, our goal is reached, for each correspondence A of the system then transforms $\mathfrak{R}'$ into itself and is a simple multiplication in it; similarly for $\mathfrak{R}''$, $\cdots$. But if this is not the case let A be a correspondence of the system which is not merely a multiplication in the sub-space $\mathfrak{R}'$. On transforming the constituent A' of A to principal axes, $\mathfrak{R}'$ is decomposed into characteristic spaces $\mathfrak{R}_1' + \mathfrak{R}_2' + \cdots$ of A', of which there are at least two. For any Hermitian matrix X of Σ we have $A'X' = X'A'$, from which it follows, as we saw above, that X' transforms each of the sub-spaces $\mathfrak{R}_1'$, $\mathfrak{R}_2'$, $\cdots$

into itself. The decomposition $\mathfrak{R}' + \mathfrak{R}'' + \cdots$ can thus be further reduced to the decomposition $(\mathfrak{R}_1' + \mathfrak{R}_2' + \cdots) + \mathfrak{R}'' + \cdots$. Proceeding in this way we finally reach our goal after at most n steps, proving:

The Hermitian forms of any system Σ can be simultaneously referred to principal axes if they all commute with one another.

The theory developed above for Hermitian correspondence is valid as it stands for unitary transformations. S being any unitary operator, a normal co-ordinate system $\mathfrak{e}_i$ can be introduced in such a way that S carries each of the fundamental vectors $\mathfrak{e}_i$ over into a multiple $\sigma_i \mathfrak{e}_i$ of itself. The characteristic numbers σ_i of S are numbers of modulus 1. In these co-ordinates the matrix of S is a diagonal matrix, the elements in the principal diagonal of which are the numbers σ_i.

The proof is quite analogous. We again start with the secular equation

$$\det(\sigma \mathbf{1} - S) = 0$$

and consider the root σ_1. There then exists a vector $\mathfrak{e}_1$ of modulus 1 which is transformed into $\sigma_1 \mathfrak{e}_1$ by the correspondence S. Supplement $\mathfrak{e}_1$ with $n - 1$ further vectors $\mathfrak{e}_2, \cdots, \mathfrak{e}_n$ so that these n vectors form a normal co-ordinate system. In these co-ordinates the matrix $\|s_{ik}\|$ of the correspondence S:

$$S\mathfrak{e}_i = \sum_k s_{ki}\, \mathfrak{e}_k$$

is again of the form

$$s_{11} = \sigma_1, \quad s_{21} = \cdots = s_{n1} = 0.$$

Since S is unitary the sum of the squares of the moduli of these elements of the first column must be unity, whence $|\sigma_1| = 1$. Similarly the sum of the squares of the moduli of the elements in the first row must also be 1:

$$|\sigma_1|^2 + |s_{12}|^2 + \cdots + |s_{1n}|^2 = 1;$$

but since $|\sigma_1|^2 = 1$ it follows that

$$s_{12} = \cdots = s_{1n} = 0.$$

The matrix S is now broken up into a 1-dimensional σ_1 and an $(n - 1)$-dimensional S' as in (5.3); the truth of the above theorem then follows immediately by induction.

The further results can be obtained in exactly the same way as above for Hermitian forms. The characteristic numbers σ_i, including their multiplicity but not their order, are uniquely determined by S, and similarly for the corresponding sub-spaces. If we wish to find a linearly independent system of characteristic vectors, the fundamental vectors of each such sub-space may be taken as forming a normal co-ordinate system. Finally, a finite or infinite set of unitary transformations can be simultaneously reduced to normal form if and only if they commute among themselves.

§ 6. Infinitesimal Unitary Transformations

A rigid body in continuous motion about a fixed point O performs an *infinitesimal rotation* in each interval $d\tau$ of time. Denoting by (dx_1, dx_2, dx_3) the infinitesimal displacement of that point of the rigid body which is at the point $P(x_1, x_2, x_3)$ at the time τ, the equations of motion of the body must be of the form

$$\dot{x}_i = \frac{dx_i}{d\tau} = \sum_k c_{ik} x_k \tag{6.1}$$

in which the coefficients c_{ik} are constants, i.e. independent of the particular point P under consideration. Employing a Cartesian co-ordinate system with O as origin, $x_1^2 + x_2^2 + x_3^2$ must remain unchanged throughout the motion; this requires that

$$\sum_i x_i \frac{dx_i}{d\tau} = 0 \quad \text{or} \quad \sum_{ik} c_{ik} x_i x_k = 0.$$

Since this equation must be satisfied identically in the x_i, the matrix $C = \|c_{ik}\|$ which characterizes the motion must be anti-symmetric: $c_{ki} = -c_{ik}$. Introducing the vector $\mathfrak{r}$ with origin at O and terminus at the point P, and the vector $\mathfrak{c} = (c_{23}, c_{31}, c_{12})$, equations (6.1) become

$$\frac{d\mathfrak{r}}{d\tau} = [\mathfrak{r}\mathfrak{c}],$$

the familiar fundamental formulæ for the kinematics of a rigid body. The square brackets denote the vector product and $\mathfrak{c}$ the vectorial angular velocity, the absolute value and direction of which give the angular velocity and direction of the axis of rotation respectively.

The continuous compounding of interest offers another example of an infinitesimal linear transformation. The interest rate being c, a real number, the increase in the capital x in time $d\tau$ is $xcd\tau$. Radioactive disintegration is the same kind of a process with negative c. The capital x, considered as a function of the time, satisfies the equation

$$\frac{dx}{d\tau} = cx \tag{6.2}$$

and consequently increases exponentially with τ. If the principal has the value x_0 at time $\tau = 0$, it will have increased to

$$x(\tau) = x_0 \cdot e^{c\tau}$$

at time τ. To obtain an alternative solution we divide, as in the method of finite differences, the time interval τ into a large number n of equal elements τ/n; x will increase by $xc\tau/n$ in each of these intervals and the capital x will accordingly be multiplied by $(1 + c\tau/n)^n$ at the end of time τ. The familiar definition

$$e^{c\tau} = \lim_{n\to\infty}\left(1 + \frac{c\tau}{n}\right)^n \tag{6.3}$$

of the exponential function follows from a comparison of these two results. But we can also solve the differential equation (6.2) by the method of successive approximations. We take as the 0th approximation the initial value x_0: $x_0(\tau) = x_0$. The $(n + 1)$st approximation is obtained from the nth by substituting the latter in place of x on the right-hand side of (6.2) and integrating:

$$x_{n+1}(\tau) = x_0 + c\int_0^\tau x_n(t)dt.$$

On carrying out this process we find

$$x_n(\tau) = x_0\left\{1 + \frac{c\tau}{1!} + \cdots + \frac{(c\tau)^n}{n!}\right\},$$

from which we obtain the familiar power series expansion

$$e^{c\tau} = 1 + \frac{c\tau}{1!} + \frac{(c\tau)^2}{2!} + \cdots \tag{6.4}$$

for the exponential function. The convergence of (6.3) and (6.4) and the identity of their limits is rigorously proved by elementary analysis.

These examples will assist in understanding the concept of an ***infinitesimal unitary transformation*** of the n-dimensional space $\mathfrak{R} = \mathfrak{R}_n$, which we now proceed to introduce. In order to avoid the use of infinitesimals we introduce a (purely fictitious) time τ and think of the infinitesimal linear correspondence which carries the vector $\mathfrak{x}$ over into $\mathfrak{x} + d\mathfrak{x}$ as taking place in the time interval $d\tau$:

$$\frac{d\mathfrak{x}}{d\tau} = C\mathfrak{x}, \quad \frac{dx_i}{d\tau} = \sum_k c_{ik}x_k.$$

(For the sake of brevity we refer to this simply as "the infinitesimal transformation C.") Since the transformation is unitary, on employing a normal co-ordinate system $\sum_i \bar{x}_i x_i$ must remain unchanged:

$$\sum_i \bar{x}_i \frac{dx_i}{d\tau} + \sum_k x_k \frac{d\bar{x}_k}{d\tau} = 0. \tag{6.5}$$

On setting

$$\frac{dx_i}{d\tau} = \sum_k c_{ik} x_k, \quad \frac{d\bar{x}_k}{d\tau} = \sum_i \bar{c}_{ki} \bar{x}_i$$

the left-hand side of (6.5) reduces to the Hermitian form

$$\sum_{i,k} (c_{ik} + \bar{c}_{ki}) \bar{x}_i x_k$$

and since it must vanish identically in the x_i we must have $c_{ik} + \bar{c}_{ki} = 0$, or the transformation C is anti-symmetric in the sense of the equation

$$c_{ik} = -\bar{c}_{ki}, \quad \tilde{C} = -C. \tag{6.6}$$

In the real domain there exists no intimate relationship between symmetric and anti-symmetric matrices, but the situation is different in the complex domain. For on setting $C = iH$ (i being the imaginary unit $\sqrt{-1}$) it follows from (6.6) that H satisfies the equation $\tilde{H} = H$, and C is consequently i times an Hermitian matrix. *In an infinitesimal unitary rotation of a vector field the velocity* $\frac{d\mathfrak{x}}{d\tau}$ *is related to* $\mathfrak{x}$ *by means of a correspondence whose matrix is i times an Hermitian matrix.* The theorem on transformation of Hermitian forms to principal axes is accordingly the limiting case of an analogous theorem on unitary transformations.

By repeated application of the infinitesimal unitary transformation

$$d\mathfrak{x} = d\tau \cdot C\mathfrak{x} \tag{6.7}$$

we obtain after time τ

$$\mathfrak{x} \to \mathfrak{x}(\tau) = U(\tau)\mathfrak{x} = e^{\tau C}\mathfrak{x} \tag{6.8}$$

where the exponential function e^A for a matrix A can be defined by either

$$\lim_{h \to \infty} \left(1 + \frac{A}{h}\right)^h$$

or the power series

$$1 + \frac{A}{1!} + \frac{A^2}{2!} + \cdots.$$

Naturally

$$U(\tau + \tau') = U(\tau)\, U(\tau').$$

Accordingly $U(\tau)$ runs through all the transformations of a 1-parameter continuous group of unitary transformations generated by the infinitesimal transformation C; the parameter τ is additive on composition. The power series is obtained by the method of successive approximations; this method can also be applied to obtain a solution in the more general case in which the infinitesimal unitary transformation C is not the same for each time element $d\tau$, i.e. in which C is a matrix $C(\tau)$ depending on the time τ. The solution of the equation

$$\frac{d\mathfrak{x}}{d\tau} = C(\tau)\mathfrak{x}$$

for this case is given by

$$\mathfrak{x}(\tau_2) = U(\tau_2\tau_1)\mathfrak{x}(\tau_1);$$

the unitary transformation $U(\tau_2\tau_1)$ which takes place in the time interval τ_1, τ_2 obeys the law of composition

$$U(\tau_3\tau_1) = U(\tau_3\tau_2)U(\tau_2\tau_1).$$

If $\mathfrak{x} = \mathfrak{x}_0$ at time $\tau = 0$, the formulæ for the successive approximations $\mathfrak{x}_l(\tau)$ are

$$\mathfrak{x}_0(\tau) = \mathfrak{x}_0; \quad \mathfrak{x}_{l+1}(\tau) = \mathfrak{x}_0 + \int_0^\tau C(t)\mathfrak{x}_l(t)dt;$$

for $U(\tau) = U(\tau\, 0)$ we obtain the infinite series $\sum_{l=0}^{\infty} U_l(\tau)$ in which

$$U_0(\tau) = \mathbf{1}; \quad U_{l+1}(\tau) = \int_0^\tau C(t)U_l(t)\,dt. \tag{6.9}$$

Written explicitly,

$$U_l(\tau) = \underset{(0 \leq t_1 \leq t_2 \leq \cdots \leq t_l \leq \tau)}{\int\int\cdots\int} C(t_1)C(t_2)\cdots C(t_l)dt_1\,dt_2\cdots dt_l.$$

The proof of the convergence of this process is readily obtained with the aid of the quantity $|A|$ associated with a matrix $A = \|a_{ik}\|$ by the equation

$$|A|^2 = \operatorname{tr}(A\tilde{A}) = \sum_{i,k}|a_{ik}|^2.$$

It follows from the well-known *Schwarz* inequality

$$|a_1b_1 + a_2b_2 + \cdots + a_nb_n|^2 \\ \leqq (|a_1|^2 + \cdots + |a_n|^2)(|b_1|^2 + \cdots + |b_n|^2) \tag{6.10}$$

that

$$|A + B| \leqq |A| + |B|$$

and that

$$|AB| \leqq |A|\,|B|.$$

The second inequality is obtained by applying (6.10) to the element

$$c_{ik} = \sum_r a_{ir} b_{rk}$$

of $C = AB$:

$$|c_{ik}|^2 \leqq \sum_r |a_{ir}|^2 \cdot \sum_r |b_{rk}|^2$$

and summing with respect to i and k. The first inequality may be stated in the form

$$\left|\int_0^\tau A(t)\,dt\right| \leqq \int_0^\tau |A(t)|\,dt.$$

for integrals. The convergence of $\sum_l U_l(\tau)$ can now be established with the aid of these auxiliary results, for we can prove that under the assumption

$$|C(t)| \leqq c \qquad (0 \leqq t \leqq \tau)$$

that

$$|U_l(\tau)| \leqq \sqrt{n} \cdot \frac{(c\tau)^l}{l!}.$$

For this is certainly true for $l = 0$, and the recursion formula (6.9) enables us to conclude that it holds for U_{l+1} if it holds for U_l. The convergence follows from this absolute convergence, for the absolute value of each component of the matrix A is certainly not greater than $|A|$.

We have only gone into these matters to reassure the reader of the legitimacy of dealing with infinitesimal quantities of the kind met here. The only thing of importance for the following is the simple relation existing between infinitesimal unitary transformations and Hermitian forms.

§ 7. Remarks on ∞-dimensional Space

The unitary spaces which appear in quantum mechanics usually have *an infinite number of dimensions.* Such a space consists of all vectors

$$\mathfrak{x} = (x_1, x_2, \cdots)$$

whose components x_i constitute an infinite sequence of numbers for which

$$\mathfrak{x}^2 = \bar{x}_1 x_1 + \bar{x}_2 x_2 + \cdots$$

converges. Within this domain addition and multiplication with numbers, as well as the construction of the scalar product of two vectors, are possible. All the axioms employed so far are satisfied, with the exception of the dimensionality axiom γ introduced in § 1.

Since the vector components $x_1, x_2, \cdots$ constitute a denumerable set, this " Hilbert space " has a denumerably infinite number of dimensions. But in addition to these, spaces of non-denumerably infinite dimensions may occur. Consider, for example, all continuous complex functions $\psi(s)$ of a real variable s of period 2π. We need not distinguish between two values of s which are congruent mod 2π, i.e. whose difference is an integral multiple of 2π; it is consequently more convenient to consider $\psi(s)$ as a function defined on the periphery of the unit circle than on the straight line. The various values of s at points on the circumference play the rôle of indices, the value $\psi(s)$ at the point s being the component of the " vector ψ " with index s. The totality of such functions $\psi(s)$ therefore constitute a linear " function space " of continuously infinite dimensions. Addition of these vectors and multiplication by a number have here the same interpretation as in the ordinary operations with functions. The square of the absolute value of the vector ψ is taken to be

$$(\psi, \psi) = \int_0^{2\pi} \bar{\psi}(s)\psi(s)ds$$

and the scalar product of two vectors ϕ and ψ as

$$(\phi, \psi) = \int_0^{2\pi} \bar{\phi}(s)\psi(s)ds.$$

A set of functions

$$\phi_1(s), \quad \phi_2(s), \quad \cdots, \quad \phi_n(s)$$

constitutes a unitary-orthogonal system of vectors if

$$\int_0^{2\pi} \bar{\phi}_i(s)\phi_k(s)ds = \delta_{ik}.$$

These vectors span an n-dimensional sub-space $\mathfrak{R}_n$ of the ∞-dimensional function space, i.e. that sub-space consisting of all vectors of the form

$$\phi(s) = x_1\phi_1(s) + x_2\phi_2(s) + \cdots + x_n\phi_n(s).$$

$x_1, x_2, \cdots, x_n$ are the components in the co-ordinate system $\phi_1, \phi_2, \cdots, \phi_n$ of the vector $\phi(s)$ in $\mathfrak{R}_n$. We have

$$(\phi, \phi) = \int_0^{2\pi} \bar{\phi}(s)\phi(s)ds = \bar{x}_1x_1 + \bar{x}_2x_2 + \cdots + \bar{x}_nx_n.$$

An arbitrary vector ψ can be broken up into a component ϕ which lies in $\Re_n$ and a component ψ' perpendicular to $\Re_n$: $\psi = \phi + \psi'$,

$$\phi(s) = \sum_{i=1}^{n} x_i\, \phi_i(s), \quad \int_0^{2\pi} \bar{\phi}_i(s)\psi'(s)ds = 0.$$

It follows from these equations that [cf. (4.3)]

$$x_i = \int_0^{2\pi} \bar{\phi}_i(s)\psi(s)ds.$$

These integrals are called the *Fourier coefficients* of the function ψ with respect to the orthogonal system ϕ_i. The orthogonal projection ϕ on $\Re_n$ cannot be longer (i.e. have greater absolute magnitude) than ψ itself; this is the content of the so-called *Bessel inequality*

$$\bar{x}_1 x_1 + \bar{x}_2 x_2 + \cdots + \bar{x}_n x_n \leqq \int_0^{2\pi} \bar{\psi}(s)\psi(s)ds. \tag{7.1}$$

In fact, since $(\phi, \psi') = 0$, $(\psi', \phi) = 0$, the "Pythagorean theorem"

$$(\psi, \psi) = (\phi, \phi) + (\psi', \psi')$$

holds.

The simplest unitary-orthogonal system in the domain of periodic functions, with which the theory of Fourier series is concerned, consists of the functions

$$\frac{1}{\sqrt{2\pi}} e(ns) \qquad [n = 0, \pm 1, \pm 2, \cdots; \; e(x) = e^{ix}]. \tag{7.2}$$

This infinite system has the property of ***completeness***; it is a complete co-ordinate system for the entire function space. The theorem that any periodic function $\psi(s)$ can be expressed as a linear combination of the functions (7.2):

$$\psi(s) = \frac{1}{\sqrt{2\pi}} \sum_{n=-\infty}^{+\infty} x_n \cdot e(ns), \qquad x_n = \frac{1}{\sqrt{2\pi}} \int_0^{2\pi} \bar{e}(ns)\psi(s)ds$$

(Fourier expansion of $\psi(s)$) is true only if certain conditions concerning the differentiability of $\psi(s)$ are fulfilled, but *any continuous function satisfies Parseval's equation*

$$\int_0^{2\pi} \bar{\psi}(s)\psi(s)ds = \sum_{n=-\infty}^{+\infty} \bar{x}_n x_n. \tag{7.3}$$

We learn from this example that *there is no essential distinction between spaces of a denumerable and of a non-denumerable infinitude of dimensions;* we have introduced into our function space a complete normal co-ordinate system (7.2) consisting of a denumerably infinite set of fundamental vectors. In an n-dimensional unitary space a system of unitary-orthogonal vectors is complete if their number is n, but not if it is less; however, such an enumeration gives no criterion for ∞-dimensional space. If we leave out a finite number of the functions (7.2) we still have an infinite set left, but the completeness of the system is destroyed thereby. The real criterion for completeness lies in the validity of the completeness relation (7.3).

We can understand the relations existing in Hilbert space by analogy with or as limiting cases of those existing in spaces of a finite number of dimensions. If we consider the values of an arbitrary periodic function $\psi(s)$ only at the points

$$s = 0 \cdot \frac{2\pi}{n}, \quad 1 \cdot \frac{2\pi}{n}, \quad \cdots, \quad (n-1) \cdot \frac{2\pi}{n}$$

and set

$$\psi\left(\frac{2\pi\nu}{n}\right) = \xi_\nu,$$

we are dealing with an n-dimensional vector space in which the components of the arbitrary vector ψ are these quantities ξ_ν $(\nu = 0, 1, \cdots, n-1)$. Let $\mathfrak{e}_\lambda$ be the vector in this space with components

$$\frac{1}{\sqrt{n}} e\left(\frac{2\pi\lambda\nu}{n}\right) \quad [\nu = 0, 1, \cdots, n-1];$$

these vectors $\mathfrak{e}_\lambda$ $(\lambda = 0, 1, \cdots, n-1)$ constitute a normal co-ordinate system for the space, relative to which the vector ξ has the components $x_0, x_1, \cdots, x_{n-1}$ which are to be calculated from

$$\xi_\nu = \frac{1}{\sqrt{n}} \sum_{\lambda=0}^{n-1} e\left(\frac{2\pi\lambda\nu}{n}\right) x_\lambda.$$

In accordance with (4.3)

$$x_\lambda = \frac{1}{\sqrt{n}} \sum_{\nu=0}^{n-1} e\left(\frac{-2\pi\lambda\nu}{n}\right) \xi_\nu$$

whence

$$\sum_{\nu=0}^{n-1} \bar{\xi}_\nu \xi_\nu = \sum_{\lambda=0}^{n-1} \bar{x}_\lambda x_\lambda.$$

By passing to the limit $n \to \infty$ we obtain the equation of *Parseval*. We do not concern ourselves here with the further considerations which may be necessary to establish a rigorous proof, but content ourselves with such reasoning by analogy.

We consider the linear correspondence or "operator" $D = \frac{1}{i}\frac{d}{ds}$ which transforms a function $\psi(s)$ in the domain of periodic functions into $\frac{1}{i}\frac{d\psi}{ds}$. $e(ns)$ is the characteristic vector (characteristic function) of this operator belonging to the characteristic number n:

$$\frac{1}{i}\frac{de(ns)}{ds} = n \cdot e(ns).$$

This operator is Hermitian; the scalar product of ϕ and $D\psi$ is the conjugate complex of that of ψ and $D\phi$, where ϕ and ψ are any two periodic functions, for by partial integration

$$\int_0^{2\pi} \bar{\phi}(s) \cdot \frac{1}{i}\frac{d\psi}{ds}ds = -\int_0^{2\pi} \psi \cdot \frac{1}{i}\frac{d\bar{\phi}}{ds}ds$$

and the right-hand side is conjugate to

$$\int_0^{2\pi} \bar{\psi} \cdot \frac{1}{i}\frac{d\phi}{ds}ds.$$

In fact, the Hermitian form

$$\frac{1}{i}\int_0^{2\pi} \bar{\psi}\frac{d\psi}{ds}ds$$

assumes the normal form

$$\sum_{n=-\infty}^{+\infty} n\bar{x}_n x_n \tag{7.4}$$

in the normal co-ordinate system whose fundamental vectors are the characteristic vectors of the operator D. The reiterated operator $DD = -\frac{d^2}{ds^2}$ appears in the theory of the vibrating string, together with the corresponding Hermitian form

$$-\int_0^{2\pi} \bar{\psi}\frac{d^2\psi}{ds^2}ds = \int_0^{2\pi} \frac{d\bar{\psi}}{ds}\frac{d\psi}{ds}ds$$

which represents the kinetic energy of the string.

We have here been dealing with a *discrete spectrum of characteristic numbers.* But in an ∞-dimensional space Hermitian forms with a *continuous spectrum* can also be constructed. Consider, for example, the function space consisting of all continuous functions $\psi(s)$ defined in the interval $-\pi \leqq s \leqq +\pi$; the square of the absolute magnitude of the "vector" ψ is then

$$(\psi, \psi) = \int_{-\pi}^{+\pi} \bar{\psi}(s)\psi(s)ds.$$

The Hermitian form

$$A[\psi] = \int_{-\pi}^{+\pi} s\bar{\psi}(s)\psi(s)\, ds \tag{7.5}$$

is already in normal form, which shows that it has as characteristic numbers all numbers between $-\pi$ and $+\pi$. The functions (7.2) again constitute a complete normal co-ordinate system in terms of which

$$\psi(s) \sim \frac{1}{\sqrt{2\pi}} \sum_{n=-\infty}^{+\infty} x_n\, e(ns).$$

Substituting this in (7.5) we find

$$A[\psi] = \sum a_{mn}\bar{x}_m x_n, \quad a_{mn} = \frac{1}{2\pi}\int_{-\pi}^{+\pi} se(-ms)e(ns)ds.$$

The evaluation of

$$\int_{-\pi}^{+\pi} s \cdot e[(n-m)s]ds$$

yields 0 when $n = m$ and by partial integration

$$\left[s \cdot \frac{e[(n-m)s]}{i(n-m)}\right]_{-\pi}^{+\pi} = 2\pi\frac{(-1)^{n-m}}{i(n-m)}$$

when $n \neq m$. The Hermitian form

$$\frac{1}{i}\sum_{n \neq m} \frac{(-1)^{n-m}}{n-m}\bar{x}_m x_n$$

has therefore as characteristic numbers all values between $-\pi$ and $+\pi$.

The characteristic vector ψ_α belonging to the characteristic value α $(-\pi \leqq \alpha \leqq +\pi)$ of $A[\psi]$ is that function which vanishes at all points $s \neq \alpha$ and is there so large that the integral of

$\bar{\psi}_\alpha \psi_\alpha$ has the value 1. Of course such a function does not really exist, but we can approximate it as closely as we wish. In order to arrive at a formulation which is mathematically rigorous for the case of continuous spectra, we must introduce in place of the idempotent Hermitian form E_α in (5.4) the idempotent form $\Delta E = \sum_{\alpha \leqq \lambda < \beta} E_\lambda$ for the entire interval $\Delta = \Delta_\alpha^\beta$ $(\alpha \leqq \lambda < \beta)$. For any given vector $\mathfrak{x}$

$$\Delta E(\mathfrak{x}) \geqq 0, \qquad \Delta_\alpha^\beta E(\mathfrak{x}) + \Delta_\beta^\gamma E(\mathfrak{x}) = \Delta_\alpha^\gamma E(\mathfrak{x}) \tag{7.6}$$

and the idempotent forms ΔE associated with two separated intervals Δ are mutually independent.

In dealing with the continuum, the sum in (5.4) is replaced by a Stieltjes integral. Consider the straight line described by the real variable λ as being covered with a substance, and let the amount of this substance on the interval Δ be denoted by Δm. We then have, in analogy to (7.6),

$$\Delta m \geqq 0, \qquad \Delta_\alpha^\beta m + \Delta_\beta^\gamma m = \Delta_\alpha^\gamma m.$$

If $\phi(\lambda)$ is a continuous function of position we can construct the integral

$$\int_0^1 \phi(\lambda) d_\lambda m. \tag{7.7}$$

An approximation to this integral can be found by dividing the entire interval $0 \leqq \lambda \leqq 1$ into small intervals Δ_i, choosing a point λ_i in Δ_i and evaluating the sum $\sum_i \phi(\lambda_i) \cdot \Delta_i m$. This sum then converges to the integral on allowing the Δ_i to approach zero. If the distribution has a continuous density

$$\lim_{\Delta\lambda \to 0} \frac{\Delta m}{\Delta \lambda} = \rho(\lambda)$$

the integral is identical with $\int_0^1 \phi(\lambda)\rho(\lambda) d\lambda$. But the Stieltjes integral (7.7) also includes the cases in which there exists no finite continuous density; in particular, it allows the existence of discrete points at which a finite amount of the substance is concentrated. If the substance is distributed over a finite number of points $\lambda = \alpha_i$ in amounts m_i, the Stieltjes integral reduces to the sum $\sum_i \phi(\alpha_i) m_i$.

We thus arrive at the following more inclusive formulation of the fundamental theorem concerning the transformation to principal axes: (1) *The Hermitian form A associates with each interval* Δ *an idempotent form* $\Delta E(\mathfrak{x})$; (2) *when two adjacent intervals* Δ_1, Δ_2 *are added together to form an interval* Δ,

$$\Delta E = \Delta_1 E + \Delta_2 E,$$

and the idempotent forms associated with separated intervals are independent ; (3) *we have*

$$\mathfrak{x}^2 = \int_{-\infty}^{+\infty} d_\lambda E(\mathfrak{x}), \qquad A(\mathfrak{x}) = \int_{-\infty}^{+\infty} \lambda \cdot d_\lambda E(\mathfrak{x}).$$

In this form the theorem is adapted to the appearance of continuous spectra of characteristic numbers, and is particularly appropriate for the purposes of quantum mechanics (cf. II, § 7). The discrete characteristic numbers lie at those points where the monotonic increasing function $\Delta^{\lambda}_{-\infty} E(\mathfrak{x}) = E(\lambda;\ \mathfrak{x})$ of λ has a discontinuity. In our example (7.5)

$$\Delta^{\beta}_{\alpha} E[\psi] = \int_{\alpha}^{\beta} \bar{\psi}(s)\psi(s)ds\ ;$$

here ψ must be taken as 0 outside the interval $(-\pi,\ +\pi)$. The evaluation in terms of the co-ordinates x_n is readily accomplished.

Consider the function space consisting of the totality of all functions $\psi(s)$ of a variable s, which assumes all values from $-\infty$ to $+\infty$, and which have a finite absolute magnitude

$$(\psi, \psi) = \int_{-\infty}^{+\infty} \bar{\psi}(s)\psi(s)ds,$$

i.e. which are " integrable square." The characteristic functions associated with the linear correspondence $\psi(s) \rightarrow \frac{1}{i}\frac{d\psi}{ds}$ are again the functions $e(\nu s)$, but the *frequency* ν can now assume *all real values.* The components of $\psi(s)$ are the quantities

$$f(\nu) = \frac{1}{\sqrt{2\pi}} \int_{-\infty}^{+\infty} \psi(s) e(-\nu s) ds.$$

Fourier's integral theorem then allows us to conclude the validity of the expansion

$$\psi(s) = \frac{1}{\sqrt{2\pi}} \int_{-\infty}^{+\infty} e(\nu s) f(\nu) d\nu$$

under certain assumptions concerning the differentiability of the function $\psi(s)$; but in any case the completeness relation [1]

$$\int_{-\infty}^{+\infty} \bar{\psi}(s)\psi(s)ds = \int_{-\infty}^{+\infty} \bar{f}(\nu)f(\nu)d\nu$$

is valid. We arrive at a somewhat different problem when we only require that the functions $\psi(s)$ be such that $\bar{\psi}(s)\psi(s)$ possess a definite mean value

$$\lim_{a\to\infty} \frac{1}{2a}\int_{-a}^{+a} \bar{\psi}(s)\psi(s)ds = (\psi, \psi);$$

this leads to the theory of *almost-periodic functions* developed by *H. Bohr.*[2] Here again the validity of the completeness relation can be established.

The theory of the characteristic numbers of Hermitian forms in infinitely many variables has been developed by *Hilbert* and *Hellinger*,[3] but it is applicable only to *bounded* forms

$$A(\mathfrak{x}) = \sum_{i,k} a_{ik}\bar{x}_i x_k,$$

i.e. forms whose values have a fixed upper bound when

$$\mathfrak{x}^2 = \sum_i \bar{x}_i x_i \leqq 1. \tag{7.8}$$

Indeed, without this assumption we cannot guarantee the convergence of $A(\mathfrak{x})$ in the entire domain (7.8); as an example, consider the form (7.4), $\sum_n n\bar{x}_n x_n$. That this form only converges in a portion of the domain (7.8) is merely another expression of the fact that not every continuous function is differentiable. The situation is more favourable for unitary forms as they satisfy the condition that they be "bounded" in consequence of their very definition; a unitary transformation is thereby to be taken as satisfying *both* of the conditions

$$U\tilde{U} = 1, \quad \tilde{U}U = 1.$$

The theorem on principal axes has been proved rigorously for bounded Hermitian and for unitary correspondences in ∞-dimensional space. A method due to *A. Wintner* [4] seems particularly appropriate for dealing with unitary correspondences; it is based on the consideration of the discrete group of all powers U^n of the given unitary transformation U, and determines the monotonic increasing function $E(\lambda\,;\mathfrak{x})$ of the real variable λ $(0 \leqq \lambda \leqq 2\pi)$ by means of the equations

$$U^n(\mathfrak{x}) = \int_0^{2\pi} e^{in\lambda}\, d_\lambda E(\lambda\,;\mathfrak{x}) \tag{7.9}$$

(the problem of trigonometric moments). *J. v. Neumann* [5] has gone furthest in dealing with linear operators for which boundedness is not postulated. In accordance with § 6 with a Hermitian form A is associated a group of unitary correspondences $e^{i\tau A} = U(\tau)$ depending on the real parameter τ and satisfying the equation

$$U(\tau + \tau') = U(\tau)U(\tau')\,; \tag{7.10}$$

the study of this group is equivalent to the study of A. It is therefore perhaps appropriate to replace this latter for ∞-dimensional space by the former, for no convergence difficulties appear in the domain of unitary transformations. We must therefore attempt to bring the operators $U(\tau)$, which are continuous functions of the real parameter τ satisfying (7.10) simultaneously into the form

$$U(\tau\,;\ \mathfrak{x}) = \int_0^{2\pi} e^{i\tau\lambda}\, d_\lambda E(\lambda\,;\mathfrak{x}). \tag{7.11}$$

This is accomplished with the aid of Wintner's method on replacing the discrete parameter n in (7.9) by the continuous parameter τ. The problem (7.11) bears the same relation to (7.9) as Fourier's integral bears to Fourier series.

In setting up a system of axioms for ∞-dimensional vector space the axioms (α), (β) of § 1 and the metric axiom (δ) of § 4 can be retained; for the proper substitute for the dimension axiom (γ) see, e.g., *v. Neumann*, "Mathematische Begründung der Quantenmechanik." [6]

The algebraic and geometric tools developed in this chapter offer a natural medium for the expression of quantum mechanics; they already hold a dominating position in the classical physics of continuous media. A masterly exposition of their mathematical content and application is found in the first part of *Courant-Hilbert's* "Methoden der mathematischen Physik," 2nd ed. (Berlin, 1930).

CHAPTER II

QUANTUM THEORY

§ 1. Physical Foundations [1]

THE magic formula

$$E = h\nu \qquad (1.1)$$

from which the whole of quantum theory is developed, establishes a universal relationship between the frequency ν of an oscillatory process and the energy E associated with such a process. The quantum of action h is one of the universal constants of nature

$$h = 6{\cdot}547 \times 10^{-27} \text{ erg secs.}$$

It was first discovered by *Planck* at the turn of the century in the laws of *black body radiation ;* that is, radiation which is enclosed in a cavity and is in thermodynamic equilibrium with matter of a definite temperature, which by emission and absorption causes an exchange of energy between the various frequencies contained in the radiation. Since this equilibrium is independent of the particular nature of the matter involved, Planck considered, as a kind of schematic matter, a system of linear oscillators of all possible frequencies. A charge oscillating with frequency ν interacts with the electromagnetic field by emitting and absorbing radiation of the same frequency. Planck assumed that the exchange of energy took place in integral multiples of an energy quantum ε ; he at first considered this assumption merely as a mathematical device, and intended to pass to the limit $\varepsilon = 0$. In order to obtain agreement with the Wien displacement law, which was derived from general thermodynamical principles, the energy quantum associated with a definite frequency ν must be taken proportional to ν: $\varepsilon = h\nu$. In this way Planck obtained his radiation formula, which is in excellent accord with observation ; according to it the amount of energy contained per unit volume in the spectral interval $\nu, \nu + d\nu$ in thermodynamic equilibrium at temperature θ is

$$u(\nu)d\nu = \frac{8\pi h\nu^3 d\nu}{c^3(e^{h\nu/k\theta} - 1)} \qquad (1.2)$$

where c is the velocity of light and k the Boltzmann constant ($\frac{3}{2}k\theta$ being the mean energy of an atom of a monatomic gas at temperature θ). On passing to the limit $h = 0$ we obtain the *Rayleigh-Jeans* radiation law

$$u(\nu) = \frac{8\pi\nu^2}{c^3} \cdot k\theta.$$

The assumption of the validity of this latter law for the entire spectrum is in gross disagreement with the facts, as it would lead to an infinite value for the total energy $\int u(\nu)d\nu$; a state of equilibrium would therefore be impossible with given finite energy.

The idea of a quantized exchange of energy, which occurs in Planck's derivation somewhat schematically and only in application to statistical thermodynamical consequences, was first seriously applied to individual atomic processes by *Einstein.* In 1905, guided by the observations of *H. Hertz, Hallwachs* and *Lenard* on the *photo-electric effect,* he enunciated the idea of a light quantum or photon as "an heuristic viewpoint concerning the generation and transformation of light"[2] according to which not only the exchange of energy between matter and radiation of frequency ν occurs in quanta of amount $h\nu$, but further, light of frequency ν can exist in the ether only in quanta of energy $h\nu$. The decisive experiments were first performed by *Millikan* ten years later. By allowing ultra-violet or X-radiation of frequency ν to fall on a metal plate electrons are released whose kinetic energy (as was already known to *Lenard*) increases with the hardness (i.e. with decrease of wave-length) of the incident radiation; the energy with which the electrons are emitted is, however, not influenced by the intensity of the radiation. The exact relation predicted by *Einstein* is

$$h\nu - P = \frac{mv^2}{2} = eV$$

where $-e$, m and v are the charge, mass and velocity of the electron, respectively. The energy $h\nu$ of the photon is transformed into kinetic energy of the electron, after subtracting from it the work P required to pull the electron out of the metal surface. If the potential difference between the metal surface and a plate placed in front of it is V' the electron current will disappear as soon as V' exceeds the critical value $V_0 = \frac{h\nu}{e}$. *Millikan* found that the potential at which the current vanished, obtained by extrapolation, was in fact exactly proportional to the frequency ν for monochromatic light of various frequencies, and that the constant of proportionality was equal to the quotient of the h obtained by *Planck* from black body radiation and the elementary quantum of electric charge e. The difference of the mean energy P for two different metals is furthermore equal to e times their contact difference of potential. The value of P, or at least its order of magnitude, is therefore known, and we find that for X-rays of a few Ångströms wave-length ($1\text{Å} = 10^{-8}$ cm.) P is negligible in comparison with $h\nu$. The equation

$$h\nu = \frac{mv^2}{2} = eV \tag{1.3}$$

governs not only the generation of secondary cathode rays by primary X-rays, but also the inverse process : the transformation at the glass wall or on the anode of the incident cathode rays into the impulse radiation first observed by *Röntgen*. If an electron which has run through the potential drop $-V$ in the X-ray tube loses its entire energy on collision, a photon of frequency ν and energy $h\nu = eV$ will spring into existence. The electron may, however, only be slowed down ; consequently ν is only the upper limit for the frequency of the impulse radiation, which will therefore consist of a continuous spectrum with a sharp limit at $\nu = \frac{eV}{h}$. The old classical theory of radiation was entirely unable to account for this most characteristic property of the impulse radiation. The frequency of the limit increases in proportion with the applied potential—and this is the exact formulation of the fact that " the higher the potential, the harder the rays " so familiar to every X-ray operator.

The observed phenomena thus confirm the hypothesis that radiation of frequency ν can be absorbed and emitted only in quanta of energy $h\nu$. This hypothesis will of course have further consequences for the theory of the structure of matter. The Planck oscillator will, for example, be unable to alter its energy continuously since it can only emit or absorb these fixed quanta of energy, and it will consequently spring to and fro on the rungs of its energy ladder, which are equally spaced at intervals $h\nu$; *ν is here the frequency of the oscillator*, a constant determined by the constitution of the oscillator. An application of the essential elements of this idea to actual atoms gave rise to the frequency rule enunciated by *Niels Bohr* (1913) :

An atom can exist only in certain discrete stationary states (" quantum states ") in which it does not radiate. Light will be emitted on transition from one state into another ; the energy which it loses in this transition, the difference $E_1 - E_2$ *of its energy in the two states, will be transformed into a photon of energy* $h\nu$, *the frequency* ν *of which is determined by the equation*

$$h\nu = E_1 - E_2. \qquad (1.4)$$

In this equation E_1, E_2 *may be any two of the discrete energy levels* ($E_1 > E_2$). *Conversely, in absorption a photon raises the atom from the energy level* E_1 *to a higher* E_2 *by giving up its energy* $h\nu$ *to the atom.*

According to classical electrodynamics an atom should continually emit radiation in consequence of the vibrations of its constituent electrons, and the frequencies of the emitted light should agree with the frequencies of the simple oscillations into which the motion of its electronic system can be resolved. But the atom will itself lose energy through this radiation, the motion of its electrons will thereby be modified and the frequencies will consequently be displaced. This entire point of view is therefore irreconcilable with one of the most fundamental physical facts: *the existence of sharp spectral lines.* On the other hand, Bohr's assumption is not only in agreement with this fact, although it offers no such detailed picture of the reaction between matter and ether as the classical theory, but contains in addition the fundamental *Ritz-Rydberg combination principle.* If we order the energy levels in an increasing series $E_0 < E_1 < E_2 < \cdots$, then in accordance with (1.4) each frequency ν is the difference of two "*terms*" $\nu_i = E_i/h$,

$$\nu(i \to k) = \nu_i - \nu_k \qquad (i > k).$$

Consequently there will occur in addition to the frequencies $\nu(i \to k)$, $\nu(k \to l)$ *the frequency*

$$\nu(i \to l) = \nu(i \to k) + \nu(k \to l) \qquad (1.5)$$

obtained from them by addition. This combination principle is valid without exception in the whole of spectroscopy, in the optical region as well as in that of X-rays, and has proved to be a valuable guide in the classification of spectra; it reduces the complex line spectra to the simpler term spectra. Unfortunately the problem is made more difficult by the fact that not all lines corresponding to possible transitions $i \to k$ need actually occur—not every term ν_i need "combine" with a given term ν_k—for the conditions of excitation may be such that certain lines have zero intensity. The *selection rules* for the allowable transitions will therefore be contained in the rules which determine the intensities of spectral lines. The combination principle, or the Bohr frequency rule, determines, so to speak, only the keyboard of the spectrum—which tones are really struck is dependent on the mode of excitation. But it will in general be possible under proper conditions of excitation, e.g. the influence of strong external electric fields, to bring out the lines which are not observed under ordinary conditions.

In the " unexcited " or normal state the atom is in the stationary state of lowest energy E_0, and consequently only the lines of the " series " $n \to 0$, of frequency $\nu_n - \nu_0$ $(n = 1, 2, \cdots)$, occur in absorption. The lowest of these $1 \to 0$ (i.e. with greatest wavelength), or more precisely the lowest which is not forbidden by the selection rules, is called the " resonance line."

The simplest atom is that of hydrogen ; in it a single electron of charge $-e$ revolves about a nucleus of opposite charge $+e$. The terms of the spectrum of atomic hydrogen are found by observation to be given by the equation

$$\frac{\nu_n}{c} = -\frac{R}{n^2} \tag{1.6}$$

where $R = 109700$ cm.$^{-1}$ is the *Rydberg constant* (spectroscopists are accustomed to give the *wave number* ν/c, the reciprocal wavelength, instead of the frequency ν). The energy levels corresponding to these frequency terms are $E_n = -\frac{Rhc}{n^2}$. To this discrete term spectrum we must add the continuous spectrum $E \geqq 0$; the additive constant in the energy is so chosen that $E = 0$ separates the hyperbolic electron orbits from the elliptic. The *Balmer series* consists of the lines $n \to 2$ with wave numbers

$$R\left(\frac{1}{4} - \frac{1}{n^2}\right) \quad (n = 3, 4, \cdots).$$

This is the oldest known series formula ; *Balmer* obtained it in 1885 by abstraction from the first four lines of the series, called H_α, H_β, H_γ, H_δ, which lie in the visible region. The lines of this series converge with increasing n to a limit with wave number $\frac{R}{4}$ $\left(\text{wave-length } \frac{4}{R} = 3650\text{Å}\right)$. $\frac{cRh}{4}$ is the work required to ionize an H-atom in the stationary state $n = 2$, i.e. the work required to remove the electron from such an atom without leaving it with kinetic energy. The continuous spectrum, arising from transitions which ionize the atom, will join on to this series limit on the short wave side. We are further acquainted with the *Lyman* series $n \to 1$ which lies in the ultraviolet and also occurs in absorption, the *Paschen* series $n \to 3$

lying in the infra-red, and finally with some members of the *Brackett* ($n \rightarrow 4$) and *Pfund* ($n \rightarrow 5$) series in the far infra-red. In order to ionize hydrogen in the normal state an amount cRh of work must be done; the corresponding "ionization potential," i.e. the potential difference an electron must traverse before it is able to ionize atomic hydrogen by means of its kinetic energy, is

$$V = \frac{cRh}{e} = 13{\cdot}53 \text{ volts.}$$

Bohr's frequency rule goes beyond the combination principle in asserting that the terms are actually energy levels, an assertion irrelevant to and not verifiable by spectroscopy. That this is, however, in fact the case is confirmed by the experiments of *Franck and Hertz* on *collision phenomena.*[3] In these experiments electrons are given an amount eV of kinetic energy by allowing them to pass through an electric field of known potential difference $- V$ and are then allowed to pass through a gas consisting of the atoms which are to be investigated with the velocity thus obtained, without further influence from external fields. The electron can give up no energy to the atom until eV is greater than the excitation energy $E_1 - E_0$ of the resonance line; if

$$E_1 - E_0 < eV < E_2 - E_0$$

then the electron can either suffer an "elastic collision," in which case it loses no energy, or it can suffer an "inelastic collision," in which case it loses an amount $E_1 - E_0$ to the atom. The electrons which have passed through the gas are of two kinds, those with kinetic energy eV and those with $eV - (E_1 - E_0)$. When the atoms which have been raised from the state 0 to the state 1 by collision with electrons fall back into the normal state they emit the resonance line and, under the above conditions, only this line. This is fully confirmed by the experiment. The kinetic energy of the emerging electrons is measured by introducing a retarding potential V'; the electrons only come through it if their energy is greater than eV'. In general the electrons possess a discrete "energy spectrum" after collision with an atom of the gas; the possible energy values are

$$eV_n' = eV \quad (E_n - E_0)$$

($n = 0, 1, 2, \cdots$, in so far as V_n' is still positive; we here disregard the possibility that a single electron may suffer more than one inelastic collision). On allowing the retarding potential V' to decrease gradually from a value which is greater than V the electron current decreases suddenly whenever V' passes through one of the values $V'_0, V'_1, \cdots$

Bohr's frequency rule reduces the determination of spectra to the problem of *obtaining the stationary states and the corresponding energy levels of an atom*, i.e. *of a mechanical system of known dynamical constitution.* The example of the linear oscillator given above and the fundamental notions of the theory of oscillations suggest the following as a general guiding principle (P) : the frequencies derived from the energy levels by means of Bohr's frequency rule shall correspond to the frequencies of the simple vibrations into which the actual motion of the atomic constituents can be resolved in accordance with the laws of dynamics. Such a resolution into simple oscillations is convincingly attainable in classical mechanics only if the system is " multiply " or " conditionally periodic," and for this case it was actually found possible to sharpen the general principle (P) into a definite rule for quantization. In the years 1913-25 the application of this quantum rule yielded a great harvest of results, and it seemed that we were in possession of the key that would unlock the mysteries of atomic processes. But the wards did not quite fit ; toward the end of this epoch its failure became more and more apparent and the physical theory was gradually reduced to a symbolic calculus of quantum numbers which had to be corrected each time a new fact was discovered. We do not wonder now that it ran such a course, but rather are surprised that it was as successful as it was !

From the beginning the quantum rules were a compromise. If a mechanical system of one degree of freedom undergoes a periodic motion the frequencies ν of the simple vibrations into which its motion can be resolved are integral multiples of a fundamental frequency ω. This frequency depends on the energy of the orbit under consideration, and this latter is restricted by the quantum rules to the discrete set E_n. The internal frequencies of the motion are therefore given by the formula

$$\nu = k \cdot \omega(n) \tag{1.7}$$

which depends on the two integers n and k. By the analogy with quantum mechanical frequencies this internal frequency (1.7) is to be ascribed to the jump $n \rightarrow (n - k)$. The fact that ν depends linearly and homogeneously on the jump k is expressed by the " *classical combination principle* "

$$\nu(n \rightarrow n - k) + \nu(n \rightarrow n - l) = \nu(n \rightarrow n - k - l) \tag{1.8}$$

in consequencie of which frequencies with the *same initial state* n will combine. But this is not in accord with the correct combination principle

$$\nu(n \to n - k) + \nu(n - k \to n - k - l) = \nu(n \to n - k - l) \quad (1.9)$$

The changes k, l in the quantum number are here the same as in (1.8), but the *final state* $n - k$ of the first frequency coincides with the *initial state* of the second; only for quantum numbers n which are large compared with k and l does the classical principle agree asymptotically with the Ritz-Rydberg combination principle. Consequently if the general principle (P) is to be satisfied without compromise our mechanics must be altered in such a way that the false combination principle (1.8) is replaced by the correct one (1.9). In 1925 *Heisenberg* discovered a way in which such an alteration can be naturally accomplished; in order to do this, however, it was necessary to give up the picture of an atom with its electronic orbits. The quantities with which the Heisenberg theory deals are only the frequencies and intensities of radiation associated with transitions between the various states of the atom.

It should be observed that the correct combination principle (1.9) is in one important respect simpler than the false one (1.8). As the formulation

$$\nu(n'' \to n') + \nu(n' \to n) = \nu(n'' \to n) \quad (1.10)$$

shows, the quantum numbers serve only as distinguishing marks or indices which do not involve a law of composition, whereas the classical formula requires the addition of quantum numbers, which are therefore numbers on a definite scale.

Another approach to quantum mechanics was discovered by *L. de Broglie* and *E. Schrödinger.*[4] This approach seems to me less cogent, but it leads more quickly to the fundamental principles of quantum mechanics and to the most important consequences for experimental science. We shall therefore follow it, since we are more concerned in giving a short but comprehensive account than in giving a complete discussion of the physical foundations. The physical, essentially statistical, interpretation of the theory, with which *Schrödinger* has not been entirely in accord, is due mainly to *M. Born.*

§ 2. The de Broglie Waves of a Particle

We consider the *undulatory character of light* as guaranteed by the phenomena of *diffraction* and *interference.* Their most decisive feature is that with them we are dealing with *the linear superposition of waves with arbitrary differences of phase.* From the mathematical standpoint, they are characterized by the fact that they involve addition and multiplication with complex numbers, and we are consequently dealing with vectors in a complex space. We can, in fact, consider a complex function $\psi(t\,;xyz)$ employed in the description of the phenomena and defined over time and space as such a vector, where each space-time point represents one dimension of a complex vector space; the differential laws for such a wave function ψ—or for several such functions simultaneously, such as the components of the electric and magnetic field strengths—are linear and homogeneous. But on the other hand the quantum phenomena which we discussed above speak just as plainly in favour of the *corpuscular nature of light.* The intensity of the monochromatic radiation employed in the production of the photoelectric effect has no influence on the velocity with which the electrons leave the metal; it influences only the frequency of this event. Even with intensities so weak that on the classical theory hours would be required before the electromagnetic energy passing through a given atom would attain to an amount equal to that of a photon, the effect begins immediately, the points at which it occurs being distributed irregularly over the entire metal plate. This constitutes a proof of the existence of photons which is no less direct than the proof that α-particles are of corpuscular nature by observing the scintillations caused by them on striking a sensitized screen. Further, if one considers the exchange of momentum in addition to that of energy in deriving the laws of black body radiation, conflict with Planck's hypothesis concerning energy quanta can be avoided only by assuming that in addition to the emission of the energy quantum $h\nu$ a quantum $h\nu/c$ of momentum is emitted in *a definite direction,* producing an equivalent reaction on the atom.[5] We here replace the continuous radiation of a spherical wave by the discontinuous emission of photons in definite directions which are irregularly distributed over the compass.

We unite the two standpoints by *retaining the linear wave equation, but considering the intensity* $\bar{\psi}\psi$ *as the relative probability that the photon appears at the point* (x, y, z) *at time* t; or, more precisely, that

$$\bar{\psi}\psi\, dxdydz \tag{2.1}$$

is the probability that at time t it will be found within the small parallelepiped with sides of length dx, dy, dz about the point

(x, y, z).* But we can only expect to arrive at a rational theory *if we deal with material particles in the same way as with photons.* This point of view was developed in the *Bose-Einstein* treatment of an atomic gas, which paralleled that employed in the theory of black body radiation ("light quant gas").[6] *Schrödinger's* researches took as their point of departure the Hamiltonian theory of mechanics, which was originally obtained by *Hamilton* himself from an analogy with geometrical optics. He argued that since we replace geometrical optics, with the aid of which interference and diffraction cannot be treated, by wave optics, it is reasonable to attempt the analogous transition in mechanics. The results amply justified the attempt. The investigations of *Davisson and Germer*, which prove the existence of interference in beams of electrons reflected from a crystal lattice, were already in progress when de Broglie published his theory. The experimental evidence that moving material particles behave in much the same way as a beam of light with respect to these phenomena is now fully established, and with no less certainty than for X-rays, by a series of further investigations by the same authors and by *G. P. Thomson*, *F. Rupp* and others.[7] The real difference between "light-like" and "electron-like" beams lies in the fact that the particles composing the latter possess *charge* and *proper mass* and can consequently be deflected by electric and magnetic fields.

A simple oscillation is one in which the function ψ, defining the state of the system, depends on the time in accordance with the law

$$\psi(t) = a \cdot e^{-i\nu t} \tag{2.2}$$

where a and ν are independent of t. [We choose as our unit of angular measure that one which proves most useful in differential calculus, for it yields the simple relation

$$\frac{1}{i}\frac{de(x)}{dx} = e(x) \tag{2.3}$$

for the fundamental trigonometric function $e^{ix} = e(x)$. The sum of the angles about a point is then 2π; it would, admittedly, be more correct from the integral standpoint to take this as 1, but then the factor 2π would appear in the differential relation. $\nu/2\pi$ is the number of oscillations in unit time; we shall not

* Just as in the classical wave theory we have an expression for the flow of energy in addition to its density, so in the more refined formulation of quantum theory we will have an expression for the probability that the photon passes through a given element of surface ("probability current") in addition to one for the probability that it be found in a given element of volume ("probability density").

hesitate, however, to use the name "frequency" for ν. *If we denote Planck's constant of action by $2\pi h$ instead of h,* and we shall throughout the present work, the fundamental formula (1.1) will still be valid in the new nomenclature.] In accordance with (2.3) the simple oscillations (2.2) are the characteristic functions of the linear Hermitian operator which carries ψ over into $-\frac{h}{i}\frac{d\psi}{dt}$; the corresponding characteristic numbers are the energies $E = h\nu$. If the dependence of a state of the system on time is described by a superposition of simple oscillations

$$\psi(t) = a_1 e^{-i\nu_1 t} + a_2 e^{-i\nu_2 t} + \cdot \cdot \cdot, \tag{2.4}$$

the energy is capable of assuming only one of the values $h\nu_1$, $h\nu_2$, $\cdot \cdot \cdot$, and we shall take the intensity $\bar{a}_r a_r = |a_r|^2$ of the oscillation of frequency ν_r in ψ as the relative probability that the energy is observed to be $h\nu_r$. The relation $E = h\nu$ is accordingly to be interpreted: *if ν is indeterminate because an entire spectrum of frequencies ν is contained in the oscillatory process, then the energy is indeterminate to the same extent; the intensities with which the various simple oscillations occur in the process measure the probabilities of the corresponding energies.* The operator $-\frac{h}{i} \cdot \frac{d}{dt}$ represents the energy:

$$H \rightarrow -\frac{h}{i}\frac{d}{dt} \tag{2.5}$$

in the following sense: *a characteristic function of* (2.5) *represents a state in which the energy assumes a definite value E with certainty. This value is the corresponding characteristic number; in an arbitrary state the components a of ψ with respect to these characteristic functions determine the relative probabilities $\bar{a}a$ of these values E.*

According to the theory of relativity energy is to be considered as the time component of a 4-vector whose spatial components constitute the linear momentum $\mathfrak{p} = (p_x, p_y, p_z)$. The fundamental metric invariant of the two vectors running from the origin to the points (t, xyz), $(t', x'y'z')$ is the scalar product

$$c^2tt' - (xx' + yy' + zz').$$

Under a Lorentz transformation, which transforms from one space-time co-ordinate system to another equally permissible one, the quantities

$$c^2t, -x, -y, -z$$

must consequently transform contragrediently to t, xyz; they are therefore the components of the vector associated with

$(t,\ x\,y\,z)$ in the space which is the dual of the 4-dimensional space-time world. Such a dual vector is given by

$$H, \ -p_x, \ -p_y, \ -p_z;$$

or, what amounts to the same thing,

$$Hdt - (p_x dx + p_y dy + p_z dz)$$

is invariant under Lorentz transformations. The same is true of the total differential operator

$$d = \frac{\partial}{\partial t}dt + \left(\frac{\partial}{\partial x}dx + \frac{\partial}{\partial y}dy + \frac{\partial}{\partial z}dz\right)$$

applied to an arbitrary function of t; x, y, z. Hence the correspondence (2.5) necessarily implies the further relations

$$p_x \to \frac{h}{i}\frac{\partial}{\partial x}, \quad p_y \to \frac{h}{i}\frac{\partial}{\partial y}, \quad p_z \to \frac{h}{i}\frac{\partial}{\partial z}, \tag{2.6}$$

which are to be given the analogous interpretation.

A homogeneous plane wave

$$\psi = a \cdot e^{i(-\nu t + \alpha x + \beta y + \gamma z)} \tag{2.7}$$

is simultaneously a characteristic function of the four mutually commutative operators (2.5), (2.6), which has as characteristic numbers

$$H = h\nu; \quad p_x = h\alpha, \quad p_y = h\beta, \quad p_z = h\gamma. \tag{2.8}$$

It represents a state in which the energy and linear momentum of the quantum possess these sharply defined values.

In classical mechanics the laws governing the motion of a particle are known as soon as we express its energy H in terms of the "canonical variables" $x\,y\,z$, $p_x p_y p_z$. In Newtonian mechanics the Hamiltonian function for a free material particle of mass m is

$$H = \frac{p_x^2 + p_y^2 + p_z^2}{2m}; \tag{2.9}$$

on employing the transition scheme developed above we obtain the corresponding wave equation

$$\frac{h}{i}\frac{\partial\psi}{\partial t} - \frac{h^2}{2m}\Delta\psi = 0 \quad \left(\Delta = \frac{\partial^2}{\partial x^2} + \frac{\partial^2}{\partial y^2} + \frac{\partial^2}{\partial z^2}\right). \tag{2.10}$$

(2.7) is a solution of this equation provided the values (2.8) of energy and linear momentum satisfy equation (2.9); in this sense (2.9) and (2.10) are equivalent. But the equation (2.10) is linear and has as its most general solution a linear superposition of simple waves (2.7); such a superposition corresponds to a state in which the energy and momentum of the particle assume their various permissible values " with a certain definite probability."

The space vector (α, β, γ) in (2.7) gives the direction of propagation of the plane wave, and the modulus of this vector is the wave number μ (the number of waves contained in 2π units of length; $2\pi/\mu$ is the wave length λ). Hence by (2.8) the absolute value p of the momentum is equal to $h\mu = \frac{2\pi h}{\lambda}$. $\frac{\nu}{\mu}$ is the *phase velocity* of the wave; in accordance with (2.9) or

$$\nu = \frac{h}{2m} \cdot \mu^2$$

it is $h\mu/2m = h\pi/\lambda m$ and depends on the wave length or frequency (*dispersion*). Since $p = mv$, where v is the velocity of the particle, the "*group velocity*" $\frac{d\nu}{d\mu} = \frac{h\mu}{m} = v$ coincides with the velocity of the particle. Experiments on diffraction and interference phenomena in electron beams, such as those performed by *Davisson and Germer*, have made it possible to test directly these relations set up by *de Broglie*.

In relativistic mechanics we have in place of (2.9) an equation which states that the square of the absolute value of the energy-momentum 4-vector is constant and equal to m^2c^2:

$$\frac{H^2}{c^2} - (p_x^2 + p_y^2 + p_z^2) = m^2c^2 \tag{2.11}$$

or

$$H = c\sqrt{m^2c^2 + (p_x^2 + p_y^2 + p_z^2)}.$$

For the transition to a wave equation it is of advantage to employ the rational form (2.11) of this expression:

$$-\frac{1}{c^2}\frac{\partial^2\psi}{\partial t^2} + \Delta\psi = \frac{m^2c^2}{h^2} \cdot \psi. \tag{2.12}$$

Here again the group velocity is equal to the velocity v of the particle, but the phase velocity is found to be c^2/v; the former is always less, the latter always more than the velocity of light. In order to return from the relativistic to the "ordinary" or Newtonian mechanics by passing to the limit $c \to \infty$, we must first replace H by $mc^2 + H$, i.e. ψ must be replaced by $e\left(-\frac{mc^2 t}{h}\right) \cdot \psi$.

The differential equation governing light waves can be obtained from (2.12) by dropping the term on the right-hand side.

Hence from the corpuscular standpoint light consists of photons or particles of proper mass 0:

$$\frac{H^2}{c^2} - (p_x^2 + p_y^2 + p_z^2) = 0.$$

In accordance with the expression (2.1) for the probability density, we are to consider as the vector in unitary system-space describing the state of the system the function ψ *in so far as it depends on the spatial co-ordinates xyz.* The integral of (2.1) with respect to the spatial co-ordinates gives the probability that the particles will be found " within the volume V at time t." Space and time must be separated from one another; the system has *at each time t* a definite state $\psi(xyz)$, which will in general vary with t. The operators which represent physical quantities must accordingly be ones which operate on an arbitrary function of the spatial co-ordinates. This requirement is satisfied by the operators (2.6) corresponding to the momentum co-ordinates, but not by differentiation with respect to time, which we have associated with the energy. We must instead consider the situation as described as follows: from the expression for the energy in terms of the canonical variables p_x, p_y, p_z we obtain the operator H which represents the energy and which operates on the function $\psi(xyz)$. The equation

$$\frac{h}{i}\frac{\partial\psi}{\partial t} + H\psi = 0$$

is then *the dynamical law* which determines the change in the state ψ in time.

The separation of space and time offers certain difficulties to the development of quantum theory from the relativistic standpoint; consequently, for the present, we base our development on the Newtonian mechanics.

Our procedure must eventually be modified in another important respect: we have here tacitly assumed, for the sake of mathematical simplicity but without physical justification, that the wave field of a material particle is described by a *scalar* quantity ψ. The modification, which is required in order to give an adequate description of the facts of spectroscopy, will be made in Chap. IV.

§ 3. Schrödinger's Wave Equation. The Harmonic Oscillator

When the particle is moving under the influence of forces the kinematic part (2.9) of the energy is augmented by the potential energy, which usually depends on the co-ordinates alone and not on the momenta. We must therefore know which Hermitian operator acting on ψ corresponds to the co-ordinate x. I assert that it is *multiplication by x*; this operator is already referred to its principal axes, its characteristic values are all real numbers x and finally $\psi(x)$, or more precisely $\psi(x)\sqrt{dx}$, is the component of the " vector " associated with the characteristic number x (we have here ignored the other co-ordinates y, z). In accordance with the statistical interpretation of the relationship between physical quantities and operators, our assertion is: the probability that x has a value between x_1 and x_2 is $\int_{x_1}^{x_2}\bar{\psi}\psi dx$; this is in agreement with the expression (2.1) for the probability density. If $V(xyz)$ is a function of position in the 3-dimensional space, e.g. the potential energy, then the physcial quantity V is represented by the operator

$$\psi \to V(xyz)\cdot\psi,$$

for the probability that V lies between V_1 and V_2 is given by the integral

$$\iiint \bar{\psi}\psi dxdydz$$

extended over that portion of space in which $V_1 \leqq V(xyz) \leqq V_2$.

The operators corresponding to x, y, z commute with each other, but the operator Q corresponding to x and the operator P corresponding to p_x do not. In fact

$$\frac{d}{dx}[x\psi(x)] - x\frac{d}{dx}\psi(x) = \psi(x)$$

or

$$PQ - QP = \frac{h}{i}\mathbf{1}$$

where the $\mathbf{1}$ on the right-hand side stands for the operator *identity*: $\psi(x) \to \psi(x)$. Because of this non-commutative relation between the operators P and Q, *p_x cannot assume a definite value with certainty when x does, and converselv.* In fact, if p_x is known to have the value $h\alpha$ with certainty, then the dependence of ψ on x is given by the factor $e^{i\alpha x}$; in consequence of this the position x of the particle is entirely indeterminate, since the probability $\bar{\psi}\psi$ of localization is the same for all points x.

If $V(x, y, z)$ is the potential energy of the field in which the particle moves, the total energy is

$$H = \frac{p_x^2 + p_y^2 + p_z^2}{2m} + V(x, y, z). \qquad (3.1)$$

We assume with *Schrödinger* that in spite of the fact that all our variables do not commute we may still apply our rules for the formulation of the wave equation; we thus obtain *Schrödinger's differential equation*

$$\frac{h}{i}\frac{\partial\psi}{\partial t} - \frac{h^2}{2m}\Delta\psi + V(xyz)\cdot\psi = 0.$$

We understand by " stationary " or " quantum states " ψ those in which the energy E has a definite value; they are characterized as solutions of the wave equation which satisfy in addition the equation [cf. (2.5)]

$$-\frac{h}{i}\frac{\partial\psi}{\partial t} = E\cdot\psi.$$

On setting $E = h\nu$, such a ψ will have the form $e^{-i\nu t}\cdot\psi$ where the new function denoted by ψ is independent of t. This function $\psi(xyz)$, which depends only on the spatial co-ordinates, satisfies the reduced equation

$$\frac{h^2}{2m}\Delta\psi + [E - V(xyz)]\psi = 0.$$

The problem is thus reduced to finding values of E and functions $\psi \neq 0$ of position which satisfy this equation and are such that the integral of $\bar{\psi}\psi$ over the entire space is finite. They are the characteristic numbers and characteristic vectors of the Hermitian operator H associated with the energy (3.1) in the function space of all functions of position ψ. *The characteristic numbers E are the possible energy levels of the particles.*

Before going any further into the interpretation of the theory we have developed, it will be well to convince ourselves that it leads to energy levels which are in agreement with the facts. The simplest example is that of the *linear oscillator;* with it we are dealing with only one co-ordinate x. The potential energy is $V(x) = \frac{a}{2}x^2$ and the total energy

$$H = \frac{1}{2}\left(\frac{p^2}{m} + ax^2\right). \tag{3.2}$$

The equation for the determination of the characteristic values E and the associated characteristic functions ψ is

$$\frac{h^2}{2m}\frac{d^2\psi}{dx^2} + \left(E - \frac{a}{2}x^2\right)\psi(x) = 0. \tag{3.3}$$

Hermitian polynomials. The solutions of this equation are expressed in terms of Hermitian polynomials. The nth Hermitian polynomial $\eta_n(x)$ is defined by the equation

$$\frac{d^n}{dx^n}(e^{-\frac{1}{2}x^2}) = (-1)^n e^{-\frac{1}{2}x^2} \cdot \eta_n(x) \; ; \tag{3.4}$$

it is of nth degree and the highest term is exactly x^n. The $\eta_n(x)$ $(n = 0, 1, 2, \cdots)$ constitute an orthogonal set of functions with the "density function" $e^{-x^2/2}$:

$$\int_{-\infty}^{+\infty} e^{-x^2/2}\, \eta_n(x)\eta_m(x)dx = 0, \quad m \neq n \; ; \tag{3.5}$$

the functions

$$\phi_n(x) = e^{-x^2/4} \cdot \eta_n(x)$$

are consequently orthogonal in the ordinary sense. To prove this we need merely to note that

$$(-1)^n \int_{-\infty}^{+\infty} \frac{d^n}{dx^n}(e^{-x^2/2}) \cdot \eta_m(x)dx$$

becomes, on integrating n times by parts,

$$\int_{-\infty}^{+\infty} e^{-x^2/2} \cdot \frac{d^n \eta_m(x)}{dx^n} dx$$

and the integrand vanishes for $m < n$. For $m = n$ we obtain

$$n\,! \int_{-\infty}^{+\infty} e^{-x^2/2} dx$$

so the equations (3.5) can be supplemented by

$$\gamma_n^2 = \int_{-\infty}^{+\infty} e^{-x^2/2}\, \eta_n^2(x)dx = n!\, \sqrt{2\pi}.$$

From (3.4) we have

$$e^{-x^2/2} \cdot \eta_{n+1}(x) = -(-1)^n \frac{d^{n+1}}{dx^{n+1}}(e^{-x^2/2})$$

and we can consider $\dfrac{d^{n+1}}{dx^{n+1}}$ as either $\dfrac{d^n}{dx^n}\left(\dfrac{d}{dx}\right)$ or $\dfrac{d}{dx}\left(\dfrac{d^n}{dx^n}\right)$. Since

$$-\frac{d}{dx}(e^{-x^2/2}) = x \cdot e^{-x^2/2}$$

and

$$\frac{d^n}{dx^n}(xe^{-x^2/2}) = x\frac{d^n}{dx^n}(e^{-x^2/2}) + n\frac{d^{n-1}}{dx^{n-1}}(e^{-x^2/2}),$$

the first of these interpretations yields the recursion formula

$$\eta_{n+1}(x) = x\eta_n(x) - n\eta_{n-1}(x). \tag{3.6}$$

From the second we find

$$-\frac{d}{dx}[e^{-x^2/2}\cdot\eta_n(x)] = e^{-x^2/2}\cdot\eta_{n+1}(x)$$

or

$$\eta_{n+1}(x) = -\frac{d\eta_n}{dx} + x\eta_n(x). \tag{3.7}$$

On subtracting the recursion formula (3.7) from (3.6) we find the simple relation

$$\frac{d\eta_n}{dx} = n\eta_{n-1}. \tag{3.8}$$

Differentiating (3.7) and substituting $(n+1)\eta_n$ for the derivative of η_{n+1} in accordance with (3.8), we obtain the differential equation

$$\frac{d^2\eta_n}{dx^2} - x\frac{d\eta_n}{dx} + n\eta_n = 0.$$

The equation for $\phi_n(\xi)$ is consequently

$$\frac{d^2\phi_n}{d\xi^2} - \frac{\xi^2}{4}\phi_n + \left(n+\frac{1}{2}\right)\phi_n = 0. \tag{3.9}$$

On going over to a new unit of length by the substitution $x = \alpha\xi$, the left-hand side of (3.3) is equal to the left-hand side of (3.9) multiplied by $h^2/2m\alpha^2$ provided

$$\frac{h^2}{2m\alpha^2}\cdot\frac{1}{4} = \frac{a\alpha^2}{2}, \qquad \frac{h^2}{2m\alpha^2}\left(n+\frac{1}{2}\right) = E.$$

Let $\omega = \sqrt{a/m}$ denote the classical frequency of the oscillator. The first of these conditions determines the new unit of length α:

$$\alpha^2 = \frac{h}{2\sqrt{am}} = \frac{\hbar}{2m\omega},$$

and the second requires that

$$E = E_n = h\omega(n+\tfrac{1}{2}). \tag{3.10}$$

It is possible to show that the $\phi_n(\xi)$ constitute a complete orthogonal system,[8] and consequently there can exist no further characteristic numbers and functions. *The oscillator possesses the discrete energy levels* (3.10) *at intervals* $h\omega$ *apart.* That the lowest energy level turns out to be $\frac{1}{2}h\omega$ instead of 0 is of itself of no significance, as we may always introduce an additive constant into the energy, although it is meaningful to assert that the least possible value of the quantity H, (3.2), is equal to $\frac{1}{2}h\omega$.

However, the wave equation not only yields the energy levels as characteristic values, but it also gives us information concerning the probability of localization by means of the characteristic functions. For convenience we now take $\alpha = \sqrt{\frac{h}{2m\omega}}$ as the unit of length. *When the oscillator is in the state described by the* n^{th} *energy level, the probability that the oscillating particle is at a distance* x *from its position of equilibrium is given by* $e^{-x^2/2} \cdot \eta_n^2(x)$. These probabilities are to be understood as relative, and refer to equal infinitesimal intervals about the points of comparison x. In particular, for the lowest energy level $n = 0$ the probability density is $e^{-x^2/2}$; we can therefore no longer say that the mass-point is at rest in the position of equilibrium, but rather *the probability of its displacement from this position is given by a Gauss error curve.* The normalized characteristic functions of (3.3) are given by

$$\psi_n(x) = \frac{1}{\gamma_n} \phi_n(x).$$

On expressing any function $\psi(x)$ of position in terms of this set

$$\psi(x) \sim \sum_{n=0}^{\infty} x_n \psi_n(x), \quad x_n = \int_{-\infty}^{+\infty} \psi(x)\psi_n(x)dx,$$

and the operator belonging to the energy H is, as we have already seen, expressed in terms of these co-ordinates ψ_n by

$$x_n \to h\omega(n + \tfrac{1}{2}) \cdot x_n.$$

In order to find the operator associated with the co-ordinate x we must express $x\psi_n(x)$ linearly in terms of the characteristic functions themselves; by (3.6) we have

$$x\phi_n = \phi_{n+1} + n\phi_{n-1}$$

whence

$$x\psi_n = \frac{\gamma_{n+1}}{\gamma_n} \psi_{n+1} + \frac{n\gamma_{n-1}}{\gamma_n} \psi_{n-1} = \sqrt{n+1}\, \psi_{n+1} + \sqrt{n}\, \psi_{n-1}.$$

The correspondence $\psi(x) \to x\psi(x)$ is thus expressed in terms of these Fourier coefficients by

$$x_n \to \sqrt{n}\, x_{n-1} + \sqrt{n+1}\, x_{n+1};$$

its matrix $\|q_{nm}\|$ contains only the elements

$$q_{n,\,n-1} = \sqrt{n}, \quad q_{n,\,n+1} = \sqrt{n+1}. \tag{3.11}$$

(On returning to the original unit of length the right-hand side must be multiplied by the factor α.) On applying the operator $\frac{d}{dx}$ to ϕ_n we obtain, in accordance with (3.8) and (3.6),

$$\frac{d\phi_n}{dx} = \tfrac{1}{2}\, n\phi_{n-1} - \phi_{n+1}$$

whence

$$\frac{d\psi_n}{dx} = \tfrac{1}{2}(\sqrt{n}\,\psi_{n-1} - \sqrt{n+1}\,\psi_{n+1}).$$

The linear Hermitian correspondence associated with the momentum $p = \frac{h}{i}\frac{d}{dx}$ is accordingly

$$x_n \to \frac{h}{2i}(-\sqrt{n}\,x_{n-1} + \sqrt{n+1}\,x_{n+1})\,;$$

its matrix $\|p_{nm}\|$ has as its only non-vanishing elements those for which $m = n \pm 1$:

$$p_{n,\,n-1} = -\frac{h}{2i}\sqrt{n}, \quad p_{n,\,n+1} = \frac{h}{2i}\sqrt{n+1}. \tag{3.12}$$

(On returning to the original unit of length these elements are to be multiplied by $1/\alpha$.—Terms with the index $n-1$ are to be omitted when $n = 0$; in fact, they automatically drop out of the above formulæ.)

§ 4. Spherical Harmonics

In order to discuss the energy levels of an electron in a spherically symmetric electrostatic field we must first discuss *spherical harmonics* and their principal properties.

1. *Definition.*—Let r denote the distance from the origin in the 3-dimensional space with co-ordinates x, y, z, and let r, θ, ϕ be polar co-ordinates with polar axis along the positive z direction:

$$x + iy = r\sin\theta\, e^{i\phi}, \quad z = r\cos\theta.$$

On setting a homogeneous polynomial u of θth degree in x, y, z equal to $r^l \cdot Y_l$, Y_l depends only on the directional co-ordinates θ, ϕ and is a function of position on the unit sphere. If u is a *harmonic function*, i.e. if it satisfies the equation $\Delta u = 0$, Y_l is said to be a *surface harmonic of degree l* and the harmonic function u itself is said to be a *spherical* (*or solid*) *harmonic of degree* l. Since in polar co-ordinates

$$\Delta u = \frac{1}{r^2}\frac{\partial}{\partial r}\left(r^2\frac{\partial u}{\partial r}\right) + \frac{1}{r^2}\Lambda u,$$

$$\Lambda u = \frac{1}{\sin\theta}\left\{\frac{\partial}{\partial\theta}\left(\sin\theta\frac{\partial u}{\partial\theta}\right) + \frac{1}{\sin\theta}\frac{\partial^2 u}{\partial\phi^2}\right\} \tag{4.1}$$

the surface harmonic Y_l satisfies the differential equation

$$\Lambda Y_l + l(l+1)Y_l = 0. \tag{4.2}$$

2. *Orthogonality.*—On applying Green's formula to the spherical harmonics $u = r^k Y_k$, $v = r^l Y_l$ on the interior of the unit sphere, we obtain the orthogonality relations

$$\int Y_k\, Y_l\, d\omega = 0, \quad k \neq l, \tag{4.3}$$

in which $d\omega = \sin\theta\, d\theta\, d\phi$ is the surface element on the unit sphere. Since the conjugate complex $\overline{Y}_k$ of a surface harmonic is also a surface harmonic, the first factor in (4.3) can be replaced by $\overline{Y}_k$.

3. *Basis.*—On writing

$$\xi = x + iy, \quad \eta = x - iy$$

the differential equation $\Delta u = 0$ becomes

$$\Delta u \equiv 4\frac{\partial^2 u}{\partial\xi\partial\eta} + \frac{\partial^2 u}{\partial z^2} = 0\,;$$

we see that a homogeneous polynomial u of degree l in ξ, η, z breaks up into harmonic polynomials $u^{(m)}$:

$$u = \Sigma u^{(m)}, \quad (m = -l, \cdots, l-1, l)$$

where $u^{(m)}$ consists of all terms in which the exponents of ξ and η have the fixed difference m. The recursion formula for the coefficients of $u^{(m)}$, which is obtained from the differential equation $\Delta u = 0$, further shows that there exists *one*, and to within a multiplicative constant *only one*, such harmonic $u^{(m)}$. Accordingly, there exist exactly $2l + 1$ linearly independent surface harmonics of degree l; we may take them to be the $Y_l^{(m)}$ defined by

$$u^{(m)} = r^l \cdot Y_l^{(m)}.$$

Writing

$$u^{(m)} = (x - iy)^{-m} \cdot P = (x + iy)^m \cdot P_*$$

and r placing

$$(x + iy)(x - iy) \quad \text{by} \quad r^2 - z^2,$$

P and P_* depend only on r^2 and z. Hence on taking $r = 1$ we have

$$Y_l^{(m)} = e^{im\phi} (\sin\theta)^{-m} \cdot P_l^{(m)} (\cos\theta). \tag{4.4}$$

For $m = -l$ we take $P = 1$, and for $m = +l$, $P_* = 1$; $P(z) = (1 - z^2)^l$ for this latter case. Since $Y_l^{(m)}$ depends on ϕ only in the factor $e^{im\phi}$,

$$\int \bar{Y}_l^{(m)} Y_l^{(m')} d\omega = 0, \quad m' \neq m. \tag{4.5}$$

This basis $Y_l^{(m)}$, in which the z-axis occupies a preferred position, is accordingly unitary-orthogonal.

4. *Completeness.*—That the totality of surface harmonics constitute a *complete orthogonal system* on the unit sphere can be proved by showing that any polynomial in x, y, z on the sphere can be written as a sum of surface harmonics. Now the general polynomial of degree l contains

$$(l + 1) + l + (l - 1) + \cdots + 1$$

arbitrary constants. But exactly this same number of linearly independent homogeneous polynomials are contained in the expression

$$r^l(Y_l + Y_{l-2} + \cdots)[= u_l + (x^2 + y^2 + z^2)u_{l-2} + \cdots], \tag{4.6}$$

for the polynomials of the form $r^l Y_l$, $r^l Y_{l-2}$, $\cdots$ are linearly independent in virtue of the orthogonality of surface harmonics. $r^l Y_l$ contains exactly $2l + 1 = (l + 1) + l$ linearly independent functions, and consequently (4.6) contains exactly

$$[(l + 1) + l] + [(l - 1) + (l - 2)] + \cdots,$$

as asserted above.

5. *Closed expressions for the surface harmonics.*—On substituting (4.4) in (4.2) we obtain the differential equation

$$(1 - z^2)\frac{d^2P}{dz^2} + 2(m - 1)z\frac{dP}{dz} + [l(l + 1) - m(m - 1)] \cdot P = 0$$

for the polynomial $P = P_l^{(m)}$ in $z = \cos\theta$. From this equation we find that $\frac{dP}{dz}$ satisfies the same differential equation on replacing m by $m - 1$; we thus obtain the recursion formula

$$P^{(m-1)}(z) = \frac{dP^{(m)}}{dz}$$

and the expression

$$P_l^{(m)}(z) = \frac{d^{l-m}}{dz^{l-m}}(1 - z^2)^l.$$

In particular, the " zonal harmonic "

$$P_l(z) \equiv P_l^{(0)}(z) = \frac{d^l(1 - z^2)^l}{dz^l}.$$

6. *Further formulæ.*—

$$\int x Y_k Y_l \, d\omega = 0 \tag{4.7}$$

unless $l - k = \pm 1$. For $x \cdot r^k Y_k$ is a polynomial of degree $k + 1$ and may, in accordance with 4, be expanded in the form $r^{k+1}(Y_{k+1} + Y_{k-1} + \cdots)$. Consequently on the unit sphere

$$x Y_k = Y_{k+1} + Y_{k-1} + \cdots \tag{4.8}$$

and the only values of $l \geqq k$ for which the integral (4.7) can have a value other than 0 is $l = k + 1$. Hence our assertion (4.7); it also follows from the above that only the first two terms can appear in (4.8).

Further, we shall also have occasion to use the differential expressions

$$L_x u = \frac{1}{i}\left(y\frac{\partial u}{\partial z} - z\frac{\partial u}{\partial y}\right), \quad L_y u, \quad L_z u, \tag{4.9}$$
$$L^2 u = L_x(L_x u) + L_y(L_y u) + L_z(L_z u)$$

in terms of polar co-ordinates. On setting in

$$du = \frac{\partial u}{\partial x}dx + \frac{\partial u}{\partial y}dy + \frac{\partial u}{\partial z}dz$$

the changes dx, dy, dz obtained by allowing ϕ to increase by $d\phi$ and holding r, θ fixed, we obtain immediately

$$L_z u = \frac{1}{i}\frac{\partial u}{\partial \phi}. \tag{4.10}$$

Similarly,

$$L_x + iL_y = e^{i\phi}\left(\frac{\partial}{\partial \theta} + i\frac{\cos\theta}{\sin\theta}\frac{\partial}{\partial \phi}\right),$$
$$L_x - iL_y = e^{-i\phi}\left(-\frac{\partial}{\partial \theta} + i\frac{\cos\theta}{\sin\theta}\frac{\partial}{\partial \phi}\right), \tag{4.10}$$
$$L^2 = -\Lambda \quad \text{[eq. (4.1)]}.$$

§ 5. Electron in Spherically Symmetric Field. Directional Quantization

Now back to physics ! Consider an electron of charge $-e$ revolving about a fixed nucleus of charge Ze situated at the origin. For $Z = 1$ we have the hydrogen atom, for $Z = 2$

singly ionized helium He^+, for $Z = 3$ doubly ionized lithium Li^{++}, etc. The potential energy is $V = -\frac{Ze^2}{r}$; we shall, however, for the present take $V(r)$ more generally as any function of the radius r. The wave equation for the determination of the energy levels is then

$$\frac{h^2}{2m}\Delta\psi + [E - V(r)]\psi = 0. \tag{5.1}$$

On expanding in terms of surface harmonics ψ becomes a sum of terms $f_l(r)Y_l$ $(l = 0, 1, 2, \cdots)$. The differential operator on the left-hand side of (5.1) sends the lth term of this sum into Y_l times

$$\frac{h^2}{2m}\left\{\frac{1}{r^2}\frac{d}{dr}\left(r^2\frac{df_l}{dr}\right) - \frac{l(l+1)}{r^2}f_l\right\} + [E - V(r)]f_l(r). \tag{5.2}$$

Consequently each individual term must satisfy the differential equation separately; we thus obtain a complete set of characteristic functions of the form

$$\psi = f_l(r)Y_l.$$

The factor $f_l(r)$ depending only on r must be such that (5.2) vanishes and $\int r^2\bar{f}_l(r)f_l(r)dr$ converges. Denoting the characteristic numbers and characteristic functions of this differential equation by

$$E_{nl}, \quad f_{nl}(r) \quad (n = 0, 1, 2, \cdots),$$

E_{nl} is a $(2l+1)$-fold energy level, as the expression $f_{nl}(r)Y_l$ contains $2l+1$ linearly independent characteristic functions associated with this single characteristic value; we may choose as a basis the functions

$$\psi_{nl}^{(m)} = f_{nl}(r) \cdot Y_l^{(m)} \quad (m = -l, \cdots, l-1, l).$$

We thus arrive at three integral quantum numbers: the "*radial quantum number*" n, the "*azimuthal quantum number*" l, and the "*magnetic quantum number*" m. The energy level depends only on the first two.

In justification of this nomenclature we determine the *angular momentum* $h\mathfrak{L}$ of the electron with components

$$hL_x = yp_z - zp_y, \cdots.$$

In quantum mechanics L_x, L_y, L_z are the operators (4.9). Hence for

$$\psi_{nl}^{(m)} = f_{nl}(r)Y_l^{(m)} = e^{im\phi} \cdot (\text{a function of } r \text{ and } \theta) \tag{5.3}$$

we have, in accordance with (4.10),

$$L_z \psi = m \cdot \psi,$$

and for the general characteristic function

$$\psi = f_{nl}(r) Y_l \tag{5.4}$$

with azimuthal quantum number l

$$L^2 \psi = l(l+1) \cdot \psi.$$

Hence in the state described by (5.4) not only the energy has a definite value E_{nl}, but also the *absolute value of the moment of momentum*

$$\boxed{\mathfrak{L}^2 = l(l+1)}. \tag{5·5}$$

The significance of the azimuthal number is that it fixes this magnitude. It is indeed remarkable that there exist states $l = 0$, $n = 0, 1, 2, \cdots$ with spherically symmetric characteristic functions $\psi = f_{n0}(r)$ for which the moment of momentum vanishes. In the states described by (5.3) not only the energy and the absolute value of the moment of momentum have definite values, but also *the z-component of the moment of momentum assumes a definite value with certainty*, for then

$$\boxed{L_z = m.}. \tag{5.6}$$

Since a *magnetic dipole moment*

$$\mathfrak{s} = \frac{-eh}{2\mu c}\mathfrak{L} \tag{5.7}$$

is associated with the angular momentum $h\mathfrak{L}$ of the revolving electron (the mass of the electron being denoted by μ whenever there is danger of confusion with the magnetic quantum number m), the influence of $\mathfrak{L}$ will be felt on subjecting the atom to a magnetic field. The existence of the *Zeeman effect* under such conditions can be traced to this cause. A fundamental experiment to observe the magnetic moment of the electron directly is due to *Stern and Gerlach.* Let a stream of one-electron atoms, which are all moving in the direction of the x-axis and are in the state (n, l) with energy level E_{nl}, be subjected to an inhomogeneous magnetic field in the direction of the z-axis. Let the x- and y-components of the magnetic field vanish in the $(x\text{-}z)$-plane, in which the beam moves, and let the z-component be a function of z alone. A magnetic dipole, the z-component of whose moment is s_z, is then acted upon by a force $\frac{dH}{dz} \cdot s_z$ in the positive z-direction. In consequence of (5.6) the atomic beam should be broken up into $2l + 1$ smaller beams by the force in the z-direction, corresponding to the various values $m = l, l - 1, \cdots, -l$ of the magnetic quantum number. On performing the experiment on silver atoms in the normal state *two* beams, corresponding to $m = \pm 1$, were observed; the value of the "Bohr magneton," the elementary magnetic moment corresponding to one unit of angular momentum, was found to agree with the value $\frac{eh}{2\mu c}$ obtained from (5.6) and (5.7). Why the unperturbed beam corresponding to $m = 0$ did not appear remained unexplained.

The older quantum theory, which employed the quantum number $k = l + 1$ with values $1, 2, \cdots$, allowed m to assume the integral values from $-k$ to $+k$; it seemed plausible to exclude the case $k = 0$, although one was thereby led into difficulties on applying the so-called "adiabatic hypothesis" to the behaviour of an atom under the influence of crossed electric and magnetic fields. In the new quantum theory no *ad hoc* hypothesis is required for this exclusion, as l can assume only the values $0, 1, 2, \cdots$. But according to either the old or the present scalar wave theory there should exist an odd number of permissible values of m for given k or l; the exclusion of the case $m = 0$ apparently required by the *Stern-Gerlach* experiment cannot be accounted for on either theory. Nor can we explain the related fact that in the anomalous Zeeman effect m may assume either an even or an odd number of values, according to the nature of the atom under consideration. Obviously something is lacking in our present scalar wave theory as well as in the older formulation; we return to this point again in Chap. IV, § 4. The older quantum theory described the situation met above as "*directional quantization*"; since the absolute value of the moment of momentum was hk and the component along the z-axis was hm, it concluded that the magnetic axis of the atom could assume only positions described by the inclination θ with the z-axis determined by the formula

$$\cos\theta = \frac{m}{k} \qquad (m = 0, \pm 1, \pm 2, \cdots, \pm k).$$

Thus in the case $k = 1$ we should expect only three possible orientations for the magnetic axis: parallel and anti-parallel to the field, which we have taken in the direction of the z-axis, and perpendicular thereto—unless we empirically exclude this latter possibility $m = 0$ because of the *Stern-Gerlach* experiment, in which case we have but two. In either case we find ourselves faced with a serious dilemma, for the direction of the z-axis is an arbitrary direction in space. In order to avoid this one then assumed that the quantization was due to the influence of the magnetic field, and consequently the preferred z-direction was interpreted physically as the *direction of the magnetic field.* But even so the difficulty is not avoided in the limiting case of vanishing magnetic field, for the directional quantization should be maintained in arbitrarily weak fields. Or stated more physically, the radiation mechanism required by the *Stern-Gerlach* effect for the orientation of the atoms, which were originally in random orientation and precessing about the z-axis, requires about 10^8 *times as long* as the greatest time consistent with the observations. The stand taken by the new quantum theory on this point is fundamentally different. The possible states (n, l) of the atom are described by the functions ψ of the $(2l + 1)$-dimensional linear family

$$\psi = f_{nl}(r)\, Y_l = \sum_{m=-l}^{+l} x_m \cdot f_{nl}(r)\, Y_l^{(m)}$$

or by the vectors of a $(2l + 1)$-dimensional space with components x_m. *The z-component of the moment of momentum, as well as the component in any arbitrary direction, is capable of assuming only the discrete values hm $(m = l, l - 1, \cdots, -l)$.* But in a state in which the z component, for example, assumes the value *hm with certainty* there is only *a certain probability* that any other component will assume *a definite one* of its possible values $h \cdot 0$, $h \cdot (\pm 1)$, $\cdots$, $h \cdot (\pm l)$. The name " directional quantization " is hardly an appropriate description of this situation.[9]

When the *electro-static central force satisfies the Coulomb law* and originates in a nucleus of charge $+ Ze$, the differential equation (5.2) for the " radial characteristic function " $f = f_{nl}(r)$ becomes

$$\left(\frac{d^2(rf)}{dr^2} - \frac{l(l+1)}{r} f\right) + \frac{2m}{h^2}(Er + Ze^2)f = 0.$$

The character of this equation is unchanged on going over to the new dependent variable v defined by $rf = e^{-\alpha r} \cdot v$:

$$\frac{d^2v}{dr^2} - 2\alpha\frac{dv}{dr} + \left\{\left(\alpha^2 + \frac{2mE}{h^2}\right) + \frac{2mZe^2}{h^2 r} - \frac{l(l+1)}{r^2}\right\}v = 0.$$

We choose α in such a way that the constant term in the coefficient of v vanishes :

$$h^2\alpha^2 = -\,2mE. \qquad (5.8)$$

We know from the general theory of linear differential equations [10] that there exist solutions of this equation in the neighbourhood of the (regular) singular point $r = 0$ in the form of a power series

$$v = \Sigma\, a_\mu r^\mu$$

in which the exponent μ begins with a certain value μ_0, which need not be an integer, and runs through the values μ_0, $\mu_0 + 1$, $\mu_0 + 2, \cdots$. On substituting this power series into the equation we find the recursion formula

$$\{\mu(\mu + 1) - l(l + 1)\}a_{\mu+1} = 2a_\mu\left(\alpha\mu - \frac{Zme^2}{h^2}\right) \tag{5.9}$$

for the coefficients a_μ. In order that it be satisfied for $\mu + 1 = \mu_0$ ($a_\mu = 0$, $a_{\mu+1} \neq 0$) we must have

$$\mu_0(\mu_0 - 1) = l(l + 1).$$

We thus have the two possibilities :

$$\mu_0 = l + 1 \quad \text{or} \quad \mu_0 = - l.$$

Considering the first possibility and taking the coefficient a_{l+1} of the lowest power as unity, all remaining coefficients can be obtained by successive applications of the recursion formula (5.9), as the denominator $\mu(\mu + 1) - l(l + 1)$ never vanishes ; let the solution thus obtained be denoted by v. The second possibility does not lead to a solution, however, as the denominator in the recursion formula for $\mu = l$ vanishes ; the second solution of the differential equation can be obtained by quadrature from the first and involves logarithmic terms.

The power series for v breaks off if for a definite exponent $\mu = \mu_0 + n$

$$\alpha\mu = \frac{Zme^2}{h^2}$$

or

$$h\alpha = \frac{Zme^2}{h(n + l + 1)}. \tag{5.10}$$

In this case f is of the form

$$e^{-\alpha r} \cdot r^l \cdot (\text{polynomial of degree } n \text{ in } r) ;$$

it is finite at $r = 0$ and the integral

$$\int r^2 \bar{f}(r) f(r) dr \tag{5.11}$$

exists, as is to be required. The corresponding characteristic numbers E are the energy levels; on writing n in place of $n + l + 1$ and solving (5.8), (5.9) for E we find

$$E_n = -\frac{Z^2 me^4}{2h^2}\frac{1}{n^2}. \tag{5.12}$$

The integer n, the ***principal*** or ***total quantum number,*** is subject to the condition $n > l$. There exist no other solutions for which the integral (5.11) converges.[8]

The energy levels depend only on the principal quantum number n; the terms for which n is a fixed number and $l = 0, 1, \cdots, n - 1$ coincide in a single degenerate term E_n of multiplicity

$$\sum_{l=0}^{n-1} (2l + 1) = n^2.$$

This theoretical result agrees with the empirical formulæ for the Balmer, Paschen, Lyman, etc., series. We find, in fact, the expression

$$-\frac{Z^2R}{n^2}, \qquad R = \frac{me^4}{4\pi h^3 c}$$

for the terms measured in wave-numbers $\left(\frac{\nu}{2\pi c} = \frac{E}{2\pi ch}\right)$. *The expression for the Rydberg constant R in terms of the fundamental constants of nature (the charge and the mass of the electron, the velocity of light and the elementary quantum of action) agrees numerically with its empirical value.* All terms and therefore all actual line frequencies ν depend on the integer Z describing the charge on the nucleus in such a way that $\sqrt{\nu}$ increases in proportion with Z. Since the X-ray terms are due to the innermost electrons, which are but slightly affected by the outer ones, we should expect to find that the hardest X-ray lines, arranged in accordance with the atomic number Z, follow this law. It was discovered by *Moseley* and gave *a conclusive proof of the fact that on going through the elements of the periodic table the charge on the nucleus increases by e from element to element.* This law uncovers with unerring certainty the holes yet remaining in the system of known elements; at present we lack but 2 (or 3) elements in the series beginning with hydrogen, $Z = 1$, and ending with uranium, $Z = 92$.

The characteristic functions associated with these energy levels, which determine the relative probabilities of the various positions of the electron, can be expressed in closed form in terms of the so-called Laguerre polynomials. The characteristic function belonging to the normal state $n = 1$, $l = 0$, is spherically symmetric : *

$$\psi = \frac{1}{\sqrt{\pi a^3}} \cdot e^{-r/a} ; \tag{5.13}$$

for hydrogen

$$a = \frac{h^2}{me^2} = 0{\cdot}532 \text{ Å} \tag{5.14}$$

(According to the older *Bohr* theory, a is the radius of the innermost electronic orbit.) a determines the order of magnitude of atomic dimensions. *In the normal state hydrogen possesses spherical symmetry* (according to the scalar wave theory—but see Chap. IV, § 8).

The radial characteristic functions $r \cdot f_{nl}(r)$ do not, however, constitute a complete orthogonal system for a given l for the full domain which we wish to consider: in addition to the discrete term spectrum (5.12) we have the continuous spectrum covering the whole region $E \geqq 0$. We go no further into this matter.[11]

§ 6. Collision Phenomena

The optical phenomena show that the quantum theory leads to the correct energy levels, but they do not lend themselves to an attempt to interpret the vector ψ in system space as a probability. Collision phenomena, which deal with the deflection of electrons or α-particles under the influence of other material bodies, are best suited for this latter purpose. The fundamental experiments of *Franck and Hertz*, as well as those of *Davisson and Germer*, belong to this latter category.

Neglecting the reaction of the moving particle on the perturbing body, the potential energy due to this latter may be taken as a given function $V(xyz)$ of position. Considering a one-dimensional problem, the energy of the moving particle is then

$$H = \frac{1}{2m} p^2 + V(x).$$

We can think of the curve $y = V(x)$ as the contour of a hill against which the particle runs. The wave equation for a

* The normalizing factor $1/\sqrt{\pi a^3}$ is calculated from

$$\iiint e^{-2r/a} dx dy dz = 4\pi \int_0^\infty e^{-2r/a} r^2 dr = \pi a^3.$$

state with given energy E is

$$\frac{h^2}{2m}\frac{d^2\psi}{dx^2} + [E - V(x)]\,\psi = 0. \tag{6.1}$$

If we neglect for the moment the perturbing field V we obtain as solutions of (6.1) the familiar de Broglie waves: ψ is a linear combination of the waves $e^{i\alpha x}$ and $e^{-i\alpha x}$ proceeding in the positive and negative directions along the x-axis, the wave number α of which is determined by

$$(h\alpha)^2 = 2mE \quad \text{or} \quad h\alpha = p.$$

Writing

$$\frac{2m}{h^2}V(x) = U(x)$$

equation (6.1) becomes

$$\frac{d^2\psi}{dx^2} + [\alpha^2 - U(x)]\,\psi = 0. \tag{6.2}$$

We now assume that as $x \to \pm\infty$, $U(x)$ behaves in such a way that the integral $\int_{-\infty}^{+\infty}|U(x)|dx$ converges; equation (6.2) then has one solution which behaves for $x \to +\infty$ asymptotically like $e^{i\alpha x}$, and another, which is linearly independent of the first, which behaves like $e^{-i\alpha x}$ in the same region.

This can most readily be seen by solving (6.2) by the method of successive approximations. Let

$$\psi = \psi_0 + \psi_1 + \psi_2 + \cdots \tag{6.3}$$

and take as the 0th approximation the function $e^{i\alpha x}$; in general ψ_{n+1} is determined in terms of ψ_n by integrating the equation

$$\frac{d^2\psi_{n+1}}{dx^2} + \alpha^2\,\psi_{n+1} = U(x)\psi_n.$$

Hence

$$\psi_{n+1}(x) = -\frac{1}{\alpha}\int_x^{\infty}\sin\alpha(x-\xi)\cdot U(\xi)\,\psi_n(\xi)\,d\xi. \tag{6.4}$$

We restrict ourselves for the moment to a region $x \geqq x_0$ such that

$$\frac{1}{\alpha}\int_{x_0}^{\infty}|U(x)|dx = g < 1.$$

If $|\psi_n(x)| \leqq a_n$ for all x, the integral (6.4) converges and we have

$$|\psi_{n+1}(x)| \leqq a_n \cdot \frac{1}{\alpha}\int_x^\infty |U(\xi)|d\xi \,;$$

we can therefore take $a_0 = 1$, $a_{n+1} = ga_n$. Then $a_n = g^n$ or

$$|\psi_n(x)| \leqq g^n \quad \text{for} \quad x \geqq x_0.$$

Consequently the series for ψ converges at least as fast as the geometric series with ratio g. It satisfies the integral equation

$$\psi(x) - \psi_0(x) = -\frac{1}{\alpha}\int_x^\infty \sin\alpha(x-\xi) \cdot U(\xi)\,\psi(\xi)\,d\xi \tag{6.5}$$

and is consequently a solution of (6.2). Since

$$|\psi(x)| \leqq 1 + g + g^2 + \cdots = \frac{1}{1-g}$$

(6.5) leads to the estimate

$$|\psi(x) - \psi_0(x)| \leqq \frac{1}{\alpha(1-g)} \cdot \int_x^\infty |U(\xi)|d\xi,$$

from which it follows that $\psi(x)$ behaves asymptotically for $x \to +\infty$ like $\psi_0(x) = e^{i\alpha x}$. Not only is $\psi \sim \psi_0$, but also $\dfrac{d\psi}{dx} \sim \dfrac{d\psi_0}{dx}$, for the equation

$$\frac{d\psi}{dx} - \frac{d\psi_0}{dx} = -\int_x^\infty \cos\alpha(x-\xi) \cdot U(\xi)\psi(\xi)d\xi$$

gives as an upper bound for the absolute value of the difference on the left-hand side the quantity

$$\frac{1}{1-g} \cdot \int_x^\infty |U(\xi)|d\xi$$

which approaches 0 as $x \to +\infty$.

The solution $\psi(x)$ which we have found in the region $x \geqq x_0$ can naturally be extended over the entire real axis by analytic continuation. Since our considerations apply just as well for $x \to -\infty$, we know that $\psi(x)$ satisfies an asymptotic equation of the form

$$\psi(x) \sim be^{i\alpha x} + b'e^{-i\alpha x} \quad \text{for} \quad x \to -\infty.$$

At the same time we must also have

$$\frac{d\psi}{dx} \sim i\alpha(be^{i\alpha x} - b'e^{-i\alpha x}).$$

$\psi(x)$ being a solution of the differential equation, $\bar{\psi}(x)$ is also :

$$\frac{d^2\psi}{dx^2} + [\alpha^2 - U(x)]\psi = 0, \quad \frac{d^2\bar{\psi}}{dx} + [\alpha^2 - U(x)]\bar{\psi} = 0.$$

Multiply the first equation by $\bar{\psi}$, the second by ψ and subtract ; we find

$$\frac{d}{dx}\left(\bar{\psi}\frac{d\psi}{dx} - \psi\frac{d\bar{\psi}}{dx}\right) = 0$$

or

$$\bar{\psi}\frac{d\psi}{dx} - \psi\frac{d\bar{\psi}}{dx} = \text{const.} \tag{6.6}$$

The determinant (6.6) has the limiting value $2i\alpha$ for $x \to +\infty$ and for $x \to -\infty$

$$2i\alpha(b\bar{b} - b'\bar{b}'),$$

whence

$$b\bar{b} - b'\bar{b}' = 1. \tag{6.7}$$

It follows from this that $b \neq 0$. On multiplying $\psi(x)$ by $1/b$ we have a solution ψ whose asymptotic behaviour is described by the equations

$$\psi(x) \sim e^{i\alpha x} + a'e^{-i\alpha x} \quad \text{for} \quad x \to -\infty,$$
$$\psi(x) \sim ae^{i\alpha x} \quad \text{for} \quad x \to +\infty \tag{6.8}$$

where $a = 1/b$, $a' = b'/b$. (6.7) is now

$$|a|^2 + |a'|^2 = 1. \tag{6.9}$$

A particle of definite energy runs against the potential energy hill from the left, i.e. *from* $x = -\infty$. *Whereas in classical mechanics the particle certainly either gets over the hill or is thrown back, according to whether its initial kinetic energy is greater or less than the maximum of* $V(x)$, *quantum mechanics states that there is a probability* $|a|^2$ *that it gets over and a probability* $|a'|^2$ *that it is thrown back.* Furthermore, these probabilities are continuous functions of the energy of the particle ; the discontinuity of the classical theory is completely broken down. If we perform the experiment successively with a large number of particles we find that they are divided into two streams, in accordance with (6.8.), proceeding in the positive and negative directions along the x-axis; the relative intensities of these are given by 1 and $|a'|^2$ for $x \to -\infty$, respectively, while for $x \to +\infty$ there exists only the positive stream of intensity $|a|^2$. Equation (7.5) thus expresses the conservation of the number of particles and *shows that we must consider the square* $|a|^2$ *of the absolute value of the amplitude a as a relative intensity or probability.*

If the integral

$$\frac{1}{\alpha} \cdot \int_{-\infty}^{+\infty} |U(x)| dx < 1$$

the solution ψ is represented throughout the whole space by the formula (6.3). In perturbation theory one is usually satisfied with the first term ψ_1. The theory of the familiar experiments of *Rutherford*, in which α-particles are allowed to fly in a given direction with given momentum into and be deflected by the field of an atom, has been developed by *Wentzel* in a similar manner.[12] The influence of the α-particle on the atom is thereby neglected; on taking it into account we are led to the theory of the experiments of *Franck and Hertz*, giving formulæ for the dispersed particles specified according to their various discrete kinetic energies and their various directions. This calculation has been carried through for hydrogen by *Born and Elsasser.*[13] A very important application of this picture of corpuscular waves "seeping" through a potential hill has been made by *G. Gamow* and *R. W. Gurney and E. U. Condon* to explain radioactive decay.[14]

§ 7. The Conceptual Structure of Quantum Mechanics

The fruitfulness of the theory has been amply established by the above applications and the examples given have served to illustrate its physical interpretation; it now seems time to set forth its general abstract formulation.

Consider a physical system of known constitution. *Each particular state, each individual case of such a system is represented by a vector* $\mathfrak{x}$ *of modulus* 1 *in a unitary system space. Each physical quantity associated with the system is represented by an Hermitian form in this space.* The fundamental question which we put to the theory is not, as in classical physics, "What value has *this* physical quantity in *this* particular case?" but rather "*What are the possible values of the physical quantity A, and what is the probability that it assumes a definite one of these values in a given case?*" The answer to this question is: *The probability* *that A assumes the value α is the value $E_\alpha(\mathfrak{x})$ of the characteristic form E_α of A associated with the value α, where the vector $\mathfrak{x}$ represents the case in question and the quantity A is represented by the Hermitian form A in the system space. The quantity represented by A is capable of assuming only those values α which are characteristic values of the form A.* In accordance with the equations

$$\mathfrak{x}^2 = \sum_\alpha E_\alpha (\mathfrak{x}), \quad A(\mathfrak{x}) = \sum_\alpha \alpha E_\alpha(\mathfrak{x})$$

the sum of the probabilities is 1 and the value $A(\mathfrak{x})$ of the form A is *the mean value or expectation of the quantity A in the state $\mathfrak{x}$.* Since all assertions concerning the probabilities in a given state $\mathfrak{x}$ are numerically unaltered when $\mathfrak{x}$ is replaced by $\varepsilon\, \mathfrak{x}$, where ε is an arbitrary complex number of modulus 1, we cannot distinguish between these two cases. The pure case or state is consequently more properly represented by the *ray* $\mathfrak{x}$ than by the *vector* $\mathfrak{x}$, and we must therefore operate in the ray field in system space rather than in the vector field.

The significance of probabilities for experimental science is that they determine *the relative frequency of occurrence in a series of repeated observations.* According to classical physics it is in principle possible to create conditions under which every quantity associated with a given physical system assumes an arbitrarily sharply defined value which is exactly reproducible whenever these conditions are the same. *Quantum physics denies this possibility.* We illustrate this by the example of directional quantization. We know conditions under which we can guarantee with practical certainty that the atoms of a hydrogen gas are in the normal state. Let us therefore assume that we can create conditions under which we can be certain that the atoms under observation are in the quantum state (n, l) with azimuthal quantum number $l = 1$ and energy E. A certain quantity L_z, which can, under these conditions, assume only the values $+ 1$, 0, or $- 1$ is associated with each direction z in space. *Stern and Gerlach* have shown us how to sharpen these conditions so that L_z takes on a definite one of these values, say $L_z = + 1$. According to the theory the utmost limit of precision is then reached. If x is another direction in space, then under these conditions which determine L_z and E only *the relative probability* that the quantity L_x assumes any one of the values $+ 1$, 0, $- 1$ can be given. Why is it impossible to go further and insure conditions under which in addition L_x takes on a definite one of the values, say 0, with certainty? Because the " measurement " of L_x, which is accomplished by separating the atoms into three classes $L_x = +1, 0, -1$, is only possible by creating conditions which destroy the homogeneity already existing with respect to L_z. Polarization of photons is obviously somewhat analogous to directional quantization of atoms. The conditions for the production of a monochromatic beam of light in a definite direction determine the energy and momentum of the photons. To each orientation s of a Nicol prism corresponds a definite quantity λ_s which is capable of assuming only the values ± 1; if $\lambda_s = +1$ the light goes through and if $\lambda_s = -1$ it does not. With the aid of such a prism we separate out the photons for which $\lambda_s = 1$ without disturbing their energy and momentum. The utmost limit of precision is then reached; a monochromatic pencil of polarized light is the most homogeneous light possible. If we now place a second Nicol of orientation σ in the path of this beam, then naturally only those photons which have $\lambda_\sigma = +1$ can pass through. But the light which we thus obtain is of the same constitution as if the first Nicol of orientation s were not used at all; the condition that all the photons have $\lambda_s = +1$ is obviously destroyed by the second Nicol.

Natural science is of a constructive character. The concepts with which it deals are not qualities or attributes which can be obtained from the objective world by direct cognition. They can only be determined by an indirect methodology, by observing their reaction with other bodies, and their implicit definition is consequently conditioned by definite laws of nature governing reactions.[15] Consider, for example, the introduction of the Galilean concept of mass, which essentially amounts to the following indirect definition: "Every body possesses a momentum, that is, a vector $m\mathfrak{v}$ having the same direction as its velocity $\mathfrak{v}$; the scalar factor m is called its mass. The momentum of a closed system is conserved, that is, the sum of the momenta of a number of reacting bodies is the same before the reaction as after it." On applying this law to the observed collision phenomena data are obtainable which allow a determination of the relative masses of the various bodies. But scientists have long held the opinion that such *constructive concepts were nevertheless intrinsic attributes of the "Ding an sich,"* even when the manipulations necessary for their determination were not carried out. *In quantum theory we are confronted with a fundamental limitation to this metaphysical standpoint.*[16]

We have already seen, toward the beginning of this chapter, that a co-ordinate x and its associated momentum p stand in a peculiar relationship to one another: the precise determination of either one of these quantities precludes the precise determination of the other. In the state represented by the wave function $\psi(x)$ $\left[\int_{-\infty}^{+\infty}\bar{\psi}\,\psi\,dx = 1\right]$ the mean values $x_0 = \langle x\rangle$ and $p_0 = \langle p\rangle$ are given by

$$\int_{-\infty}^{+\infty} x\,\bar{\psi}(x)\,\psi(x)dx \quad \text{and} \quad \frac{h}{i}\int_{-\infty}^{+\infty}\bar{\psi}\,\frac{d\psi}{dx}dx.$$

No loss of generality is incurred by taking these mean values as zero; the first can be made to vanish by replacing x by $x - x_0$ or $\psi(x)$ by $\psi(x + x_0)$ and the second by replacing $\psi(x)$ by $e\left(-\frac{p_0 x}{h}\right)\cdot\psi(x)$. The mean values $(\Delta x)^2$, $(\Delta p)^2$ of $(x - x_0)^2$, $(p - p_0)^2$ are then given by

$$(\Delta x)^2 = \int_{-\infty}^{+\infty} x^2\bar{\psi}(x)\psi(x)dx,$$

$$(\Delta p)^2 = -\,h^2\int_{-\infty}^{+\infty}\bar{\psi}(x)\,\frac{d^2\psi}{dx^2}dx = h^2\int_{-\infty}^{+\infty}\frac{d\bar{\psi}}{dx}\,\frac{d\psi}{dx}dx.$$

From these expressions the general inequality

$$\Delta p\cdot\Delta x \geqq \tfrac{1}{2}h$$

can readily be obtained (I am indebted to *W. Pauli* for this remark); the less the uncertainty in x, the greater the uncertainty in p, and conversely.*

In general the conditions under which an experiment is performed will not even guarantee that all the individuals constituting the system under observation are in the same "state," as represented in the quantum theory by a ray in system space. This is, for example, the case when we only take care that all the atoms are in the quantum state (n, l) without undertaking to separate them, with respect to m by means of the *Stern-Gerlach* effect. In order to apply quantum mechanics it is therefore necessary to set up a criterion which will enable us to determine whether the given conditions are sufficient to insure such a "*pure state.*" We say that the conditions $\mathfrak{C}'$ effect a greater homogeneity than the conditions $\mathfrak{C}$ if (1) every quantity which has a sharp, reproducible value under $\mathfrak{C}$ has the same definite

* Cf. Appendix 1 at the end of the book.

value under $\mathfrak{C}'$ and if (2) there exists a quantity which is strictly determinate under $\mathfrak{C}'$ but not under $\mathfrak{C}$. The desired criterion is obviously this: *The conditions $\mathfrak{C}$ guarantee a pure state if it is impossible to produce a further increase in homogeneity.* (This maximum of homogeneity was obtained in classical physics only when all quantities associated with the system had definite values.)

In the pure state represented by the vector $\mathfrak{a} = (a_i)$, a quantity $\mathcal{Q}$ represented by the Hermitian matrix $Q = \|q_{ik}\|$ has the expectation or mean value

$$\langle Q \rangle = \sum_{i,\,k} a_k \bar{a}_i q_{ik}.$$

The numbers

$$a_{ik} = a_i \bar{a}_k \tag{7.1}$$

are the components of a positive definite Hermitian form A of trace 1, i.e.

$$|(\mathfrak{a}\mathfrak{x})|^2 = \sum_i a_i \bar{x}_i \cdot \sum_i \bar{a}_i x_i.$$

(Positive definite is to be understood here in the weakened sense $A(\mathfrak{x}) \geqq 0$.) It is to be noted that $\langle Q \rangle$ depends linearly and homogeneously on the quantity $\|q_{ik}\|$ under consideration:

$$Q = \operatorname{tr}(AQ). \tag{7.2}$$

If a *statistical aggregate* **A** is created by subjecting a large number of individuals of the physical system under observation to the conditions $\mathfrak{C}$, then the mean value of a physical quantity Q will be given by (7.2) where A is a certain positive definite Hermitian form of trace 1 which is characteristic for the aggregate—even if the conditions $\mathfrak{C}$ do not guarantee maximum homogeneity. The reason for this is that (7.2) is still correct if we mix statistical aggregates, each of which does possess maximum homogeneity, in any proportions; any statistical case may indeed be considered as a mixture of pure states. As *J. v. Neumann* has remarked, this formula (7.2) can be derived from the simple axioms [17]:

1. If P, Q are physical quantities and λ a real number, then $\langle \lambda P \rangle = \lambda \langle P \rangle$, $\langle P + Q \rangle = \langle P \rangle + \langle Q \rangle$.

2. If the quantity Q is capable of assuming only positive values (i.e. if the form Q is positive definite), then $\langle Q \rangle \geqq 0$.

3. If Q is a pure number, i.e. if it is independent of all physical conditions, then $\langle Q \rangle = Q$.

Assuming not only that any physical quantity Q is represented by an Hermitian form, but also that conversely any

Hermitian form represents some quantity associated with the system, it follows from (1) that

$$\langle Q\rangle = \sum_{i,k} a_{ki}q_{ik},$$

where the coefficients a_{ki} are independent of Q. (We shall return to this assumption in Chap. IV, § 9.) The matrix $A = \|a_{ik}\|$ must be Hermitian since $\langle Q\rangle$ is always real. On bringing A into the normal form $\Sigma\alpha_i x_i\bar{x}_i$ (2) requires for the special Hermitian forms of the type $Q = \Sigma q_i x_i \bar{x}_i$ that $\Sigma\alpha_i q_i \geqq 0$ for arbitrary non-negative values q_i; consequently $\alpha_i \geqq 0$ and A is positive definite.

The probability that in the statistical aggregate **A** the quantity Q assumes the value κ is

$$w = \mathrm{tr}\,(AE_\kappa) \tag{7.3}$$

where E_κ is the idempotent form associated with the characteristic number κ.

We can also distinguish " pure states " among general statistical aggregates, " mixed states," by the fact that they cannot be obtained by mixing two or more different statistical aggregates. This corresponds to the theorem that an Hermitian matrix A of the form (7.1) is not expressible as the sum $B + C$ of two positive definite Hermitian forms B and C which are not merely multiples of A. This can be readily proved on taking the vector $\mathfrak{a} = (a_i)$ as one of the co-ordinate axes in system space. The positive definite Hermitian forms A with unit trace, i.e. the statistical aggregates, constitute *a convex region* $\mathfrak{S}$ in the sense that with A and B their " centre of mass " $\lambda A + \mu B$ (λ, μ arbitrary positive numbers whose sum is unity) belongs also to $\mathfrak{S}$. A point of $\mathfrak{S}$ which cannot be considered as such a centre of mass of two points of $\mathfrak{S}$ distinct from the point in question is called, following *Minkowski*, an " *extreme point.*" [18] $\mathfrak{S}$ is the " *convex core* " of the class $\mathfrak{E}$ of all extreme points, i.e. it is the smallest convex domain which includes all the points of $\mathfrak{E}$. We cannot dispense with a single extreme point of $\mathfrak{S}$; if we leave out but a single point of $\mathfrak{E}$ the entire convex core shrinks together. We may accordingly characterize *the pure states as the " extremes " among all the possible statistical aggregates.*

It is often convenient to dispense with the normalization tr $A = 1$; (7.3) then gives the relative rather than the absolute probabilities. The simplest statistical aggregate is that one characterized by the unit Hermitian form with matrix $\mathbf{1}$; it represents *total ignorance.* In thermo-dynamics the important rôle is played by the *canonical aggregate* $A = e^{-H/k\theta}$; H is here the Hermitian form which represents the energy, k the *Boltzmann* constant and the number θ the temperature.[19]

§ 8. The Dynamical Law. Transition Probabilities

Having considered the general probability laws of the quantum theory, we now turn to the dynamical law governing the change in the state $\mathfrak{x}$ of a physical system during an interval dt of time. The dynamical law states that this change is effected by the infinitesimal unitary operator $-\frac{idt}{h}\cdot H$, where H is the Hermitian form which represents the energy:

$$\boxed{\frac{h}{i}\frac{d\mathfrak{x}}{dt}+H\mathfrak{x}=0}. \tag{8.1}$$

The peculiar significance of the energy in quantum mechanics is due to its appearance in the dynamical law. We also consider this law as a fundamental axiom of quantum theory of universal validity. For the matrix X:

$$x_{ik}=x_i\bar{x}_k,$$

which characterizes a statistical aggregate of the pure state described by the vector $\mathfrak{x}=(x_i)$ [cf. eq. (7.2)], we obtain the equation

$$\frac{h}{i}\frac{dX}{dt}=XH-HX \tag{8.2}$$

on applying (8.1) and taking into account the fact the H is Hermitian. This same equation also governs the change in time of a statistical aggregate X for a mixed state.[20]

For the integration of (8.1) it is convenient to choose as our co-ordinate system the characteristic vectors of H; the corresponding characteristic numbers E_n are the energy levels. We call this particular system the ***Heisenberg co-ordinate system***, as *Heisenberg* tacitly employed it in his fundamental paper on quantum mechanics. This Heisenberg co-ordinate system is in general not uniquely determined; the essential point is the decomposition of the system space $\mathfrak{R}$ into the characteristic sub-spaces $\mathfrak{R}'=\mathfrak{R}(E')$, $\mathfrak{R}''=\mathfrak{R}(E'')$, $\cdots$ associated with the various characteristic numbers E', E'', $\cdots$. The states represented by vectors $\mathfrak{x}$ in such a characteristic space are called ***quantum*** or ***stationary states***; in them the energy has a sharply defined value. The cases in which H possesses only discrete characteristic numbers include "conditionally periodic motion," the only ones for which the older quantum theory could be formulated. The nomenclature and symbolism employed in the following is adapted to discrete characteristic spectra, but this by no means precludes the possibility that the spectrum is entirely or partly continuous. Equation (8.1) becomes, on resolving it into components with respect to Heisenberg's co-ordinate system,

$$\frac{h}{i}\frac{dx_n}{dt} + E_n x_n = 0$$

and has as solution

$$x_n(t) = x_n \cdot e^{-i\nu_n t} \qquad (E_n = h\nu_n). \tag{8.3}$$

This is an explicit formulation of the unitary transformation $\mathfrak{x} \to \mathfrak{x}(t) = U(t)\mathfrak{x}$ which the state vector $\mathfrak{x}$ undergoes in time t. Since $|x_n(t)|^2$ is constant, *the probabilities for the various energy values do not change in the course of time.* The finite law

$$X(t) = U(t)XU^{-1}(t) \tag{8.4}$$

for the dependence of the statistical state $X(t)$ on the time t is fully equivalent to the differential law (8.2).

The mean value $q = q(t)$ of the physical quantity represented by the fixed Hermitian operator Q:

$$q(t) = \text{tr}\,[X(t) \cdot Q]$$

can, on taking into account the symmetry properties of the trace, be written also in the form

$$q(t) = \text{tr}\,[X \cdot Q(t)]$$

where

$$Q(t) = U^{-1}(t)QU(t). \tag{8.5}$$

Consequently the situation can be described either by considering Q as fixed for all time and the statistical state $X(t)$ as varying with the time in accordance with the law (8.4)—and this is the fundamental stand taken by quantum mechanics—or we can take the initial state X as representing the state of the system for all time and allow the operator $Q(t)$ representing the quantity Q to vary with time in accordance with the law (8.5). This latter interpretation lends itself to comparison with classical mechanics. (8.5) is equivalent to the differential law

$$\frac{h}{i}\frac{dQ}{dt} = HQ - QH, \tag{8.6}$$

for in virtue of (8.2) and (8.6)

$$\frac{dq}{dt} = \text{tr}\left(\frac{dX}{dt} \cdot Q\right) = \text{tr}\left(X \cdot \frac{dQ}{dt}\right).$$

In particular, the quantity Q is constant in time, i.e. *the probabilities associated with it do not change in course of time, if the Hermitian form Q which represents it commutes with H.*

In Heisenberg's co-ordinate system equation (8.5) becomes

$$q_{mn}(t) = q_{mn} \cdot e^{-i(\nu_n - \nu_m)t}. \tag{8.7}$$

The matrix $Q(t)$ is thus expressed in terms of components performing simple oscillations with frequencies $\nu_m - \nu_n$. The corresponding amplitude is q_{mn}. On going over from the mth to the nth stationary state the system loses an amount $h(\nu_m - \nu_n)$ of energy; if this energy is radiated as light, its frequency is given by

$$\nu_{mn} = \nu_m - \nu_n. \tag{8.8}$$

Classical mechanics collects together all the transitions from a fixed level m to all possible levels $n = 1, 2, \cdots$ into a single state of motion, the motion of the system in the mth quantum state, whose harmonic components have the corresponding transition frequencies $\nu_{m1}, \nu_{m2}, \cdots$. For any quantity A it therefore associates a constant amplitude a_{mn} with the transition $m \rightarrow n$. But in classical mechanics (for systems with one degree of freedom) we have

$$\nu_{mn} = k \cdot \omega(n), \quad k = m - n,$$

instead of equation (8.8). On multiplying the two Fourier series A, B

$$\sum_k a_k \cdot e^{ik\omega t} \quad \text{and} \quad \sum_k b_k \cdot e^{ik\omega t}$$

we obtain the Fourier series C with coefficients

$$c_k = \sum a_r b_s \quad (r + s = k).$$

Accordingly classical mechanics associates with the quantity $C = AB$ the amplitudes

$$c_{mn} = \sum a_{m,\, m-r} \cdot b_{m,\, m-s} \quad (r + s = m - n), \tag{8.9}$$

whereas quantum mechanics assigns to it the amplitudes

$$c_{mn} = \sum_t a_{mt} b_{tn} = \sum_r a_{m,\, m-r} \cdot b_{m-r,\, n}. \tag{8.10}$$

The difference between these two results lies in the fact that in (8.9) both factors a, b have the first index m in common, whereas in (8.10) the first index of b is the same as the last index of a. *This is in exact analogy with the difference between the " classical " and the correct Ritz-Rydberg combination principle.* This was Heisenberg's starting-point; the correct combination principle indicates the pertinent fact that the rule (8.9) for the multiplication of amplitudes must be replaced by (8.10). Admittedly such multiplication is not commutative, and it collects together amplitudes which the older model assigned to different orbits.

We denote $|a_{mn}|^2$ as the ***intensity*** *of the quantity* A *in the transition* $m \to n$. When multiple energy levels occur ("*degeneracy*") only the sum $\sum |a_{mn}|^2$, extended over all indices m for which $E_m = E'$ and all indices n for which $E_n = E''$, has an invariantive significance; in such a case this sum is taken as the intensity of A in the transition $E' \to E''$. If A_* is that portion of A in which $\mathfrak{R}(E')$ intersects $\mathfrak{R}(E'')$ the sum defined above is the trace of $A_* \tilde{A}_*$.

Consider an atom with one or more electrons and let $\mathfrak{r}$ be the vector from the nucleus to a representative electron. Then $\mathfrak{q} = e\mathfrak{r}$, or in case there is more than one electron the sum $\mathfrak{q} = \sum e\mathfrak{r}$, extended over the various electrons, is the electric dipole moment of the atom. In classical electrodynamics *the intensity of the light of frequency* ν *emitted by the atom* is calculated from the amplitude $\mathfrak{q}(\nu)$ of the harmonic components of $\mathfrak{q}$ with the same frequency ν in the following manner.† The rate at which energy flows through a surface element do at the point P, whose distance from the atom at O is large compared with the wave-length, is given by

$$|\mathfrak{q}^{\perp}(\nu)|^2 \cdot \frac{\nu^4}{2\pi c^3} d\omega,$$

where $\mathfrak{q}^{\perp}$ is the component of $\mathfrak{q}$ perpendicular to $\overrightarrow{OP}$ and $d\omega$ is the solid angle subtended at O by do. We have further assumed that the wave-length under consideration is large compared with the radius of the atom. Since each photon of frequency ν carries with it energy $h\nu$, we postulate that this law is to be taken over into quantum theory as follows: the probability that an atom in state n goes over into state n' in unit time and emits a photon of frequency ν, whose direction lies within the solid angle $d\omega$, is given by

$$|q^{\perp}_{nn'}|^2 \cdot \frac{\nu^3}{2\pi h c^3} d\omega. \tag{8.11}$$

We thus arrive at a definite rule for the calculation of the intensities of the lines emitted by the atom. The fact that we can now make such a prediction indicates a distinct superiority of the new theory over the old. *In particular, the transition* $n \to n'$ *does not occur if the corresponding coefficient in the Hermitian form*

† By this we mean that the terms $\mathfrak{q}(\nu)e^{i\nu t} + \bar{\mathfrak{q}}(\nu)e^{-i\nu t}$ occur in the harmonic analysis of $\mathfrak{q}$.

for $\mathfrak{q}$ *is zero.* This constitutes the general *selection rule.* The connection between the state of polarization of the emitted light and the direction of oscillation of the electric moment is also carried over into quantum theory. But a real derivation of our intensity rule can naturally only be obtained by considering the question of interaction between the atom and the ether; see § 13.

Examples: 1. *The Oscillator.*

The Hermitian form

$$\int_{-\infty}^{+\infty} x\, \bar{\psi}(x)\, \psi(x)\, dx,$$

representing the co-ordinate x of the oscillating particle has, as we have already found [(3.11)], the coefficients

$$\left. \begin{aligned} q_{nn'} &= 0 \quad \text{if} \quad n' \neq n \pm 1\,; \\ q_{n,\,n-1} = \sqrt{\frac{hn}{2m\omega}}, &\quad q_{n,\,n+1} = \sqrt{\frac{h(n+1)}{2m\omega}} \end{aligned} \right\} \tag{8.12}$$

with respect to Heisenberg's co-ordinate system, in which the energy is referred to its principal axes. We thus obtain *the selection rule* $n \rightarrow n \pm 1$; *the quantum number* n *can only change by* ± 1, *the oscillator then absorbing or emitting a photon of frequency* $\nu = \omega$ *and energy* $h\omega$, *in accordance with* (3.10). The selection rule makes it clear why no higher harmonics are excited in the simple oscillator. We have also found that the matrix $\|p_{nn'}\|$, which represents the linear momentum in Heisenberg's co-ordinate system, is given by (3.12)

$$\left. \begin{aligned} p_{n,\,n-1} = -\frac{1}{i}\sqrt{\frac{h\,m\,\omega\,n}{2}}, &\quad p_{n,\,n-1} = \frac{1}{i}\sqrt{\frac{h\,m\,\omega(n+1)}{2}}, \\ p_{nn'} = 0 \quad &\text{for} \quad n' \neq n \pm 1 \end{aligned} \right\} \tag{8.13}$$

2. *Electron in spherically symmetric field.*

The result (4.7) for surface harmonics yields *the selection rule*

$$l \rightarrow l \pm 1 \tag{8.14}$$

for the azimuthal quantum number l; for $l = 0$ only the transition $0 \rightarrow 1$ is possible. On introducing the magnetic quantum number m as in § 4, the characteristic functions $\psi_{nl}^{(m)}$ depend on the meridian angle ϕ about the z-axis only in the multiplicative factor $e^{im\phi}$; here

$$x \pm iy = r \sin\theta \cdot e^{\pm i\phi}, \quad z = r\cos\theta.$$

In order to obtain the dependence of the matrices $q_x + iq_y$, $q_x - iq_y$, q_z on the transition $m \rightarrow m'$ we must evaluate the integral

$$\int_0^{2\pi} e(a\phi)\, e(-m\phi)\, e(m'\phi)\, d\phi,$$

where $a = 1, \ -1, \ 0$, respectively. The integral vanishes unless $m' + a = m$. *The only components of $q_x + iq_y$ which do not vanish are those corresponding to the transitions $m \rightarrow m - 1$ in which the magnetic quantum number decreases by* 1; *for $q_x - iq_y$, $m \rightarrow m + 1$; for q_z, $m \rightarrow m$.*

This last selection rule cannot be obtained from the spectra themselves as long as the terms corresponding to different values of m ($|m| \leqq l$) coincide. But these terms are broken up into their various components by a homogeneous magnetic field in the direction of the z-axis (*Zeeman effect*). On "longitudinal" observation of the light emitted in the z-direction we find instead of the one line $(n, l) \rightarrow (n', l')$ several left- and right-circularly polarized components, the former of which arise from the transitions $m \rightarrow m - 1$ and the latter from $m \rightarrow m + 1$. On "transverse" observation, e.g. along the y-axis, we find two transverse linearly polarized lines arising from $m \rightarrow m \pm 1$, and in addition a longitudinally (i.e. along the z-axis) polarized line corresponding to the transition $m \rightarrow m$. (Polarization as here used means the direction of oscillation of the electric dipole, and therefore the direction of the electric field strength.)

In the *term spectrum of the alkali elements*, which is, however, typical in this respect, even for the more complicated spectra of the other elements, we distinguish between several series by means of the letters $s, p, d, f, g, \cdots$. Each series consists of infinitely many terms which we number in the direction of increasing frequency by the integer n. It is found convenient to let n run from 1 on in the s-series, from 2 on in the p-series, from 3 in the d-series, etc. The values of the terms ns, np, $nd, \cdots$ are then given by the "hydrogen-like" formula

$$-\frac{R}{(n + \kappa)^2},$$

in which $\kappa = \kappa_s, \kappa_p, \kappa_d, \cdots$ is a correction term depending but slightly on n, the numerical value of which but rarely exceeds 1/2 and is very close to 0 for high series ($f, g, \ldots$). *Only terms lying in neighbouring series combine to produce a line*, i.e. an s-term combines only with a p-term, p only with s and d, d with p and f, etc. In particular, the transitions $np \rightarrow 1s$ give rise to the *principal series*, which also appears in absorption, $nd \rightarrow 2p$ to the lines of the *diffuse series*, $ns \rightarrow 2p$ to the *sharp series*, and $nf \rightarrow 3d$ to the *Bergmann series*.[21]

The alkalies A are univalent, i.e. in chemical reactions only *one* electron, the valence electron, plays a rôle; the others, together with the nucleus, constitute an inert *closed shell.* It is therefore reasonable to assume that the optical spectra of the alkalies are caused by quantum jumps involving only this valence electron, while the core A^+ remains in its normal state. We have seen above that hydrogen in the normal state is represented by a spherically symmetric wave function ψ; we therefore assume, disregarding the reaction of the valence electron on the core, that this feature of the core being " closed " is to be expressed by ascribing *spherical symmetry* to it.* We have then to deal with the problem of an electron in a spherically symmetric field, which we have already discussed above. In accordance with the empirical combination principle and the theoretical selection rule for the azimuthal quantum number l, the $s, p, d, f, \cdots$ terms are to be taken as having $l = 0, 1, 2, 3, \cdots$ respectively. n then runs from $l + 1$ on in the series with azimuthal quantum number l, as in hydrogen.**

§ 9. Perturbation Theory

The problem with which *perturbation theory* is concerned is the following: Let the energy H consist of two terms $\mathsf{H} = H + \varepsilon W$, the second of which, the *perturbation term* εW, is small compared with the first; this we express by the " infinitesimal " numerical constant ε, of which powers higher than the first are to be neglected. Assume that the quantum problem for the " unperturbed system " with energy H has already been solved, so that the Hermitian form H has already been brought into normal (diagonal) form, and let $\mathfrak{R}', \mathfrak{R}'', \cdots$ be the characteristic spaces of H with characteristic numbers $E', E'', \cdots$. The problem is to find the solution of the equations for the " perturbed system " with energy H.

In order to illustrate the typical difference between degenerate and non-degenerate systems we first consider the system space as 2- instead of ∞-dimensional; then

$$\mathsf{H} = \begin{Vmatrix} E_1 & 0 \\ 0 & E_2 \end{Vmatrix} + \varepsilon W.$$

* Why He and not H is the first closed atom is only to be understood as the result of a profound modification of wave mechanics; see Chap. IV.

** Concerning the introduction of the "true quantum number" for elements other than hydrogen, see Chap. IV, § 10.

If $E_1 \neq E_2$ the unitary transformation which brings H into diagonal form differs from the identity only by terms of order ε. Consequently the probabilities $|x_1|^2$, $|x_2|^2$ that in the pure state $\mathfrak{x}$ H has the values E_1, E_2 will change only by amounts of the same relative order ε; they remain constant to the same approximation with which εW may be neglected in comparison with H. But the situation is quite different for degenerate systems, for which $E_1 = E_2 = E$, for the principal axes of H are then indeterminate and this arbitrariness is expressed in the "instability" of the system under the influence of a perturbation. We set up that normal co-ordinate system $\mathfrak{e}_1'$, $\mathfrak{e}_2'$ in which W assumes the diagonal form; the co-ordinate vectors are then also characteristic vectors of H, since $E_1 = E_2$. But these vectors can obviously differ arbitrarily from the original co-ordinate vectors $\mathfrak{e}_1$, $\mathfrak{e}_2$, whereas the energies $h\nu_1'$, $h\nu_2'$ can only differ from E by a term of order ε. On returning to the original co-ordinate system we have

$$x_1 = a_{11} \cdot e(-\nu_1't) + a_{12} \cdot e(-\nu_2't),$$
$$x_2 = a_{21} \cdot e(-\nu_1't) + a_{22} \cdot e(-\nu_2't),$$

where $\mathfrak{a}_1 = (a_{11}, a_{21})$, $\mathfrak{a}_2 = (a_{12}, a_{22})$ are two mutually perpendicular vectors whose directions coincide with those of $\mathfrak{e}_1'$, $\mathfrak{e}_2'$. The probabilities for the two states $\mathfrak{e}_1$, $\mathfrak{e}_2$ vary periodically in time with the small beat frequency $\nu_2' - \nu_1'$ (resonance between states $\mathfrak{e}_1$, $\mathfrak{e}_2$). *Quantum states with the same energy are therefore in resonance with one another.* The magnitudes of the components of $\mathfrak{x}$ in the characteristic spaces $\mathfrak{R}'$, $\mathfrak{R}''$, $\cdots$, i.e. the probabilities for the various numerically different values of H remain approximately constant under a small perturbation, but this is not the case for the absolute values $|x_n|$ of the individual components x_n resolved along the axes of an arbitrary Heisenberg co-ordinate system of the unperturbed system.

In accordance with the foregoing we can formulate the perturbation problem in two forms: I. Determine the change, due to the perturbation, in those states in which the energy H of the unperturbed system is determinate. This formulation has a sound physical interpretation if we consider the perturbation as acting during a time interval t_1, t_2. We then find *how the probabilities for the various quantum states change under the influence of the perturbation.*[22] II. Determine the quantum states and energy levels of the perturbed system, i.e. the characteristic values and characteristic spaces of H. *We ask in particular how the terms are broken up and displaced under the perturbation.* We consider II first.

We first decompose the Hermitian form W into two parts: $W_0 + V$. To the first belong those portions of W in which a characteristic space $\mathfrak{R}', \mathfrak{R}'', \cdots$ of H intersects itself, and to V those in which two different characteristic spaces intersect. If the characteristic values of H have but finite multiplicity the problem of bringing W', that part of W in which $\mathfrak{R}'$ intersects itself, into diagonal form deals only with the space $\mathfrak{R}'$ of a finite number of dimensions. If $\mathfrak{R}'$ is not simply a one-dimensional space, the resonance phenomena mentioned above will appear. The co-ordinate system, consisting of characteristic vectors of H, is now more precisely specified, for now W_0 also appears as a diagonal matrix; let E_n be the characteristic values of the $H + \varepsilon W_0 = \mathsf{H}_0$ so obtained. The single term value E' associated with $\mathfrak{R}'$ has in general been resolved into as many different characteristic values E_n of H_0 as there are dimensions in the sub-space $\mathfrak{R}'$.

The remainder $V = \|v_{mn}\|$ of the matrix is such that $v_{mn} = 0$ if the characteristic values E_m, E_n of H are equal. The infinitesimal unitary rotation

$$\delta x = \varepsilon \cdot Cx, \quad C = \|c_{mn}\|,$$

of order ε transforms H into $\mathsf{H} + \delta\mathsf{H}$ where

$$\delta\mathsf{H} = \varepsilon(\mathsf{H}C - C\mathsf{H}) \sim \varepsilon(HC - CH).$$

On choosing this transformation in such a way that $\delta\mathsf{H} = -\varepsilon V$, $\mathsf{H} = \mathsf{H}_0 + \varepsilon V$ goes over into H_0; this can be accomplished by choosing $c_{mn} = 0$ if $E_m = E_n$ and

$$c_{mn} = \frac{v_{mn}}{E_n - E_m}$$

otherwise. *The characteristic values* E_n *of* H_0 *are therefore the energy levels of the perturbed system of energy* H *if we neglect terms of order* ε^2.

W_0 can be considered as the time mean of the perturbation terms, averaged over the motion of the unperturbed system. For by (8.7) the mean value of the element $a_{mn}(t)$ of the matrix $A(t)$, which represents an arbitrary physical quantity of the system, is a_{mn} or 0, according as $\nu_m = \nu_n$ or not. In statistics angular brackets are often used to denote the mean value of a quantity; we may therefore write

$$W_0 = \langle W \rangle, \quad \mathsf{H}_0 = \langle \mathsf{H} \rangle.$$

The solution of II naturally provides an answer to the question I. But it is more convenient to employ *the method of variation of constants* for the calculation of the effect of the perturbation over a limited time interval—the smaller the constant ε, the longer we may take this time interval to be. Assume that at time $t = 0$ the system is in the quantum state 0 and that the perturbation begins to act at this time; we ask for the probability that the system will be found in the state n at time t. That is, we seek that solution of the equations

$$-\frac{1}{i}\frac{dx_n}{dt} = \nu_n x_n + \frac{\varepsilon}{h}\sum_m W_{nm} x_m \quad (n = 0, 1, 2, \cdots)$$

which reduces to

$$x_0 = 1, \quad x_1 = x_2 = \cdots = 0$$

at time $t = 0$. Writing

$$x_n = \xi_n \cdot e(-\nu_n t)$$

the equations for ξ_n are

$$-\frac{1}{i}\dot{\xi}_n = \frac{\varepsilon}{h}\sum_m W_{nm}\xi_m e^{i(\nu_n - \nu_m)t};$$

for $\varepsilon = 0$, $\dot{\xi}_n = 0$. Neglecting terms of order ε^2, we can take the initial conditions

$$\xi_0 = 1, \quad \xi_1 = \xi_2 = \cdots = 0$$

as the 0th approximation; on substituting these values in the equation we obtain as the first approximation

$$\xi_n = \frac{\varepsilon W_{n0}}{h}\,\frac{e^{-i(\nu_0 - \nu_n)t} - 1}{\nu_0 - \nu_n} \quad (\nu_n \neq \nu_0).$$

On setting $\nu_0 - \nu_n = \nu$, the desired probability is

$$|x_n|^2 = |\xi_n|^2 = 2\left(\frac{\varepsilon}{h}\right)^2 \frac{1 - \cos(\nu t)}{\nu^2} \cdot |W_{n0}|^2$$

$$= \frac{2[1 - \cos(\nu t)]}{h^2\nu^2}|\mathsf{H}_{0n}|^2. \quad (9.1)$$

It is to be noted that in accordance with this result the probability of transition from state 0 *to state* n *is determined by* $|\mathsf{H}_{0n}|^2$. In the case of resonance ($\nu_n = \nu_0$) the transition probability increases at first with the square of t:

$$|x_n|^2 = \left(\frac{\varepsilon}{h}\right)^2 \cdot |W_{n0}|^2 t^2.$$

§ 10. The Problem of Several Bodies. Product Space

A physical system consisting of *two particles* of masses m, m', co-ordinates $x\,y\,z;\ x'\,y'\,z'$ and linear momenta $\mathfrak{p}, \mathfrak{p}'$, has as its Hamiltonian function

$$H = \frac{1}{2m}(p_x^2 + p_y^2 + p_z^2) + \frac{1}{2m'}(p_x'^2 + p_y'^2 + p_z'^2)$$

$$+ V(x\,y\,z;\ x'\,y'\,z'), \quad (10.1)$$

where V is the potential energy. We assume, as in the older physics of central forces, that we are here dealing with an *action at a distance* so that the potential energy depends only on the simultaneous positions of the two particles. This assumption naturally breaks down when, in accordance with the theory of relativity, we take into account the finite velocity of propagation of the disturbance, which requires the introduction of a *field*. The wave function ψ of the system will depend on all six co-ordinates $x\,y\,z$; $x'\,y'\,z'$ in addition to t ; the operators corresponding to these functions in the domain of such functions ψ are multiplication by $x, \cdots; \; x', \cdots$, and to the linear momenta correspond the derivatives $\frac{h}{i}\frac{\partial}{\partial x}, \cdots; \; \frac{h}{i}\frac{\partial}{\partial x'}, \cdots$

From (10.1) we then obtain the wave equation

$$-\frac{h}{i}\frac{\partial\psi}{\partial t} + \frac{h^2}{2m}\Delta\psi + \frac{h^2}{2m'}\Delta'\psi - V\cdot\psi = 0. \tag{10.2}$$

We must ask for *the probability that the one particle is to be found at a point P and, simultaneously, the other is to be found at a point P'*. The probability density is accordingly to be computed for a 6-dimensional space with co-ordinates $x\,y\,z$; $x'\,y'\,z'$. Indeed, the wave field is not to represent directly occurrences taking place in physical space, but is to determine the appearance at definite positions or with definite energies and momenta ; there is consequently nothing absurd in the fact that its medium is this abstract 6-dimensional configuration space.

In order to be independent of the special procedure by which the scalar wave mechanics puts together two systems $\mathfrak{a}$, $\mathfrak{b}$ to form a single system $\mathfrak{c}$, as suggested by this example involving the Hermitian forms representing the co-ordinates and momenta of the two systems, we must first discuss the *multiplication of spaces* from a purely mathematical standpoint.

With each vector $\mathfrak{x} = (x_i)$ in a space $\mathfrak{R}$ of m dimensions and each vector $\mathfrak{y} = (y_k)$ in a space $\mathfrak{S}$ of n dimensions there is associated a vector $\mathfrak{z} = \mathfrak{x} \times \mathfrak{y}$ with components

$$z_{ik} = x_i y_k \tag{10.3}$$

in an $m \cdot n$-dimensional space $\mathfrak{T} = \mathfrak{R} \times \mathfrak{S}$, the ***product space.*** The components are here numbered by means of the *index pair* $(i\,k) = l$. The totality of vectors $\mathfrak{z} = \mathfrak{x} \times \mathfrak{y}$ do not themselves constitute a linear manifold, but their linear combinations fill the entire product space $\mathfrak{T}$. With the linear correspondences A in $\mathfrak{R}$ and B in $\mathfrak{S}$:

$$x'_{i'} = \sum_i a_{i'i}\, x_i, \quad y'_{k'} = \sum_k b_{k'k}\, y_k$$

is associated a linear correspondence $C = A \times B$ in $\mathfrak{T}$:

$$x'_{i'}\, y'_{k'} = \sum_{i,k} a_{i'i}\, b_{k'k}\, x_i\, y_k,$$

or

$$z'_{l'} = \sum_{l} c_{l'l} z_l, \quad c_{l'l} = a_{i'i} b_{k'k} \quad [l = (ik),\ l' = (i'k')].$$

Naturally, to this multiplication corresponds the law of composition

$$(A \times B)(A_1 \times B_1) = (AA_1 \times BB_1),$$

where A, A_1 are correspondences of $\mathfrak{R}$ on itself and B, B_1 are correspondences in $\mathfrak{S}$. A co-ordinate system in $\mathfrak{R}$ and one in $\mathfrak{S}$ together determine a co-ordinate system in $\mathfrak{T}$; if the co-ordinate system in $\mathfrak{R}$ is subjected to the transformation A and that in $\mathfrak{S}$ to the transformation B, then the co-ordinate system associated with them in $\mathfrak{T}$ undergoes the transformation $A \times B$. In accordance with the equation

$$d(x_i y_k) = dx_i \cdot y_k + x_i \cdot dy_k,$$

to the infinitesimal correspondence H in $\mathfrak{R}$, $\mathcal{J}$ in $\mathfrak{S}$ corresponds the infinitesimal correspondence

$$(H \times \mathbf{1}_s) + (\mathbf{1}_r \times \mathcal{J}) \tag{10.4}$$

in $\mathfrak{T}$, where $\mathbf{1}_r$, $\mathbf{1}_s$ denote the unit matrices in $\mathfrak{R}$, $\mathfrak{S}$, respectively. All of the foregoing is applicable to arbitrary vector spaces. When $\mathfrak{R}$ and $\mathfrak{S}$ are both unitary spaces, then $\mathfrak{T}$ is also, for by (10.3)

$$\sum \bar{z}_l z_l = \sum \bar{x}_i x_i \cdot \sum \bar{y}_k y_k$$

is an invariant if $\sum \bar{x}_i x_i$, $\sum \bar{y}_k y_k$ are; $A \times B$ is unitary if A and B are.

Accordingly, two physical systems $\mathfrak{a}$ and $\mathfrak{b}$ are compounded to form a total system $\mathfrak{c}$ as follows. The system space $\mathfrak{T}$ of $\mathfrak{c}$ is $\mathfrak{R} \times \mathfrak{S}$, where $\mathfrak{R}$ is the system space of $\mathfrak{a}$ and $\mathfrak{S}$ of $\mathfrak{b}$. Let the arbitrary physical quantity α in $\mathfrak{R}$ be represented by the Hermitian form A; on replacing all these forms A by $A \times \mathbf{1}_s$, where $\mathbf{1}_s$ is the unit form in an arbitrary space $\mathfrak{S}$, there exist between these latter exactly the same relations as between the A, so that from the solution of a quantum problem in $\mathfrak{R}$ there arises a solution for the corresponding problem in $\mathfrak{R} \times \mathfrak{S}$, but there exists no real distinction between the two. In the system $\mathfrak{c}$ obtained by composition we have therefore to associate the Hermitian form $A \times \mathbf{1}_s$ with a quantity α of $\mathfrak{a}$ and $\mathbf{1}_r \times B$ with β of $\mathfrak{b}$, where A, B are the forms associated with α, β in $\mathfrak{R}$, $\mathfrak{S}$, respectively. The totality of quantities of the composite system $\mathfrak{c}$ is obtained by starting from the quantities belonging to the component systems $\mathfrak{a}$ and $\mathfrak{b}$ and multiplying and adding them together in all possible ways. The quantities α of $\mathfrak{a}$ commute with the quantities β of $\mathfrak{b}$, for

$$(A \times \mathbf{1}_s)(\mathbf{1}_r \times B) = A \times B = (\mathbf{1}_r \times B)(A \times \mathbf{1}_s).$$

We refer to the content of these last two sentences when we say that $\mathfrak{c}$ *consists of two kinematically independent parts* $\mathfrak{a}$ *and* $\mathfrak{b}$.

The two systems are *dynamically independent* if the energy H of the composite system is the sum of the energies $H^{(1)}$, $H^{(2)}$ of the partial systems:

$$H = (H^{(1)} \times \mathbf{1}) + (\mathbf{1} \times H^{(2)}).$$

The infinitesimal unitary correspondence $\frac{dt}{ih} \cdot H$ in the total system space is then that one which is due to the infinitesimal unitary correspondences $\frac{dt}{ih} \cdot H^{(1)}$, $\frac{dt}{ih} \cdot H^{(2)}$ in the two original system spaces [(10.4)]. If $H^{(1)}$ and $H^{(2)}$ are both in diagonal form, then H is also, and the characteristic numbers are given by

$$E_l = E_i^{(1)} + E_k^{(2)} \quad \text{or} \quad \nu_l = \nu_i^{(1)} + \nu_k^{(2)} \quad [l = (ik)]$$

If we have a pure state for the total system which is represented by the vector $\mathfrak{c}$ of absolute value 1 and components c_{ik}, and if $Q = \|q_{ii'}\|$ is an arbitrary quantity in $\mathfrak{a}$, then the expectation of Q in the pure state $\mathfrak{c}$ is

$$\langle Q \rangle = \Sigma q_{ii'} \delta_{kk'} \bar{c}_{ik} c_{i'k'}.$$

This has the form (7.2) with

$$A = \|a_{ii'}\| = \|\sum_k \bar{c}_{ik} c_{i'k}\|.$$

$A(\mathfrak{x})$ is the Hermitian form

$$\sum_k \left| \sum_i c_{ik} \bar{x}_i \right|^2$$

in $\mathfrak{R}$. But we see from this that we are not dealing with a pure state in $\mathfrak{a}$, for $a_{ii'}$ will not in general have the form $a_i \bar{a}_{i'}$. *Conditions which insure a maximum of homogeneity within* $\mathfrak{c}$ *need not require a maximum in this respect within the partial system* $\mathfrak{a}$. Furthermore: *if the state of* $\mathfrak{a}$ *and the state of* $\mathfrak{b}$ *are known, the state of* $\mathfrak{c}$ *is in general not uniquely specified*, for a positive definite Hermitian form $\|a_{ik,\, i'k'}\|$ in the product space, which describes a statistical aggregate of states $\mathfrak{c}$, is not uniquely determined by the Hermitian forms

$$\sum_k a_{ik,\, i'k}, \quad \sum_i a_{ik,\, ik'}$$

to which it gives rise in the spaces $\mathfrak{R}$, $\mathfrak{S}$. In this significant sense quantum theory subscribes to the view that "*the whole is greater than the sum of its parts*," which has recently been raised to the status of a philosophical creed by the Vitalists and the Gestalt Psychologists.

The kinematically independent parts into which a system can be resolved need not be spatially separated, nor need they even refer to different particles. We can, for example, resolve a single particle, whose physical quantities can all be expressed in terms of x, y, z; p_x, p_y, p_z, into three partial systems with fundamental quantities x, $p_x \mid y$, $p_y \mid z$, p_z. For quantities which belong to different partial systems, for example a quantity which can be expressed in terms of x, p_x alone and one which is in terms of y, p_y alone, commute with each other in the sense of matrix multiplication.

In the *perturbation theory* we are usually concerned with a system which consists of two kinematically independent parts and which are almost dynamically independent. Disregarding the interaction εW for the moment, let $h\nu_n$ and $h\rho_r$ be the energy levels of the two parts, so that $h(\nu_n + \rho_r)$ are the energy levels of the unperturbed total system. On writing in equation (9.1) $s = (n, r)$ in place of 0 and $s' = (n', r')$ in place of n, whence

$$\nu = (\nu_n + \rho_r) - (\nu_{n'} + \rho_{r'}) = \nu_{nn'} + \rho_{rr'};$$
$$\nu_{nn'} = \nu_n - \nu_{n'}, \quad \rho_{rr'} = \rho_r - \rho_{r'},$$

we find as the probability that the total system goes over from the state s to the state s' during time t:

$$2\left(\frac{\varepsilon}{h}\right)^2 \frac{1 - \cos(\nu_{nn'} + \rho_{rr'})t}{(\nu_{nn'} + \rho_{rr'})^2} \cdot |W(nr, n'r')|^2. \qquad (10.5)$$

The probability that the first system will be found in the state n' after time t, the total system having been in the state $s = (nr)$ originally, is obtained from (10.5) by summation with respect to r'.

§ 11. Commutation Rules. Canonical Transformations

The development of wave mechanics in §§ 1-3 went beyond the general scheme of §§ 7 and 8 in that it employed certain specific Hermitian forms to represent the co-ordinates and momenta of the particle. We are now interested in seeing how this can be formulated in an invariant manner, without recourse to any special co-ordinate system in system space.

For the Hermitian forms q, p representing a rectangular

co-ordinate and its associated momentum we postulate the ***commutation rule***

$$pq - qp = \frac{h}{i}\mathbf{1}. \tag{11.1}$$

If the system has only one degree of freedom, these two quantities appear as *canonical variables* in classical mechanics. All physical quantities of the system are then functions of p and q; in order to avoid complications we restrict ourselves to *polynomials* f in p and q, and assume, in particular, that the Hamiltonian function H has this form. What are we to understand by the *derivatives* f_p and f_q of f with respect to p and q in this domain in which p and q are not commutative in multiplication? We should in any case require that differentiation with respect to q should obey the following postulates:

(1) $p_q = 0$, $q_q = 1$;
(2) $(f + g)_q = f_q + g_q$ and $(\alpha f)_q = \alpha \cdot f_q$, where α is a number;
(3) $(fg)_q = f_q \cdot g + f \cdot g_q$.

We see immediately that these conditions uniquely determine the derivative of a polynomial f, unless they happen to lead to contradictions. But that they do *not* lead to contradictions can be seen from the fact that they are obeyed by the definition

$$ih \cdot f_q = fp - pf. \tag{11.2}$$

(1) follows immediately from the commutation rule (11.1), and the linearity (2) of the process is evident. (3) is proved by the formula

$$(fg)p - p(fg) = f(gp - pg) + (fp - pf)g$$

which involves only the distributive and associative character of matrix multiplication. Similarly we can show that

$$- ih \cdot f_p = fq - qf. \tag{11.2}$$

The fundamental dynamical law gives us the equation (8.6):

$$\frac{h}{i}\frac{df}{dt} = Hf - fH$$

for any Hermitian form f. On applying this equation to p and q—which obviously suffices to establish the corresponding result for any polynomial f of p and q—and comparing it with the formulæ (11.2) applied to the particular function H, we are led to the familiar *Hamiltonian equations of classical mechanics*:

$$\frac{dq}{dt} = H_p, \quad \frac{dp}{dt} = - H_q. \tag{11.3}$$

It is a universal trait of quantum theory to retain all the relations of classical physics; but whereas the latter interpreted these relations as conditions to which the values of physical quantities were subject in all individual cases, the former interprets them as conditions on the quantities themselves, or rather on the Hermitian matrices which represent them. This is the more significant formulation which the new quantum theory has given Bohr's correspondence principle.

The commutation rule (11.1) is of a rather remarkable nature. It is entirely impossible for matrices in a space of a finite number of dimensions, and it alone precludes the possibility that in an ∞-dimensional space q (or p) have only a discrete spectrum of characteristic numbers. For on referring q to its principal axes

$$q = \|q_{mn}\|, \quad q_{nn} = q_n, \quad q_{mn} = 0 \quad (m \neq n)\,; \; p = \|p_{mn}\|,$$

the left side of the commutation rule has the components $p_{mn}(q_n - q_m)$; hence the main diagonal consists of nothing but zeros! The question arises as to whether it can be concluded from (11.1) alone that the forms representing q and p can always be given the form

$$\int_{-\infty}^{+\infty} x\,\bar{\psi}(x)\,\psi(x)\,dx, \quad \int_{-\infty}^{+\infty} \bar{\psi}(x) \cdot \frac{h}{i}\frac{d\psi}{dx}\,dx$$

for an arbitrary vector ψ with components $\psi(x)$ on employing an appropriate co-ordinate system in system space. We shall see in Chap. IV, § 15, that, on introducing a certain irreducibility condition, this is in fact the case.

On taking into account the three space co-ordinates q_α and their associated linear momenta p_α $(\alpha = 1, 2, 3)$, we have in place of the one commutation rule (11.1) the following: [23]

$$\left.\begin{aligned} p_\alpha p_\beta - p_\beta p_\alpha = 0, \quad q_\alpha q_\beta - q_\beta q_\alpha = 0 \text{ for all } \alpha, \beta\,; \\ p_\alpha q_\beta - q_\beta p_\alpha = \frac{h}{i}\delta_{\alpha\beta}, \quad \delta_{\alpha\beta} = \begin{cases} 1 & (\alpha = \beta) \\ 0 & (\alpha \neq \beta). \end{cases} \end{aligned}\right\} \tag{11.4}$$

The same commutation rules apply to the case in which we have several particles, the only difference being that then α runs through 6, 9, · · · values, according to the number of particles, instead of 3. These commutation rules are the necessary and sufficient condition that the dynamical law, which governs the time rate of change of the state vector $\mathfrak{x}$ in system space, leads to the Hamiltonian equations for the "canonical variables" q_α, p_α representing the co-ordinates and associated momenta of the various particles composing the physical system—whatever the dependence of the Hamiltonian function H on these quantities may be.

In classical mechanics *the Hamiltonian equations are invariant with respect to canonical transformations.*[24] In a system of f degrees of freedom the transition from a set of variables q_α, p_α, describing the state to a set q'_α, p'_α ($\alpha = 1, 2, \cdots, f$) is a ***canonical transformation*** if the difference

$$\Sigma p'_\alpha dq'_\alpha - \Sigma p_\alpha dq_\alpha \tag{11.5}$$

is a total differential. If, for example, the q_α are subjected to a transformation

$$q_\alpha = \phi_\alpha(q'_1 \cdots q'_f)$$

among themselves, the p_α must transform as the components of a " covariant vector " in q-space in order that the whole be a canonical transformation (" extended point transformation ") :

$$p'_\alpha = \sum_\beta \frac{\partial \phi_\beta}{\partial q'_\alpha} \cdot p_\beta.$$

Perhaps the simplest canonical transformation is that in which the rôles of q and p are interchanged :

$$p'_\alpha = -q_\alpha, \quad q'_\alpha = p_\alpha.$$

The canonical transformations constitute a group [cf. III, § 1]. For the identity, i.e. the transition from (p, q) to (p, q), is a canonical transformation ; the inverse $(p', q') \rightarrow (p, q)$ of a canonical transformation $(p, q) \rightarrow (p', q')$ is also canonical ; and from the canonical transformations $(p, q) \rightarrow (p', q')$, $(p', q') \rightarrow (p'', q'')$ it follows that the resultant transformation $(p, q) \rightarrow (p'', q'')$ is also canonical, for if

$$\Sigma p'_\alpha dq'_\alpha - \Sigma p_\alpha dq_\alpha, \quad \Sigma p''_\alpha dq''_\alpha - \Sigma p'_\alpha dq'_\alpha$$

are total differentials their sum

$$\Sigma p''_\alpha dq''_\alpha - \Sigma p_\alpha dq_\alpha$$

is also.

An *infinitesimal* canonical transformation is one in which p', q' differ infinitely little from p, q. We can consider it as an infinitesimal deformation of the $2f$-dimensional (p, q)-space which takes place in the infinitesimal time interval $\varepsilon = \delta t$. We introduce the components δp, δq of the displacement vector by means of the equations

$$p'_\alpha - p_\alpha = \varepsilon \cdot \delta p_\alpha, \quad q'_\alpha - q_\alpha = \varepsilon \cdot \delta q_\alpha \cdot$$

Since (11.5) must be a total differential,

$$\Sigma p'_\alpha \, dq'_\alpha + \Sigma q_\alpha \, dp_\alpha = dT \tag{11.6}$$

must also ; in our case T must differ only infinitesimally from $\Sigma p_\alpha q_\alpha$. We may therefore write

$$T = \sum_\alpha p_\alpha q'_\alpha - \varepsilon S ;$$

considering S as a function of p_α and q'_α we have, in accordance with (11.6),

$$p'_\alpha = p_\alpha - \varepsilon \frac{\partial S}{\partial q'_\alpha}, \quad q_\alpha = q'_\alpha - \varepsilon \frac{\partial S}{\partial p_\alpha}$$

or

$$\delta p_\alpha = - \frac{\partial S}{\partial q_\alpha'}, \quad \delta q_\alpha = \frac{\partial S}{\partial p_\alpha}. \tag{11.7}$$

Since we may legitimately neglect terms of order ε^2, we may identify q'_α with q_α on the right-hand side of these equations. We call S *the generating function of the infinitesimal canonical transformation.*

In accordance with the Hamiltonian equations, the state of a system, represented by a point (p, q) in (p, q)-space, goes over into a state $(p + dp, q + dq)$ during time dt. If we follow this transition for *all possible initial states* (p, q) we obtain an infinitesimal deformation of the space whose points represent the state of the system. *The Hamiltonian equations assert that this deformation is an infinitesimal canonical transformation with generating function* $H \cdot dt$. It follows from this without any calculation that these equations have a significance which is independent of any particular choice of canonical variables.

Now in quantum theory the Hamiltonian equations (11.3) assert that the state vector $\mathfrak{x}$ in system space undergoes the infinitesimal unitary rotation

$$d\mathfrak{x} = - \frac{i \, dt}{h} \cdot H\mathfrak{x}, \tag{8.1}$$

so the infinitesimal canonical transformation of the quantities p, q is here obtained by subjecting the argument $\mathfrak{x}$ in the Hermitian forms representing them to the infinitesimal rotation

$$\varepsilon \, . \, \delta\mathfrak{x} = - \frac{i\varepsilon}{h} \cdot S\mathfrak{x}.$$

We find that the increments of the quantities p_α, q_α are in fact

$$\varepsilon \, . \, \delta q_\alpha = \frac{i\varepsilon}{h}(Sq_\alpha - q_\alpha S), \quad \varepsilon \, . \, \delta p_\alpha = \frac{i\varepsilon}{h}(Sp_\alpha - p_\alpha S),$$

and, in virtue of the commutation relations (11.4), this agrees exactly with (11.7). On generating a finite canonical transformation by the successive application of an infinity of infinitesimal ones we arrive at the result that *the unitary correspondences of system space on itself in quantum theory :*

$$\mathfrak{x}' = U\mathfrak{x}$$

correspond to the canonical transformations of classical mechanics ; more precisely, only those for which the matrix U is expressible in terms of the matrices p, q, but we may for the present pass over the question as tc whether every matrix U can be obtained, or at least arbitrarily closely approximated, in this way. Since the commutation rules (11.4) remain unchanged under rotations of the normal co-ordinate system, they are valid for an arbitrary set of canonical variables. This is also evident from the fact that they are the conditions that the dynamical law (8.1) lead to the Hamiltonian equations

$$\frac{dq_\alpha}{dt} = \frac{\partial H}{\partial p_\alpha}, \quad \frac{dp_\alpha}{dt} = -\frac{\partial H}{\partial q_\alpha}. \tag{11.3}$$

The general procedure for the quantum mechanical treatment of a physical system suffers from the disagreeable fact that the expression for the energy in terms of the canonical variables must be taken from the classical model, and in addition the transition to quantum mechanics is even then not unique, for the model offers no means of telling whether a monomial such as p^2q is to be interpreted as p^2q, pqp, qp^2 or a linear combination of all three [cf. IV, § 14]. The provisional character of such a procedure is clear, but the results so far obtained seem to justify the hope that the path we have entered upon will lead to a unique formulation of the laws governing the *actual physical phenomena.* We need then concern ourselves longer with the general mechanical scheme.

§ 12. Motion of a Particle in an Electro-magnetic Field. Zeeman Effect and Stark Effect

Let the spatial co-ordinates $x\,y\,z$ now be denoted by $x_1\,x_2\,x_3$ and the time t by x_0. If ϕ is the scalar and $c\,\mathfrak{A}$ the vector potential of the electro-magnetic field, then in the theory of relativity

$$(-\phi, \mathfrak{A}_x, \mathfrak{A}_y, \mathfrak{A}_z) = (\phi_0, \phi_1, \phi_2, \phi_3)$$

are the components of a vector in the space dual to the 4-dimensional world. Let

$$F_{\alpha\beta} = \frac{\partial \phi_\beta}{\partial x_\alpha} - \frac{\partial \phi_\alpha}{\partial x_\beta};$$

F_{10}, F_{20}, F_{30} are the components of the electric field strength $\mathfrak{E}$, $c(F_{23}, F_{31}, F_{12})$ the components of the magnetic field strength $\mathfrak{H}$. Denoting the components of the velocity of a particle by v_1, v_2, v_3, its proper time is

$$ds = \sqrt{dt^2 - (dx_1^2 + dx_2^2 + dx_3^2)/c^2}$$
$$= dt\sqrt{1 - v^2/c^2} \quad (v^2 = v_1^2 + v_2^2 + v_3^2).$$

With the world vector $u^\alpha = \dfrac{dx_\alpha}{ds}$ is associated the dual u_α with components

$$u_r = u^r \quad (r = 1, 2, 3), \quad u_0 = -c^2 u^0.$$

The invariant equations of motion for a particle of mass m and charge $-e$ are

$$\frac{d(mu_\alpha)}{ds} = -e \sum_{\beta=0}^{3} F_{\alpha\beta} u^\beta$$

or

$$\frac{d(mu_i)}{dt} = -e\Big(F_{i0} + \sum_{k=1}^{3} F_{ik} v_k\Big) \quad (i = 1, 2, 3). \tag{12.1}$$

The right-hand side is in fact the ponderomotive force

$$-e\Big(\mathfrak{E} + \frac{1}{c}[\mathfrak{v}\mathfrak{H}]\Big).$$

These equations arise from the Hamiltonian function

$$H = e\phi_0 + c\sqrt{m^2c^2 + \sum_{i=1}^{3}(p_i + e\phi_i)^2}, \tag{12.2}$$

in which $x_1 x_2 x_3$; $p_1 p_2 p_3$ are the canonical variables. In fact, the Hamiltonian equations

$$v_i \equiv \frac{dx_i}{dt} = \frac{\partial H}{\partial p_i} = \frac{c(p_i + e\phi_i)}{\sqrt{}}$$

yield

$$p_i + e\phi_i = mu_i;$$

in the remaining equations

$$\frac{dp_i}{dt} = -\frac{\partial H}{\partial x_i} = -e\left\{\frac{\partial \phi_0}{\partial x_i} + c\sum_{k=1}^{3} \frac{\partial \phi_k}{\partial x_i} \frac{p_k + e\phi_k}{\sqrt{}}\right\}$$

the left-hand side is

$$= \frac{d(mu_i)}{dt} - e\left\{\frac{\partial \phi_i}{\partial x_0} + \sum_{k=1}^{3} \frac{\partial \phi_i}{\partial x_k} \cdot v_k\right\}.$$

But this is the desired equation (12.1):

$$\frac{d(mu_i)}{dt} = -e\left\{\Big(\frac{\partial \phi_0}{\partial x_i} - \frac{\partial \phi_i}{\partial x_0}\Big) + \sum_{k=1}^{3}\Big(\frac{\partial \phi_k}{\partial x_i} - \frac{\partial \phi_i}{\partial x_k}\Big)v_k\right\}.$$

The negative energy $-H$ is the time component p_0 of the dual vector whose space components are the components of linear momentum $\mathfrak{p} = (p_1, p_2, p_3)$, so the equation (12.2) can be written in the rational form

$$\frac{1}{c^2}(p_0 + e\phi_0)^2 - \sum_{i=1}^{3} (p_i + e\phi_i)^2 = m^2c^2.$$

From this we obtain the simple rule: *The influence of an electromagnetic field on a particle of charge $-e$ can be expressed by replacing p_α by $p_\alpha + e\phi_\alpha$ in the equations of motion for a free particle.*

On going over to quantum theory p_α becomes the operator $\frac{h}{i}\frac{\partial}{\partial x_\alpha}$ and is contragradient to the 4-dimensional displacement dx_α, as is seen from the equation

$$d\psi = \sum_\alpha \frac{\partial \psi}{\partial x_\alpha} dx_\alpha.$$

Our rule is now: *On introducing a field of potential ϕ_α*

$$\frac{\partial}{\partial x_\alpha} \textit{ must be replaced by } \frac{\partial}{\partial x_\alpha} + \frac{ie}{h}\phi_\alpha \tag{12.3}$$

in the wave equation of the particle. Only $\tilde{\psi}\psi$ has a simple physical significance; it is therefore to be assumed that the laws which govern ψ remain invariant on replacing ψ by $e^{i\lambda}\cdot\psi$, where λ is any real function of position in space-time. On the other hand, in the classical theory of the electro-magnetic field only the field strengths, and not the potentials, have an objective significance, i.e. the laws are invariant on replacing ϕ_α by $\phi_\alpha - \frac{\partial\mu}{\partial x_\alpha}$, where μ is also an arbitrary function of the x_α. On examining our wave equation for these invariantive properties we find that it is not invariant under each of them separately, but that there must exist a certain relation between λ and μ. *The field equations for the potentials ψ and ϕ of the material and electromagnetic waves are invariant under the simultaneous replacement of*

$$\psi \textit{ by } e^{i\lambda}\cdot\psi \quad \text{and} \quad \phi_\alpha \textit{ by } \phi_\alpha - \frac{h}{e}\frac{\partial\lambda}{\partial x_\alpha};$$

here λ is an arbitrary function of the space-time co-ordinates. This "*principle of* **gauge invariance**" is quite analogous to that previously set up by the author, on speculative grounds, in order to arrive at a unified theory of gravitation and electricity.[25] But I now believe that this gauge invariance does not tie together electricity and gravitation, but rather *electricity and matter* in the manner described above. We shall discuss this principle more thoroughly in Chap. IV; its significance and its interpretation will then be more apparent.

On passing to the limit $c \to \infty$ in (12.2), after separating out the factor mc^2, we return to ordinary mechanics:

$$H = e\phi_0 + \frac{1}{2m}\sum_i (p_i + e\phi_i)^2.$$

On neglecting terms which are quadratic in the ϕ_i, we find, in addition to the kinetic energy $\sum_i p_i^2/2m$, the potential

$$V = -e\phi + \frac{e}{m}(\mathfrak{p}\mathfrak{A}). \tag{12.4}$$

We have already made use of the first part, that due to the electric field, in § 5. If we have, in addition to the field originating in the nucleus, a homogeneous electro-static field in the direction of the z-axis and of strength F, for which $\phi = -F \cdot z$, it adds the perturbation term

$$W = eF \cdot z$$

to the energy. A homogeneous static magnetic field $\mathfrak{H}$ is obtained from the vector potential $c\mathfrak{A} = \frac{1}{2}[\mathfrak{H}\mathfrak{r}]$, $\mathfrak{r} = (x, y, z)$; this adds to the energy the perturbation term

$$\frac{e}{2mc}(\mathfrak{p}[\mathfrak{H}\mathfrak{r}]) = \frac{e}{2mc}([\mathfrak{r}\mathfrak{p}]\mathfrak{H}),$$

i.e.

$$\boxed{W = \frac{eh}{2mc}(\mathfrak{H}\mathfrak{L})}\,. \tag{12.5}$$

Zeeman Effect.—If the homogeneous magnetic field strength, of magnitude $|\mathfrak{H}|$, is in the direction of the z-axis, the perturbation term is

$$W = ho \cdot L_z, \quad o = \frac{e|\mathfrak{H}|}{2\mu c}. \tag{12.6}$$

On choosing the characteristic functions $\psi_{nl}^{(m)}$ as our co-ordinate system in the system space of the functions ψ, W, as well as the energy of the unperturbed atom, is in diagonal form; in the state defined by nl, m it has the value

$$ho \cdot m. \tag{12.7}$$

The components $(nl, m) \rightarrow (n'l', m')$, consistent with the selection rule for m, into which the line with frequency $\nu = \frac{1}{h}(E_{nl} - E_{n'l'})$ is broken up give rise to but *three lines:* one corresponding to all the transitions $m \rightarrow m$, which is linearly polarized in the direction of the z-axis and is undisplaced; one which is circularly polarized perpendicular to the z-axis, the frequency ν of which is displaced by $+ o$ $(m \rightarrow m - 1)$; and one which is circularly polarized in the opposite sense, with frequency $\nu - o$ instead of ν $(m \rightarrow m + 1)$. This *normal Zeeman effect* is found only in the so-called singlet lines.

Stark Effect.—In accordance with the general perturbation theory, the displacement and resolution of terms in the presence of a homogeneous electric field is determined, to terms of first order, by the matrix

$$eF \cdot \langle z \rangle.$$

In consequence of the selection rule $l \rightarrow l \pm 1$, $\langle z \rangle = 0$, unless accidentally all energy levels whose azimuthal quantum numbers differ by 1 coincide. Ignoring this exceptional case, *we should expect to find no* 1st *order perturbation effect increasing linearly with the field strength F (linear Stark effect)*, but only a quadratic effect, which is much smaller. This is in agreement with the experimental data on alkali atoms. Hydrogen is, however, degenerate, since for it energy levels with the same principal quantum number n and $l = 0, 1, \cdots, n - 1$ coincide. The calculations for this case have been carried out by *Schrödinger* and compared with experiment.[26]

§ 13. Atom in Interaction with Radiation

Following *Jeans*, black body radiation is mathematically equivalent to a system of infinitely many oscillators. Maxwell's equations for the free ether are

$$\operatorname{div} \mathfrak{H} = 0, \quad \operatorname{curl} \mathfrak{E} + \frac{1}{c}\frac{\partial \mathfrak{H}}{\partial t} = 0;$$

$$\operatorname{div} \mathfrak{E} = 0, \quad \operatorname{curl} \mathfrak{H} - \frac{1}{c}\frac{\partial \mathfrak{E}}{\partial t} = 0.$$

In order to simplify the relations, we assume that the walls of the radiation cavity of volume V are *reflecting;* then $\mathfrak{E}$ is perpendicular to the walls at the boundaries of the cavity. Since the black body is at rest it is of no particular advantage to carry through the calculation in a relativistically invariant manner; we may therefore normalize the vector potential

$c\mathfrak{A}$ in such a way that the scalar potential vanishes. We then have $\mathfrak{E} = -\frac{\partial \mathfrak{A}}{\partial t}$ and the equations in the first row are satisfied by $\mathfrak{H} = c \cdot \text{curl}\, \mathfrak{A}$; the equations in the second row become

$$\text{div}\, \mathfrak{A} = 0, \quad \Delta \mathfrak{A} - \frac{1}{c^2} \frac{\partial^2 \mathfrak{A}}{\partial t^2} = 0.$$

On the boundary $\mathfrak{A}$ is normal to the walls. Let the characteristic numbers and characteristic functions of the equations

$$\Delta \mathfrak{A} + \frac{\nu^2}{c^2} \mathfrak{A} = 0, \quad \text{div}\, \mathfrak{A} = 0,$$

with the boundary condition that $\mathfrak{A}$ is there normal, be denoted by

$$\nu = \rho_\alpha \ (\geqq 0), \quad \mathfrak{A}_\alpha \quad [\alpha = 1, 2, 3, \cdots],$$

normalized in accordance with

$$\int_V (\mathfrak{A}_\alpha \mathfrak{A}_\beta) dV = 4\pi \delta_{\alpha\beta}.$$

On setting

$$\mathfrak{A} = \sum_\alpha q^\alpha \cdot \mathfrak{A}_\alpha,$$

where the coefficients q^α depend on time but not on position, we find for them the equations

$$\frac{d^2 q^\alpha}{dt^2} + \rho_\alpha^2 q^\alpha = 0.$$

Introducing $\frac{dq^\alpha}{dt} = p^\alpha$ in addition to the q^α, this equation is that for an oscillator with Hamiltonian function

$$H_\alpha = \frac{1}{2}(p^\alpha)^2 + \frac{1}{2}\rho_\alpha^2 (q^\alpha)^2;$$

we readily find on applying

$$\mathfrak{E} = -\sum_\alpha p^\alpha \mathfrak{A}_\alpha, \quad \mathfrak{H} = c \sum_\alpha q^\alpha \cdot \text{curl}\, \mathfrak{A}_\alpha$$

that the energy of the radiation field is in fact given by

$$H = \frac{1}{8\pi} \int_V (\mathfrak{E}^2 + \mathfrak{H}^2) dV = \sum_\alpha H_\alpha;$$

with this we have proved the theorem due to *Jeans*. For high frequencies ρ there are approximately

$$\frac{V \rho^2 d\rho}{\pi^2 c^3} \tag{13.1}$$

modes of oscillation in the frequency interval ρ, $\rho + d\rho$.[27] We are interested above all in the limiting case of an infinitely large cavity; the spectrum then becomes continuous and our formula for the density of frequencies becomes exact.

On *quantizing* this mechanical system of infinitely many oscillators [28] in accordance with the theory of the oscillator (§ 3) and the process of composition (§ 10—but cf. remark on p. 109), we find as possible quantum states s, each of which is characterized by the fact that in it there is associated with each index α an integer $n_\alpha \geqq 0$. In this quantum state

$$H_\alpha = h\rho_\alpha\left(n_\alpha + \frac{1}{2}\right),$$

or, on choosing the additive constant in the energy in such a way that the lowest energy value which the black body radiation is capable of assuming is 0,

$$H_\alpha = n_\alpha \cdot h\rho_\alpha, \quad H = \sum_\alpha n_\alpha \cdot h\rho_\alpha.$$

In the language of photons this means that when the cavity is in the state s it contains n_α photons of each kind α. The matrix element

$$q^\alpha_{ss'} \quad [s = n_1, n_2, \cdots, n_\alpha, \cdots; \; s' = n'_1, n'_2, \cdots, n'_\alpha, \cdots]$$

vanishes unless all the equations

$$n'_1 = n_1, \quad n'_2 = n_2, \quad n'_3 = n_3, \cdots$$

hold with the exception of $n'_\alpha = n_\alpha$, which is to be replaced by

$$n'_\alpha = n_\alpha + 1 \quad \text{or} \quad n'_\alpha = n_\alpha - 1.$$

In the first case we have, by eq. (8.12),

$$q^\alpha_{ss'} = \sqrt{\frac{h(n_\alpha + 1)}{2\rho_\alpha}} \quad \text{(Emission)}, \qquad (13.2)$$

and in the second

$$q^\alpha_{ss'} = \sqrt{\frac{h\,n_\alpha}{2\rho_\alpha}} \quad \text{(Absorption)}. \qquad (13{\cdot}2)$$

The first transition $s \rightarrow s'$ consists in a photon of kind α springing into being, the second in the disappearance of one such photon. It follows from the above that in a transition for which $q^\alpha_{ss'} \neq 0$ all other $q^\beta_{ss'}$ must vanish.

Let an atom with fixed nucleus and electric dipole moment $\mathfrak{q}$ *interact with the radiation field.* Differentiate the quantum states of the atom from one another by means of the index n and denote the corresponding energies by $h\nu_n$; then $\mathfrak{q} = \|\mathfrak{q}_{nn'}\|$.

A quantum state of the total system consisting of both atom and radiation is characterized by the quantum numbers

$$n\,;\ n_1\,,\ n_2\,,\ \cdots,\ n_\alpha\,,\ \cdots.$$

The effect of the radiation on the atom is, in accordance with eq. (12.4) of the preceding paragraph, given to a first approximation by the perturbation term

$$\varepsilon W = (\dot{\mathfrak{q}}\mathfrak{A}).$$

It can be shown that the addition of such a term to the Hamiltonian function of the total system will, according to classical theory, not only indicate an influence exerted on the atom by the radiation field, but will also modify the equations of Maxwell in a way which indicates that the motion of the electrons in the atom affects the radiation field. The perturbation term will accordingly call forth emission as well as absorption. To a sufficient approximation we may take for $\mathfrak{A}$ its value *at the point occupied by the nucleus, provided we restrict ourselves to radiation whose wave-length is large compared with the dimensions of the atom.* We now have

$$\varepsilon W = \sum_\alpha (\dot{\mathfrak{q}}\,\mathfrak{A}_\alpha) q^\alpha. \tag{13.3}$$

From this it follows than an element $\varepsilon \cdot W_{ns,\,n's'}$ can only differ from 0 if s and s' are such that all $n'_\beta = n_\beta$ with the exception of a single one n'_α, which must equal $n_\alpha \pm 1$. Then only the αth term contributes to the sum (13.3), and we have

$$\varepsilon W_{ns,\,n's'} = (\dot{\mathfrak{q}}_{nn'}\,\mathfrak{A}_\alpha) \cdot q^\alpha_{ss'}. \tag{13.4}$$

Bohr's frequency condition, which asserts that the emission or absorption of a photon in state α with energy $h\rho_\alpha$ is associated with a quantum jump of the atom in which an amount $\pm h(\nu_n - \nu_{n'}) = h\rho_\alpha$ of energy is lost or won, need by no means be satisfied here. The finite cavity has its own frequencies ρ_α, and may therefore be in no position to take up the frequencies associated with the quantum jumps of the atom. This is true in principle, but as a matter of fact, as we shall see, *Bohr's frequency condition is fulfilled to a very close approximation in the overwhelming majority of all transitions;* and this is more and more the case the larger the cavity is.

Let the atom be in the state n and the radiation in the state $s = \{n_\alpha\}$. We set

$$\sum h\, n_\alpha\, \rho_\alpha = V \cdot U(\rho)d\rho, \tag{13.5}$$

where the sum on the left is to be extended over those indices α for which ρ_α lies between ρ and $\rho + d\rho$; hence $U(\rho)d\rho$ is the energy density of the radiation contained in the frequency range ρ, $\rho + d\rho$. In accordance with (10.5), the probability that the atom will find itself in the state n' after time t is given by

$$\frac{2}{h^2}\sum_{s'}\frac{1-\cos(\nu_{nn'}+\rho_{ss'})t}{(\nu_{nn'}+\rho_{ss'})^2}\cdot|\varepsilon W_{ns,\,n's'}|^2. \tag{13.6}$$

The contribution to this sum due to the cases in which a photon is emitted is, in accordance with equations (13.2), (13.4), given by

$$\frac{2}{h^2}\sum_{\alpha}\frac{1-\cos(\nu_{nn'}-\rho_\alpha)t}{(\nu_{nn'}-\rho_\alpha)^2}\cdot\frac{h(n_\alpha+1)}{2\rho_\alpha}\,|(\dot{\mathfrak{q}}_{nn'}\mathfrak{A}_\alpha)|^2, \tag{13·6$_e$}$$

and that for absorption by

$$\frac{2}{h^2}\sum_{\alpha}\frac{1-\cos(\nu_{nn'}+\rho_\alpha)t}{(\nu_{nn'}+\rho_\alpha)^2}\cdot\frac{hn_\alpha}{2\rho_\alpha}\,|(\dot{\mathfrak{q}}_{nn'}\mathfrak{A}_\alpha)|^2. \tag{13.6$_a$}$$

Consider first the case in which the term level $\nu_{n'}$ is *higher* than ν_n; $\nu_{nn'} = \nu_n - \nu_{n'} = -\nu$ is then negative. We now collect together all those terms α in the sum (13·6$_a$) for which ρ_α lies between ρ and $\rho + d\rho$. Since the position of the atom is not exactly fixed—even in consequence of the variations caused by the emission of photons—we may, for small wave-lengths, replace $\mathfrak{A}_\alpha^2$ by its mean value $4\pi/V$ as given by the normalizing equation $\int\mathfrak{A}_\alpha^2 dV = 4\pi$, and we may also assume that all directions are equally probable for $\mathfrak{A}_\alpha$. The square $|(\mathfrak{A}_\alpha\,\mathfrak{q})|^2$ of the scalar product of $\mathfrak{A}$ with a fixed vector $\mathfrak{q}$ has then the mean value $\frac{4\pi}{3V}\cdot|\mathfrak{q}|^2$. (13.6$_a$) then becomes

$$\frac{1-\cos(\rho-\nu)t}{(\rho-\nu)^2}\cdot\frac{4\pi}{3}\frac{|\dot{\mathfrak{q}}_{nn'}|^2}{V}\cdot\frac{\sum h\,n_\alpha\,\rho_\alpha}{2\rho^2}.$$

On introducing (13.5) the sum (13.6$_a$) may, to a good approximation, be replaced by the integral

$$\frac{4\pi}{3}\frac{|\dot{\mathfrak{q}}_{nn'}|^2}{h^2}\cdot\int\frac{1-\cos(\rho-\nu)t}{(\rho-\nu)^2}\cdot\frac{U(\rho)d\rho}{\rho^2}.$$

Essentially the only elements which contribute to the value of this integral, for a time t large in comparison with the duration $1/\nu$ of an oscillation, are those for which ρ lies near to ν. On developing

$$\frac{U(\rho)}{\rho^2} = \frac{U(\nu)}{\nu^2} + \cdots$$

in powers of $\rho - \nu$, the first term in the expansion contributes

$$\frac{U(\nu)}{\nu^2}\cdot t\int_{-\infty}^{+\infty}\frac{1-\cos x}{x^2}\,dx = \pi t\cdot\frac{U(\nu)}{\nu^2} \tag{13.7}$$

to the integral; all others are to be neglected. Similarly the entire amount (13.6_e) due to emission is negligible, for its denominator $(\rho + \nu)^2$ vanishes nowhere. This means that the transition is almost invariably associated with the absorption of a photon whose frequency lies very close to ν. The probability that the atom will appear in the higher state n' after lapse of time t increases in proportion with t; the factor

$$\frac{4\pi^2}{3}\frac{U(\nu)}{(h\nu)^2}\cdot|\dot{q}_{nn'}|^2 = \frac{4\pi^2}{3h^2}\cdot U(\nu)|q_{nn'}|^2$$

is *the probability that the transition $n \to n'$ take place in unit time.*

This formula was obtained for the case in which the state n' possessed a higher energy level than n. In the reverse case only the sum (13.6_e) due to emissions contributes an appreciable amount. We now put $\nu_{nn'} = \nu_n - \nu_{n'} = \nu$ and obtain the same formula with this difference: in place of n_α we now have $n_\alpha + 1$, or in place of the sum (13.5) the sum

$$\sum_\alpha h(n_\alpha + 1)\rho_\alpha = \sum hn_\alpha\rho_\alpha + \sum h\rho_\alpha.$$

The first is $V \cdot U(\rho)d\rho$, and we denote the second by $V \cdot u(\rho)d\rho$. This latter is equal to $(h\rho)$ times the number of modes of vibration of the cavity within the frequency interval $\rho, \rho + d\rho$; hence by (13.1)

$$V \cdot u(\rho)d\rho = V \cdot \frac{h\rho^3 d\rho}{\pi^2 c^3}, \quad \boxed{u(\nu) = \frac{h\nu^3}{\pi^2 c^3}}.$$

The probability that the atom drop from state n into the lower state n' in unit time is given by

$$\frac{4\pi^2}{3h^2}\cdot[U(\nu) + u(\nu)]|q_{nn'}|^2.$$

The additional term $u(\nu)$ is characteristic for *spontaneous emission.* When the radiation is not enclosed in a black body, i.e. when there is no radiation density $U(\nu)$, *the probability that the atom drop from the state n to the lower state n' in unit time,* emitting thereby a photon whose frequency lies in the immediate neighbourhood of $\nu = \nu_n - \nu_{n'}$, *is*

$$\frac{4\nu^3}{3hc^3}\cdot|q_{nn'}|^2.$$

This agrees with the formula obtained by integrating (8.11) over all directions. The probability that the atom jump from the level n into a *higher* level n' $(\nu_{n'} > \nu_n)$ under these same conditions is zero.

In the energy field of the black body radiation we find not only absorption, but also "*stimulated emission*," both of which are proportional to the energy density $U(\nu)$. On setting

$$\boxed{A_{nn'} = \frac{4\pi^2}{3h^2}|\mathfrak{q}_{nn'}|^2}, \tag{13.8}$$

the probability for a jump from state n to a higher state n' in unit time is

$$\boxed{w_{nn'} = A_{nn'} \cdot U(\nu)} \quad (\nu = \nu_{n'} - \nu_n), \tag{13.9}$$

and the probability for the inverse jump, the drop from n' to n, is

$$\boxed{w_{n'n} = A_{n'n}[U(\nu) + u(\nu)]}. \tag{13.9}$$

Since $||\mathfrak{q}_{nn'}||$ is an Hermitian matrix,

$$\boxed{A_{nn'} = A_{n'n}}. \tag{13.10}$$

If there are a number of atoms in the radiation field and the whole system is in a steady state, then on the average as many atoms must make the jump $n \rightarrow n'$ in unit time as make the inverse jump $n' \rightarrow n$. On denoting the number of atoms in the state n by N_n, these considerations are expressed in the condition

$$A_{nn'} \cdot N_n U(\nu) = A_{n'n} \cdot N_{n'}[U(\nu) + u(\nu)]$$

or

$$\frac{N_n}{N_{n'}} = 1 + \frac{u(\nu)}{U(\nu)}. \tag{13.11}$$

The probability coefficients $A_{nn'} = A_{n'n}$ have entirely disappeared—or rather, *almost* entirely, for the equation is valid only under the assumption that $A_{nn'} \neq 0$ or $\mathfrak{q}_{nn'} \neq 0$, i.e. the transition $n \rightleftarrows n'$ is not to be forbidden by the selection rules. But for such a system in thermal equilibrium N_n must, as shown by *Boltzmann*, be proportional to

$$e^{-E_n/k\theta} = e^{-h\nu_n/k\theta},$$

where θ is the temperature and k the Boltzmann constant. Equation (13.11) then becomes

$$e^{h(\nu_{n'} - \nu_n)/k\theta} = 1 + \frac{u(\nu)}{U(\nu)}$$

or *the Planck radiation formula:*

$$U(\nu) = \frac{u(\nu)}{e^{h\nu/k\theta} - 1};$$

this formula is valid for all frequencies ν whose energies can be exchanged by the absorbing and emitting atoms in accordance with Bohr's frequency condition.[29]

We have thus finally returned to the historical origin of the quantum theory. We must now add three remarks concerning this treatment, due to Dirac, of energy exchange between matter and radiation. In the first place, it is able to explain the fact that *the spectral lines are not sharp, but possess a natural breadth.*[30] Secondly, we must inquire what causes this difference between absorption and emission, processes which are transformed into each other on changing the direction of time. Indeed, the fundamental mechanical and field laws are invariant under the transformation $t \to -t$! The answer is that *this difference is due to the preferential direction in time involved in the application of the theory of probability;* we assume a fixed initial state and calculate, with the aid of transition probabilities, the distribution over the various states at a *later* time, not the distribution which would result from the equations for an *earlier* time. If no assumption is made concerning this preferential direction, t should be replaced by $|t|$ in (13·7). And finally, the fact that we have here treated Maxwell's equations as classical equations of motion, and as such have subjected them to the process of quantization, may give rise to serious doubts—for in our general formulation Maxwell's equations are already the quantum-theoretic wave equations for the photon! But we shall see in Chap. IV, § 11, that this method is in fact the correct one to employ in order to go from *one* corpuscle to *an indefinite number* of corpuscles. For since the number of photons must remain indefinite—as a photon can, in contrast to an electron, spring into being or disappear—the method of composition described in § 10 is not applicable to them.

CHAPTER III

GROUPS AND THEIR REPRESENTATIONS

§ 1. Transformation Groups

THE concept of a *group*, one of the oldest and most profound of mathematical concepts, was obtained by abstraction from that of a *group of transformations*.[1]

A point-field, a domain of elements which we call points, on which the transformations operate, underlies the transformations. This point-field may be either the totality of a finite number of individually exhibited elements or an infinite set, in particular a continuum such as space or time. A *mapping* or **correspondence** S of the point-field on itself is determined by a law which associates with each point p of the field a point p' as image: $p \rightarrow p' = Sp$; two correspondences Sp and Tp are identical if for all points p the two image points Sp and Tp coincide. If the point-field contains a finite number of elements the correspondence S can be defined by giving explicitly the image point for each point p; for infinite sets, however, the association is only possible by giving the *law* of the function S.

Among such correspondences there is a particular one which associates with each point p the point p itself: $p \rightarrow p$; it is called the ***identity*** I. Two correspondences can be applied successively: if the first sends the arbitrary point p into $p' = Sp$, the second p' into $p'' = Tp'$, then the correspondence resulting from the composition of the two is defined by the association $p \rightarrow p'' = T(Sp)$ and is denoted by TS (read from right to left !). The resultant correspondence depends on the order of the two factors S and T. In order that composition be possible it is essential that the correspondences are ones which map the point-field on itself, and not on another point-field.

We shall restrict ourselves to *one-to-one correspondences* S: the image points $p' = Sp$ associated with p shall always be distinct, and each given point p' shall appear as the image of one (and only one) of the points p. Consequently such a one-to-one correspondence $S: p \to p'$ determines a second, the *inverse* $S^{-1}: p' \to p$ of S, which just cancels it :

$$S'(Sp) = p, \quad S(S'p') = p' \quad \text{or}$$
$$S^{-1}S = I, \qquad SS^{-1} = I.$$

The inverse of S^{-1} is again S and the identity I is its own inverse. The resultant TS of two one-to-one correspondences S, T is itself one-to-one, and its inverse is $(TS)^{-1} = S^{-1}T^{-1}$ —for on inverting the correspondences $p \to p' \to p''$ there results $p'' \to p' \to p$. Henceforth we shall consider only those correspondences, also called *transformations* or *substitutions*, which are one-to-one. In this domain we have, in accordance with what has been said, the two fundamental operations of inversion and composition.

Examples.—1. Let the point-field consist of n elements exhibited individually ; bring them into a particular order by numbering them with the integers

$$1, 2, \cdot \cdot \cdot, n. \tag{1.1}$$

This numbering consists in a one-to-one reciprocal relation between the elements of the point-field and the integers or possible " positions " q in the series (1.1). A permutation consists in the transition from one such arrangement to another. If we wish to operate in space we may think of the positions as fixed compartments into which the movable elements can be laid, or, conversely, we may think of the elements as fixed and shift the movable numbers about. With each permutation is associated a one-to-one correspondence $p \to p'$ which tells which element p' occupies, after the exchange, the position previously held by p. Insofar as the method of numbering is considered as left to convention, the permutation is nothing more than this one-to-one correspondence. The concept is to be understood in this way when we are concerned with the composition or successive application of permutations.

2. A kinematical example of a group is offered by the motions of a space-filling substance, in particular those of a rigid body. The positions or numbers of the preceding example are here represented by the material points and the point-field is the space itself. The one-to-one correspondence $p \to p'$ connects the initial with the final state : that material point which originally covered the spatial point p is taken to the point p' by the motion. Congruent correspondences of space on to itself will also be briefly referred to as " motions " in the geometrical sense.

The concept of a ***group of transformations*** is now readily formulated. We understand by it any system $\mathfrak{G}$ of transformations of a given point-field, which is closed in the sense of the following conditions:

1. It contains the identity;
2. If S belongs to $\mathfrak{G}$, then its inverse S^{-1} does also;
3. The resultant TS of any two transformations S, T of $\mathfrak{G}$ is also a transformation of $\mathfrak{G}$.

As examples we name the group of all $n!$ permutations of n things, the congruent mappings or " motions " of 3-dimensional Euclidean space, all homogeneous linear transformations in n variables with non-vanishing determinants (affine correspondence of an n-dimensional vector space) and the group of unitary transformations in n dimensions.

If the point p goes over into p' by means of a transformation of the group $\mathfrak{G}$, then p' is said to be equivalent to p (with respect to the group $\mathfrak{G}$). The same concept is applied when we are considering instead of a point p a figure consisting of points. Expressed in these terms, the three requirements for a group are nothing other than the three axioms of equality:

1. p is equivalent to p;
2. If p' is equivalent to p, then p is equivalent to p';
3. If p' is equivalent to p and p'' to p', then p'' is equivalent to p.

According to *Klein's* Erlanger Program [2] any geometry of a point-field is based on a particular transformation group $\mathfrak{G}$ of the field; figures which are equivalent with respect to $\mathfrak{G}$, and which can therefore be carried into one another by a transformation of $\mathfrak{G}$, are to be considered as the same. In Euclidean geometry this rôle is played by the group of congruency transformations, consisting of the motions referred to above, and in affine geometry by the group of affine transformations, etc. The group expresses the specific isotropy or homogeneity of the space; it consists of all one-to-one " isomorphic correspondences " of the space on itself, i.e. those transformations which leave undisturbed all objective relations between points of the space which can be expressed geometrically. The symmetry of a particular figure in such a space is described by a sub-group of $\mathfrak{G}$ consisting of all transformations of $\mathfrak{G}$ which carry the figure over into itself. The art of ornamental tiling, which was perfected by the Egyptians, contains implicitly considerable knowledge of a group-theoretic nature; we here find, perhaps, the oldest fragment of mathematics in human culture. But only recently have we been able to formulate clearly the formal principles of this art; attempts in this direction were already made by *Leonardo da Vinci*, who sought to give a general and systematic account of the various types of symmetry possible in a building. But the most wonderful symmetrical structures are exhibited in crystals, the symmetry of which is described by those congruency transformations of Euclidean space which bring the atomic lattices of the crystal into coincidence with themselves. The most important application of group theory to natural science heretofore has been in this field.

The following considerations fit naturally into the present discussion. Let the point-field **M** on which the transformations S of the group $\mathfrak{G}$ operate be mapped on the point-field **N** by means of the one-to-one correspondence $A : p \rightarrow q$; the case in which the correspondence serves to introduce new numbering or new co-ordinates is of particular importance. Through this correspondence A of **M** on **N** the transformation S of **M** becomes a transformation T of **N**; in the particular case mentioned above T is simply a description of the transformation S in the new co-ordinates. It is evident that to the composition of transformations S corresponds the composition of the corresponding transformations T of **N** and that a group $\mathfrak{G}$ of transformations S goes over into a group $\mathfrak{H}$ of transformations T. The relation between these two transformations is

$$T = ASA^{-1}, \tag{1.2}$$

for if we denote the transformation S by $p \rightarrow p'$ and if q, q' are the points of **N** associated with p, p' by A, then the transformation $q \rightarrow q'$ of **N** is effected by

$$q \rightarrow p \rightarrow p' \rightarrow q'.$$

We may also write $\mathfrak{H} = A\mathfrak{G}A^{-1}$. In particular, these considerations apply when **N** and **M** are the same point-field.

§ 2. Abstract Groups and their Realization

An arbitrary number of transformations of a given point-field on to itself can be applied successively; we are of course not restricted to merely two. But when we perform this process step by step it is automatically reduced to a succession of compositions of transformations taken two at a time:

$$ABC \cdots = A[B(C \cdots)].$$

This possibility of performing an extended composition in steps involving but two transformations at a time shows that the associative law

$$(AB)C = A(BC)$$

holds for any three transformations A, B, C.

The *structure of a transformation group* is obtained from it by abstraction when we allow the transformations themselves to degenerate into elements of an immaterial nature, retaining only their individuality and the rules in accordance with which two given transformations are composed, in a given order, to form a third. In accordance with what has been said such composition necessarily obeys the associative law. Perhaps it also obeys other universal laws, but since we have at present no indication of this we attempt a formulation of the abstract structure of the group by means of the following definitions :

An ***abstract group*** *is a system of elements within which a law of composition is given such that by means of it there arises from any two (the same or different) elements a, b of the group, taken in this order, an element ba. The following conditions shall thereby be satisfied :*

1. *The associative law* $c(ba) = (cb)a$;
2. *There shall exist an element* $\mathbf{I}$, *the unit element, which leaves an arbitrary element a unaltered on composition with it :*

$$\mathbf{I}a = a\mathbf{I} = a.$$

3. *To each element a shall exist an inverse* a^{-1} *which yields on composition with it the unit element* $\mathbf{I}$:

$$aa^{-1} = a^{-1}a = \mathbf{I}.$$

Such an abstract group is not to be confused with its *realization by transformations*, i.e. by one-to-one correspondences of a given point-field. *A* ***realization*** *consists in associating with each element a of the abstract group a transformation T(a) of the point-field in such a way that to the composition of elements of the group corresponds composition of the associated transformations :*

$$T(ba) = T(b)T(a). \tag{2.1}$$

It follows from this that to the unit element $\mathbf{I}$ corresponds the identity I and to inverse elements a, a^{-1} correspond inverse transformations :

$$T(a^{-1}) = T^{-1}(a). \tag{2.2}$$

The first assertion follows from the particular case

$$T(a)T(\mathbf{I}) = T(a)$$

of (2.1) by left-handed composition with the reciprocal of the transformation $T(a)$; (2.2) is then contained in (2.1) as the particular case $b = a^{-1}$. The realization is said to be ***faithful*** when to distinct elements of the group correspond distinct transformations :

$$T(a) \neq T(b) \text{ when } a \neq b.$$

In accordance with the fundamental equation (2.1) the necessary and sufficient condition for "faithfulness" is that $T(a)$ shall be the identity *only if* a is the unit element. For if a, b are two elements of the group it then follows from $T(a) = T(b)$, i.e.

$$T(a)T^{-1}(b) = T(a)T(b^{-1}) = T(ab^{-1}) = I$$

that under these conditions $ab^{-1} = \mathbf{1}$, i.e. $a = b$. If the abstract group is obtained from a transformation group $\mathfrak{G}$ by abstraction, then conversely $\mathfrak{G}$ is a faithful realization of it.

In the study of transformation groups we always deal with two manifolds, the structureless point-field and the manifold of group elements, the structure of which is expressed by the law of composition. The original problem thus resolves itself into two; the examination of the various group structures possible and the examination of the possibility of obtaining realizations of the given abstract group by transformations of a given point-field. The historical development of the subject has shown that it is advantageous to effect this division into two problems; they are of fundamentally different character and require fundamentally different mathematical equipment for their discussion.

In accordance with our method of introducing the abstract group, which we henceforth refer to simply as the group, it serves merely to give the structure of the group; the nature of its elements is immaterial. This abstraction from the nature of the elements is expressed mathematically by the concept of *isomorphism*. If we have two groups $\mathfrak{g}$, $\mathfrak{g}'$ and there is associated with each element a of $\mathfrak{g}$ an element a' of $\mathfrak{g}'$ in a one-to-one way: $a \rightleftarrows a'$, such that

$$(ba)' = b'a', \tag{2.3}$$

then the two groups are said to be ***simply isomorphic***. Simply isomorphic abstract groups offer no means of distinguishing one from the other. The concept of isomorphism can, of course, be applied to transformation groups. Two isomorphic transformation groups can be considered as faithful representations of *one and the same* abstract group. A group may be isomorphic with itself; it is then said to be *automorphic*. Such an automorphism occurs when $\mathfrak{g}$ and $\mathfrak{g}'$ coincide, i.e. when a one-to-one reciprocal association $a \rightleftarrows a'$ satisfying the condition (2.3) is established between the elements of the group $\mathfrak{g}$.

The question arises whether or not every abstract group possesses a faithful realization. If this were not the case the concept of an abstract group as developed above would be too broad—there would exist, in addition to the associative law, other purely formal laws for the composition of transformations which are satisfied by every transformation group. Conversely, a proof of the realizability of any abstract group would tell us that all that can be said about the formal laws for the composition of transformations is contained in our conditions (1) to (3). We can, in fact, construct a faithful realization of any abstract group $\mathfrak{g}$ by taking as the point-field the group manifold itself and letting correspond to each element a of the group the transformation

$$s \to s' = as$$

of the group manifold on to itself. This "*left-translation*" $\mathfrak{t}_a$ is obviously a one-to-one reciprocal transformation which has as inverse the transformation $s = a^{-1}s'$. If a and b are distinct elements the corresponding transformations $\mathfrak{t}_a$, $\mathfrak{t}_b$ are distinct, for they allow the unit element $\mathbf{I}$ to correspond to the distinct elements a, b respectively. If we perform in succession two left-translations

$$s \to s' = as, \quad s' \to s'' = bs'$$

the resulting transformation is, in consequence of the associative law,

$$s \to s'' = b(as) = (ba)s.$$

Consequently the left-translations constitute in fact a faithful realization of the abstract group. However, the right-translations behave otherwise, for if we denote the mapping $s \to s' = sa$ of the group manifold on itself by $\mathfrak{t}^*(a)$, we find instead of (2.1) the equation

$$\mathfrak{t}^*(ba) = \mathfrak{t}^*(a)\mathfrak{t}^*(b).$$

§ 3. Sub-groups and Conjugate Classes

A ***sub-group*** $\mathfrak{g}'$ of a given abstract group $\mathfrak{g}$ is a set of elements contained in $\mathfrak{g}$ which itself fulfils the characteristic group conditions: the unit element $\mathbf{I}$ belongs to $\mathfrak{g}'$, with a belongs also a^{-1} and with a, b also ba. These three conditions can be reduced to the one: if a, b are any two elements of $\mathfrak{g}'$, then ba^{-1} also belongs to $\mathfrak{g}'$. We assume, of course, that the partial system consists not merely of the element $\mathbf{I}$, but the other limiting case, in which $\mathfrak{g}'$ coincides with $\mathfrak{g}$, shall be included under the concept of a sub-group.

Examples are readily found. In the group of Euclidean motions are contained, for example, the group of rotations (which leaves one point, the centre, fixed) and the group of translations. The unitary transformations constitute a sub-group of the complete group of all homogeneous linear transformations; the even permutations a sub-group of the group of all permutations. If we are dealing with a transformation group $\mathfrak{G}$, all those transformations of $\mathfrak{G}$ which leave a particular point p fixed (i.e. which carry p over into itself) constitute a sub-group $\mathfrak{G}_p$. Instead of a point p the fixed element may be any figure composed of points; the transformations of the sub-group must either leave the figure as a whole fixed (i.e. they must carry each point of the figure over into another such) or the more restrictive condition that they leave each point of the figure fixed. We can also obtain sub-groups of $\mathfrak{G}$ by employing invariant functions instead of invariant figures. If $\psi(p)$ is any function of position on the point-field with elements p we associate with the transformation $S: p \to p'$ the function ψ' defined by $\psi'(p') = \psi(p)$ and say that it is obtained from ψ by the transformation S. If $p' = Sp$, $p'' = Tp'$, the equations

$$\psi(p) = \psi'(p') = \psi''(p'')$$

show that the composition of the transitions $\psi \to \psi'$ and $\psi' \to \psi''$ associated with S and T result in the transition $\psi \to \psi''$ associated with TS. Now consider all transformations S of $\mathfrak{G}$ which carry $\psi(p)$ over into itself, i.e. for which $\psi(Sp) = \psi(p)$ is an identity in p; they constitute a sub-group $\mathfrak{H}$ of $\mathfrak{G}$, and $\psi(p)$ is an *invariant* of $\mathfrak{H}$. In this way we can separate out the rotations from the homogeneous linear transformations by requiring the invariance of the unit quadratic form. The sub-groups contained in a finite group $\mathfrak{g}$, which is described by exhibiting each of its elements and giving explicitly the result of composition of each two, can be obtained by inspection.

There is associated with each element a of the group $\mathfrak{g}$ a *cyclic sub-group* denoted by (a):

$$\cdots, a^{-2}, \quad a^{-1}, \quad a^0 = \mathbf{1}, \quad a, \quad a^2, \cdots, \tag{3.1}$$

the elements a^n of which are defined inductively by the equations

$$a^0 = \mathbf{1}, \quad a^{n+1} = a^n a.$$

These elements constitute in fact a group, for n and m being any integral exponents we have

$$a^{n+m} = a^n a^m.$$

(a) is the smallest sub-group which contains a, i.e. its elements are common to all sub-groups of $\mathfrak{g}$ which contain a. The elements of the set (3.1) can either be distinct or—and this latter must be the case if $\mathfrak{g}$ is a finite group—they must repeat themselves after a cycle of h terms: $\mathbf{1}, a, a^2, \cdots, a^{h-1}$ are distinct but $a^h = \mathbf{1}$. h is called the *order* of the element a.

The **order** *of a finite group* is the number of its elements; accordingly, the order of an element a agrees with the order of the cyclic sub-group (a) generated by a. A group is said to be **commutative** or **Abelian** if composition of its elements obeys the rule $ba = ab$. Cyclic groups are therefore Abelian.

If a runs through the sub-group $\mathfrak{h}$ of $\mathfrak{g}$ the associated (left-) translations $\mathfrak{t}_a$ constitute a group of transformations which is simply isomorphic with $\mathfrak{h}$, the point-field of which is the group manifold. We say that two elements s, s' which are equivalent with respect to this transformation group are (left-)equivalent with respect to $\mathfrak{h}$ and express this situation by the notation "$s' \equiv s$ with respect to $\mathfrak{h}$"; the condition for it is that $s' = as$ where a is an element of $\mathfrak{h}$. In this way the elements of $\mathfrak{g}$ are divided into sets of elements which are equivalent to $\mathfrak{h}$. If the number of such sets is finite, it is called the **index** of $\mathfrak{h}$ in $\mathfrak{g}$. If $\mathfrak{g}$ is a finite group the number of elements in each of these sets is given by the order of $\mathfrak{h}$, for different translations $\mathfrak{t}_a$ send s into different elements: $as \neq bs$ if $a \neq b$. *The order of* $\mathfrak{h}$ *is accordingly a divisor of the order of* $\mathfrak{g}$, *and the quotient of these two is the index of* $\mathfrak{h}$.

The considerations at the end of § 2 above, which were developed for groups of transformations, suggest a second realization of the abstract group $\mathfrak{g}$. We associate with the element a the correspondence

$$s \rightarrow s' = asa^{-1} \tag{3.2}$$

of the group manifold on itself. This correspondence, which we call the "*conjugation*" $\mathfrak{k}_a$, is reciprocal one-to-one, and has as inverse $s = a^{-1}s'a$. The law of composition is obeyed, for from

$$s \rightarrow s' = asa^{-1}, \quad s' \rightarrow s'' = bs'b^{-1}$$

we obtain the product

$$s'' = basa^{-1}b^{-1} = (ba)s(ba)^{-1}.$$

Two elements s, s' of $\mathfrak{g}$ are said to be **conjugate** if they are equivalent with respect to the group of all conjugations. Accordingly, the whole group is divided into *classes*, any element of one of which is conjugate to any other element of the same class. When we speak of classes within a group without a more explicit description we mean these conjugate classes.

The realization of $\mathfrak{g}$ by the group of conjugations is in general a "contracted" rather than a faithful realization. In particular, the conjugation $\mathfrak{k}_a$ coincides with the identity if a commutes with all elements s of the group. The totality of all such elements a is called the **central** of the group; it is obviously an Abelian sub-group of $\mathfrak{g}$. But this disadvantage of conjugation over translation is offset by an advantage; conjugation is an isomorphic correspondence within the group itself which leaves the unit element invariant and which associates with each sub-group $\mathfrak{h}$ of $\mathfrak{g}$ another such, the conjugate sub-group $a\mathfrak{h}a^{-1}$. These facts, which are expressed by the equation

$$a(st)a^{-1} = (asa^{-1})(ata^{-1}),$$

were already contained implicitly in the considerations at the end of § 1. $\mathfrak{h}$ is said to be a ***self-conjugate*** or ***invariant sub-group*** if it coincides with all its conjugate sub-groups.

The importance of this last concept is best seen in the following:

Theorem. If $\mathfrak{h}$ *is an invariant sub-group and* $\equiv$ *denotes equivalence with respect to it, then it follows from*

$$s' \equiv s, \; t' \equiv t \quad \text{that} \quad s't' \equiv st. \tag{3.3}$$

To prove this we note that $s' = as$, $t' = bt$ (a, b in $\mathfrak{h}$) yield

$$s't' = asbt = (ac)(st). \tag{3.4}$$

for $c = sbs^{-1}$ belongs to $\mathfrak{h}$ with b. Since ac lies in $\mathfrak{h}$ our assertion is proven. It is readily seen that the invariantive nature of $\mathfrak{h}$ is necessary as well as sufficient for the validity of (3.3). In dealing with an invariant sub-group $\mathfrak{h}$ we need not distinguish between right and left equivalence with respect to $\mathfrak{h}$—indeed, the above proof was based on this fact.

We may, if we like, consider equivalent elements as not differing from one another (by application of the principle of definition by abstraction); but by thus allowing equivalent elements to fall together the group property of $\mathfrak{g}$ is, in general, forfeited. In accordance with the above theorem it still remains, however, if $\mathfrak{h}$ is an invariant sub-group. The group obtained from $\mathfrak{g}$ by identifying all elements which are equivalent with respect to $\mathfrak{h}$ is called the ***factor group*** $\mathfrak{g}/\mathfrak{h}$; its order is the index of the invariant sub-group $\mathfrak{h}$ of $\mathfrak{g}$.

These concepts are of assistance in examining the way in which a group may be " contracted " on setting up a realization. Let the transformation $T(a)$ of a given point-field on itself correspond to the element a of the abstract group $\mathfrak{g}$ in the realization under consideration. Then $T(a) = T(a')$ if and only if a' is obtained from a by composition with an element e (i.e. $a' = ea$) for which $T(e)$ is the identity. Such elements e obviously constitute a sub-group $\mathfrak{h}$ of $\mathfrak{g}$, for it follows from

$$T(e) = I, \quad T(e') = I \quad \text{that} \quad T(ee') = T(e)T(e') = I.$$

$\mathfrak{h}$ is, in fact, an invariant sub-group, for if $T(e)$ is the identity, the same is true of

$$T(aea^{-1}) = T(a)T(e)T^{-1}(a) = T(a)T^{-1}(a).$$

In any realization of an abstract group $\mathfrak{g}$ *by a group of transformations the elements of a certain invariant sub-group* $\mathfrak{h}$ *of* $\mathfrak{g}$ *correspond to the identical transformation; two different elements will be associated with the same transformation if and only if they are equivalent with respect to* $\mathfrak{h}$. *The group of transformations is consequently a faithful realization of the factor group* $\mathfrak{g}/\mathfrak{h}$.

§ 4. Representation of Groups by Linear Transformations

On requiring that the transformations which are to serve as a realization of a given abstract group $\mathfrak{g}$ be linear and homogeneous we arrive at a problem which is most fruitful from the mathematical standpoint and which is at the same time of greatest importance for quantum mechanics; we then speak of a ***representation,*** instead of a realization, of the group.[3] An n-dimensional representation of $\mathfrak{g}$, or a representation of *degree* n, consists in associating with each element s of the group an affine transformation $U(s)$ of the n-dimensional vector space $\mathfrak{R} = \mathfrak{R}_n$ in such a way that these transformations obey the law of composition

$$U(s)U(t) = U(st). \tag{4.1}$$

We then say that s *induces* the transformation $U(s)$ in the representation space $\mathfrak{R}$. On choosing a definite co-ordinate system in $\mathfrak{R}$ each transformation $U(s)$ is represented by a square matrix of n rows and columns, the determinant of which does not vanish. On replacing the original co-ordinate system by another, obtained from it by the transformation A, the correspondence which was formerly represented by the matrix $U(s)$ is now represented by the matrix $AU(s)A^{-1}$. Consequently if the association $s \to U(s)$ is a representation, the association

$$s \to AU(s)A^{-1}$$

is obviously also one; this latter representation is said to be ***equivalent*** to the former. They are essentially the same, differing only in the choice of the co-ordinate system in terms of which they are described.

Examples.—A representation in one dimension consists in assigning to each element s of the group a non-vanishing number $\chi(s)$ in such a way that

$$\chi(st) = \chi(s)\,\chi(t). \tag{4.2}$$

In particular, $\chi(\mathbf{I}) = 1$. A most trivial 1-dimensional representation is obtained by assigning to each s the number 1: $\chi(s) = 1$. This special case is called the *identical representation.*

Consider next the so-called ***symmetric group,*** the group $\pi = \pi_f$ of all $f!$ permutations of f things. The association

$$s \to \delta_s = \pm 1,$$

according as s is an even or an odd permutation, defines a 1-dimensional representation, the "alternating" representation of the group π. For the character δ_s, which distinguishes between the even and the odd permutations, satisfies the equation

$$\delta_{st} = \delta_s \cdot \delta_t.$$

Let $\mathfrak{g}$ be a finite cyclical group of order h; the elements s are then

$$\mathbf{I},\ a,\ a^2,\ \cdot\cdot\cdot,\ a^{h-1}$$

and $a_h = \mathbf{I}$. Consider the 1-dimensional representation $s \to \chi(s)$ in which $\chi(a) = \varepsilon$. The condition (4.2) for a representation then tells us that to the elements s of this series correspond

$$1,\ \varepsilon,\ \varepsilon^2,\ \cdot\cdot\cdot,\ \varepsilon^{h-1},$$

and that to a^h corresponds ε^h. Hence $\varepsilon^h = 1$; ε must therefore be an hth root of unity and the law defining the representation is $a^r \to \varepsilon^r$ $(r = 0, 1, 2, \cdot\cdot\cdot)$. Conversely, when ε is an arbitrary h^{th} root of unity this association defines a 1-dimensional representation of $\mathfrak{g}$. We have thus obtained a complete survey of all possible 1-dimensional representations of a cyclical group.

The only example of a multi-dimensional representation which we offer at this time is the following trivial one. If $\mathfrak{g}$ is itself a group of linear transformations of an n-dimensional vector space $\mathfrak{R}$, then the association $s \to s$ defines an n-dimensional representation of $\mathfrak{g}$. This example implies more than one might at first sight imagine. We have in fact to do the following: we first obtain the structure of the group $\mathfrak{g}$ by abstraction from the group of linear transformations and then return to the original realization by means of the correspondence $s \to s$ between an element s of the abstract group on the one hand and the linear transformation s on the other.

The concept of ***equivalence*** has a more general significance than that discussed above. It may refer to an arbitrary system Σ of linear correspondences U of the n-dimensional vector space $\mathfrak{R}$. We need not assume that these correspondences possess an inverse (i.e. that they have a non-vanishing determinant), nor need we assume that they are associated with the elements s of a group, as is the case with representations. On expressing the set of correspondences U in terms of a new co-ordinate system each matrix U goes over into the matrix $U' = AUA^{-1}$; the system Σ is transformed into the equivalent system Σ' consisting of the U'. A is here a fixed non-singular matrix.

Consider a correspondence U of $\mathfrak{R}$ on to itself. A linear sub-space $\mathfrak{R}'$ of $\mathfrak{R}$ is said to be ***invariant** under* U if the vectors of $\mathfrak{R}'$ are transformed into vectors of $\mathfrak{R}'$ by U. If $\mathfrak{R}'$ is invariant then the space $\mathfrak{R}$ (mod. $\mathfrak{R}'$) obtained by projecting $\mathfrak{R}$ with respect to $\mathfrak{R}'$ is also invariant (cf. I, § 2, in particular Fig. 1). $\mathfrak{R}'$ being invariant, U gives rise to a correspondence U' of $\mathfrak{R}'$ on to itself; we say that U *induces* U' in $\mathfrak{R}'$. Similarly for the space obtained by projection. We now pass from a single correspondence U to a system Σ of correspondences. $\mathfrak{R}'$ is said to be *invariant under* Σ if it is invariant under each correspondence U of Σ. Describing $\mathfrak{R}$ in terms of a co-ordinate system which is adapted to the invariant sub-space $\mathfrak{R}'$, all matrices U of the system Σ reduce simultaneously to the form illustrated in Fig. 1, p. 8. Σ is called ***irreducible*** if $\mathfrak{R}$ contains no sub-space, other than $\mathfrak{R}$ itself and the space 0 consisting only of the vector 0, which is invariant under Σ. We shall have occasion to reduce $\mathfrak{R}$ in such a way that each constituent separated off is irreducible under a given system Σ. This requires the construction of a series of sub-spaces

$$0, \mathfrak{R}_1, \mathfrak{R}_2, \cdots, \mathfrak{R}_r = \mathfrak{R}, \tag{4.3}$$

beginning with 0 and ending with $\mathfrak{R}$, in which each member is contained in the preceding one and is such that $\mathfrak{R}_i$ (mod. $\mathfrak{R}_{i-1}$) is irreducible. Naturally $\mathfrak{R}_i$ shall actually be larger than $\mathfrak{R}_{i-1}$, not merely coincide with it. The implications of this reduction are most readily seen in terms of the matrices U of the correspondences of the system Σ on adapting the co-ordinate system to the " composition series " (4.3), i.e. by choosing first a co-ordinate system in $\mathfrak{R}_1$, then supplementing it with additional fundamental vectors in order to obtain a co-ordinate system for $\mathfrak{R}_2, \mathfrak{R}_3, \cdots$ in turn.

Σ is said to be ***completely reducible*** if $\mathfrak{R}$ can be decomposed into two sub-spaces $\mathfrak{R} + \mathfrak{R}'$, each of which are invariant under Σ and such that neither of them consists merely of the vector 0. This concept of *complete reducibility* is more exacting than that of mere *reducibility*. On describing $\mathfrak{R}$ in terms of a co-ordinate system which is adapted to this decomposition, each matrix U of Σ assumes the form illustrated in Fig. 2, p. 9. We are then faced with the problem of decomposing $\mathfrak{R}$ (or Σ) into constituents, none of which is completely reducible, i.e. of decomposing $\mathfrak{R} = \mathfrak{R}_1 + \cdots + \mathfrak{R}_k$ into invariant sub-spaces, none of which is completely reducible.

We often find that reducibility implies complete reducibility, i.e. that in many cases we have the theorem: If $\mathfrak{R}'$ is an invariant sub-space of $\mathfrak{R}$, a second invariant sub-space $\mathfrak{R}''$ can be found such that $\mathfrak{R}$ is completely reducible (with respect to Σ) into $\mathfrak{R}' + \mathfrak{R}''$. We shall soon see that this is actually the case when $\mathfrak{R}$ is a unitary space and Σ is a system of unitary transformations.

It was shown in Chap. I, § 3, that if the system Σ is reducible, then the system Σ^* of "transposed" correspondences of the dual space on itself is also reducible. If $\mathfrak{H}: s \to U(s)$ is an n-dimensional representation of the group $\mathfrak{g}$ the transposed $U^*(s)$ do not constitute a representation; it is readily seen, however, that on employing instead the contragredient correspondences

$$\breve{U}(s) = [U^*(s)]^{-1}$$

we do obtain a representation $s \to \breve{U}(s)$ of the dual vector space. This we call the ***contragredient*** *representation* $\breve{\mathfrak{H}}$.

§ 5. Formal Processes. Clebsch-Gordan Series

Continuous groups offer what are perhaps the simplest examples of the theory of representations. We consider in particular the group $\mathfrak{c} = \mathfrak{c}_n$ of all linear and homogeneous transformations s in n variables $x_1, x_2, \cdots, x_n$ with non-vanishing determinants; we consider each set of values x_i as a vector in an n-dimensional vector space $\mathfrak{r} = \mathfrak{r}_n$. The classical theory of invariants, first developed in England about the middle of the last century, concerned itself in particular with the representations of $\mathfrak{c}$ induced on the coefficients of arbitrary forms in the variables x_i. A quadratic form in these variables is a linear combination of the $n(n+1)/2$ linearly independent products $x_i x_k$; under the influence of a linear transformation s of the x_i these products undergo a linear transformation $[s]_2$, and the correspondence $s \to [s]_2$ is obviously a representation $[\mathfrak{c}]^2$ in $n(n+1)/2$ dimensions of the group $\mathfrak{c}$. The transformation s of the variables x_i sends the arbitrary quadratic form

$$\Sigma a_{ik} x_i x_k$$

into a quadratic form

$$\Sigma a'_{ik} x'_i x'_k$$

in the new variables, where the coefficients a'_{ik} are obtained from the a_{ik} by a certain linear transformation s_2 associated with s; s_2 is obviously contragredient to $[s]_2$. The quadratic form characterized by a fixed set of $n(n+1)/2$ coefficients a_{ik} may therefore be considered as a vector in a space of this number of dimensions, and the transformation s of the variables x_i induces the transformation s_2 in this space. The space thus defined by the totality of n-ary quadratic forms is thus the point-field for a group of linear homogeneous transformations which constitute a representation of the group $\mathfrak{c}$.

We may in the same way deal with cubic, quartic, $\cdots$, f-ic forms. The totality of monomials of order f are contained in the formula

$$x_1^{f_1} x_2^{f_2} \cdots x_n^{f_n} \tag{5.1}$$

where the f_i are non-negative integers whose sum

$$f_1 + f_2 + \cdots + f_n = f.$$

They constitute the substratum of a representation $[\mathfrak{c}]^f$ in

$$\begin{bmatrix} n \\ f \end{bmatrix} = \frac{n(n+1)\cdots(n+f-1)}{1 \cdot 2 \cdots f}$$

dimensions.

But we can exhibit representations of $\mathfrak{c}$ which are formally yet simpler than these arising from the theory of forms. Let (x_i) and (y_i) be two arbitrary vectors in our n-dimensional space $\mathfrak{r}$ and consider the products $x_i y_k$. On subjecting the x_i and the y_i to the same transformation s of $\mathfrak{c}$ (transition to a new co-ordinate system) the n^2 products undergo a certain linear transformation $s \times s$ associated with s and the correspondence $s \to s \times s$ is an n^2-dimensional representation $(\mathfrak{c})^2$ of $\mathfrak{c}$. Now a system of numbers $F(i, k)$, depending on two indices i, k which run through the values $1, 2, \cdots, n$, is said to be a tensor of second order if under the influence of a transformation s of $\mathfrak{r}$ the $F(i, k)$ undergo the same transformation as the products $x_i y_k$ of the components of two arbitrary vectors $\mathfrak{x}$, $\mathfrak{y}$ of $\mathfrak{r}$. Hence the tensors of order 2 are the substratum of the representation $(\mathfrak{c})^2$ of $\mathfrak{c}$. $(\mathfrak{c})^2$ contains the representation $[\mathfrak{c}]^2$ which is induced in the sub-space of *symmetric* tensors of order 2; the tensor with components $F(i, k)$ being symmetric if $F(ik) = F(ki)$.

In geometry the *anti-symmetric* tensors, i.e. tensors whose components satisfy the condition $F(ik) = -F(ki)$, play a more important rôle than the symmetric ones. In particular, two arbitrary vectors (x_i), (y_i) define a surface element with components

$$x\{ik\} = x_i y_k - x_k y_i;$$

of these quantities but $n(n-1)/2$ are linearly independent, say those for which $i < k$. On subjecting the components x_i of the vector $\mathfrak{x}$ and the components y_i of the vector $\mathfrak{y}$ to the same linear transformation s, the components of the surface element defined by them undergo an $n(n-1)/2$-dimensional linear transformation $\{s\}_2$. $s \to \{s\}_2$ is a representation $\{\mathfrak{c}\}^2$ whose substratum is the totality of anti-symmetric tensors of order 2. Hence the representation $(\mathfrak{c})^2$ is reduced into the representations $[\mathfrak{c}]^2$ and $\{\mathfrak{c}\}^2$, for any tensor $F(ik)$ can obviously be written

$$F(ik) = \frac{1}{2}[F(ik) + F(ki)] + \frac{1}{2}[F(ik) - F(ki)],$$

i.e. in a unique manner as the sum of its symmetric and anti-symmetric parts. That this reduction is correct is further borne out by the fact that the dimensionalities satisfy

$$n^2 = \frac{n(n+1)}{2} + \frac{n(n-1)}{2}.$$

Similarly three arbitrary vectors $\mathfrak{x}$, $\mathfrak{y}$, $\mathfrak{z}$ determine a 3-dimensional element of volume with components

$$x\{ikl\} = \begin{vmatrix} x_i & x_k & x_l \\ y_i & y_k & y_l \\ z_i & z_k & z_l \end{vmatrix}. \tag{5.2}$$

These elements constitute the substratum of a representation $\{\mathfrak{c}\}^3$ in

$$\begin{Bmatrix} n \\ 3 \end{Bmatrix} = \frac{n(n-1)(n-2)}{1 \cdot 2 \cdot 3}$$

dimensions. Continuing in this way we can construct 4-, 5-, $\cdots$, n-dimensional elements; this process must cease with n-rowed determinants, for a determinant of the form (5.2) with more than n rows must necessarily vanish identically.

We shall see that the representations of $\mathfrak{c}$ whose substrata are the symmetric and anti-symmetric tensors of order f are irreducible, and shall in fact solve the general problem of effecting the complete reductions of $(\mathfrak{c})^f$, the representation induced by $\mathfrak{c}$ in the space of all tensors of order f, into its irreducible constituents (Chap. V).

The tensor concept really depends on the $\times$-multiplication introduced in II, § 10. If the m variables x_i undergo a transformation A and the n variables y_k a transformation B, then the mn products $x_i y_k$ undergo a transformation $A \times B$. Considering the x_i as the components of an arbitrary vector $\mathfrak{x}$ in an m-dimensional space $\mathfrak{R}_m$ and the y_k as the components of

$\mathfrak{y}$ in $\mathfrak{R}_n$, the products $x_i \cdot y_k$ may be considered as the components of a vector $\mathfrak{x} \times \mathfrak{y}$ in an mn-dimensional vector space $\mathfrak{R}_m \times \mathfrak{R}_n$. Hence two representations

$$\mathfrak{H} : s \to U(s), \quad \mathfrak{H}' : s \to U'(s) \tag{5.3}$$

of $\mathfrak{g}$ in m, n-dimensions, respectively, give rise to a new mn-dimensional representation which we denote by $\mathfrak{H} \times \mathfrak{H}'$:

$$\mathfrak{H} \times \mathfrak{H}' : s \to U(s) \times U'(s). \tag{5.4}$$

This presents a general method of obtaining a new representation $\mathfrak{H} \times \mathfrak{H}'$ from two given representations $\mathfrak{H}$, $\mathfrak{H}'$.

Denoting the representation $s \to s$ of the linear group $\mathfrak{c}$ for the moment by $(\mathfrak{c})$, the representations of $\mathfrak{c}$ whose substrata are the tensors of order 2, 3, $\cdots$ are then $(\mathfrak{c}) \times (\mathfrak{c}) = (\mathfrak{c})^2$, $(\mathfrak{c}) \times (\mathfrak{c}) \times (\mathfrak{c}) = (\mathfrak{c})^3, \cdots$.

We should, perhaps, have discussed the ***addition*** $+$ of two representations before discussing their ***multiplication*** $\times$. Consider the variables x_i and y_k as the components of a single vector $\mathfrak{z}$ in an $(m + n)$-dimensional vector space; when the x_i are subjected to the transformation A and the y_k to the transformation B these $m + n$ variables undergo a certain transformation (A, B). Hence we obtain from (5.3) the representation

$$\mathfrak{H} \times \mathfrak{H}' : s \to [U(s), U'(s)]$$

in $m + n$ dimensions. The inverse of this process is complete reduction, as discussed above: $\mathfrak{H} + \mathfrak{H}'$ is completely reducible into the components $\mathfrak{H}$ and $\mathfrak{H}'$.

Another important formal method is the following: Any representation Γ in N-dimensions of the linear group $\mathfrak{c}_n$ in n-dimensions may be used to construct an N-dimensional representation of any abstract group $\mathfrak{g}$ from an n-dimensional representation $\mathfrak{H}$ of the same. Γ associates with the linear transformation u in n-dimensional space a linear transformation U in N dimensions, so if $\mathfrak{H} : s \to u$ is an n-dimensional representation of the group $\mathfrak{g}$ with elements s, then

$$s \to u \to U$$

is an N-dimensional representation $s \to U$ of $\mathfrak{g}$ which we may denote by $\Gamma(\mathfrak{H})$. To this is due the importance of the representations of the linear group for the general theory of representations. For example, take Γ to be the representation of $\mathfrak{c}$ whose substratum is the dual space, the space of all tensors of order 2, of the symmetric or anti-symmetric tensors of order 2, etc.; we then obtain from the representation $\mathfrak{H}$ of the abstract group $\mathfrak{g}$ the representation $\breve{\mathfrak{H}}$, $\mathfrak{H} \times \mathfrak{H}$, $[\mathfrak{H} \times \mathfrak{H}]$, $\{\mathfrak{H} \times \mathfrak{H}\}$, etc.

The three most important formal processes are (1) addition, (2) $\times$-multiplication, and (3) the Γ process. The first two generate a new representation from one or two given representations, the third a new one from a given representation. The first two are completely circumscribed, but the third contains a general method, for Γ may be any representation of the linear group $\mathfrak{c}_n$.

If $\mathfrak{g}'$ is a sub-group of $\mathfrak{g}$, then any representation $\mathfrak{H} : s \to U(s)$ of $\mathfrak{g}$ contains a representation of $\mathfrak{g}'$; we need only let the element s run through the sub-group $\mathfrak{g}'$! This too may be considered as a formal process (4) which generates a representation of $\mathfrak{g}'$ from a given representation of $\mathfrak{g}$.

The $\times$-multiplication occurs in yet another connection. Given two groups $\mathfrak{g}$, $\mathfrak{g}'$, we can consider the pairs (s, s'), the first member s of which is an element of $\mathfrak{g}$ and the second s' an element of $\mathfrak{g}'$, as the elements of a new group $\mathfrak{g} \times \mathfrak{g}'$, the ***direct product*** of $\mathfrak{g}$ and $\mathfrak{g}'$, obeying the multiplication law

$$(s, s')(t, t') = (st, s't').$$

The order of $\mathfrak{g} \times \mathfrak{g}'$ is the product of the orders of $\mathfrak{g}$ and $\mathfrak{g}'$. If $\mathfrak{H} : s \to U(s)$ is an n-dimensional representation of $\mathfrak{g}$ and $\mathfrak{H}' : s' \to U'(s')$ an n'-dimensional representation of $\mathfrak{g}'$, then

$$(s, s') \to U(s) \times U'(s') \tag{5.5}$$

is obviously a representation in nn' dimensions of the group $\mathfrak{g} \times \mathfrak{g}'$; we denote it by $\mathfrak{H} \boldsymbol{\times} \mathfrak{H}'$ (with a boldface $\boldsymbol{\times}$). This construction may be broken up into two steps. First introduce the representation

$$(s, s') \to U(s)$$

of $\mathfrak{g} \times \mathfrak{g}'$; there is no reason why we should not designate it by the same letter $\mathfrak{H}$ as the representation $s \to U(s)$ of $\mathfrak{g}$—we are accustomed to calling the function $f(x)$, considered as a function of the two variables x, y, by the same letter as the function $f(x)$ of the single variable x. $U(s)$ and $U'(s')$ are thus to be considered as functions of the same variable pair (s, s'), and then the representation $\mathfrak{H} \boldsymbol{\times} \mathfrak{H}'$ of $\mathfrak{g} \times \mathfrak{g}'$ may be obtained by ordinary $\times$-multiplication from $\mathfrak{H}$ and $\mathfrak{H}'$. The differentiation between boldface $\boldsymbol{\times}$ and ordinary $\times$ is accordingly purely pedantic.

Examples. Unimodular Group in Two Dimensions

Let $\mathfrak{g} = \mathfrak{c} = \mathfrak{c}_2$ consist of all linear transformations s of two variables x, y:

$$x' = ax + by, \quad y' = cx + dy, \tag{5.6}$$

whose determinant $ad - bc = 1$ ("unimodular" linear transformations *). A homogeneous polynomial in x, y of order f is a linear combination of the $f + 1$ monomials

$$x^f, x^{f-1} y, \cdots, x y^{f-1}, y^f. \tag{5.7}$$

Under the influence of s they undergo a linear transformation which we denoted above by $[s]_f$; they constitute the substratum of a representation $[\mathfrak{c}]^f : s \rightarrow [s]_f$ in $f + 1$ dimensions which we now denote by $\mathfrak{C}_f$. $\mathfrak{C}_f$ is, although we have yet to prove it, irreducible.

We can restrict ourselves within $\mathfrak{c}$ to the sub-group $\mathfrak{c}_1$ of "*principal*" transformations which transform each of the variables separately:

$$x' = ax, \quad y' = \frac{1}{a} y, \tag{5.8}$$

where $a \neq 0$ is an arbitrary constant. $\mathfrak{c}_1$ is Abelian. This transformation multiplies the monomials of the set (5.7) by

$$a^f, a^{f-2}, \cdots, a^{-(f-2)}, a^{-f}.$$

On associating the number a^r with the element (5.8) of $\mathfrak{c}_1$ we obtain a 1-dimensional representation which we denote for the moment by $\mathfrak{C}^{(r)}$; here r can be any fixed integral exponent. We have just seen that the irreducible representation $\mathfrak{C}_f$ of $\mathfrak{c}_2$ is completely reduced on restricting ourselves to the sub-group $\mathfrak{c}_1$ into $f + 1$ one-dimensional representations $\mathfrak{C}^{(r)}$ with $r = f$, $f - 2, \cdots, -f$. This is an example of the process (4).

As an example of multiplication and addition we consider the problem of reducing the product $\mathfrak{C}_f \times \mathfrak{C}_g$ of the two representations $\mathfrak{C}_f$, $\mathfrak{C}_g$ of $\mathfrak{c}$ into its irreducible components. The result is contained in the formula

$$\boxed{\mathfrak{C}_f \times \mathfrak{C}_g = \sum_v \mathfrak{C}_v} \tag{5.9}$$

where v runs through the series

$$\boxed{v = f + g, f + g - 2, \cdots, |f - g|} \tag{5.10}$$

without repetition, decreasing by 2 from term to term. This equation is essentially identical with the *Clebsch-Gordan* series which plays such an important rôle in the theory of invariants of binary forms. We shall see in the succeeding chapters that it may justly be considered as the fundamental mathematical

* $\mathfrak{c}_n$ will usually denote the group of all non-singular linear transformations in n-dimensions; it will however occasionally be used to denote the more restricted unimodular group, in which case the restriction will be explicitly stated.

formula for the classification of atomic spectra and for the theory of the valence bond.

The proof consists in showing that

$$\mathfrak{C}_f \times \mathfrak{C}_g = \mathfrak{C}_{f+g} + (\mathfrak{C}_{f-1} \times \mathfrak{C}_{g-1}), \tag{5.11}$$

for (5.9) then follows by mathematical induction and the fact that obviously

$$\mathfrak{C}_f \times \mathfrak{C}_0 = \mathfrak{C}_f.$$

A new co-ordinate system for the representation space of $\mathfrak{C}_f$ is obtained by replacing the basis (5.7) of homogeneous polynomials of order f by another basis. In this sense we can say that the polynomials of order f constitute the substratum of the representation $\mathfrak{C}_f$. The substratum of the representation $\mathfrak{C}_f \times \mathfrak{C}_g$ is then the totality of polynomials

$$\Phi = \Phi(x\,y\,;\,\xi\,\eta)$$

depending on the components of two arbitrary vectors $(x\,y)$, $(\xi\,\eta)$, homogeneous and of order f in the first, and homogeneous and of order g in the second; we write the total order $f + g = h$. The Φ are thus linear combinations of the $(f+1)(g+1)$ monomials

$$x^i\,y^k \cdot \xi^\iota\,\eta^\kappa \quad \text{where} \quad i + k = f,\; \iota + \kappa = g. \tag{5.12}$$

Both vectors are transformed cogrediently under the same transformation s, (5.6). The problem consists in completely reducing the space of the polynomials Φ into two sub-spaces $(\Phi)_0$ and $(\Phi)'$ which are the substrata of the representations $\mathfrak{C}_h$ and $\mathfrak{C}_{f-1} \times \mathfrak{C}_{g-1}$ respectively. We first discuss the structure of these two sub-spaces.

$(\Phi)_0$. Expand

$$(\alpha x + \beta y)^f(\alpha\xi + \beta\eta)^g = \alpha^h \cdot \phi_0 + \binom{h}{1}\alpha^{h-1}\beta \cdot \phi_1 + \cdots + \beta^h \cdot \phi_h \tag{5.13}$$

in powers of the undetermined coefficients α, β. The $\phi_i = \phi_i(xy\,;\,\xi\eta)$ are special polynomials of the type Φ and span the sub-space $(\Phi)_0$. We must now show that this sub-space is invariant under the transformation (5.6) of the variables; i.e. that $\phi'_i = \phi_i(x'y'\,;\,\xi'\eta')$ is a linear combination of the $\phi_i = \phi_i(xy\,;\,\xi\eta)$. It is clear that if this is the case then $\mathfrak{c}$ induces the representation $\mathfrak{C}_h$ in $(\Phi)_0$, for on identifying the two vectors

$$\xi = x, \quad \eta = y \tag{5.14}$$

ϕ_i becomes

$$\phi_i(xy\,;\,xy) = x^{h-i} \cdot y^i.$$

Hence we are certain *a priori* that the $h + 1$ functions ϕ_i are linearly independent.

In order to arrive at the desired proof we replace x, y in (5.13) by

$$x' = ax + by, \quad y' = cx + dy,$$

and in the same way ξ, η by

$$\xi' = a\xi + b\eta, \quad \eta' = c\xi + d\eta.$$

Now note that $\alpha x' + \beta y'$ is the linear form

$$(\alpha a + \beta c)x + (\alpha b + \beta d)y = Ax + By$$

in x and y; hence

$$(\alpha x' + \beta y')^f(\alpha\xi' + \beta\eta')^g = (Ax + By)^f(A\xi + B\eta)^g,$$

and by (5.13)

$$\sum_i \binom{h}{i} \alpha^{h-i}\beta^i \cdot \phi_i' = \sum_i \binom{h}{i} A^{h-i}B^i \cdot \phi_i.$$

On replacing A, B on the right-hand side of this equation by

$$A = \alpha a + \beta c, \quad B = \alpha b + \beta d,$$

and equating coefficients of $\alpha^{h-i}\beta^i$, we obtain ϕ_i' as a linear combination of the ϕ_k.

$(\Phi)'$. The substratum of the representation $\mathfrak{C}_{f-1} \times \mathfrak{C}_{g-1}$ consists of the polynomials

$$\Psi = \Psi(xy;\ \xi\eta)$$

of order $f - 1$ in (x, y) and of order $g - 1$ in (ξ, η). They are not polynomials of type Φ; in order to increase the order in the components of each vector by 1 we replace each such Ψ by

$$\Phi = (x\eta - y\xi) \cdot \Psi.$$

The factor thus introduced in no way affects the representation.

The last step in the proof consists in showing that the total space of polynomials Φ is completely reducible into these two sub-spaces; i.e. in showing that any polynomial Φ can be written in the form

$$\Phi = (a_0\phi_0 + a_1\phi_1 + \cdots + a_h\phi_h) + (x\eta - y\xi)\Psi \qquad (5.15)$$

with unique constant coefficients a_i. (The development in terms of powers of the determinant $x\eta - y\xi$ obtained from this by induction is the Clebsch-Gordan series.) First, the dimensionalities are correct, for

$$(f + 1)(g + 1) = (f + g + 1) + fg.$$

Hence it suffices to show that the various terms in (5.15) are linearly independent, i.e. that an expression of the form (5.15), in which Ψ is a polynomial of order $f - 1$ in (x, y) and of order $g - 1$ in (ξ, η), can vanish only if Ψ vanishes identically and if all the coefficients a_i are zero. The proof is extremely simple. We first let $(\xi\eta) = (xy)$ as in (5.14), then the equation $\Phi = 0$ becomes

$$a_0 x^h + a_1 x^{h-1} y + \cdots + a_h y^h = 0$$

identically in x and y; hence $a_i = 0$. Having established this we return to the two sets of variables xy; $\xi\eta$ and obtain the equation

$$(x\eta - y\xi)\Psi = 0,$$

from which it follows that $\Psi = 0$—in an algebraic identity for polynomials we may always remove a factor, such as $x\eta - y\xi$, which does not vanish identically.

Our formula (5.9) also holds for the group $\mathfrak{c}$ of *all* linear transformations of x, y with non-vanishing determinant. We must then interpret $\mathfrak{C}_v$, $v = h - 2l$ in (5.9) as that representation whose substratum is the totality of homogeneous polynomials of order v in x and y multiplied by $(x\eta - y\xi)$. In other words, the new $\mathfrak{C}_v$ differs from the old in that the transformation of the $(v + 1)$-dimensional representation space corresponding to s in the representation $\mathfrak{C}_v$ is to be multiplied by the lth power of the determinant $ad - bc$.

$\mathfrak{C}_f \times \mathfrak{C}_g$ is a representation of $\mathfrak{c}_2 \times \mathfrak{c}_2$, the group consisting of pairs (s, s') whose members s and s' run independently through the entire group $\mathfrak{c}_2$. On introducing the restriction that s' is the element $\bar{s}$ obtained from s by replacing the coefficients of the linear transformation s by their conjugate complex, $\mathfrak{C}_f \times \mathfrak{C}_g$ becomes a representation $\mathfrak{C}_{f,g}$ of $\mathfrak{c}_2$, the substratum of which may be taken as the monomials

$$x^i y^k \cdot \bar{x}^\iota \bar{y}^\kappa \quad (i + k = f, \quad \iota + \kappa = g)$$

of order f in (x, y) and order g in $(\bar{x}, \bar{y})$. It can be shown that $\mathfrak{C}_{f,g}$ is also irreducible.

§ 6. The Jordan-Hölder Theorem and its Analogues

Perhaps the most fundamental theorem of mathematics is that on which the concept of *cardinal numbers* depends. Let the members of a finite set of objects distinguished by marks $a, b, c \cdots$ be exhibited individually in this order and associated with the symbols $1, 2, \cdots n$. The theorem then states that the "number" n is independent of the order in which the objects are exhibited. The proof of this theorem is of considerable mathematical interest and offers the simplest example of the type of proof employed in establishing the Jordan-Hölder theorem. A new enumeration consists in associating the symbol 1 with any one of the objects, the symbol 2 with any one of the remaining objects, etc., until the entire set is exhausted, the last object receiving the symbol n'. We assert that $n' = n$.

The proof is divided into two steps. (1) If in the new enumeration the symbol 1 is associated with the same object a as in the old, our theorem for the series from 1 to n is reduced to that for the series from 1 to $n - 1$. This is immediately evident on discarding the object a and reducing by one the symbols associated with the objects $b, c, \cdots$ in the new as well as in the old enumeration. (2) If, on the other hand, the symbol 1 is associated with one of the other objects $b, c, \cdots$ then in the new enumeration the object a is associated with some symbol i contained in the series $2, 3, \cdots, n'$. We now introduce a third enumeration which enables us to make the transition between the first and the second by interchanging the symbols 1 and i in the second enumeration. The number n' is obviously unaltered by this process. But we have now introduced an equivalent enumeration in which the object a is associated with the same symbol 1 as in the original and have reduced the general case to the one considered in (1) above. The proof of the theorem then follows immediately by the method of mathematical induction.

As an auxiliary result of these fundamental considerations we have the theorem that any permutation can be obtained by the successive application of transpositions.

The *Jordan-Hölder* theorem is concerned with an abstract group $\mathfrak{g}$. An invariant sub-group $\mathfrak{g}'$ of $\mathfrak{g}$ which does not coincide with $\mathfrak{g}$ itself is said to be *maximal* if there exists no invariant sub-group of $\mathfrak{g}$—except $\mathfrak{g}'$ and $\mathfrak{g}$—containing $\mathfrak{g}'$. The factor group $\mathfrak{g}/\mathfrak{g}'$ is then ***simple***, i.e. it contains no invariant sub-group with the exception of itself and that consisting only of the unit element $\mathbf{I}$. As was recognized by *Galois*, the so-called ***composition series***

$$\mathfrak{g}_0 = \mathfrak{g}, \mathfrak{g}_1, \mathfrak{g}_2, \cdots, \mathfrak{g}_{r-1}, \mathfrak{g}_r = \mathbf{I} \tag{6.1}$$

is of fundamental importance for the solution of algebraic equations. This series begins with $\mathfrak{g}$ and ends with $\mathbf{I}$, and each member is a maximal invariant sub-group of the preceding member. We assume that the composition series terminates; this is naturally the case for finite groups, as the order necessarily decreases from term to term. The successive factor groups

$$\mathfrak{g}/\mathfrak{g}_1, \mathfrak{g}_1/\mathfrak{g}_2, \cdots, \mathfrak{g}_{r-1}/\mathfrak{g}_r = \mathfrak{g}_{r-1} \tag{6.2}$$

are simple. The Jordan-Hölder theorem asserts that *the structure of these factor groups, except for the order in which they appear, is uniquely determined by* $\mathfrak{g}$.

Consider, therefore, a second composition series

$$\mathfrak{g}'_0 = \mathfrak{g},\ \mathfrak{g}'_1,\ \mathfrak{g}'_2,\ \cdots$$

of the same group $\mathfrak{g}$; it is to be compared with the "standard series" (6.1). The proof of the fact that this new series also contains exactly $r + 1$ terms and that the corresponding factor groups are, except for the order in which they occur, isomorphic with the factor groups (6.2) is again accomplished in two steps.

(1) If the two second members $\mathfrak{g}'_1$, $\mathfrak{g}_1$ coincide, the theorem for the group $\mathfrak{g}$, whose standard series contains $r + 1$ members, is reduced to the corresponding theorem for the group $\mathfrak{g}_1$, whose standard series contains but r members.

(2) If $\mathfrak{g}_1$ and $\mathfrak{g}'_1$ do not coincide we construct the intersection $\mathfrak{h}$ of $\mathfrak{g}_1$ and $\mathfrak{g}'_1$, i.e. the set consisting of all elements common to the two. $\mathfrak{h}$ is then an invariant sub-group of $\mathfrak{g}'_1$ and, as we shall prove, $\mathfrak{g}'_1/\mathfrak{h}$ is isomorphic with $\mathfrak{g}/\mathfrak{g}_1$. That two elements s, t of $\mathfrak{g}$ are equivalent with respect to $\mathfrak{g}_1$, i.e. that they belong to the same "set," is expressed by the equation $t = a_1 s$ where a_1 is in $\mathfrak{g}_1$. If s and t are at the same time elements of the sub-group $\mathfrak{g}'_1$, then a_1 is also in $\mathfrak{g}'_1$ and consequently it is an element of $\mathfrak{h}$. We may therefore consider as the elements of $\mathfrak{g}'_1/\mathfrak{h}$ those sets in $\mathfrak{g}$ which contain an element of $\mathfrak{g}'_1$. The elements contained in these classes then constitute an invariant sub-group $\mathfrak{H}$ of $\mathfrak{g}$ containing both $\mathfrak{g}_1$ and $\mathfrak{g}'_1$, and $\mathfrak{g}'_1/\mathfrak{h}$ is simply isomorphic with $\mathfrak{H}/\mathfrak{g}_1$. But since $\mathfrak{g}'_1$ is maximal either $\mathfrak{H} = \mathfrak{g}$ or $\mathfrak{H} = \mathfrak{g}'_1$. The second case implies that $\mathfrak{g}_1$ is contained in $\mathfrak{g}'_1$, and since it is maximal it must coincide with $\mathfrak{g}'_1$, contrary to assumption. Hence $\mathfrak{H}$ coincides with $\mathfrak{g}$ and our assertion is proved. The intersection $\mathfrak{h}$ of $\mathfrak{g}_1$ and $\mathfrak{g}'_1$ depends *symmetrically* on both, whence $\mathfrak{g}/\mathfrak{g}'_1$ and $\mathfrak{g}_1/\mathfrak{h}$ are also simply isomorphic.

We now proceed as follows. We construct a composition series for $\mathfrak{h}$, which we denote simply by $\mathfrak{h}, \cdots$, and compare the following four composition series of $\mathfrak{g}$:

$$\begin{array}{llll} \mathfrak{g}, & \mathfrak{g}_1, & \mathfrak{g}_2, & \cdots \\ \mathfrak{g}, & \mathfrak{g}_1, & \mathfrak{h}, & \cdots \\ \mathfrak{g}, & \mathfrak{g}'_1, & \mathfrak{h}, & \cdots \\ \mathfrak{g}, & \mathfrak{g}'_1, & \mathfrak{g}'_2 & \cdots \end{array}$$

The comparison of the first and second series is reduced to case (1). The second and third series agree from the member $\mathfrak{h}$ on, and the two foregoing factor groups

$$\mathfrak{g}/\mathfrak{g}_1,\quad \mathfrak{g}_1/\mathfrak{h}$$

are, as we have seen, *simply isomorphic with*

$$\mathfrak{g}/\mathfrak{g}'_1, \quad \mathfrak{g}'_1/\mathfrak{h}$$

on interchanging their order. The comparison between the third and fourth series is again reduced to the case (1). The proof of the theorem for composition series containing $r + 1$ members is thus reduced to the proof of the corresponding theorem for series with but r members, and since it obviously holds for $r = 2$ (i.e. for simple groups) the method of mathematical induction establishes its general validity.

The close methodological agreement between the construction involved in the proof of this theorem and that involved in the proof of the independence of the cardinal number of a set of the order in which the objects are enumerated is immediately evident.

E. Noether[4] has given a generalization of the Jordan-Hölder theorem which is of importance for us. A correspondence $s \to s' = As$ of the group on itself is said to be ***automorphic*** if multiplication is invariant under it, i.e. if $(st)' = s't'$—we here neither assume that different elements s generate different elements s' nor that for a given element s' there exists an element s such that $s \to s'$ in virtue of the automorphism. Let Σ be a system of such automorphic correspondences of $\mathfrak{g}$. We now admit only sub-groups of $\mathfrak{g}$ which are invariant under Σ, i.e. sub-groups whose elements are carried over by all operations of the system Σ into elements of the same sub-group. We say that two such " allowed " sub-groups $\mathfrak{g}_1$ and $\mathfrak{g}_2$ have the same structure if we can set up a one-to-one simple isomorphic correspondence between the elements of the one and the elements of the other in such a way that every operation A of the system Σ sends corresponding elements of the two sub-groups over into corresponding elements. The Jordan-Hölder theorem still holds under this modification; its proof can be aken over unaltered.

The vectors of an n-dimensional vector space $\mathfrak{R}$ constitute an Abelian group whose multiplication is the addition $+$ of vectors. We must for the moment supplement addition by the operation of multiplication of a vector by an arbitrary number; hence the concepts and theorems applying to vector space are not truly specializations of the concepts and theorems of Abelian groups, but there exists a thorough-going analogy between the two. Indicating this analogy between a group (on the left) and vector space (on the right) by $\sim$ we have, for example, sub-group $\sim$ linear sub-space, automorphism $\sim$ linear correspondence. Indeed, a linear sub-space is a system $\mathfrak{R}'$ of vectors such that with $\mathfrak{x}$ and $\mathfrak{y}$ their sum $\mathfrak{x} + \mathfrak{y}$ and the product $\lambda\mathfrak{x}$ by an arbitrary number λ also belong to $\mathfrak{R}'$, and a correspondence $\mathfrak{x} \to \mathfrak{x}' = A\mathfrak{x}$ is linear if it sends $\mathfrak{x} + \mathfrak{y}$ and $\lambda\mathfrak{x}$ over into $\mathfrak{x}' + \mathfrak{y}'$ and $\lambda\mathfrak{x}'$, respectively. Every "sub-group" is here invariant, as we are dealing with Abelian groups. If $\mathfrak{R}'$ is a sub-space of $\mathfrak{R}$ the space $\mathfrak{R}$ (mod. $\mathfrak{R}'$) obtained by projecting $\mathfrak{R}$ with respect to $\mathfrak{R}'$ is the exact analogue of a factor group. A composition series consists of a sequence of spaces each member of which is a linear sub-space of the preceding one and has one less dimension. The last member is the space 0, consisting of the vector 0 alone, and the number of members in the series is 1 greater than the dimensionality n. The Jordan-Hölder theorem is here valid but trivial.

On the other hand, this theorem is of considerable importance on going over to Noether's generalization. Consider a system Σ of linear correspondences of the vector space $\mathfrak{R}$ on itself; the terms invariant, equivalent, reduction shall in the following refer to this system. Two invariant sub-spaces $\mathfrak{R}_1$ and $\mathfrak{R}_2$ are similar or equivalent if a one-to-one linear correspondence $\mathfrak{x}_1 \rightleftarrows \mathfrak{x}_2$ can be set up between the vectors of the one and the vectors of the other in such a way that any operation A of the system sends corresponding vectors over into corresponding vectors. On reading the series (4.3) established in § 4 backwards, we have the exact analogue of the composition series: each member of the series is followed by a maximal sub-space which is invariant under Σ. (The possibility of constructing the composition series in increasing as well as decreasing order is due to the fact that the addition of vectors is commutative.) Furthermore, we can obtain the concepts and theorems relating to a system Σ of correspondences as genuine special cases of those of group theory, and not merely as analogues, by supplementing the system Σ with all similarity transformations, i.e. by all correspondences of the form $\mathfrak{x} \to \mathfrak{x}' = \lambda\mathfrak{x}$ representing multiplication by an arbitrary number λ. The Jordan-Hölder-Noether theorem now states: Given a second composition series

$$0, \mathfrak{R}_1', \mathfrak{R}_2', \cdots, \mathfrak{R}, \tag{6.3}$$

the corresponding projection spaces

$$\mathfrak{R}_1', \mathfrak{R}_2' \text{ (mod. } \mathfrak{R}_1'), \mathfrak{R}_3' \text{ (mod. } \mathfrak{R}_2'), \cdots$$

are equivalent to the projection spaces (4.3)

$$\mathfrak{R}_1, \mathfrak{R}_2 \text{ (mod. } \mathfrak{R}_1), \mathfrak{R}_3 \text{ (mod. } \mathfrak{R}_2), \cdots$$

of the original series, taken in a suitable order. The number of members is, of course, the same in both. The reader is advised to reconstruct the proof of this theorem by carrying through the proof of the Jordan-Hölder theorem step by step for this case.

In particular, if the system Σ consists of the transformations $U(s)$ associated with the various elements s of a group in a representation $\mathfrak{H} : s \rightarrow U(s)$, our result yields the

Uniqueness theorem: The irreducible representations separated off from $\mathfrak{H}$ *by successive reduction are completely determined by* $\mathfrak{H}$, *except for the order in which they occur, considering equivalent representations as the same. In particular, the complete reduction of* $\mathfrak{H}$ *into irreducible components is unique, always considering equivalent representations as the same.*

§ 7. Unitary Representations

For the case in which the representation space $\mathfrak{R}$ is unitary and the correspondences $U(s)$ of $\mathfrak{R}$ on itself, associated with the element s of the group under consideration, are also unitary, certain of the concepts introduced above are to be modified accordingly. Two representations

$$s \rightarrow U(s), \quad s \rightarrow U'(s) = AU(s)A^{-1},$$

are to be considered as equivalent only if A is unitary, i.e. if it is a transformation from one normal co-ordinate system in $\mathfrak{R}$ to another such. If $\mathfrak{R}'$ is a sub-space of $\mathfrak{R}$ a unitary-orthogonal co-ordinate system can be set up in $\mathfrak{R}'$ and supplemented by additional fundamental vectors to form a complete unitary-orthogonal co-ordinate system for the entire space $\mathfrak{R}$: every sub-space of a unitary space is *per se* unitary. Invariance and reduction remain as before, but we allow only those decompositions of $\mathfrak{R}$ into two sub-spaces $\mathfrak{R}_1 + \mathfrak{R}_2$ in which $\mathfrak{R}_1$, $\mathfrak{R}_2$ are perpendicular. For a system of unitary correspondences *reducibility implies complete reducibility and we have the theorem: If* $\mathfrak{R}'$ *is invariant with respect to* Σ *then* $\mathfrak{R}$ *may be broken up into* $\mathfrak{R}' + \mathfrak{R}''$ *in such a way that* $\mathfrak{R}''$ *is also invariant under* Σ. We need merely to define $\mathfrak{R}''$ as the space defined by all vectors perpendicular to $\mathfrak{R}'$. The theorem naturally holds for the case in which Σ is a system of *infinitesimal* unitary correspondences or, what amounts to the same, a system of Hermitian forms. The theorem developed in the preceding section proves that these irreducible components are uniquely determined, in the sense of (unitary) equivalence, to within a permutation.

Examples

(1) *The Unitary Group in Two Dimensions*

The group $\mathfrak{c} = \mathfrak{c}_2$ of linear transformations in two dimensions contains the sub-group $\mathfrak{u} = \mathfrak{u}_2$ of unitary transformations. Hence the representation $\mathfrak{C}_f$ of $\mathfrak{c}$ obtained in § 5 is also a representation of $\mathfrak{u}$. This representation is not unitary as it stands, but it can readily be made unitary by a slight change. The transformation of $\mathfrak{C}_f$ corresponding to the unitary transformation s of the co-ordinates x, y is that induced by s on the monomials

$$x_n = x^i y^k \quad (i + k = f) \tag{7.1}$$

of order f. For purposes of symmetry we label these co-ordinates with the index $n = i - k$ which runs through the values $f, f - 2, \cdots, -f$. This is also desirable because on restricting ourselves to the sub-group of " principal transformations "

$$x \to \varepsilon x, \quad y \to \frac{1}{\varepsilon} y$$

x_n is multiplied by the factor ε^n. We now employ, instead of (7.1), the variables

$$x_n = \frac{x^i y^k}{\sqrt{i\,!\,k\,!}} \tag{7.2}$$

obtained from them by multiplication with a constant. The representation $\mathfrak{C}_f$ of $\mathfrak{u}$ will then be unitary, as follows from the equation

$$\frac{1}{f!}(x\bar{x} + y\bar{y})^f = \Sigma \frac{x^i y^k \cdot \bar{x}^i \bar{y}^k}{i\,!\,k\,!} = \Sigma x_n \bar{x}_n.$$

We call $\mathfrak{C}_f$ even or odd according as f is even or odd. The even representations associate the identity **1** with the reflection

$$x' = -x, \quad y' = -y,$$

and the odd associate with it the transformation $-\mathbf{1}$. $\mathfrak{C}_f$ is also irreducible when considered as a representation of $\mathfrak{u}$, and on letting f assume the values $0, 1, 2, \cdots$ they form *a complete system of inequivalent irreducible representations of* $\mathfrak{u}$. The proof of these assertions, which we employ heuristically in the following, will be given in Chapter V. On writing a homogeneous polynomial of order f in the variables x, y in the form

$$\Sigma \bar{a}_n x_n$$

the coefficients a_n transform under the influence of a unitary transformation s like the components of a vector in the representation space of $\mathfrak{C}_f$.

The complete reduction

$$(\mathfrak{C}_f \times \mathfrak{C}_g) = \mathfrak{C}_{f+g} + (\mathfrak{C}_{f-1} \times \mathfrak{C}_{g-1})$$

was accomplished by breaking up the space of the "polynomials Φ" into two invariant sub-spaces $(\Phi)_0$ and $(\Phi)'$. We must now verify that these two sub-spaces are mutually orthogonal in the unitary sense. A general polynomial Φ may be written

$$\Sigma \bar{a}_{n\nu} x_n \xi_\nu \qquad \begin{pmatrix} n = f, f-2, \cdots, -f \\ \nu = g, g-2, \cdots, -g \end{pmatrix}$$

where the x_n are given by (7.2) and the ξ_ν are the corresponding monomials

$$\xi_\nu = \frac{\xi^\iota \eta^\kappa}{\sqrt{\iota!\,\kappa!}} \qquad (\iota + \kappa = g,\ \iota - \kappa = \nu).$$

Two such polynomials Φ with coefficients $a_{n\nu}$, $b_{n\nu}$ are orthogonal if

$$\Sigma \bar{a}_{n\nu} b_{n\nu} = 0.$$

The polynomial $x_f \xi_g$, whose highest coefficients $a_{fg} = 1$ while all others vanish, is to within a constant factor $x^f \cdot \xi^g$ and is obviously perpendicular to all polynomials $(\Phi)'$, for in all these latter the coefficient of $x^f \xi^g$ vanishes. But under the unitary transformation

$$s : x' = \alpha x + \beta y, \quad y' = -\bar{\beta} x + \bar{\alpha} y, \tag{7.3}$$

where $\alpha\bar{\alpha} + \beta\bar{\beta} = 1$, $x^f \xi^g$ goes into

$$(\alpha x + \beta y)^f (\alpha \xi + \beta \eta)^g. \tag{7.4}$$

Since $(\Phi)'$ and the orthogonality of polynomials are both invariant under the unitary transformation s, (7.4) is also orthogonal to $(\Phi)'$ and, with the help of the definition (5.12) of $(\Phi)_0$, it follows from this that *all* polynomials of $(\Phi)_0$ are unitary-orthogonal to those of $(\Phi)'$.

(7.3) is the most general unimodular unitary transformation. This is derived in the same way as the familiar formula for the orthogonal transformations of two variables with unit determinant in plane analytical geometry. On writing the coefficients

$$\alpha = \kappa + i\lambda, \quad \beta = -\mu + i\nu \tag{7.5}$$

in terms of their real and imaginary parts we see that each such transformation is characterized by four real parameters κ, λ, μ, ν, the sum of whose squares is 1. The composition of two transformations s: $(\kappa, \lambda, \mu, \nu)$ is accomplished in terms of these parameters by *Hamilton's quaternion multiplication*; this latter led to the vector calculus.

(2) *Unitary Groups in n-Dimensions*

The totality of tensors of order f is the substratum of an n^f-dimensional unitary representation $(\mathfrak{u})^f$ of the group $\mathfrak{u} = \mathfrak{u}_n$, for on denoting the components of an arbitrary tensor by $F(i_1 i_2 \cdots i_f)$ the sum

$$\sum_{(i_1, \cdots, i_f)} |F(i_1 i_2 \cdots i_f)|^2 \tag{7.6}$$

is a unitary invariant. On restricting ourselves to the $\begin{Bmatrix} n \\ f \end{Bmatrix}$-dimensional linear manifold of anti-symmetric tensors we take as the variables in tensor space those components $F(i_1 i_2 \cdots i_f)$ for which $i_1 < i_2 < \cdots < i_f$. The sum (7.6) for these components only is, however, equal to the *complete* sum (7.6) divided by $f\,!$; hence the representation $\{\mathfrak{u}\}^f$ of $\mathfrak{u}$, whose substratum consists of all anti-symmetric tensors, is unitary. The situation is somewhat different for symmetric tensors. The most general symmetric tensor of order f transforms like $\mathfrak{x} \times \mathfrak{x} \times \cdots \times \mathfrak{x}$ (f terms), i.e. we may for the present purpose set

$$F(i_1 i_2 \cdots i_f) = x_{i_1} x_{i_2} \cdots x_{i_f}. \tag{7.7}$$

We write the monomial on the right in the form

$$x_1^{f_1} x_2^{f_2} \cdots x_n^{f_n} \tag{5.1}$$

as before; f_r is the number of times the index r appears in the series $i_1, i_2, \cdots, i_f$. In this sense we write the components of a symmetrical tensor

$$F(i_1 i_2 \cdots i_f) = \phi(f_1, f_2, \cdots, f_n).$$

The sum (7.6) becomes in this case

$$\sum \frac{f\,!}{f_1\,! f_2\,! \cdots f_n\,!} |\phi(f_1, f_2, \cdots, f_n)|^2,$$

extended over all integral $f_r \geqq 0$ for which $f_1 + f_2 + \cdots + f_n = f$. The coefficient indicates how often the term $|F(i_1 i_2 \cdots i_f)|^2$ occurs in the sum in consequence of the fact that its value is unchanged on permuting the indices. We must therefore consider the quantities

$$\frac{\phi(f_1, f_2, \cdots, f_n)}{\sqrt{f_1\,! f_2\,! \cdots f_n\,!}}$$

as independent components of an arbitrary symmetric tensor of order f in order to obtain a *unitary* representation $[\mathfrak{u}]^f$.

The truth of this assertion follows from the fact that the special tensor (7.7) satisfies the equation

$$\frac{1}{f!}(x_1\bar{x}_1 + \cdots + x_n\bar{x}_n)^f = \Sigma \frac{x_1^{f_1}\cdots x_n^{f_n}\cdot \bar{x}_1^{f_1}\cdots \bar{x}_n^{f_n}}{f_1!\cdots f_n!}. \qquad (7.8)$$

We have already seen in I, § 5 that a normal co-ordinate system can be so chosen that a commutative system Σ of unitary correspondences is completely reduced to a set of 1-dimensional systems. *The only irreducible unitary representations of an Abelian group are accordingly* 1-*dimensional.* For it follows from

$$U(s)U(t) = U(st) \qquad (4.1)$$

and the Abelian character of the group that the unitary matrices $U(s)$ associated with the elements s are commutative.

If $\mathfrak{H}$ and $\mathfrak{H}'$ are unitary representations, then $\mathfrak{H} + \mathfrak{H}'$, $\mathfrak{H} \times \mathfrak{H}'$ are also.—The first *fundamental problem* for a given group $\mathfrak{g}$ is *to find a complete system of inequivalent irreducible unitary representations of* $\mathfrak{g}$, for then any unitary representation of $\mathfrak{g}$ can be obtained by the addition of these irreducible representations. The second *fundamental problem* is *to reduce the product* $\mathfrak{H} \times \mathfrak{H}'$ *of two irreducible representations* $\mathfrak{H}$, $\mathfrak{H}'$ *of* $\mathfrak{g}$ *into its irreducible components ; or better* (after having solved the first problem), *to determine how often each of the irreducible representations occurs in this product.*

We illustrate these problems on the example offered by rotation groups, which are of particular importance in quantum physics.

§ 8. Rotation and Lorentz Groups

(a) *The Group of Rotations in the Plane*

We describe the 2-dimensional plane by a complex co-ordinate x. The rotations of the plane are then given by

$$x \to x' = \varepsilon x, \qquad (8.1)$$

where $\varepsilon = e^{i\phi}$ is a constant with unit modulus. (The rotations of the real 2-dimensional plane thus coincide with the unitary transformations of a *single* complex variable.) The angle of rotation ϕ determines the rotation completely, but it is of course only determined mod. 2π by the rotation. The angle of rotation behaves additively on composition : the rotation ϕ followed by the rotation ϕ' results in the rotation $\phi + \phi'$. This rotation group is accordingly a one-parameter continuous Abelian group. We obtain a 1-dimensional representation $\mathfrak{D}^{(m)}$ of our rotation group $\mathfrak{d} = \mathfrak{d}_2$ by associating with the element ε, (8.1), the linear correspondence

$$x \to x' = \varepsilon^m \cdot x = e^{im\phi} \cdot x, \tag{8.2}$$

where m is any fixed integer. I assert that the $\mathfrak{D}^{(m)}$, m running through all integral values, constitute a complete system of irreducible unitary representations of $\mathfrak{d}_2$. This can be seen as follows.

Any irreducible representation is necessarily 1-dimensional: it associates with the rotation ϕ a number $\chi(\phi)$ of absolute value 1 such that

$$\chi(\phi + \phi') = \chi(\phi) \cdot \chi(\phi').$$

We assume that our representation is continuous; then $\chi(\phi)$ is a continuous function of ϕ with period 2π. First, $\chi(0) = 1$. We write $\chi(\phi) = e^{i\lambda}$ and determine $\lambda(\phi)$ uniquely by the requirements that $\lambda(0) = 0$ and that $\lambda(\phi)$ shall be a *continuous* function of ϕ. We then have

$$\lambda(\phi + \phi') = \lambda(\phi) + \lambda(\phi'), \tag{8.3}$$

for the right- and left-hand sides of this equation could at most differ by an integral multiple of 2π, but as it is written both sides agree for $\phi' = 0$ and vary continuously with ϕ'. (8.3) satisfies the condition $\lambda(0) = 0$ and we obtain from it the further equations

$$\lambda(-\phi) = -\lambda(\phi), \quad \lambda(h\phi) = h \cdot \lambda(\phi), \tag{8.4}$$

where h is any integer. On replacing ϕ in the second of these equations by ϕ/h we obtain

$$\lambda\left(\frac{\phi}{h}\right) = \frac{1}{h}\lambda(\phi). \tag{8.5}$$

It follows immediately from (8.4), (8.5) that for every rational number k/h (k, h integers)

$$\lambda\left(\frac{k}{h}\phi\right) = \frac{k}{h}\lambda(\phi). \tag{8.6}$$

In accordance with our assumptions $\lambda(2\pi)$ is an integral multiple $2m\pi$ of 2π. On setting $\phi = 2\pi$ in (8.6) we obtain the equation $\lambda(\phi) = m\phi$ for all ϕ which are rational fractions of 2π; the continuity requirement then allows us to assert its validity for all real values of the argument ϕ.

The simple equation

$$\mathfrak{D}^{(m)} \times \mathfrak{D}^{(m')} = \mathfrak{D}^{(m+m')}$$

is here valid.

Consider the function $f(p)$ on the unit circle in the complex x plane. If the point p goes over into the point p' under the rotation $\mathfrak{e}$, the function f goes into a function f' which is defined by the equation

$$f'(p') = f(p).$$

The transition $f \to f'$ is a linear correspondence in the ∞-dimensional space of functions $f(p)$ and is associated with the rotation $\mathfrak{e}$; this obviously defines an ∞-dimensional representation of the rotation group $\mathfrak{d}_2$, which we denote by $\mathfrak{K}$. $\mathfrak{K}$ is unitary if we take as the square of the absolute value of a "vector" f the integral of $|f(p)|^2$ with respect to the element of arc dp on the unit circle. The fact that any function (satisfying suitable conditions) on the unit circle can be developed in a Fourier series means that in the reduction of $\mathfrak{K}$ into its irreducible components each of the 1-dimensional representations $\mathfrak{D}^{(m)}$ occurs once and only once. More precisely, this reduction is to be interpreted with regard to the completeness relation.

(*b*) *The Group of Rotations in* 3-*dimensional Space*

We consider the functions $f = f(P)$ on the unit sphere as the vectors of an ∞-dimensional unitary space whose metric is given by $\int |f(P)|^2 d\omega$; $d\omega$ is the surface element of the sphere over which the integration is to be extended. If the point P goes over into $P' = sP$ under the rotation s, the function f goes over into the function f' defined by $f'(P') = f(P)$. The surface harmonics Y_l of degree l [cf. II, § 4] obviously span a $(2l + 1)$-dimensional sub-space $\mathfrak{R}_l$ which is invariant under the totality of transitions $f \to f'$ induced in function space by the various elements s of the rotation group $\mathfrak{d} = \mathfrak{d}_3$—here again we speak of this representation as $\mathfrak{K}$. They are consequently the substratum of a certain representation $\mathfrak{D}_l$ of $\mathfrak{d}$ which is induced in $\mathfrak{R}_l$ by $\mathfrak{d}$. On choosing a definite direction as that of the z-axis we may, as in II, § 4, take the set

$$Y_l^{(m)} \quad (m = l, l-1, \cdots, -l)$$

as a basis for the surface harmonics of degree l. We then have a *unitary* representation, and the sub-spaces $\mathfrak{R}_l$ corresponding to the various values 0, 1, 2, $\cdots$ of l are mutually perpendicular in the unitary sense (orthogonality properties of surface harmonics). $\mathfrak{d}$ contains the 2-dimensional rotation group $\mathfrak{d}_2$—e.g. as the sub-group of rotations about the z-axis. The structure of $Y_l^{(m)}$ shows that on restricting $\mathfrak{d}_3$ to this sub-group $\mathfrak{d}_2$ the representation $\mathfrak{D}_l$ is reduced into the 1-dimensional representations $\mathfrak{D}^{(m)}$ for which $m = l, l-1, \cdots, -l$. The fact that any function on the unit sphere possesses a unique expansion in terms of surface harmonics means that on reducing $\mathfrak{K}$ into its irreducible components each of the representations $\mathfrak{D}_l$, $l = 0, 1, 2, \cdots$, occurs exactly once. This reveals the true significance of surface harmonics; they are characterized by the fundamental symmetry properties here developed, and the solution of the potential equation in polar co-ordinates is merely an accidental approach to their theory.

Rotations are orthogonal transformations of three variables x, y, z. If we wish to include with the proper rotations with determinant $+1$ also the improper ones with determinant -1 —" augmented rotation group $\mathfrak{d}'$ "—this can be done by introducing the reflection

$$i : x' = -x, \quad y' = -y, \quad z' = -z \tag{8.7}$$

in the origin. Its reiteration ii is the identity, and it commutes with all rotations. The matrix corresponding to it in the representation defined by the surface harmonics of degree l is the $(2l+1)$-dimensional matrix $(-\mathbf{1})^l$, for the surface harmonics of degree l are homogeneous polynomials of degree l in x, y, z. We can thus obtain two representations $\mathfrak{D}_l^+$, $\mathfrak{D}_l^-$ of the augmented rotation group from the representation $\mathfrak{D}_l$ of proper rotations; these two coincide with $\mathfrak{D}_l$ for proper rotations, but in the first the matrix associated with the reflection i is $+\mathbf{1}$ whereas in the second it is $-\mathbf{1}$. We call this ± 1 the ***signature*** of the representation. Hence in the ∞-dimensional representation $\mathfrak{K}$ of the augmented group $\mathfrak{d}'$ each $\mathfrak{D}_l$ occurs once with signature $(-1)^l$, but not with the opposite signature. Although we are not as yet in a position to prove it, the $\mathfrak{D}_l$ $(l = 0, 1, 2, \cdots)$ constitute a complete system of inequivalent irreducible (single-valued) representations of the rotation group $\mathfrak{d}$, and the $\mathfrak{D}_l^+$, $\mathfrak{D}_l^-$ together constitute such a system for the augmented rotation group $\mathfrak{d}'$.

Now consider the unitary function space of all functions $f(P)$ in 3-dimensional space for which the integral $|f|^2$ over all space is finite. Let the representation induced in this space by rotations s, in which the transition from f to the transformed function $f' = sf$ is associated with s, be denoted by $\mathfrak{E}$. Each function $f(P)$ can be expanded in a series of terms of the form $\phi(r) \cdot Y_l$. Choose a complete orthogonal system $\phi_1(r), \phi_2(r), \cdots$ in the domain of functions $\phi(r)$ of the radius r, in the sense of the equations

$$\int_0^\infty r^2 \bar{\phi}_m(r)\phi_n(r)dr = \delta_{mn}.$$

The functions of the form $\phi_n(r) \cdot Y_l$ then constitute a $(2l + 1)$-dimensional sub-space $\mathfrak{R}_{nl}$ which is invariant under rotations and in which $\mathfrak{C}$ induces the representation $\mathfrak{D}_l$. Different $\mathfrak{R}_{nl}$ are mutually unitary-orthogonal. Each $\mathfrak{D}_l$ then appears in $\mathfrak{C}$ infinitely often, its various occurrences being distinguished by the "radial quantum number" n. Consider the analysis of single electron spectra given in Chap. II, § 5, in the light of these mathematical developments. We then see that the azimuthal quantum number l is of purely group-theoretic significance, whereas the radial quantum number n refers to the dynamical situation, for the manner in which the orthogonal system $\phi_n(r)$ is to be chosen is determined by the dynamical differential equation.

The proper rotations of 3-dimensional Euclidean space about the origin of Cartesian co-ordinates x, y, z, i.e. the real orthogonal transformations with determinant $+ 1$, are most easily represented by a stereographic projection of the unit sphere about the origin on to the equatorial plane $z = 0$, the south pole of the sphere being the centre of projection. If the point $(x', y', 0)$ be the image on the plane of the point (x, y, z) on the sphere and we write $\zeta = x' + iy'$, the formulæ for the projection are

$$x + iy = \frac{2\zeta}{1 + \zeta\bar{\zeta}}, \quad x - iy = \frac{2\bar{\zeta}}{1 + \zeta\bar{\zeta}}, \quad z = \frac{1 - \zeta\bar{\zeta}}{1 + \zeta\bar{\zeta}}.$$

But it is preferable to introduce the two homogeneous complex co-ordinates ξ, η in place of ζ by means of the equation $\zeta = \eta/\xi$; the south pole $\xi : \eta = 0 : 1$ is then included. We then have

$$x + iy : x - iy : \quad z \quad : \quad 1 \quad = \\ 2\eta\bar{\xi} \; : \; 2\xi\bar{\eta} \; : \xi\bar{\xi} - \eta\bar{\eta} : \xi\bar{\xi} + \eta\bar{\eta}.$$

Accordingly each *unitary* transformation

$$\sigma : \xi' = \alpha\xi + \beta\eta, \quad \eta' = \gamma\xi + \delta\eta$$

of the co-ordinates ξ, η corresponds to a rotation s of the sphere, the points of which are represented by the rays $\xi : \eta$ of 2-dimensional unitary space. Since, as is readily seen, any point and tangential direction through it on the sphere can be carried over into any other such configuration on the sphere by means of such rotations, we obtain in this way *all* rotations. Since we are only concerned with the ratios of the coefficients α, β, γ, δ, the arbitrary factor of proportionality may be chosen in such a way that the determinant of the transformation is 1. Nevertheless this normalization is somewhat artificial as the correspondence is still double-valued, for on multiplying the coefficients of the unitary transformation by -1, i.e. on going over from σ to $-\sigma$, the normalization is unaffected. Hence to each element σ, (7.4), of the *unimodular* unitary group $\mathfrak{u}$ corresponds a rotation $s : \sigma \to s$ under which the co-ordinates $x + iy$, $x - iy$, z transform like

$$2\eta\bar{\xi}, \quad 2\xi\bar{\eta}, \quad \xi\bar{\xi} - \eta\bar{\eta}, \tag{8.8}$$

or

$$x \sim \eta\bar{\xi} + \xi\bar{\eta}, \quad y \sim \frac{1}{i}(\eta\bar{\xi} - \xi\bar{\eta}), \quad z \sim \xi\bar{\xi} - \eta\bar{\eta}. \tag{8.9}$$

(The symbol $\sim$, which we occasionally employ, means that the expression on the left transforms like the one on the right.) We obtain in this way all rotations, each one exactly twice. The rotations about the z-axis are obtained from the " principal transformations "

$$\xi' = \varepsilon\xi, \quad \eta' = \frac{1}{\varepsilon}\eta$$

of $\mathfrak{u}$. In fact, on setting $\varepsilon = e^{i\omega} = e(\omega)$ the angle of rotation about the z-axis is $\phi = -2\omega$. In virtue of the correspondence $\sigma \to s$ the rotations in 3-dimensions constitute a representation of the group $\mathfrak{u}$; and, conversely, the association $s \to \sigma$ is a representation of the group $\mathfrak{d} = \mathfrak{d}_3$ of 3-dimensional rotations by $\mathfrak{u}$, although this representation is double-valued. In virtue of this correspondence $s \to \sigma$ any representation $U(\sigma)$ of $\mathfrak{u}$ yields a representation of $\mathfrak{d}_3$ (" Γ process," § 5) ; $\mathfrak{C}_v$ may thus be thought of as a representation of $\mathfrak{d}_3$, in which case we write it $\mathfrak{D}_j$, where $j = \frac{1}{2}v$. The (" even ") $\mathfrak{D}_j$ with integral j are single-valued, those with half-integral (i.e. half an *odd* integer) j are double-valued. On restricting the group $\mathfrak{d}_3$ to the sub-group $\mathfrak{d}_2$ of rotations about the z-axis $\mathfrak{D}_j$ is reduced into the $2j + 1$ one-dimensional representations $\mathfrak{D}^{(m)}$ $(m = j, j - 1, \cdots, -j)$. To show this we first note that the substratum of our representation $\mathfrak{D}_j$ consists of the monomials (7.2)

$$x(m) = \frac{\xi^i\eta^k}{\sqrt{i!\,k!}} \quad (i + k = 2j, \quad i - k = 2m),$$

where m runs through the values $j, j-1, \cdots, -j$. The transformation induced on these variables by a rotation ϕ about the z-axis is accordingly

$$x(m) \to e(-m\phi) \cdot x(m).$$

The representation $\sigma \to s$ of $\mathfrak{u}$ is itself contained among the representations $\mathfrak{D}_j$ of $\mathfrak{u}$ constructed above; it is, in fact, $\mathfrak{D}_1$. To show this we note that if (ξ, η), (ξ', η') be subjected to the same transformation σ of $\mathfrak{u}$, then the determinant $\xi\eta' - \eta\xi'$, as well as $\xi\bar{\xi} + \eta\bar{\eta}$, is invariant. Consequently $(\bar{\xi}, \bar{\eta})$ transform cogrediently to $(\eta', -\xi')$, or as $(\eta, -\xi)$; hence

$$x + iy \sim \eta^2, \quad x - iy \sim -\xi^2, \quad z \sim \xi\eta. \tag{8.10}$$

The representations $\mathfrak{D}_j$ with integral j are identical with those obtained above as the representations induced on surface harmonics of order j, for each polynomial in x, y, z of degree j is, in virtue of (8.10), equivalent to a form of order $2j$ in ξ, η.

If we wish to augment $\mathfrak{u} = \mathfrak{u}_2$ in a manner paralleling the augmentation of $\mathfrak{d} = \mathfrak{d}_3$ by the improper rotation i (reflection in the origin) we must consider it as an abstract group rather than a group of linear transformations in two variables. Denote the element corresponding to i by ι and the elements of the original $\mathfrak{u}$ by σ as before. We define the augmented $\mathfrak{u}'$ as the totality of elements of the types σ and $\iota\sigma$; ι must naturally obey the multiplication laws

$$\iota\sigma = \sigma\iota, \quad \iota\iota = 1.$$

$\mathfrak{C}_v^+$ and $\mathfrak{C}_v^-$ are then those representations of $\mathfrak{u}'$ which coincide with $\mathfrak{C}_v$ for elements of the restricted group $\mathfrak{u}$ and which associate with the element ι the unit matrix $+\mathbf{1}$ and its negative $-\mathbf{1}$, respectively. The sign $\pm$ is again called the signature. The representation $\mathfrak{C}_2^-$ associates the augmented rotation group $\mathfrak{d}'_3$ with $\mathfrak{u}'$.

(c) *The Lorentz Group*

Let the 3-dimensional Euclidean space be referred to homogeneous projective co-ordinates x_α ($\alpha = 0, 1, 2, 3$) defined by

$$x = \frac{x_1}{x_0}, \quad y = \frac{x_2}{x_0}, \quad z = \frac{x_3}{x_0}.$$

The equation of the unit sphere is then

$$-x_0^2 + x_1^2 + x_2^2 + x_3^2 = 0 \tag{8.11}$$

and the formulæ for the stereographic projection considered above become

$$\left.\begin{aligned} x_0 &= \bar{\xi}\xi + \bar{\eta}\eta, & x_1 &= \bar{\xi}\eta + \bar{\eta}\xi \\ x_2 &= \frac{1}{i}(\bar{\xi}\eta - \bar{\eta}\xi), & x_3 &= \bar{\xi}\xi - \bar{\eta}\eta \end{aligned}\right\}. \tag{8.12}$$

On subjecting ξ, η to an *arbitrary* linear transformation σ the x_α undergo a corresponding *real* linear transformation s which leaves the equation (8.11) invariant. If the absolute value of the determinant of σ is 1, we can readily show that the form

$$-x_0^2 + x_1^2 + x_2^2 + x_3^2 \tag{8.13}$$

is itself invariant under the corresponding s, and that the determinant of s is $+1$.

We now consider $x_0 = ct$, x_1, x_2, x_3 as the co-ordinates of space-time; (8.11) is then the equation of the light-cone, the generators of which are the possible paths for a beam of light. In the restricted theory of relativity normal co-ordinate systems for space-time are connected with each other by arbitrary *Lorentz transformations*, i.e. by any real linear transformation which leaves the form (8.13) invariant and which does not interchange past and future. Lorentz transformations constitute a group, the "complete Lorentz group," and this group describes the homogeneity of the 4-dimensional world. This group consists of "*positive*" and "*negative*" transformations, i.e. transformations with determinants $+1$ and -1, respectively. The first constitute the "restricted Lorentz group," from which the complete group is obtained by introducing in addition the spatial reflection

$$x_0 \to x_0, \quad x_\alpha \to -x_\alpha \quad (\alpha = 1, 2, 3). \tag{8.14}$$

Under the restricted group right and left, as well as past and future, are fundamentally different. Since the expression for x_0 in (8.12) is positive definite, we may state the result obtained above in the form: *any linear transformation of ξ, η, with determinant of absolute value* 1, *induces a positive Lorentz transformation s in the x_α.* Transformations σ which differ only by a factor $e^{i\lambda}$ of absolute value 1 give rise to the same s. The correspondence $\sigma \to s$ is naturally a representation.

The question of whether *every* positive Lorentz transformation s can be obtained in this way arises immediately. That this is in fact the case can be seen from general continuity considerations, for the positive Lorentz transformations constitute a single connected continuum. But it is also easily proved by elementary methods. Since we have seen in (*b*) above that the rotations of space s are obtained from the unitary transformations σ, we need only to examine the Lorentz transformation

$$(x_0 + x_3) \to a^2(x_0 + x_3), \quad (x_0 - x_3) \to \frac{1}{a^2}(x_0 - x_3),$$
$$x_1 \to x_1, \quad x_2 \to x_2,$$

affecting the time axis, where a is a real non-vanishing constant. But this transformation is obtained from the unimodular σ:

$$\xi \to a\,\xi, \quad \eta \to \frac{1}{a}\eta.$$

Returning to the general case, the correspondence $s \to \sigma$ is a 2-dimensional representation of the restricted Lorentz group. But σ is determined by s only to within the arbitrary " gauge factor " $e^{i\lambda}$; we may therefore normalize it by the condition that the determinant of σ shall itself be unity, not merely its absolute value. Even so, σ remains double-valued, for $-\sigma$ satisfies the normalizing condition as well as σ. This representation $s \to \sigma$ contains the representation of the rotation group considered in (*b*) on allowing s to run through the subgroup of spatial rotations contained in the restricted Lorentz group.

The expressions (8.12) are Hermitian forms with matrices

$$S_0 = \begin{Vmatrix} 1 & 0 \\ 0 & 1 \end{Vmatrix}, \quad S_1 = \begin{Vmatrix} 0 & 1 \\ 1 & 0 \end{Vmatrix}, \quad S_2 = \begin{Vmatrix} 0 & -i \\ i & 0 \end{Vmatrix}, \quad S_3 = \begin{Vmatrix} 1 & 0 \\ 0 & -1 \end{Vmatrix} \tag{8.15}$$

Hence if $\mathfrak{x}$ denotes the one-columned matrix with elements ξ, η equations (8.12) may be written

$$x_\alpha = \tilde{\bar{\mathfrak{x}}}\, S_\alpha\, \mathfrak{x}. \tag{8.16}$$

On replacing ξ, η by $\bar{\eta}$, $-\bar{\xi}$ the x_α undergo the spatial reflection (8.14). That is *one* way of including the negative Lorentz transformations. But if we require that the corresponding transformation of ξ, η be linear, we must introduce in addition to $\mathfrak{x} = (\xi, \eta)$ a second pair $\mathfrak{x}' = (\xi', \eta')$ which undergoes the transformation σ' contragredient to $\bar{\sigma}$. Then

$$(\bar{\eta}, -\bar{\xi}) \sim (\xi', \eta') \text{ to within the factor } \bar{d},$$
$$(\eta, -\xi) \sim (\bar{\xi}', \bar{\eta}') \text{ to within the factor } d,$$

where d is the determinant of σ. Defining

$$S'_0 = S_0, \quad S'_\alpha = -S_\alpha \quad (\alpha = 1, 2, 3),$$

the quantities

$$x'_\alpha = \tilde{\bar{\mathfrak{x}}}' S'_\alpha\, \mathfrak{x}'$$

undergo the same transformation s as (8.16), provided the absolute value of the determinant of σ is 1. The same is true for any linear combination of the two, e.g. $x_\alpha + x'_\alpha$. Hence the quantities

$$x_\alpha = \tilde{\mathfrak{x}} S_\alpha \mathfrak{x} + \tilde{\mathfrak{x}}' S'_\alpha \mathfrak{x}' \tag{8.17}$$

undergo the given positive Lorentz transformation s when ξ, η are subjected to a certain transformation σ and simultaneously ξ', η' to the transformation σ' contragredient to $\bar{\sigma}$. *Furthermore, they undergo the transformation* (8.14) *on interchanging the two pairs* $\mathfrak{x}$, $\mathfrak{x}'$, i.e. on subjecting the four variables to the transformation

$$T: \xi \to \xi', \quad \eta \to \eta'; \quad \xi' \to \xi, \quad \eta' \to \eta. \tag{8.18}$$

The expression

$$\bar{\xi}\xi' + \bar{\eta}\eta'$$

is invariant in virtue of the transformation law of ξ', η' defined above. To obtain an expression which is also invariant under the interchange (8.18) we must add to the above the expression obtained from it by this interchange:

$$(\bar{\xi}\xi' + \bar{\eta}\eta') + (\bar{\xi}'\xi + \bar{\eta}'\eta). \tag{8.19}$$

It will be found advantageous to denote the column consisting of the four elements (ξ, η; ξ', η') by a single letter $\mathfrak{x}$. Let that linear transformation of these four variables which transforms ξ, η in accordance with S_α and ξ', η' in accordance with S'_α be denoted simply by S_α: (8.17) then becomes

$$x_\alpha = \tilde{\mathfrak{x}} S_\alpha \mathfrak{x}. \tag{8.16'}$$

We must now ask to what extent the linear transformation σ of the four variables $\mathfrak{x}$ is determined by the requirement that it induce a given (positive or negative) Lorentz transformation s of the Hermitian forms x_α. It suffices for this purpose to inquire what transformations of the $\mathfrak{x}$ induce the identity on the variables x_α. The only transformations of this latter kind are those which multiply ξ, η with a common factor $e^{i\lambda}$ of absolute value 1 and at the same time ξ', η' with any factor $e^{i\lambda'}$ (independent of the first) of absolute value 1. But σ can be more precisely specified by the requirement that (8.19), i.e. $\tilde{\mathfrak{x}} T \mathfrak{x}$, be also invariant. The two arbitrary "gauge factors" $e^{i\lambda}$, $e^{i\lambda'}$ must then coincide: *the substitution σ is then determined to within a factor $e^{i\lambda}$.*

Our analysis reduces the problem of the representations of the Lorentz group to the corresponding problem for the unimodular linear group $\mathfrak{c}_2$.

§ 9. Character of a Representation

The *trace* of a linear correspondence A, i.e. the sum of the elements in the principal diagonal of the matrix A, is an invariant under transformations of co-ordinates which is of particular importance. The trace $\chi(s)$ of the correspondence $U(s)$ associated with the element s of the group $\mathfrak{g}$ in a representation $\mathfrak{H}$ of $\mathfrak{g}$ is called the ***group characteristic***, or, in order to avoid assigning yet another meaning to this second word, which has already appeared in another important connection in quantum mechanics, simply the ***character*** *of the representation* $\mathfrak{H}$. *Equivalent representations have the same character*; the name is so chosen because the converse of this theorem is true within wide limits. Since $U(\mathbf{I}) = \mathbf{1}$, the value of the character $\chi(\mathbf{I})$ for the unit element is equal to the dimensionality of the representation.

It follows from the equations

$$U(asa^{-1}) = U(a)U(s)U(a^{-1}) = U(a)U(s)U^{-1}(a)$$

that the matrices $U(s)$ and $U(asa^{-1})$ differ only in their orientation and consequently have the same trace:

$$\chi(asa^{-1}) = \chi(s).$$

Now s and asa^{-1} are any two conjugate elements of the group $\mathfrak{g}$, i.e. they belong to the same *class* of conjugates in the sense of § 3. We speak of a function $f(s)$ on the group manifold which has the same value for all elements s belonging to the same class as a *class function*; such a function can at most allow us to distinguish between different classes, but not between elements of the same class. The distinguishing feature of class functions can also be expressed in the equation

$$f(st) = f(ts).$$

The validity of this equation for $f = \chi$ follows from

$$U(st) = U(s)U(t), \quad U(ts) = U(t)U(s)$$

and the fact that the trace of the matrix AB is equal to the trace of BA.

The character $\chi(s)$ of a unitary representation: $U(s^{-1}) = \tilde{U}^*(s)$, satisfies the equation

$$\chi(s^{-1}) = \bar{\chi}(s). \tag{9.1}$$

We shall say that the characters of irreducible representations are *primitive*. Any unitary representation $\mathfrak{H}$ can be reduced into its irreducible components, and the normal co-ordinate system in the corresponding sub-spaces can be so chosen that two irreducible constituents are equal if they are equivalent. If in this sense

$$\mathfrak{H} = m\mathfrak{h} + m'\mathfrak{h}' + \cdots, \tag{9.2}$$

where $\mathfrak{h}, \mathfrak{h}', \cdots$ are inequivalent irreducible representations and $m, m' \cdots$ are the numbers of times they occur in $\mathfrak{H}$, then the character X of $\mathfrak{H}$ is expressed in terms of the characters $\chi, \chi', \cdots$ of $\mathfrak{h}, \mathfrak{h}', \cdots$ by the equation

$$\mathrm{X}(s) = m\chi(s) + m'\chi'(s) + \cdots. \tag{9.3}$$

From an n-dimensional representation $\mathfrak{H} : s \to U(s)$, with the character $\chi(s)$, and an n'-dimensional $\mathfrak{H}' : s \to U'(s)$ of character $\chi'(s)$ we can construct the (nn')-dimensional representation $\mathfrak{H} \times \mathfrak{H}'$. The elements in the principal diagonal of $U(s) \times U'(s)$ are obtained by multiplying all elements in the principal diagonal of $U(s)$ by those in the principal diagonal of $U'(s)$: *the character of* $\mathfrak{H} \times \mathfrak{H}'$ *is consequently* $\chi(s)\,\chi'(s)$. Again, if $\mathfrak{H}$ is a representation of the group $\mathfrak{g}$, $\mathfrak{H}'$ a representation of the group $\mathfrak{g}'$, then the representation $\mathfrak{H} \times \mathfrak{H}'$ of $\mathfrak{g} \times \mathfrak{g}'$ has the character ζ defined by

$$\zeta(s, s') = \chi(s)\,\chi'(s'), \tag{9.4}$$

where s runs through the elements of $\mathfrak{g}$ and s' those of $\mathfrak{g}'$.

We need not distinguish between a 1-dimensional representation and its character; the character satisfies the simple equation (4.2). This holds, for example, for the characters $e(m\phi)$, eq. (8.2), of the rotation group $\mathfrak{d}_2$.

By the theorem on the transformation of unitary correspondences to principal axes, each element of the group $\mathfrak{u} = \mathfrak{u}_2$ is conjugate to a principal element, i.e. an element of the form

$$\begin{Vmatrix} \varepsilon & 0 \\ 0 & \frac{1}{\varepsilon} \end{Vmatrix}, \quad |\varepsilon| = 1 \tag{9.5}$$

The characteristic values ε, $1/\varepsilon$ are determined to within the order in which they appear. Introducing the angle ω by the equation $\varepsilon = e(\omega)$, ω characterizes a *class* of conjugate elements of $\mathfrak{u}$; we are only concerned with ω mod. 2π, and furthermore the class $-\omega$ coincides with the class ω. Since for any representation $\mathfrak{C}$ of $\mathfrak{u}$ the character $\chi(s)$ depends only on the class of the element s, it suffices to calculate it for elements of the form (9.5). It must be a periodic function of the angle ω with period 2π, and it must furthermore be an even function of ω; its value for $\mathfrak{C}_f$ is

$$\chi_f = \varepsilon^f + \varepsilon^{f-2} + \cdots + \varepsilon^{-f} = \frac{\varepsilon^{f+1} - \varepsilon^{-(f+1)}}{\varepsilon - \varepsilon^{-1}}. \tag{9.6}$$

The characters of the representations considered in the other examples of the preceding section are just as readily calculated.

§ 10. Schur's Lemma and Burnside's Theorem

Lemma (10.1).[5] *Assumption.* Let Σ be an irreducible system of linear correspondences of an m-dimensional vector space $\mathfrak{r}$ on to itself, and Ω such a system of an n-dimensional vector space $\mathfrak{S}$. A linear correspondence A shall satisfy the equation

$$\Sigma A = A\Omega \tag{10.2}$$

in the following double sense: for each U of Σ there shall exist a V of Ω such that

$$UA = AV, \tag{10.3}$$

and conversely for each V of Ω there shall exist a U of Σ such that this relation is fulfilled.

Assertion. Either $A = 0$ or $m = n$ and $\det A \neq 0$; in the latter case Σ and Ω are equivalent.

Proof. We first make use of the assumption that Σ is irreducible in connection with equation (10.2) in the first sense. Considering the kth column

$$a_{1k}, a_{2k}, \cdots, a_{mk}$$

of A as a vector $\mathfrak{a}^{(k)}$, equation (10.3) asserts that the vector $U\mathfrak{a}^{(k)}$ associated with $\mathfrak{a}^{(k)}$ through the correspondence U is a linear combination of the vectors $\mathfrak{a}^{(k)}$, specifically that

$$U\mathfrak{a}^{(k)} = \sum_h v_{hk}\,\mathfrak{a}^{(h)}, \quad V = \|v_{hk}\|.$$

Consequently the sub-space of $\mathfrak{r}$ spanned by the n vectors $\mathfrak{a}^{(k)}$ is invariant under Σ. But because of the assumption that Σ is irreducible either $\mathfrak{a}^{(k)} = 0$, $A = 0$, or the $\mathfrak{a}^{(k)}$ span the entire space $\mathfrak{r}$, in which case m of them are linearly independent; this latter is possible only if $n \geqq m$. That our conclusion contains two possibilities is due to the fact that the concept of irreducibility contains such an alternative.

The second part of the assumption can be given a simple geometrical interpretation on going over to the transposed matrices: Ω^* is irreducible and for each V^* of Ω^* there exists a U^* of Σ^* such that

$$V^*A^* = A^*U^*.$$

The reasoning employed in the first part of the theorem allows us to conclude: either $A^* = 0$ or $m \geqq n$. We summarize the results thus far obtained in the statement: Either $A = 0$ or

$m = n$; in the latter case the $m = n$ columns $\mathfrak{a}^{(k)}$ of A are linearly independent, i.e. the determinant of A does not vanish. But then U and V are determined uniquely by the relation (10.3) and Σ and Ω are equivalent.

In formulating these results it is desirable to consider the case of equivalence separately:

I. *If the two irreducible systems Σ, Ω are inequivalent,* (10.2) *can only be satisfied by $A = 0$.*

II. *If Σ is an irreducible system a correspondence A commutes with all correspondences U of the system Σ:*

$$UA = AU \tag{10.4}$$

if and only if A is a multiple of the unit matrix $\mathbf{1}$.

Assertion II follows from the lemma proved above by elementary methods and the fundamental theorem of algebra. For by the latter there exists a number α such that $\det(A - \alpha\mathbf{1}) = 0$, and since $A' = A - \alpha\mathbf{1}$ satisfies (10·4) for all U if A does, we conclude that since $\det A' = 0$ we must have $A' = 0$.

Applied to representations, our results are:

Fundamental Theorem (10.5). I. *If $s \to U(s)$, $s \to V(s)$ are two inequivalent irreducible representations of a group* $\mathfrak{g}$, *the equation*

$$U(s)A = AV(s)$$

can be satisfied by no matrix A which is independent of s, except $A = 0$.

II. *A matrix A which is independent of s and which satisfies the equation*

$$U(s)A = AU(s)$$

for all s is necessarily a multiple of the unit matrix $\mathbf{1}$.

If there exists a matrix A which satisfies $U(s)A = AU(s)$ identically in s and which is not merely a multiple of the unit matrix $\mathbf{1}$, the argument employed above supplies us with a constructive process for the reduction of the representation $s \to U(s)$ with the aid of A.

We now consider an application of these important results, which are fundamental for the entire theory of representations, in order to prove a theorem due to *Burnside*. Let Σ be a multiplicative system, i.e. if U, U' are two correspondences in Σ then the product UU' is also a correspondence in Σ. This concept is somewhat wider than that of a group; we need not require that U possess an inverse—its determinant may be 0.

Burnside's Theorem (10.6).[6] *In an irreducible multiplicative* *system Σ of linear correspondences $U = \|u_{ik}\|$ of an n dimensional vector space on to itself the components u_{ik} are linearly independent.*

This asserts that the only matrix L which satisfies the equation

$$\operatorname{tr}(UL) = \sum_{i,k} l_{ki} u_{ik} = 0$$

for all matrices U of the system is $L = 0$. Contrary to the assertion, we assume there exist non-vanishing matrices satisfying this equation; such matrices we shall call *L-matrices.* It is of course possible that every L-matrix whose first column

$$l_{11}, \quad l_{21}, \quad \cdots, \quad l_{n1}$$

vanishes must itself vanish. But in any case we can find a definite column index h with the following properties: there exist non-vanishing L-matrices whose first $h - 1$ columns vanish and are such that if the hth column also vanishes then necessarily $L = 0$. We shall call L-matrices whose first $h - 1$ columns vanish *special L-matrices.* They constitute a linear family of $m \leqq n$ dimensions; we denote a basis for this family by

$$L^{(1)}, \quad L^{(2)}, \quad \cdots, \quad L^{(m)}.$$

The hth column of a special L-matrix will be written $\mathfrak{l}$.

Since Σ is multiplicative the equation

$$\operatorname{tr}(U'UL) = 0$$

is satisfied by each L-matrix, where U, U' are arbitrary correspondences of the system Σ. With L, UL is also an L-matrix; obviously it is a special L-matrix if L is. Each of the matrices

$$UL^{(1)}, \quad UL^{(2)}, \quad \cdots, \quad UL^{(m)}$$

is therefore a linear combination of $L^{(1)}, \cdots, L^{(m)}$ and each of the vectors $U\mathfrak{l}^{(1)}, \cdots, U\mathfrak{l}^{(m)}$ is a linear combination of the vectors $\mathfrak{l}^{(1)}, \cdots, \mathfrak{l}^{(m)}$. Accordingly the vectors $\mathfrak{l}^{(1)}, \cdots, \mathfrak{l}^{(m)}$ span a non-vanishing sub-space which is invariant under all the correspondences U, and in consequence of the irreducibility assumed above it follows that $m = n$ and the vectors $\mathfrak{l}^{(1)}, \cdots, \mathfrak{l}^{(n)}$ span the entire n-dimensional space. The basis $L^{(1)}, \cdots, L^{(n)}$ of the family of special L-matrices can be chosen in such a way that $\mathfrak{l}^{(1)}, \cdots, \mathfrak{l}^{(n)}$ are the fundamental vectors of the space; $\mathfrak{l}^{(1)}$ is then the column $(1, 0, 0, \cdots, 0)$, etc. Since then

$$U\mathfrak{l}^{(r)} = u_{1r}\mathfrak{l}^{(1)} + \cdots + u_{nr}\mathfrak{l}^{(n)} \tag{10.7}$$

we must also have

$$UL^{(r)} = u_{1r}L^{(1)} + \cdots + u_{nr}L^{(n)}. \tag{10.8}$$

We now consider an arbitrary column, say the kth, of L. (This is of course of no interest if $k < h$, for the first $h - 1$ columns vanish.) Suppressing the second index k, we now let $\mathfrak{l} = (l_1, \cdots, l_n)$ denote the kth column of L. Then in accordance with (10.8), equation (10.7) holds for the present $\mathfrak{l}$, i.e. the kth instead of the hth column of L. Introducing for the moment the matrix

$$\Lambda = \begin{Vmatrix} l_1^{(1)} & \cdots & l_1^{(n)} \\ \cdots & \cdots & \cdots \\ l_n^{(1)} & \cdots & l_n^{(n)} \end{Vmatrix},$$

consisting of the kth columns of $L^{(1)}, \cdots, L^{(n)}$, we may write (10.7) as the matrix equation

$$U\Lambda = \Lambda U.$$

But it follows from this that Λ must be a multiple of the unit matrix, i.e.

$$l_i^{(r)} = \lambda \cdot \delta_i^r, \quad \delta_i^r = \begin{cases} 1 & (r = i) \\ 0 & (r \neq i) \end{cases};$$

or, returning to the original notation by adding the column index k,

$$l_{ik}^{(r)} = \lambda_k \cdot \delta_i^r.$$

Here we have, by the foregoing, $\lambda_1 = \cdots = \lambda_{h-1} = 0$, $\lambda_h = 1$.

The equation

$$\text{tr}\,(UL^{(r)}) = 0$$

becomes

$$\sum_{k=1}^{n} u_{kr}\lambda_k = 0, \quad (r = 1, \cdots, n), \tag{10.9}$$

i.e. all correspondences of the system Σ^* carry the vector λ with components $(\lambda_1, \lambda_2, \cdots, \lambda_n)$ over into the null-vector. In consequence of the irreducibility of Σ this vector must therefore vanish, which is in contradiction with the equation $\lambda_h = 1$; Burnside's theorem then follows by ***reductio ad absurdum***.—If we know that the unit matrix is contained in the system Σ, as is the case for a representation, we can conclude that $\lambda_i = 0$ by taking U in (10.9) as the unit matrix.

Reducibility requires that on employing an appropriate co-ordinate system all matrices U of the system Σ have an entire rectangle of vanishing elements and consequently implies a *system* of homogeneous linear relations between the components u_{ik} of a very special kind. Burnside's theorem states that if there exists no system of homogeneous linear relations of this special kind, then there exists no linear dependence at all. The real reason for this remarkable fact is of course to be found in the assumption that Σ is closed with respect to multiplication.

If our system Σ consists of an irreducible representation which associates with the elements s of the group $\mathfrak{g}$ the matrix $U(s)$, we see from Burnside's theorem that the components of $U(s)$ are linearly independent. The method developed above can readily be extended to prove the same for the components of two or more inequivalent irreducible representations $U(s)$, $U'(s)$, $\cdots$.[7] From this it follows that in particular *there can exist no linear dependences between their characters* $\chi(s)$, $\chi'(s)$, $\cdots$. Any unitary representation $\mathfrak{H}$ can be reduced into irreducible components; the character of $\mathfrak{H}$ is expressed in terms of the characters of these irreducible representations by (9.3). Since $\chi(s)$, $\chi'(s)$ are linearly independent the coefficients m, m', $\cdots$, which give the number of times the irreducible representations $\mathfrak{h}$, $\mathfrak{h}'$, $\cdots$ appear in $\mathfrak{H}$, are uniquely determined. This constitutes a new indirect proof of the following result, which has already been proved in § 6 in a more general and more elementary way: *The irreducible representations into which $\mathfrak{H}$ can be reduced, as well as the number of times they occur, are uniquely determined by $\mathfrak{H}$, no distinction being made between equivalent representations.* Two unitary representations $\mathfrak{H}_1$ and $\mathfrak{H}_2$ are obviously equivalent if every irreducible representation which is contained in the one is contained in the other the same number of times. Hence if $\mathfrak{H}_1$ and $\mathfrak{H}_2$ are inequivalent the character of $\mathfrak{H}_1$ cannot be the same as the character of $\mathfrak{H}_2$ because of the linear independence of the primitive characters: *a unitary representation is uniquely determined by its character alone*, and its character may be used as a unique name for the representation itself. We here go no further into these extensions of Burnside's theorem, which are due to *Frobenius* and *I. Schur*, as we shall obtain the same results by a more profound method in the next section under assumptions which are more restrictive but which are sufficient for our purposes.

We mention only one consequence. $\mathfrak{H}$, $\mathfrak{H}'$ being representations of the groups $\mathfrak{g}$, $\mathfrak{g}'$, respectively, then $\mathfrak{H} \times \mathfrak{H}'$ is an irreducible representation of $\mathfrak{g} \times \mathfrak{g}'$. Indeed, there can exist no homogeneous linear relation with constant coefficients $c_{ik,\iota\kappa}$ between the components $u_{ik}(s)u'_{\iota\kappa}(s')$ of $U(s) \times U'(s')$ except the trivial one $c = 0$. For on applying Burnside's theorem for the irreducible system $\mathfrak{H}$ we have

$$\sum_{\iota,\kappa} c_{ik,\iota\kappa}\, u'_{\iota\kappa}(s') = 0,$$

and on applying it again for $\mathfrak{H}'$ we must have $c_{ik,\iota\kappa} = 0$.

§ 11. Orthogonality Properties of Group Characters

If the abstract group $\mathfrak{g}$ *is finite, then any representation* $\mathfrak{H} : s \rightarrow U(s)$ *is equivalent to a unitary one.* To show this take any positive definite Hermitian form, e.g. the unit form, subject it to all transformations $U(s)$ of $\mathfrak{H}$ and sum over s. We thus obtain a positive definite Hermitian form H which is invariant under each of the transformations $U(s)$. Now choose the co-ordinate system in such a way that H becomes the unit form; then $U(s)$, expressed in terms of these co-ordinates, is unitary. This same method of summation over the elements of the group gives rise to the fundamental orthogonality relations.

Let $\mathfrak{H} : s \rightarrow U(s)$, $\mathfrak{H}' : s \rightarrow U'(s)$ be two inequivalent irreducible representations of the finite group $\mathfrak{g}$, the former being g-dimensional and the latter g'-dimensional. We write

$$U(s) = ||u_{ik}(s)||, \quad U'(s) = ||u'_{\iota\kappa}(s)||,$$
$$U'^{-1}(s) = ||\check{u}'_{\iota\kappa}(s)||.$$

For a unitary representation $\mathfrak{H}'$

$$\check{u}'_{\kappa\iota}(s) = \bar{u}'_{\iota\kappa}(s).$$

If A is an arbitrary matrix with g rows and g' columns then obviously the sum

$$\sum_t U(t)AU'^{-1}(t) = B, \tag{11.1}$$

taken over all elements t of $\mathfrak{g}$, is invariant in the sense that

$$U(s)BU'^{-1}(s) = B. \tag{11.2}$$

In fact, the left-hand side of (11.2) becomes, in virtue of the fact that $s \rightarrow U(s)$ is a representation of $\mathfrak{g}$,

$$\sum_\tau U(\tau)AU'^{-1}(\tau),$$

where $\tau = st$, s being fixed and t running through all elements of the group. We therefore obtain equation (11.2) or

$$U(s)B = BU'(s).$$

In accordance with the fundamental theorem (10.5) it follows from this that $B = 0$, i.e.

$$\sum_t \sum_{k,\kappa} u_{ik}(t)a_{k\kappa}\check{u}'_{\kappa\iota}(t) = 0.$$

Writing s in place of t and remembering that the $a_{k\kappa}$ are arbitrary numbers, we obtain the $g^2 \cdot g'^2$ equations

$$\sum_s u_{ik}(s)\check{u}'_{\kappa\iota}(s) = 0,$$

or, in dealing with unitary representations,

$$\sum_s u_{ik}(s)\bar{u}'_{\iota\kappa}(s) = 0. \tag{11.3}$$

Taking the single irreducible representation $s \to U(s)$ instead of the two inequivalent representations $\mathfrak{H}$, $\mathfrak{H}'$, we find by the same argument that the square matrix

$$U(s)AU^{-1}(s) = B,$$

found from an arbitrary square matrix A, must satisfy the

$$U(s)B = BU(s).$$

This requires, however, that B be a multiple of the unit matrix $\mathbf{1}$, i.e.

$$\sum_s \sum_{k,\kappa} u_{ik}(s) a_{k\kappa} \breve{u}_{\kappa\iota}(s) = \alpha \cdot \delta_{i\iota}.$$

the number α depends on the matrix A, the dependence being of course linear and homogeneous. Taking as A that matrix which has as its only non-vanishing element $a_{k\kappa} = 1$, we obtain the equation

$$\sum_s u_{ik}(s)\breve{u}_{\kappa\iota}(s) = \alpha_{\kappa k} \delta_{i\iota}. \tag{11.4}$$

Now $||\breve{u}_{\iota\kappa}(s)||$ is the matrix reciprocal to $||u_{\iota\kappa}(s)||$:

$$\sum_s \breve{u}_{\kappa i}(s) u_{ik}(s) = \delta_{\kappa k}.$$

On taking $\iota = i$ in (11.4) and summing over $i = 1, 2, \cdots, g$ we find that

$$h \cdot \delta_{\kappa k} = g\alpha_{\kappa k},$$

where h is the order of the group $\mathfrak{g}$.

Expressing the sum $\sum_s$ in terms of the mean value $\mathfrak{M} = \frac{1}{h} \sum_s$, our results may be written in the form

$$\mathfrak{M}\{u_{ik}(s)\bar{u}_{\iota\kappa}(s)\} = \begin{cases} \frac{1}{g} & \text{for } i = \iota,\ k = \kappa \\ 0 & \text{otherwise} \end{cases} \tag{11.5}$$

for any irreducible unitary representation $\mathfrak{H} : s \to U(s)$ and

$$\mathfrak{M}\{u_{ik}(s)\bar{u}'_{\iota\kappa}(s)\} = 0 \tag{11.6}$$

for any two inequivalent irreducible unitary representations $s \to U(s)$, $s \to U'(s)$. *The components of one or more inequivalent irreducible unitary representations constitute a unitary-orthogonal set of functions on the group manifold.*

It follows from these fundamental orthogonality relations that *the components* $u_{ik}(s)$, $u'_{\iota\kappa}(s)$, $\cdots$ *are linearly independent.* Since the number of linearly independent functions of an argument s which assumes but h values cannot be greater than h we must have

$$g^2 + g'^2 + \cdots \leqq h.$$

On the left-hand side of this equation occur the squares of the degrees of any inequivalent irreducible representation of $\mathfrak{g}$.

We obtain the orthogonality properties of the characters by writing $k = i$, $\kappa = \iota$ in (11.5), (11.6) and summing over these indices :

Any primitive character satisfies the equation

$$\mathfrak{M}\{\chi(s)\bar{\chi}(s)\} = 1, \tag{11.7}$$

and the characters $\chi(s)$, $\chi'(s)$ *of any two inequivalent irreducible representations satisfy*

$$\mathfrak{M}\{\chi'(s)\bar{\chi}(s)\} = 0. \tag{11.7'}$$

The primitive characters of inequivalent representations constitute a normal orthogonal set of functions. They are consequently linearly independent, and from this follow all the consequences discussed in the previous section. In particular, a representation of $\mathfrak{g}$ can be unambiguously described by its character, no distinction being made between equivalent representations. The number of times m the irreducible χ occurs in the representation $\mathbf{X}$ is, following (9.3), given by

$$m = \mathfrak{M}\{\mathbf{X}(s)\bar{\chi}(s)\}, \tag{11.8}$$

and we have

$$\mathfrak{M}\{\mathbf{X}(s)\bar{\mathbf{X}}(s)\} = m^2 + m'^2 + \cdot \cdot \cdot.$$

This last equation offers a simple criterion for the irreducibility of a given representation in terms of its character χ : *it is necessary and sufficient that the mean value of* $\chi\bar{\chi} = |\chi|^2$*—which is in any case integral—be unity.*

Since the characters are class functions we are in dealing with them concerned with an argument which runs through the K different classes of $\mathfrak{g}$; there can therefore be no more than K linearly independent class functions. *Hence a finite group can have no more inequivalent irreducible representations than classes.*

Whereas the general concept of a representation seemed at first to open up limitless possibilities, we now see that all representations are constructed from primitive ones and that the number of possible primitive representations is confined within narrow limits. The further content of the general theory of representations can be stated in the theorem that *the sets of functions, the orthogonality of which we have shown above, are complete orthogonal systems.* The primitive characters constitute a complete orthogonal system in the domain of class functions, i.e. there exist exactly K inequivalent irreducible representations. The components of a complete system of K

inequivalent irreducible representations constitute a complete orthogonal system for the totality of functions defined on the group manifold, or

$$h = g^2 + g'^2 + \cdots,$$

where the sum on the right is extended over such a complete system and $g, g', \cdots$ are the dimensionalities of the individual irreducible representations.

§ 12. Extension to Closed Continuous Groups

The theory developed in the preceding sections cannot be extended to arbitrary groups, but it is applicable *mutatis mutandis* to a group whose elements constitute a *continuous closed manifold of a finite number of dimensions.* Just as the immediate neighbourhood of a point on a surface constitutes a plane, so the immediate neighbourhood of a point p_0 on an r-dimensional continuous manifold constitutes an r-dimensional *linear* manifold and the line elements from p_0 to neighbouring points p define an r-dimensional linear vector space. We assume that the infinitesimal elements of our group $\mathfrak{g}$ (i.e. those elements in the neighbourhood of the unit element $\mathbf{I}$), or rather the infinitesimal vectors leading to them from $\mathbf{I}$, constitute such an r-dimensional vector space, the "tangential space" to $\mathfrak{g}$ at $\mathbf{I}$. The concept of an infinitesimal rotation will be familiar to the reader from the kinematics of rigid bodies, as well as the fact that these infinitesimal rotations in 3-dimensional space constitute a 3-dimensional linear family—in n-dimensional space an $[n(n-1)/2]$-dimensional family. The multiplication of two infinitesimal elements of the group is then expressed by the *addition* of the corresponding vectorial line elements in the tangential space.

A parallelepiped which will serve as a volume element in the neighbourhood of $\mathbf{I}$ is defined by r linearly independent line elements, and its volume is given as usual by the absolute value of the determinant of the components of these r vectors. This volume element is, of course, not entirely independent of the choice of a co-ordinate system in the tangential space, but the transformation to a new co-ordinate system only multiplies the volumes of all such elemental volumes in the neighbourhood of $\mathbf{I}$ by a constant numerical factor. These volumes are therefore determined to within the choice of a unit of measure; more than this we can hardly require.

On extending the theory developed in the preceding section to continuous groups integration replaces summation, and it is therefore necessary to be able to measure volumes on the entire group manifold of $\mathfrak{g}$. With the aid of the foregoing volume elements in the neighbourhood of $\mathbf{I}$ can be measured and compared immediately with each other, and the same is true for the volume elements at any other point of the group manifold. The only difficulty lies in carrying the unit of volume from the point $\mathbf{I}$ to any other point a. Examination of the argument of § 11 reveals that the measurement of volume must have the following invariantive properties: the volume of an arbitrary element must be unaltered by a left-translation of the group manifold which transforms the general element t into $\tau = at$. But this requirement just suffices to specify the process uniquely. Consider the volume element at a which arises from an elemental volume at $\mathbf{I}$ by the left-translation which throws $\mathbf{I}$ into a; *per definitionem the volumes of these two elements shall be the same.* On carrying the volume element from a to b by means of the translation $t' = (ba^{-1})t$ the equation $t' = b(a^{-1}t)$ shows that with this definition of volume the volumes of the elements so obtained at a and b are equal.

We further assume that our continuous group manifold is *closed*—in the sense, for example, that the surface of a sphere is a closed manifold in contrast with a Euclidean plane, which is open. This guarantees that we shall be able to integrate continuous functions of position on the group manifold over the entire manifold. We now choose the unit of volume in such a way that the volume of the entire manifold $\mathfrak{g}$ is 1; the integrals are then mean values. We naturally require that the components of $U(s)$ in a representation $s \rightarrow U(s)$ are continuous functions of the element s of $\mathfrak{g}$. *The laws* (11.5), (11.6), (11.7), (11.7′) *and all consequences obtained from them in* § 11 *are then valid for irreducible representations of the continuous group* $\mathfrak{g}$ *and their characters.*[8]

The theory would be extraordinarily restricted if *the measure of volume, which we have introduced in such a way that it is invariant under left-translations, were not automatically invariant under* (1) *right-handed translations:* $s \rightarrow s' = sa$ *and* (2) *inversion:* $s \rightarrow s' = s^{-1}$. The first of these properties will be established by showing that the volume of a volume element at $\mathbf{I}$ is unchanged on taking it to a by a left-translation and returning it to $\mathbf{I}$ by a right-translation. Obviously each infinitesimal element δs of the group then undergoes the linear transformation A:

$$\delta s \rightarrow \delta' s = a \cdot \delta s \cdot a^{-1},$$

i.e. the conjugation $\mathfrak{k}_a$ associated with the element a. Such linear transformations in the r-dimensional vector-space of the infinitesimal elements of the group constitute a representation $a \to A$ of the abstract group $\mathfrak{g}$. *Since* $\mathfrak{g}$ *is closed, each A must be "absolute-unimodular,"* i.e. the determinant of A must have the absolute value 1; and this in turn allows us to conclude that the definition of transportation of volumes by either left- or right-translations leads to the same result. To prove this consider the element a and its powers $a^2, a^3, \cdots$. Since the group manifold $\mathfrak{g}$ is closed, the infinite set $a, a^2, a^3, \cdots$ on $\mathfrak{g}$ possesses a point of condensation b, i.e. an infinite set of exponents n can be found such that as n runs through this set a^n converges to b. To the elements a^n and b correspond the conjugations A^n and B, respectively, and in virtue of the continuity assumed above det (A^n) converges to det (B) as n runs through the chosen set. Now since det (B) is a finite non-vanishing number, and since, if the absolute value of the determinant of A differed from 1, det (A^n) would tend toward 0 or ∞, we may conclude the truth of the above assertion. This also enables us to prove the truth of (2), invariance under inversion. For inversion sends the element δs at $\mathfrak{l}$ into $-\delta s$, and this transformation is absolute-unimodular. Now send one of two inverse volume elements at $\mathfrak{l}$ to a by a left-translation and the other to a^{-1} by a right-translation; we thus obtain volume elements at a and a^{-1} which go into each other by the inversion $s \to s' = s^{-1}$. Since both left- and right-translations conserve volumes, these two volume elements have the same volume.

Examples of the Orthogonality Properties

We have already found the primitive characters for the *group of rotations* $\mathfrak{d}_2$ *of a circle into itself*: $e(m\phi)$, $m = 0, \pm 1, \pm 2, \cdots$, where ϕ is the angle of rotation. They constitute, in fact, a unitary-orthogonal set of functions:

$$\int_0^{2\pi} e(m\phi)\,\bar{e}(m'\phi)\,d\phi = \begin{cases} 2\pi & (m = m') \\ 0 & (m \neq m') \end{cases}.$$

If there existed further irreducible representations their characters would necessarily be orthogonal to all of these; but this is impossible, for the functions $e(m\phi)$, where m takes on all integral values, already constitute a complete orthogonal system. We have, however, already shown by a more direct method (§ 8), which did not involve Parseval's equation, that the system of primitive characters $e(m\phi)$ was complete. It is therefore natural to consider Parseval's equation as the simplest case of the general group-theoretic completeness theorem mentioned in § 11.

The character of the representation $\mathfrak{C}_f$ of the *2-dimensional unitary unimodular group* $\mathfrak{u} = \mathfrak{u}_2$ is given by (9.6). Writing

$$\varepsilon = e(\omega), \quad \Delta = \varepsilon - \varepsilon^{-1} = 2i \sin \omega, \quad \frac{1}{2\pi} \Delta\bar{\Delta} d\omega = d\sigma,$$

we have

$$\int_{\omega=0}^{\pi} \chi_f \chi_g \, d\sigma = \begin{cases} 1 & (f = g) \\ 0 & (f \neq g) \end{cases}. \tag{12.1}$$

This leads us to suspect that $d\sigma$ is the volume of that portion of the group manifold occupied by those elements σ of the group whose angles of rotation lie between ω and $\omega + d\omega$. [The total volume of the group manifold is then

$$\frac{1}{2\pi} \int_0^{\pi} \Delta\bar{\Delta} d\omega = 1.]$$

If this is correct, (12.1) are the orthogonality relations predicted by the general theory, and the equation

$$d\sigma = \frac{1}{2\pi} \Delta\bar{\Delta} d\omega$$

defines the density of the various classes of the group. In the last chapter we shall actually carry through the determination of volume and verify these results.

If there were yet another irreducible representation, with character χ, then $\xi = \Delta \cdot \chi$ would be an odd periodic function of ω with period 2π which would be orthogonal to all the functions $\xi_f = \Delta \cdot \chi_f$, i.e. to the functions

$$\sin \omega, \quad \sin 2\omega, \quad \sin 3\omega, \quad \cdots.$$

But these latter are already a complete orthogonal set for the domain of odd periodic functions, and consequently *the* $\mathfrak{C}_f$ $(f = 0, 1, 2, \cdots)$ *constitute a complete system of irreducible representations of the group* $\mathfrak{u}$. A direct proof, which is independent of Parseval's equation, is also to be found in Chap. V, § 16—indeed, it is there carried through for $\mathfrak{u}_n$ in an arbitrary number n of dimensions.

The Clebsch-Gordan series

$$\chi_f \chi_g = \chi_{f+g} + \chi_{f+g-2} + \cdots + \chi_{|f-g|} \tag{12.2}$$

for the characters χ_f is readily verified. If we know on general grounds that the character of a representation specifies it uniquely, this equation can be used as a proof of the reducibility of $\mathfrak{C}_f \times \mathfrak{C}_g$ into irreducible components with characters as on the right. Since the characters are much more readily handled than the representations themselves this principle offers a very powerful method for obtaining assertions concerning representations. Let $f \geqq g$ and multiply equation (12.2), which is to be verified, by Δ:

$$\xi_f \chi_g = \sum_v \xi_v \quad (v = f + g, f + g - 2, \cdots, f - g).$$

The product of

$$\xi_f = \varepsilon^{f+1} - \varepsilon^{-(f+1)} \quad \text{with} \quad \chi_g = \varepsilon^g + \varepsilon^{g-2} + \cdots + \varepsilon^{-g}$$

is the difference of two sums; the one is

$$\varepsilon^{f+g+1} + \varepsilon^{f+g-1} + \cdots + \varepsilon^{f+1-g},$$

the exponent decreasing by 2 from term to term, and the other is obtained from this one by replacing all exponents by their negative. Hence the product is in fact

$$\sum \{\varepsilon^{2v+1} - \varepsilon^{-(2v+1)}\}, \quad v = f + g, f + g - 2, \cdots, f - g.$$

The representations $\mathfrak{C}_f^+$, $\mathfrak{C}_f^-$ $(f = 0, 1, 2, \cdots)$ constitute a complete set of inequivalent irreducible representations of the augmented group $\mathfrak{u}_2'$. To establish this we first note that in an irreducible representation of $\mathfrak{u}'$ the matrix associated with the element ι must be a multiple of the unit matrix, for it commutes with the irreducible system of matrices constituting the representation. Furthermore, $\iota\iota = \mathsf{I}$, so this matrix can only be $+1$ or -1. Since the matrix associated with ι is a multiple of the unit matrix, and since the extension of $\mathfrak{u}$ to $\mathfrak{u}'$ involves the addition of a *single* element ι, the representation must remain irreducible on restricting the group $\mathfrak{u}'$ to the sub-group $\mathfrak{u}$. Hence every irreducible representation of $\mathfrak{u}_2'$ is obtained by supplementing the irreducible representations of $\mathfrak{u}_2$ by the association

$$\iota \to +1 \quad \text{or} \quad \iota \to -1.$$

If $\mathfrak{H}$, $\mathfrak{H}'$ run independently through complete systems of inequivalent irreducible representations of the two (finite or closed continuous) groups $\mathfrak{g}$, $\mathfrak{g}'$, respectively, then the $\mathfrak{H} \times \mathfrak{H}'$ constitute a complete system of inequivalent irreducible representations for the direct product $\mathfrak{g} \times \mathfrak{g}'$. To prove this we note that since the primitive characters $\chi(s)$ of $\mathfrak{g}$ constitute a complete orthogonal system for class functions of the element s which runs through $\mathfrak{g}$ and the primitive characters $\chi'(s')$ of $\mathfrak{g}'$ do the same for $\mathfrak{g}'$, the totality of the products $\chi(s) \cdot \chi'(s')$ constitute a complete orthogonal system for the class functions of the element (s, s') which runs through the group $\mathfrak{g} \times \mathfrak{g}'$.

The representations $\mathfrak{C}_{f,g}$ introduced in § 5 constitute a complete system of irreducible representations of $\mathfrak{c}_2$ when f, g run independently through the numbers 0, 1, 2, · · ·; we here only mention this fact without going further into it.

§ 13. The Algebra of a Group

We return for the present to finite groups. In order to be able to express the completeness theorem we associate with each function $x(s)$ on the group manifold of the finite group $\mathfrak{g}$ its " Fourier coefficient matrix," the ***group matrix***.

$$X = \sum_s x(s) U(s), \tag{13.1}$$

where $\mathfrak{H} : s \to U(s)$ is a representation of $\mathfrak{g}$. The trace of X,

$$\xi = \sum_s x(s)\chi(s), \tag{13.2}$$

is the Fourier coefficient of $x(s)$ with respect to the character $\chi(s)$ of $\mathfrak{H}$. It is here desirable to consider the function $x(s)$ as a *single quantity* $\boldsymbol{x}$ *in the group domain ;* each element s of the group is a dimension in " group space " and the number $x(s)$ is the s-component of the quantity $\boldsymbol{x}$. We may express the quantities themselves symbolically in the form

$$\boldsymbol{x} = \sum_s x(s) \cdot \boldsymbol{s}. \tag{13.3}$$

The matrix X is associated with the quantity $\boldsymbol{x}$ in the representation $\mathfrak{H} : \boldsymbol{x} \to X$ in $\mathfrak{H}$. *Addition of " group quantities " and multiplication of them by a number* are introduced in the usual way : $\boldsymbol{x} + \boldsymbol{y}$ has the components $x(s) + y(s)$ and $\alpha\boldsymbol{x}$ the components $\alpha \cdot x(s)$. Group quantities consequently behave like vectors in an h-dimensional space, where h is the order of the group. The following definition of *multiplication of two arbitrary group quantities* $\boldsymbol{x}$ *and* $\boldsymbol{y}$ is suggested by (13.3) :

$$\boldsymbol{z} = \boldsymbol{xy} = \sum_{t, t'} x(t)y(t')\boldsymbol{tt'} = \sum_s z(s) \cdot \boldsymbol{s}$$

where

$$z(s) = \sum_{tt'=s} x(t)y(t'). \tag{13.4}$$

This last equation, in which the sum is to be extended over all pairs of elements t, t' whose product is s, defines the product $\boldsymbol{z}$ of the quantities $\boldsymbol{x}$ and $\boldsymbol{y}$. We denote this product by $\boldsymbol{xy}$ and its components by $xy(s)$; this is not to be confused with $x(s) \cdot y(s)$, the ordinary product of the two numbers $x(s)$, $y(s)$. Addition and multiplication of group quantities parallel addition and multiplication of the group matrices associated with them by (13.1). Indeed, the product of

$$X = \sum_s x(s)U(s), \quad Y = \sum_s y(s)U(s)$$

is given by

$$Z = XY = \sum_{t,t'} x(t)y(t')U(tt') = \sum_s z(s)U(s),$$

where $z(s)$ is defined by (13.4)

The operations to which the group quantities may be subjected: (1) addition, (2) multiplication with a number, and (3) multiplication with one another, satisfy the usual laws of ordinary algebra with two important exceptions: *multiplication is not commutative* and *division is not in general possible*, i.e. the equation $\boldsymbol{ax} = \boldsymbol{b}$ for given $\boldsymbol{a} \neq 0$ and $\boldsymbol{b}$ may have no unique solution or even no solution at all. But there does exist a quantity $\mathbf{1}$ having the properties of unity: $\mathbf{1a} = \boldsymbol{a}\mathbf{1} = \boldsymbol{a}$ for every quantity $\boldsymbol{a}$; its components all vanish with the exception of the one associated with $s = \mathbf{I}$, which is 1. A domain of quantities as described above is called an ***algebra***,[9] and the "group quantities" are the ***elements*** *of the algebra;* care must be taken not to confuse these with the *elements of the group* (cf. V, § 5). The association $\boldsymbol{x} \to X$ in the representation $\mathfrak{H}$ satisfies the conditions:

1. $\mathbf{1} \to \mathbf{1}$, to the element $\mathbf{1}$ corresponds the unit matrix $\mathbf{1}$;
2. if $\boldsymbol{x} \to X$, $\boldsymbol{y} \to Y$ and α is a number, then

$$\boldsymbol{x} + \boldsymbol{y} \to X + Y, \quad \alpha\boldsymbol{x} \to \alpha X, \quad \boldsymbol{xy} \to XY.$$

A representation $\mathfrak{H}$ of the group is the same as *a realization or* "***representation***" *of the algebra of the group by matrices* such that these conditions are satisfied. Actually all we have done here is this: we have gone over from the matrices $U(s)$ associated with the individual elements of the group to the linear manifold of matrices for which they constitute a basis.

What characterizes an element $\boldsymbol{a}$ of the algebra whose components $a(s)$ define a class function? We have in general

$$ax(s) = \sum_t a(st)x(t^{-1}), \quad xa(s) = \sum_t a(ts)x(t^{-1}),$$

and a class function satisfies the equation

$$a(st) = a(ts).$$

Hence such an $\boldsymbol{a}$ is characterized by the fact that it commutes with all elements $\boldsymbol{x}$ of the algebra: $\boldsymbol{ax} = \boldsymbol{xa}$. Employing a term carried over from group theory to algebra we may say: *those elements whose components depend only on the class of conjugate group elements to which the argument s belongs constitute the* **central** *of the algebra.*

We are interested only in unitary representations $s \to U(s)$. For such a representation the Hermitian conjugate of (13.1) is

$$\tilde{X} = \sum_s \bar{x}(s)\tilde{U}(s) = \sum_s \bar{x}(s)U(s^{-1}) = \sum_s \bar{x}(s^{-1})U(s).$$

Hence on defining the conjugate $\tilde{\boldsymbol{x}}$ of the element $\boldsymbol{x}$ by $\tilde{x}(s) = \bar{x}(s^{-1})$, Hermitian conjugate matrices are associated with conjugate elements in a unitary representation; this characterizes unitary representations. An element will be said to be real if it coincides with its conjugate. We have seen that the character $\chi(s)$ of a unitary representation satisfies this condition $\chi(s) = \bar{\chi}(s^{-1})$.

Let $\mathfrak{H}$ be a g-dimensional irreducible unitary representation of $\mathfrak{g}$. $C = ||c_{ik}||$ being a given g-dimensional matrix, the element $\boldsymbol{c}$ of the algebra defined by

$$c(s) = \sum_{ik} c_{ik} \cdot \frac{g}{h}\bar{u}_{ik}(s) = \frac{g}{h}\,\mathrm{tr}\,[C\tilde{U}(s)]$$

is such that $\boldsymbol{c} \to C$ in $\mathfrak{H}$; this is readily verified with the aid of the orthogonality relations. *Hence in the correspondence* $\boldsymbol{x} \to X$ *X runs through all g-dimensional matrices.* We denote the quantity with components $\frac{g}{h}\,\bar{u}_{ik}(s)$ by $\boldsymbol{e}_{ik}$. The set H of all elements of the form

$$\sum_{i,k} c_{ik}\,\boldsymbol{e}_{ik},$$

where the coefficients c_{ik} are arbitrary, is naturally closed with respect to the operations of addition and multiplication by a number. But the product of two elements in H is again an element in H; indeed, if $\boldsymbol{c}$ is in H and $\boldsymbol{x}$ is an *arbitrary* element of the algebra both $\boldsymbol{cx}$ and $\boldsymbol{xc}$ are also in H. We express this situation in a terminology paralleling that of the theory of groups: H *is an invariant sub-algebra of the algebra Γ of all group quantities.* To prove these assertions we first note that the definition (13.1), together with the condition that $s \to U(s)$ be a representation, yields the equation

$$XU(s^{-1}) = \sum_t x(t)U(ts^{-1}),$$

or, on replacing $U(s^{-1})$ by $\tilde{U}(s)$,

$$X\tilde{U}(s) = \sum_t \tilde{U}(st^{-1})x(t). \tag{13.5}$$

Multiplying on the left by $C = ||c_{ik}||$ and constructing the trace we find

$$\frac{g}{h}\operatorname{tr}[(CX)\tilde{U}(s)] = \sum_t c(st^{-1})x(t) = cx(s),$$

whence $\boldsymbol{y} = \boldsymbol{cx}$ is in H :

$$\boldsymbol{cx} = \sum_{i,k} y_{ik}\,\boldsymbol{e}_{ik} \tag{13.6}$$

and the matrix

$$||y_{ik}|| = CX. \tag{13.7}$$

In the same way we can show that if $\boldsymbol{c}$ belongs to H then $\boldsymbol{xc}$ does also. If

$$\boldsymbol{x} \to X = ||x_{ik}|| \text{ in } \mathfrak{H}$$

we call

$$\sum x_{ik}\,\boldsymbol{e}_{ik}$$

the component of $\boldsymbol{x}$ *in* H. In accordance with (13.6), (13.7) this component is the product of $\boldsymbol{x}$ with

$$\boldsymbol{\varepsilon} = \boldsymbol{e}_{11} + \boldsymbol{e}_{22} + \cdots + \boldsymbol{e}_{gg};$$

it is $\boldsymbol{\varepsilon x} = \boldsymbol{x\varepsilon}$. $\boldsymbol{\varepsilon}$ is a real element belonging to the central of the group algebra Γ, with components $\frac{g}{h}\cdot\bar{\chi}(s)$; it is "*idempotent*," i.e. it satisfies the equation $\boldsymbol{\varepsilon\varepsilon} = \boldsymbol{\varepsilon}$. In particular, the product of two elements

$$\boldsymbol{a} = \sum a_{ik}\,\boldsymbol{e}_{ik}, \quad \boldsymbol{b} = \sum b_{ik}\,\boldsymbol{e}_{ik}$$

of H with coefficient matrices A, B, is the quantity $\boldsymbol{ab}$ in H with the coefficient matrix AB. $\boldsymbol{\varepsilon}$ is the **1**, the "*modulus*," or "*principal unit*," of the sub-algebra H since $\boldsymbol{\varepsilon x} = \boldsymbol{x\varepsilon} = \boldsymbol{x}$ when $\boldsymbol{x}$ is in H. The algebra H is identical with the algebra of all g-dimensional matrices ("*simple matric algebra*"). The "units" $\boldsymbol{e}_{ik}$ satisfy the equations

$$\boldsymbol{e}_{ir}\,\boldsymbol{e}_{rk} = \boldsymbol{e}_{ik}, \quad \boldsymbol{e}_{ir}\,\boldsymbol{e}_{sk} = 0 \text{ for } r \neq s. \tag{13.8}$$

The central of the sub-algebra H consists only of the multiples of its modulus $\boldsymbol{\varepsilon}$.

An irreducible representation $\mathfrak{H}' : s \to U'(s) = ||\,u'_{\iota\kappa}(s)\,||$ of dimensionality g' which is not equivalent to $\mathfrak{H}$ yields another invariant sub-algebra H′ consisting of all elements of the form

$$\boldsymbol{c}' = \sum_{\iota,\kappa} c'_{\iota\kappa}\boldsymbol{e}'_{\iota\kappa}, \quad ||\,c'_{\iota\kappa}\,|| = C'.$$

The components of $\boldsymbol{e}'_{\iota\kappa}$ are $\frac{g'}{h}\bar{u}'_{\iota\kappa}(s)$. It follows from the orthogonality relations existing between inequivalent representations that $\boldsymbol{c}' \rightarrow 0$ in the representation $\mathfrak{H}$. If $\boldsymbol{c}$ is in H, then, by applying (13.6) for $\boldsymbol{x} = \boldsymbol{c}'$, $\boldsymbol{cc}' = \boldsymbol{y}$ is also, but since then $X = 0$ (13.7) yields $\boldsymbol{y} = 0$; the two sub-algebras are independent in the sense that the product of an element in one with an element in the other is always 0. Hence the "units" satisfy

$$\boldsymbol{e}_{ik}\,\boldsymbol{e}'_{\iota\kappa} = 0. \tag{13.9}$$

The modulus

$$\boldsymbol{\varepsilon}' = \sum_{\iota} \boldsymbol{e}'_{\iota\iota}$$

of H' satisfies $\boldsymbol{\varepsilon\varepsilon}' = \boldsymbol{\varepsilon}'\boldsymbol{\varepsilon} = 0$ in addition to $\boldsymbol{\varepsilon}'\boldsymbol{\varepsilon}' = \boldsymbol{\varepsilon}'$.

If $a(s)$ is a class function, $\boldsymbol{a}$ belongs to the central of Γ and if $\boldsymbol{a} \rightarrow A$ in the g-dimensional irreducible representation $\mathfrak{H}$ then the matrix A commutes with all matrices X. *Hence A is a multiple of the unit matrix:* $A = \frac{\alpha}{g}\mathbf{1}$. By (13.2) we find that the trace α of A is *

$$\alpha = \sum_{s} a(s)\chi(s).$$

In this way the entire theory of representations can be translated into the language of modern algebra. This leads to a greater freedom of operation and is preferable for the expression of the completeness theorem. The orthogonality relations between $u_{ik}(s)$, $u'_{\iota\kappa}(s)$, $\cdots$ lead to Bessel's inequality

$$g \cdot \mathrm{tr}\,(X\tilde{\bar{X}}) + \cdots \leqq h \cdot \sum_{s} x(s)\bar{x}(s), \tag{13.10}$$

where X in the sum on the left is the matrix (13.1) associated with $x(s)$ in the g-dimensional irreducible representation $\mathfrak{H}$ and the sum is taken over any set of inequivalent irreducible representations $\mathfrak{H}, \cdots$. This inequality is obtained by expressing the fact that the mean value of $z(s)\,\bar{z}(s)$ is non-negative (cf. I, § 7), where $\boldsymbol{z}$ is that element obtained from $\boldsymbol{x}$ on subtracting from $\boldsymbol{x}$ its components in $\mathsf{H}, \cdots$:

$$\boldsymbol{z} = \boldsymbol{x} - (\sum x_{ik}\,\boldsymbol{e}_{ik} + \cdots) = \boldsymbol{x} - (\boldsymbol{x\varepsilon} + \cdots).$$

Since the characters constitute an orthogonal system we also have the Bessel inequality

$$\xi\bar{\xi} + \cdots \leqq h \cdot \sum_{s} x(s)\bar{x}(s) \tag{13.11}$$

* Cf. also Appendix 2 at the end of the book.

where ξ is defined by (13.2). *The completeness theorem asserts that in both cases the equality sign holds when the sum is extended over a complete system of inequivalent irreducible representations, where in* (13.10) $x(s)$ *is any function on the group manifold and in* (13.11) *any class function. The second relation is a special case of the first, since for class functions* $X = \frac{\xi}{g}\mathbf{1}$.

If the abstract group $\mathfrak{g}$ is a finite continuous group which is closed in the sense of § 12, instead of a finite group as above, the sums must be replaced by integrals; the measure of volume on the group manifold is introduced as in § 12. We then have in place of (13.1), (13.4):

$$X = \int x(s)\, U(s) ds,$$

$$xy(s) = \int x(st^{-1}) y(t) dt = \int x(t) y(t^{-1}s) dt.$$

The modulus $\mathbf{1}$ of the algebra must have as components the values of a function $1(s)$ which vanishes everywhere on the group manifold except at the point $s = \mathbf{1}$ and must there be so large that $\int 1(s) ds = 1$. Such a function does not exist, but we can construct functions approximating these conditions arbitrarily close.

The completeness relations assert that any element $\boldsymbol{x}$ of the algebra of a finite group $\mathfrak{g}$ is the sum of its components in the totality of sub-algebras associated with a complete system of inequivalent irreducible representations. The group algebra Γ is thus reduced to a set of independent simple matric algebras. It suffices to prove this theorem for $\boldsymbol{x} = \mathbf{1}$:

$$\mathbf{1} = \boldsymbol{\varepsilon} + \boldsymbol{\varepsilon}' + \cdots = (\boldsymbol{e}_{11} + \cdots + \boldsymbol{e}_{gg}) + \cdots, \quad (13.12)$$

for on multiplying this by $\boldsymbol{x}$ it follows for all elements $\boldsymbol{x}$. These assertions cannot be carried over to continuous groups in the form here stated; we must hold to the formulation (13.10) (with $=$ instead of $\leqq$) containing an arbitrary function $x(s)$. We go into the proof of these results in Chap. V, where all the results of this section will be derived anew and discussed in detail from another more profound point of view.

§ 14. Invariants and Covariants

We first discuss briefly the classical concept of an ***invariant***. Consider, for example, the group $\mathfrak{c} = \mathfrak{c}_2$ of homogeneous linear transformations of two variables ξ, η with unit determinant. Let

$$a\xi^2 + 2b\xi\eta + c\eta^2$$

be an arbitrary quadratic form in the two variables. The "discriminant" $ac - b^2$ is an invariant, for the discriminants of two forms which are such that either goes into the other on transforming ξ, η by some element of $\mathfrak{c}$ have the same value. We may have, instead of one arbitrary quadratic form, one or more arbitrary forms f, ϕ, $\cdots$ of given orders, n, ν, $\cdots$. An invariant is a rational integral function I of the coefficients of these forms which is homogeneous in the coefficients of each of the forms f, ϕ, $\cdots$ and which has the same value on replacing these coefficients by the coefficients of the forms f', ϕ', $\cdots$ into which f, ϕ, $\cdots$ are transformed by an arbitrary transformation σ of $\mathfrak{c}$ affecting the variables ξ, η.

The coefficients $a_0, a_1, \cdots, a_n$ of an arbitrary form of order n in the variables ξ, η undergo a certain linear transformation on subjecting the variables to a transformation σ of $\mathfrak{c}$, and the correspondence between σ and this transformation constitutes a *representation* of the group $\mathfrak{c}$. The same is true for the totality of monomials

$$a_0^{r_0} a_1^{r_1} \cdots a_n^{r_n} \quad (r_0 + r_1 + \cdots + r_n = r)$$

of order r in these coefficients. A homogeneous polynomial I of order r in the a_i is a linear combination of these monomials. We thus see that if I is of given degrees r, ρ, $\cdots$ in the coefficients of the arbitrary forms f, ϕ, $\cdots$ it is a linear combination of quantities which constitute the substratum of a definite representation of $\mathfrak{c}$; this representation is known as soon as we have given the orders n, ν, $\cdots$ of the forms f, ϕ, $\cdots$ in the variables ξ, η and the degrees r, ρ, $\cdots$ of the invariant I in the arbitrary coefficients of f, ϕ, $\cdots$. Discarding the all too special formal algebraic assumptions involved in the "classical" concept of an invariant, and which the theory of invariants has from the beginning attempted to outgrow by generalizations in various directions, we may express the concept in modern group-theoretic language as follows:

Let $\mathfrak{H} : s \to U(s)$ be a given representation of an abstract group $\mathfrak{g}$ in an n-dimensional representation space $\mathfrak{R}$ with variables x_i; a linear form in the x_i is said to be an invariant in the representation space $\mathfrak{R}$ of $\mathfrak{H}$ if it is unchanged under all the transformations $U(s)$. If I_1, I_2, $\cdots$ are invariants in the representation space of $\mathfrak{H}$, then any linear combination $\alpha_1 I_1 + \alpha_2 I_2 + \cdots$ of them with constant coefficients α_1, α_2, $\cdots$ is also an invariant. The most important problem arising here is naturally that concerning the number m of linearly independent invariants in the given representation space. If $y_1, y_2, \cdots y_m$ constitute such a complete set of linearly independent invariants, and if we choose as co-ordinates in $\mathfrak{R}$ these m quantities and $n - m$ further linear forms $y_{m+1}, \cdots, y_n$ such that the two sets together constitute a complete system of linearly independent linear forms in $\mathfrak{R}$, the transformation $U(s)$ is, in terms of the variables y,

$$\begin{aligned} y'_1 &= y_1, \cdots, y'_m = y_m; \\ y'_{m+1} &= u_{m+1,1}(s)\, y_1 + \cdots + u_{m+1,n}(s)\, y_n, \\ &\cdots\cdots\cdots \\ y'_n &= u_{n1}(s)\, y_1 \qquad + \cdots + u_{nn}(s)\, y_n. \end{aligned}$$

If we are dealing with a unitary representation the y's can be so chosen that they define a normal co-ordinate system; $\mathfrak{H}$ is then reduced into m times the 1-dimensional identical representation $y' = y$ and an $(n - m)$-dimensional representation. Hence the problem of finding the number of linearly independent invariants in the representation space $\mathfrak{R}$ reduces to finding how often the identical representation with the character 1 is contained in the given $\mathfrak{H}$. But by formula (11.8) the solution of this problem is given by

$$m = \mathfrak{M}\{\chi(s)\}, \tag{14.1}$$

or: *the mean value of the character χ of $\mathfrak{H}$, which is always a non-negative integer, gives the number of linearly independent invariants in the representation space of $\mathfrak{H}$.*

The formula (14.1) answers the principal question arising in the linear invariant theory, and we now proceed to an extremely brief discussion of the algebraic invariant theory. Let $\mathfrak{G}, \mathfrak{H}, \cdots$ be representations of the same abstract group $\mathfrak{g}$ in the spaces with variables $x_i, y_k, \cdots$. We consider rational integral functions $I(x_i, y_k, \cdots)$ which are homogeneous in the variables x_i, homogeneous in the variables y_k, etc. If on subjecting $x, y, \cdots$ to those linear transformations corresponding to the same arbitrary group element s in the representations $\mathfrak{G}, \mathfrak{H}, \cdots$ I remains unchanged, then it is said to be *a rational integral invariant of the system* $[\mathfrak{G}, \mathfrak{H}, \cdots]$ *of representations.* If the orders $p, q, \cdots$ of the function I in the variables $x_i, y_k, \cdots$ are given, the problem reduces to the one discussed above; for the monomials in these variables which are homogeneous of order p in the x_i, homogeneous of order q in the y_k, $\cdots$ constitute the substratum of a representation obtained in a certain way from $\mathfrak{G}, \mathfrak{H}, \cdots$. But if we consider simultaneously invariants of all possible orders belonging to the system $[\mathfrak{G}, \mathfrak{H}, \cdots]$ we are confronted with new problems. The most important of these, which is answered in the affirmative by the so-called *fundamental theorem of the theory of invariants* is: Do there exist a finite number of invariants such that all others can be expressed rationally and integrally in terms of them? This involves the question of algebraic, rather than linear, dependence between the invariants. We only mention this higher branch of the theory of invariants, and do not go into it further, as it bears no direct relation to quantum mechanics.[10]

In addition to invariants or scalars, ***covariant linear quantities***, such as vectors and tensors, play an important rôle in physics. Let $\mathfrak{g}$ be the group of all linear transformations between the normal co-ordinate systems in space or in space-time, e.g. the 3-dimensional group of Euclidean rotations or the group of Lorentz transformations, and let $\mathfrak{H} : s \rightarrow U(s)$ be an n-dimensional representation of $\mathfrak{g}$. *A covariant quantity of kind $\mathfrak{H}$ is an entity having n components $a_1, a_2, \cdots, a_n$ relative to any given co-ordinate system for the variables of the transformation group $\mathfrak{g}$ and which is such that on going over to a new co-ordinate system by means of the transformation s of $\mathfrak{g}$ the new components a_i are obtained from the old by the corresponding transformation $U(s)$ of $\mathfrak{H}$.* If $\mathfrak{H}$ is irreducible such a quantity is said to be *primitive* or *simple*. *Physical quantities are generally simple.* Thus, for example, the entity whose components are the electro-magnetic field strengths in the 4-dimensional world is described as an "anti-symmetric tensor of order 2" rather than merely as a "tensor of order 2"; we shall see in Chap. V, § 4, that it is therefore a simple quantity. The reduction of a representation into its irreducible constituents implies the reduction of the corresponding kind of quantities into simple quantities. It would appear that the only simple quantities with which we deal are *tensors* which are characterized by certain *symmetry conditions* in addition to their *order*. We shall prove this theorem for the complete linear group $\mathfrak{c}$ and for its unitary sub-group $\mathfrak{u}$ in Chap. V; it asserts that all representations of $\mathfrak{c}$ (or $\mathfrak{u}$) can be obtained by reduction from the powers $\mathfrak{c}, (\mathfrak{c})^2, (\mathfrak{c})^3, \cdots$ and that the irreducible constituents of $(\mathfrak{c})^f$ are obtained by imposing certain symmetry conditions.

We must accordingly generalize the problem of the linear theory of invariants in the following manner. Consider two unitary representations $\mathfrak{h} : \sigma \rightarrow s$, $\mathfrak{H} : \sigma \rightarrow S$ of the abstract group $\mathfrak{g}$ with elements σ; let their dimensionalities be n, N and let $\mathfrak{h}$ be irreducible. *We wish to determine all covariant quantities of kind $\mathfrak{h}$ in the representation space of $\mathfrak{H}$.* Calling the variables in this representation space x_i, which undergo the transformation S under the influence of σ, such a quantity I has n components $I_1, I_2, \cdots, I_n$ which are linearly independent linear forms in the variables x_i. When the x_i undergo the transformation S the n linear forms I_α go over into new ones which are obtained from the I_α (in which the variables x_i have been transformed in accordance with S) by means of the transformation s of $\mathfrak{h}$. If there exist two or more covariant quantities

$$I = (I_1, I_2, \cdots, I_n), \quad I' = (I'_1, I'_2, \cdots, I'_n), \cdots$$

of the kind $\mathfrak{h}$ in the representation space of $\mathfrak{H}$, then any linear combination $\alpha I + \alpha' I' + \cdots$ with constant coefficients α is again a quantity of the same kind. *We ask for the number m of linearly independent quantities of this kind. The answer is that m is equal to the number of times the irreducible representation $\mathfrak{h}$ is contained in $\mathfrak{H}$.* Hence if χ, $\mathbf{X}$ are the characters of $\mathfrak{h}$, $\mathfrak{H}$, we have

$$m = \mathfrak{M}\{\mathbf{X}(s)\bar{\chi}(s)\}. \tag{14.2}$$

In order to prove this statement we choose the co-ordinate system x_i in the representation space of $\mathfrak{H}$ in such a way that the matrices of $\mathfrak{H}$ are reduced into their irreducible constituent sub-matrices, the m representations $\mathfrak{h} : \mathfrak{h}' = \mathfrak{h}'' = \cdots = \mathfrak{h}^{(m)} = \mathfrak{h}$ being separated out first. The remaining constituents $\mathfrak{h}^{(m+1)}$, $\cdots$ are inequivalent to $\mathfrak{h}$. Denote the variables in the corresponding invariant sub-spaces by

$$x'_1, \cdots, x'_n; \; x''_1, \cdots, x''_n; \; \cdots; \; x_1^{(m)}, \cdots, x_n^{(m)}; \; \cdots$$

The matrix S is completely reduced into the sub-matrices $s' = s, \cdots, s^{(m)} = s$; $s^{(m+1)}, \cdots$ arranged along the principal diagonal. Let

$$\left.\begin{aligned} y_1 &= a_{11} x_1 + \cdots + a_{1N} x_N, \\ &\cdot \quad \cdot \quad \cdot \quad \cdot \quad \cdot \quad \cdot \\ y_n &= a_{n1} x_1 + \cdots + a_{nN} x_N \end{aligned}\right\}$$

be a covariant quantity of the kind $\mathfrak{h}$. We can write this in the form $y = Ax$ in terms of the column x of the N variables x_i, the column y of the n variables y_α and the matrix $A = \|a_{\alpha i}\|$. The requirement that I be a quantity of kind $\mathfrak{h}$ means that when x is replaced by $x' = Sx$, y goes over into $y' = sy$, or

$$sy = ASx, \quad sAx = ASx, \quad sA = AS. \tag{14.3}$$

Corresponding to the reduction of x-space into irreducible sub-spaces, the matrix A of the correspondence of x-space on y-space is reduced into matrices $A', \cdots, A^{(m)}$; $A^{(m+1)}, \cdots$ consisting of the first n rows, $\cdots$, the mth set of n rows, $\cdots$, $\cdots$ of A. Equation (14.3) then becomes

$$sA' = A's, \; \cdots, \quad sA^{(m)} = A^{(m)}s; \; sA^{(m+1)} = A^{(m+1)}s^{(m+1)}, \; \cdots.$$

It follows from the fundamental theorem (10.5) on representations that $A', \cdots, A^{(m)}$ are all multiples of the n-dimensional unit matrix and that the remaining $A^{(m+1)}, \cdots$ are all zero. But this is just our assertion that $y = (y_1, y_2, \cdots, y_n)$ is a linear combination of the m quantities

$$x' = (x_1', x_2', \cdots, x_n'),$$
$$\cdot \quad \cdot \quad \cdot \quad \cdot \quad \cdot \quad \cdot$$
$$x^{(m)} = (x_1^{(m)}, x_2^{(m)}, \cdots, x_n^{(m)})$$

of the kind $\mathfrak{h}$.

§ 15. Remarks on Lie's Theory of Continuous Groups of Transformations

In § 12 we made use of the concept of *infinitesimal elements of a group* in order to establish a method of measuring volume on a continuous group manifold. We here discuss this concept in detail for the 3-dimensional group $\mathfrak{d}$ of rotations in Euclidean space.[11] This group serves to describe the mobility of a body in Euclidean space, one point O of which is fixed in space. Each possible position of the body can be considered as arising from any given initial position by an operation of $\mathfrak{d}$. A material substance distributed throughout the space or any portion of it moves as a rigid body about O if the position of each of its elements at a given moment is associated with its initial position by means of a correspondence belonging to $\mathfrak{d}$. This is the description of the motion of such a rigid body which compares the position in any moment directly with the initial position, ignoring the intermediate states which it has assumed in going from the one into the other. But it seems more natural to consider it in terms of a continuous motion in which the position of the body undergoes an infinitesimal rotation from moment to moment, so that the motion as a whole is the integration of a series of infinitesimal operations of $\mathfrak{d}$. On employing an auxiliary variable t in order to avoid the use of infinitesimals and thinking of this parameter as time, the velocity field $dx = \dot{x}$, $dy = \dot{y}$, $dz = \dot{z}$ of an infinitesimal rotation is defined by [cf. I, § 6]

$$dx = bz - cy, \quad dy = cx - az, \quad dz = ay - bx, \tag{15.1}$$

where the constants a, b, c are independent of position (x, y, z). These velocity fields, which obviously constitute a 3-dimensional linear manifold, are the infinitesimal elements of $\mathfrak{d}$; they are the "vectors" which define the linear space tangent to the group manifold at the point which represents the unit element I. The continuous motion of a rigid body about O is characterized by the fact that at each moment its velocity field belongs to the 3-parameter linear family (15.1). We may take as a basis of this family the three elements D_x, D_y, D_z obtained by choosing

$$a = 1, b = 0, c = 0\ ;\ a = 0, b = 1, c = 0\ ;\ a = 0, b = 0, c = 1.$$

We call these "the infinitesimal rotations about the x-, y- and z-axes." *S. Lie* was the first to undertake a systematic study of the construction of transformation groups from their infinitesimal elements. In fact, once they are known all the substitutions of the continuous group can be generated by integration, i.e. by successive application of such infinitesimal elements—at least, all those which belong to the same connected "sheet" as the identity. (Example: the proper orthogonal transformations can be obtained from the infinitesimal ones, but not the improper transformations with determinant -1).

In general, consider a continuous r-parameter transformation group $\mathfrak{G}$, and let the group manifold be described in terms of the parameters $s_1, s_2, \cdots, s_r$ in the neighbourhood of the unit point, at which they vanish. A portion of the group manifold is thereby mapped in a one-to-one continuous manner on a neighbourhood of the origin in the r-dimensional number space of the parameters s. Let the n-dimensional point-field of the transformations be described in terms of co-ordinates $x_1, x_2, \cdots, x_n$ in the neighbourhood of the point under consideration, and let the correspondence $x \to x'$:

$$x'_i = \phi_i(x_1 x_2 \cdots x_n\ ;\ s_1, \cdots, s_r)$$

be associated with the element $(s_1, s_2, \cdots, s_r)$ of the abstract group in its realization by the transformation group. The infinitesimal transformation $x \to x + dx$ obtained by assigning the infinitesimal increments ds to the parameters s in the neighbourhood of $s = 0$ is given by

$$dx_i = \left(\frac{\partial \phi_i}{\partial s_1}\right) ds_1 + \cdots + \left(\frac{\partial \phi_i}{\partial s_r}\right) ds_r\ ; \qquad (15.2)$$

the parentheses indicate that the differential quotients are to be computed for $s_1 = 0, \cdots, s_r = 0$. We postulate a material substance which fills the point-field and which is capable of executing those and only those motions in which the positions of its elements at an arbitrary moment t' are obtained from their positions at time t by a transformation of $\mathfrak{G}$. Again its motion can be more simply described as the result of successive deformations corresponding to infinitesimal operations (15.2) of our group; the velocity field must at any time have the form

$$\dot{x}_i = \left(\frac{\partial\phi_i}{\partial s_1}\right)\sigma_1 + \cdot\cdot\cdot + \left(\frac{\partial\phi_i}{\partial s_r}\right)\sigma_r, \tag{15.3}$$

where $\sigma_1, \cdot\cdot\cdot, \sigma_r$ are constants independent of position. This r-dimensional linear family constitutes the *infinitesimal group of motions of our substance.* It is to be observed that the application of these infinitesimal processes to our transformation group presupposes that the functions ϕ_i are differentiable with respect to s at the point $s = 0$. In the theory of abstract groups the point-field is the group manifold itself and we take as a realization (left-)translation. In the neighbourhood of the unit element $s = 0$, $t = 0$ we have, as law of composition,

$$(st)_\alpha = \psi_\alpha(s_1 \cdot\cdot\cdot s_r;\ t_1 \cdot\cdot\cdot t_r) \quad [\alpha = 1, \cdot\cdot\cdot, r].$$

The introduction of a measure of volume in § 12 presupposes that the functions ψ_α are, for sufficiently small t, differentiable with respect to the s at the point $s = 0$, and that for sufficiently small s they are differentiable with respect to t at $t = 0$.

The composition of infinitesimal elements of the group is expressed by addition of the parameters σ introduced by (15.3). It might therefore appear as if the infinitesimal elements of an r-parameter continuous group need satisfy no condition other than that they constitute a linear family. However, that is not the case; there are further "*integrability conditions*" to be satisfied. The example of a sphere which rolls without slipping on a horizontal table shows that the possible positions of a body whose infinitesimal motions have but three degrees of freedom can nevertheless constitute a 5-dimensional manifold. The integrability conditions we are seeking, which involve second order derivatives, guarantee that this situation does not arise. We obtain these conditions on expressing the fact that the commutator $sts^{-1}t^{-1}$ of two infinitesimal elements s, t of the group also is an element of the group. This commutator converges to $\mathbf{I}$ as s approaches the unit element $\mathbf{I}$, whatever t may be, and similarly as $t \to \mathbf{I}$ for arbitrary s. The commutator of the two infinitesimal *linear* correspondences A and B:

$$dx = Ax, \quad d'x = Bx$$

is the infinitesimal correspondence $AB - BA$; to show this we note that the equation

$$\mathsf{A}(s)\mathsf{B}(t) = \Gamma(s, t)\mathsf{B}(t)\mathsf{A}(s)$$

leads, on writing

$$\mathbf{A}(0) = \mathbf{1},\ \mathbf{B}(0) = \mathbf{1},\ \left(\frac{d\mathbf{A}}{ds}\right)_{s=0} = A,\ \left(\frac{d\mathbf{B}}{dt}\right)_{t=0} = B,$$

$$\lim_{s,t\to 0} \frac{\Gamma(s, t) - \mathbf{1}}{s \cdot t} = \left(\frac{d^2\Gamma}{ds\,dt}\right)_{s,t=0} = C,$$

to the equation

$$C = AB - BA.$$

Our main purpose in mentioning these matters is to prepare the ground for an understanding from general principles of the commutation rules satisfied by the three infinitesimal rotations D_x, D_y, D_z:

$$\begin{Vmatrix} 0 & 0 & 0 \\ 0 & 0 & -1 \\ 0 & 1 & 0 \end{Vmatrix}, \quad \begin{Vmatrix} 0 & 0 & 1 \\ 0 & 0 & 0 \\ -1 & 0 & 0 \end{Vmatrix}, \quad \begin{Vmatrix} 0 & -1 & 0 \\ 1 & 0 & 0 \\ 0 & 0 & 0 \end{Vmatrix}. \tag{15.4}$$

They are, as is readily shown,

$$\left.\begin{array}{c} D_x D_y - D_y D_x = D_z, \quad D_y D_z - D_z D_y = D_x, \\ D_z D_x - D_x D_z = D_y. \end{array}\right\} \tag{15.5}$$

We could, of course, take the unimodular unitary group $\mathfrak{u}_2$ in two dimensions as fundamental, instead of the group $\mathfrak{d}_3$ of rotations. We denote the two variables which undergo the transformations σ of the unitary group by ξ, η as in § 8. In consequence of the correspondence $\sigma \to s$, which was established there by means of a stereographic projection, the 3-dimensional rotation group now appears as a representation of $\mathfrak{u}_2$. We can take as a basis for the 3-parameter linear manifold of infinitesimal operators of $\mathfrak{u}_2$ the three particular operators—

$$\left.\begin{array}{llll} \frac{1}{2i} S_x: & d\xi = \frac{1}{2i}\eta, & d\eta = \frac{1}{2i}\xi; \\ \frac{1}{2i} S_y: & d\xi = -\frac{1}{2}\eta, & d\eta = \frac{1}{2}\xi; \\ \frac{1}{2i} S_z: & d\xi = \frac{1}{2i}\xi, & d\eta = -\frac{1}{2i}\eta; \end{array}\right\} \tag{15.6}$$

here, in agreement with (8.15),

$$S_x = \begin{Vmatrix} 0 & 1 \\ 1 & 0 \end{Vmatrix}, \quad S_y = \begin{Vmatrix} 0 & -i \\ i & 0 \end{Vmatrix}, \quad S_z = \begin{Vmatrix} 1 & 0 \\ 0 & -1 \end{Vmatrix}.$$

They are the infinitesimal transformations of $\mathfrak{u}_2$ corresponding to the three infinitesimal transformations D_x, D_y, D_z of $\mathfrak{d}_3$ in virtue of the correspondence $\sigma \rightarrow s$; that this is in fact the case is readily seen from (8.10) or

$$x \sim -\xi^2 + \eta^2, \quad y \sim \frac{1}{i}(\xi^2 + \eta^2), \quad z \sim 2\xi\eta.$$

Given any representation $\mathfrak{H} : \sigma \rightarrow U(\sigma)$ of $\mathfrak{u}_2$, its infinitesimal operators with matrices

$$\frac{1}{i}\,(M_x, M_y, M_z)$$

corresponding to the infinitesimal operators (15.6) in $\mathfrak{u}_2$ satisfy the same equations (15.5) as the D_x, D_y, D_z:

$$M_x M_y - M_y M_x = iM_z, \cdot \cdot \cdot. \tag{15.7}$$

The matrices M_x, M_y, M_z are of course Hermitian. For reasons which will appear in the following chapter we call these the components of *moment of momentum* (or *angular momentum*) $\mathfrak{M}$ of the representation $\mathfrak{H}$, and

$$\mathfrak{M}^2 = M^2 = M_x^2 + M_y^2 + M_z^2$$

the square of the magnitude of the moment of momentum. If $\mathfrak{H}$, $\mathfrak{H}'$ are two representations with angular momenta $\mathfrak{M}$, $\mathfrak{M}'$ then, in accordance with the general formula II, (10.4), which governs the composition of infinitesimal operators by $\times$-multiplication, the representation $\mathfrak{H} \times \mathfrak{H}'$ has as moment of momentum $(\mathfrak{M} \times \mathbf{1}) + (\mathbf{1} \times \mathfrak{M}')$.

We next calculate the moment of momentum $\mathfrak{M}_j$ of the irreducible representation $\mathfrak{C}_f = \mathfrak{D}_j$ $(j = f/2)$ of $\mathfrak{u}_2$. It will be found more convenient to employ in place of $\frac{1}{2i} S_x$, $\frac{1}{2i} S_y$, the transformations

$$\left.\begin{aligned} \frac{1}{2}(S_x + iS_y):\ & d\xi = \eta, \quad d\eta = 0 \\ \frac{1}{2}(S_x - iS_y):\ & d\xi = 0, \quad d\eta = \xi \end{aligned}\right\}. \tag{15.8}$$

In general

$$d(\xi^r\eta^s) = r\,\xi^{r-1}\,\eta^s\,d\xi + s\,\xi^r\,\eta^{s-1}\,d\eta,$$

and on substituting in this the variables

$$x(m) = \frac{\xi^r\eta^s}{\sqrt{r!\ s!}} \quad (r + s = 2j, \quad r - s = 2m)$$

of the representation space of $\mathfrak{D}_j$, we find that the three infinitesimal transformations of $\mathfrak{u}_2$ defined by (15.6), (15.8) induce in this space the transformations

$$\left.\begin{aligned}\frac{1}{2}(S_x + iS_y) : dx(m) &= \sqrt{r(s+1)}\, x(m-1)\\ &= \sqrt{(j+m)(j-m+1)}\, x(m-1),\\ \frac{1}{2}(S_x - iS_y) : dx(m) &= \sqrt{s(r+1)}\, x(m+1)\\ &= \sqrt{(j-m)(j+m+1)}\, x(m+1),\\ S_z : dx(m) &= \frac{r-s}{2}\, x(m) = m\, x(m).\end{aligned}\right\}$$

Hence

$$\left.\begin{aligned}(M_x + iM_y)(m, m-1) &= \sqrt{(j+m)(j-m+1)},\\ (M_x - iM_y)(m, m+1) &= \sqrt{(j-m)(j+m+1)},\\ M_z(m, m) &= m.\end{aligned}\right\} \qquad (15.9)$$

All other components (m, m') vanish. M^2 is a multiple of the unit matrix in $\mathfrak{R}_j$:

$$M^2 = j(j+1),$$

for it follows from

$$\begin{aligned}(M_x + iM_y)(M_x - iM_y) &= M_x^2 + M_y^2 - i(M_xM_y - M_yM_x)\\ &= M_x^2 + M_y^2 + M_z\end{aligned}$$

that

$$M^2 = (M_x + iM_y)(M_x - iM_y) - M_z + M_z^2,$$

and from this and (15.9) that

$$M^2(m, m) = (j+m)(j-m+1) - m + m^2 = j(j+1).$$

If on reducing an arbitrary representation $\mathfrak{H}$ the irreducible representation $\mathfrak{D}_j$ is found to occur exactly g_j times, then M^2 has $j(j+1)$ as a $[(2j+1)g_j]$-fold characteristic number and M_z has the characteristic number m with multiplicity

$$\sum_j g_j \qquad (j = |m|, |m| + 1, \cdots).$$

From this we again see that the multiplicity g_j with which $\mathfrak{D}_j$ occurs in the reduction of $\mathfrak{H}$ is uniquely determined by $\mathfrak{H}$. These infinitesimal operations can be used to give a relatively elementary constructive proof of the fact that the $\mathfrak{D}_j$ are the only irreducible representation of $\mathfrak{u}_2$.[12]

§ 16. Representation by Rotations of Ray Space

In quantum theory the representations take place in system space; but this is to be considered as a ray rather than a vector space, for a pure state is represented by a ray rather than a vector. Two unitary transformations U and εU which differ only by a numerical factor ε of absolute magnitude 1 are consequently to be considered as the same, $U \simeq \varepsilon U$, for they determine the same rotation of the ray field. In a "*ray representation*," which associates with each element s of the abstract group $\mathfrak{g}$ a unitary rotation $U(s)$ of the rays of n-dimensional representation space, the gauge factor $\varepsilon(s)$ may be taken arbitrarily for each unitary matrix $U(s)$; if $\mathfrak{g}$ is a continuous group we choose it, however, in such a way that $U(s)$ depends continuously on s. The condition for a representation is now only

$$U(s)U(t) \simeq U(st), \tag{16.1}$$

i.e.

$$U(s)U(t) = \delta(s, t)U(st), \tag{16.2}$$

where $\delta(s, t)$ is a numerical factor, of modulus 1, depending on s and t. If by change of gauge $U(s)$ is replaced by $\varepsilon(s)U(s)$, $\delta(s, t)$ is replaced by

$$\varepsilon(st)\varepsilon^{-1}(s)\varepsilon^{-1}(t)\delta(s, t).$$

In the equation

$$X = \sum_s x(s)U(s),$$

defining the connection between the components $x(s)$ of an element $\boldsymbol{x}$ of the algebra of the group and the group matrix X which represents it, the $x(s)$ are also dependent on the gauge and are sent into $\varepsilon(s)x(s)$ on the change of gauge defined by $U(s) \rightarrow \varepsilon(s)U(s)$. In order that the multiplication law for two elements $\boldsymbol{x}, \boldsymbol{y}$ shall, as we require, parallel the multiplication of the matrices which represent them we must define

$$xy(s) = \sum_{tt'=s} \delta(t, t')\, x(t)\, y(t') \tag{16.3}$$

in terms of the chosen gauge. The condition

$$\bar{x}(s^{-1}) = x(s)$$

for a real element $\boldsymbol{x}$ is only appropriate if the gauge is so chosen that $U(s^{-1})$ is the matrix reciprocal to $U(s)$. The algebra of the group is to be adapted in this way to the ray representation under consideration, whereas in dealing with "vector representations" it is uniquely determined by the law of composition of the group alone.[13]

Examples.

I. The 1-dimensional representations are now entirely uninteresting, for any 1-dimensional matrix $\simeq \mathbf{1}$. But *under certain circumstances Abelian groups may possess multi-dimensional unitary ray representations*, whereas any irreducible unitary vector representation of an Abelian group is necessarily of degree 1.

We first investigate the simplest example, a finite cyclical group (a) of order h, consisting of the elements

$$\mathbf{1},\ a,\ a^2,\ \cdot\cdot\cdot,\ a^{h-1}\ (a^h = \mathbf{1}).$$

Let the element a correspond to the unitary matrix A in the ray representation; then $A^h = \alpha\mathbf{1}$ is necessarily a multiple of the unit matrix. Since α is of modulus 1 we may change the gauge in such a way that A goes into $A/\sqrt[h]{\alpha}$; then $A^h = \mathbf{1}$ and the correspondence $a^k \to A^k$ is a vector representation of the cyclical group. Hence by introducing an appropriate change of gauge the ray representation can be made into a vector representation, $\delta(s, t)$ being then 1.

II. The simplest example of an Abelian group which gives rise to multi-dimensional irreducible ray representations must consequently be non-cyclic. Consider the group consisting of the four elements $\mathbf{1}$, a, b, c with the multiplication table

$$\begin{gathered} a^2 = b^2 = c^2 = \mathbf{1}, \\ bc = cb = a, \quad ca = ac = b, \quad ab = ba = c. \end{gathered} \tag{16.4}$$

A ray representation $\mathfrak{B}$ is given by

$$U(\mathbf{1}) = \begin{Vmatrix} 1 & 0 \\ 0 & 1 \end{Vmatrix}, \quad U(a) = \begin{Vmatrix} 0 & 1 \\ 1 & 0 \end{Vmatrix}, \quad U(b) = \begin{Vmatrix} 0 & -i \\ i & 0 \end{Vmatrix}, \quad U(c) = \begin{Vmatrix} 1 & 0 \\ 0 & -1 \end{Vmatrix} \tag{16.5}$$

The normalization is here chosen in such a way that

$$U^2(a) = U(a)U(a^{-1}) = \mathbf{1}$$

and similarly for $\mathbf{1}$, b, c. The algebra defined by (16.3) for this representation is non-commutative in spite of the Abelian nature of the group; it is the *algebra of complex quaternions*. On denoting the elements of this algebra by

$$\boldsymbol{x} = \kappa\,\mathbf{1} + \lambda\,\boldsymbol{a} + \mu\,\boldsymbol{b} + \nu\,\boldsymbol{c},$$

the "units" $\mathbf{1}$, $\boldsymbol{a}$, $\boldsymbol{b}$, $\boldsymbol{c}$ have the same multiplication table as the corresponding matrices U:

	$\mathbf{1}$	$\boldsymbol{a}$	$\boldsymbol{b}$	$\boldsymbol{c}$
$\mathbf{1}$	$\mathbf{1}$	$\boldsymbol{a}$	$\boldsymbol{b}$	$\boldsymbol{c}$
$\boldsymbol{a}$	$\boldsymbol{a}$	$\mathbf{1}$	$i\boldsymbol{c}$	$-i\boldsymbol{b}$
$\boldsymbol{b}$	$\boldsymbol{b}$	$-i\boldsymbol{c}$	$\mathbf{1}$	$i\boldsymbol{a}$
$\boldsymbol{c}$	$\boldsymbol{c}$	$i\boldsymbol{b}$	$-i\boldsymbol{a}$	$\mathbf{1}$

(The product $\boldsymbol{xy}$ occupies the intersection of the row $\boldsymbol{x}$ with the column $\boldsymbol{y}$.)

The "real" quantities are those for which all components κ, λ, μ, ν are real. Since in the calculus of quaternions $\mathbf{I}$, $i\boldsymbol{a}$, $i\boldsymbol{b}$, $i\boldsymbol{c}$ are taken as the fundamental units, they are those whose scalar component κ is real and whose vectorial components λ/i, μ/i, ν/i are purely imaginary.

III. *The group* $\mathfrak{u} = \mathfrak{u}_2$ *of unitary transformations* σ *in two dimensions with determinant* 1. Consider a representation $\sigma \rightarrow U(\sigma)$ by rotations in n-dimensional ray space. On changing the gauge in such a way that $U(\sigma)$ goes into

$$U(\sigma) : \sqrt[n]{\det U(\sigma)}, \tag{16.6}$$

the determinant of the new $U(\sigma)$ is 1. The only possible difficulty consists in the fact that the nth root

$$\varepsilon(\sigma) = \sqrt[n]{\det U(\sigma)} \tag{16.7}$$

is multiple-valued. It is "*locally*" single-valued, i.e. if we have chosen a definite one ε_0 of the n values for the point $\sigma = \sigma_0$, we can uniquely determine the root $\varepsilon(\sigma)$ in a sufficiently small neighbourhood of σ_0 in such a way that it depends continuously on σ and goes over into ε_0 for $\sigma = \sigma_0$. Hence we can *continue* the determination of the root for $\sigma = \sigma_0$ in a unique manner along a path in the group manifold, starting in σ_0. The only question is whether $\varepsilon(\sigma)$ returns to its original value when we allow σ to describe a closed path. *This is to be answered in the affirmative, since the group manifold of* $\mathfrak{u}$ *is simply connected* in the sense that any closed curve can be drawn together into a point by a continuous deformation. For in accordance with equation (7.5) the elements of the group are mapped in a one-to-one continuous manner on the quadruple $(\kappa\lambda\mu\nu)$ of real numbers which are subject to the condition

$$\kappa^2 + \lambda^2 + \mu^2 + \nu^2 = 1.$$

Hence the group manifold has the same topological properties as a 3-dimensional sphere in 4-dimensional space. These considerations thus show that the nth root (16.7) is broken up into n single-valued continuous functions over the entire group manifold. The method of proof here employed, which is of fundamental importance in the whole of mathematics, is perhaps best known to the reader in the proof of *Cauchy's* integral theorem ; it follows from the fact that the integral of an analytic function is locally single-valued, that it is single-valued *in the large if the region in which we are operating is simply connected.*

The result of our topological considerations showed that the formula (16.6) defines n single-valued continuous functions $U(\sigma)$. One of them is such that in it $U(\mathbf{I})$ is the unit matrix ;

we henceforth denote it alone by $U(\sigma)$. On writing the equation

$$U(\sigma)U(\tau) = \delta(\sigma, \tau)U(\sigma\tau) \qquad (16.8)$$

for $\tau = \mathbf{I}$, and taking into account the fact that $U(\mathbf{I}) = \mathbf{1}$, we find $\delta(s, \mathbf{I}) = 1$. On forming the determinant of both sides of (16.8) we obtain the equation

$$1 = [\delta(\sigma, \tau)]^n.$$

$\delta(\sigma, \tau)$ is consequently an nth root of unity which depends continuously on τ for fixed σ and which reduces to 1 for $\tau = \mathbf{I}$; hence it is identically equal to 1, and (16·8) becomes

$$U(\sigma)U(\tau) = U(\sigma\tau).$$

Consequently the only ray representations of $\mathfrak{u}_2$ *are also vector representations*, and our considerations show that *this theorem is valid for any continuous group whose elements constitute a simply connected manifold.* On going over to the 3-dimensional rotation group $\mathfrak{d}_3$ by stereographic projection, all $\mathfrak{D}_j$, even those with half-integral j, are single-valued when considered as ray representations. Any single-valued continuous ray representation of $\mathfrak{d}_3$ is reducible into irreducible constituents, and the only irreducible ray representations are the $\mathfrak{D}_j$ $(j = 0, 1/2, 1, 3/2, \cdots)$ obtained earlier in the chapter. But $\mathfrak{d}_3$ is not simply connected; we must resort to a two-sheeted covering surface, similar to a Riemannian surface but without cuts or branch points, which is simply connected. This accounts for the fact that there exist irreducible ray representations of $\mathfrak{d}_3$ which may be *single-* or *double-valued* vector representations, but there cannot exist multiple-valued representations of higher degree.

I have been able to prove the same theorem for the n-dimensional rotation group $(n \geqq 3)$.[14] This means that there exist *two* closed continuous motions (i.e. motions which lead back to the initial state) of a rigid body, which is free to rotate about a fixed point 0, such that any other closed motion can be continuously deformed into one of the two. One of these may be taken as *rest*, and the other is such that it cannot be continuously deformed into rest.

CHAPTER IV

APPLICATION OF THE THEORY OF GROUPS TO QUANTUM MECHANICS

A. The Rotation Group

§ 1. The Representation Induced in System Space by the Rotation Group

IN accordance with III, § 8, we can interpret the theory of a single electron in a spherically symmetric electrostatic field, as developed in II, § 5, in the following manner. A rotation of physical space, i.e. an orthogonal transformation from the Cartesian co-ordinates xyz into $x'y'z'$, induces a unitary transformation $U(s) : \psi \rightarrow \psi'$ defined by

$$\psi'(x'y'z') = \psi(xyz) \tag{1.1}$$

in the system-space $\mathfrak{R}$ of the electron, the vectors of which are the wave functions $\psi(xyz)$ describing the state of the electron. The correspondence $s \rightarrow U(s)$ is a definite representation $\mathfrak{E}$, of infinitely many dimensions, of the rotation group $\mathfrak{d}_3$. This representation $\mathfrak{E}$ can be reduced into its irreducible constituents $\mathfrak{D}_l$, and it is found that each $\mathfrak{D}_l$ with integral l occurs an infinite number of times. The total system-space $\mathfrak{R}$ is correspondingly decomposed into mutually orthogonal sub-spaces $\mathfrak{R}(nl)$; $\mathfrak{R}(nl)$ has $2l + 1$ dimensions and the rotation group induces the representation $\mathfrak{D}_l$ in it. If we introduce in addition the improper rotations ($\mathfrak{d}_3'$) $\mathfrak{D}_l$ always appears in $\mathfrak{E}$ with the signature $(-1)^l$. The ∞-dimensional sub-spaces $\sum\limits_n \mathfrak{R}(nl)$ associated with the various values of l are uniquely determined, but their further decomposition into the summands $\mathfrak{R}(nl)$ is quite arbitrary. In particular, this can be done in such a way that the *energy* of the states composing $\mathfrak{R}(nl)$ has a definite value $E(nl)$.

We now calculate the operators induced in system-space by the *infinitesimal* rotations of physical space. Denoting the increase $\psi'(xyz) - \psi(xyz)$ by $d\psi$, equation (1.1) becomes

$$d\psi + \left(\frac{\partial\psi}{\partial x}dx + \frac{\partial\psi}{\partial y}dy + \frac{\partial\psi}{\partial z}dz\right) = 0$$

for the infinitesimal rotation s which sends

$$x,\ y,\ z \text{ into } x' = x + dx, \quad y' = y + dy, \quad z' = z + dz.$$

Taking as s the three infinitesimal rotations D_x, D_y, D_z in turn [III, (15.4)] and writing the corresponding infinitesimal unitary operators in the form

$$d\psi = \frac{1}{i}(L_x,\ L_y,\ L_z)\psi,$$

we find

$$L_x = \frac{1}{i}\left(y\frac{\partial}{\partial z} - z\frac{\partial}{\partial y}\right), \ \cdot \cdot \cdot. \tag{1.2}$$

$h\mathfrak{L}$ is accordingly the moment of momentum [cf. II, (4.9)].

On going over from *one* electron to two, the vectors of system space are the functions $\psi(x_1y_1z_1\,;\ x_2y_2z_2)$ of the Cartesian coordinates of both electrons. The unitary transformation $U : \psi \rightarrow \psi'$ induced in system-space by the rotation s is now defined by the equation

$$\psi'(x_1'y_1'z_1'\,;\ x_2'y_2'z_2') = \psi(x_1y_1z_1\,;\ x_2y_2z_2),$$

where $x_1'y_1'z_1'$ and $x_2'y_2'z_2'$ are obtained from $x_1y_1z_1$ and $x_2y_2z_2$ by the *same* orthogonal transformation s. This situation can be described as follows: The state space $\mathfrak{R}^2$ of the system consisting of two electrons is $\mathfrak{R} \times \mathfrak{R}$ and the representation $\mathfrak{E}^2$ induced in it is $\mathfrak{E} \times \mathfrak{E}$.

This representation is, as we see, determined by the kinematical constitution of the system alone, and is in no way influenced by the dynamical relationships; the rule for $\times$-multiplication for the induced representation on composition of partial systems presupposes only kinematical, not dynamical, independence of the partial systems.

We can, without further trouble, formulate the situation discussed above in terms of the general scheme of quantum mechanics in a manner which is independent of the particular assumptions of *Schrödinger's* scalar wave theory. This is all the more important since it has all along seemed doubtful whether the matter waves could be described in terms of a single state function ψ. We set up an analogy between the *actual* displacement of the state of the system in time and the *virtual* change produced by an arbitrary rotation of space. The transition from time t to time t' changes the (arbitrary) state $\mathfrak{x}$ at time t into a state $\mathfrak{x}'$ at time t', obtained from $\mathfrak{x}$ by a unitary transformation U corresponding to a displacement of the time axis which sends t over into t'. The displacements along the time axis constitute a one-parameter continuous group which is isomorphic with the group of transformations U associated with them in system-space. The former group is generated from the infinitesimal displacement $t \rightarrow t + dt$, and it therefore suffices to give the infinitesimal unitary operator

$$d\mathfrak{x} = \frac{dt}{ih} \cdot H\mathfrak{x}$$

associated with it in system-space. We called the Hermitian operator H the *energy*.

On subjecting the physical system (or the spatial co-ordinate system in terms of which it is described) to a virtual rotation s, the state $\mathfrak{x}$ goes over into another state $\mathfrak{x}'$. Since nothing intrinsic to the system is changed thereby and since the state space $\mathfrak{R}$ is linear and unitary, the transition $U(s) : \mathfrak{x} \rightarrow \mathfrak{x}'$ associated with s must also be linear and unitary. As in the case of the group of actual displacements in time, this group of virtual rotations in space must induce a certain representation $\mathfrak{R}$ in the system-space $\mathfrak{R}$; this latter is more properly to be considered as a ray, rather than a vector, space. But if we go over from the rotation group to the unimodular unitary group $\mathfrak{u}_2$ (or $\mathfrak{u}_2'$) by stereographic projection (III, § 8) and take this latter as fundamental, it is, in accordance with III, § 16, not necessary to distinguish between ray and vector representations. The group of proper rotations can be generated from its infinitesimal operations, and we may take as a basis for these the infinitesimal rotations D_x, D_y, D_z about the x-, y-, and z-axis. It then suffices to know the infinitesimal unitary transformations

$$d\mathfrak{x} = \frac{1}{i}(M_x, M_y, M_z)\,\mathfrak{x}$$

which they induce in system space. We call the real physical quantities of the system which are represented by the Hermitian operators M_x, M_y, M_z the x-, y-, z-components of the ***moment of momentum*** $\mathfrak{M}$. In order to express them in terms of the usual units they must, as was also the case with the energy, be multiplied by the quantum of action h. *The moment of momentum plays the same rôle with respect to the virtual rotations of space as the energy with respect to the actual displacements in time.*

One argument for the appropriateness of our definition of moment of momentum is that in the case of the *Schrödinger* theory it leads to the usual formulæ of classical mechanics. As a further justification we prove the general theorem that the moment of momentum so defined is constant in time. We saw in II, § 8, that the necessary and sufficient condition that the physical quantity represented by the Hermitian operator A be constant in time was that A commute with the Hermitian operator H induced by the infinitesimal displacement of time. In exactly the same way we can show that the commutativity of A with M_x, M_y, M_z constitutes the necessary and sufficient condition that the quantity represented by A remains unaltered under the virtual proper rotations of space, i.e. that A is a *scalar* with respect to these rotations. Now the energy is a scalar, hence

$$HM_x - M_x H = 0, \cdots.$$

But, on the other hand, these equations assert that M_x, M_y, M_z are constant in time.

The infinitesimal rotations generate only the group of proper rotations; in order to obtain the complete orthogonal group we must supplement them with the reflection i in the origin, or extend the group $\mathfrak{u}_2$ to the group $\mathfrak{u}_2'$ by the addition of the element ι (III, § 8). ι will induce a unitary operator I in system space which commutes with all $U(s)$, in particular with the moment of momentum $\mathfrak{M} = (M_x, M_y, M_z)$, and which satisfies the equation $II = \mathbf{1}$; this shows that I is Hermitian, as well as unitary. A quantity A which is unchanged by reflection must commute with I; hence, in particular, the energy H must commute with I. The physical quantity represented by I, which we call the ***signature***, is constant in time, as it commutes with H. It has, in common with all quantities arising in group theory which are not associated with infinitesimal operators, no analogue in classical mechanics.

We reduce the total system-space into invariant sub-spaces with respect to the group of displacements in time; such an invariant sub-space is carried over into itself by the generating infinitesimal operation $d\mathfrak{x} = \frac{1}{i}H\mathfrak{x}$. Since we are here dealing with a one-parameter Abelian group, or with a single operator H, this reduction can be carried to the point in which all the constituent sub-spaces are 1-dimensional. The states contained in one of these invariant sub-spaces we call *quantum states*.

We now proceed in exactly the same manner to reduce the representation $\mathfrak{R}$ induced in system space by the group of rotations into its irreducible constituents $\mathfrak{D}_j$. We make use of the fact that these are known to us *a priori*; only the number of times they appear in $\mathfrak{R}$ depends on the particular representation $\mathfrak{R}$. (Of course, we have not as yet shown that the $\mathfrak{D}_j$ really constitute a complete system of irreducible representations of $\mathfrak{d}_3$, and it may seem risky to apply the process of reduction to the ∞-dimensional representation $\mathfrak{R}$. This procedure can, however, be justified on the basis of the fact that $\mathfrak{d}_3$ is a closed group. But in the final formulation of quantum mechanics it will not be necessary to base our conclusions on such general considerations, as the reduction into $\mathfrak{D}_j$ will be obtained by elementary means.) The entire system-space $\mathfrak{R}$ is thus decomposed into sub-spaces $\mathfrak{R}_j$, $\mathfrak{R}'_{j'}$, $\cdots$ such that $\mathfrak{R}_j$ is of dimensionality $2j + 1$ and the representation induced in it by the group $\mathfrak{u}_2$ is $\mathfrak{D}_j$. On adapting the co-ordinate system in system-space to this decomposition the variables fall into classes

$$x(m) \quad (m = j, j-1, \cdots, -j); \qquad x'(m') \quad (m' = j', j'-1, \cdots, -j'); \cdots;$$

under the influence of an arbitrary transformation σ of $\mathfrak{u}_2$, applied to the variables ξ, η the co-ordinates of system-space transform in accordance with the law

$$x(m) \sim \frac{\xi^i \eta^k}{\sqrt{i!\,k!}} \quad (i + k = 2j,\ i - k = 2m).$$

With the reduction of $\mathfrak{R}$ or $\mathfrak{N}$ is associated the reduction of the angular momentum $\mathfrak{M}$; in the sub-space $\mathfrak{R}_j$ the components of $\mathfrak{M}$ are given by III, (15.9), from which it follows that the square M^2 of the moment of momentum has there the fixed value $j(j + 1)$. (It is evident from general considerations that M^2 must be a multiple of the unit matrix in $\mathfrak{R}_j$, for it is a scalar and must therefore commute with all the operators of the irreducible representation $\mathfrak{D}_j$.) If the state of the system is represented by a vector lying in $\mathfrak{R}_j$, the z-component of its moment of momentum is capable of assuming the values $m = j$, $j - 1, \cdots, -j$; the z-component naturally only apparently occupies a preferred status, due to the fact that the co-ordinates in $\mathfrak{R}_j$ were chosen in a manner which differentiated the z-axes from the others. That M_z, M^2 can *a priori* assume only discrete values m, $j(j + 1)$ is essentially due to the fact that the rotation group is closed; since the group of displacements in time is open, the analogous result for the energy need not in general hold. In this connection we wish to emphasize again that the operator H depends on the dynamical relationships existing in the system, whereas the representation $\mathfrak{N}$ induced by the group of rotations is determined only by the kinematical situation (number of elementary particles, etc.). The signature I also assumes a definite one of its values ± 1 in each sub-space $\mathfrak{R}_j$. For lack of a better name we call the states which lie in the sub-space $\mathfrak{R}_j$, which is invariant under the group of rotations, ***"simple" states*** of ***inner quantum number*** j. We must be prepared to find that j may here assume half-integral as well as integral values, in contrast with the *Schrödinger* theory.

On uniting two kinematically independent systems, with system-spaces $\mathfrak{R}$, $\mathfrak{R}'$ in which the rotation group induces the representations $\mathfrak{N}$, $\mathfrak{N}'$, the total system has as system-space $\mathfrak{R} \times \mathfrak{R}'$, in which the representation $\mathfrak{N} \times \mathfrak{N}'$ is induced. In particular, the moment of momentum of the total system is

$$(\mathfrak{M} \times \mathbf{1}) + (\mathbf{1} \times \mathfrak{M}')$$

where $\mathfrak{M}$ and $\mathfrak{M}'$ are the angular momenta of the two partial systems. The theorem that *the moment of momentum behaves additively with respect to composition* is contingent only on the assumption that the parts are kinematically independent, whereas the corresponding theorem for energy applies only if they are dynamically independent, i.e. in the absence of interaction between the parts. This difference is based on the fact that whereas the energy represents that *actual* change of state in the course of time, the moment of momentum represents the *virtual* change associated with a fictitious rotation. We reduce $\mathfrak{R}$, $\mathfrak{R}'$ into the invariant irreducible sub-spaces $\mathfrak{R}_j$, $\mathfrak{R}'_{j'}$ respectively, i.e into the simple states of the two partial systems having inner quantum numbers, j, j'. The Clebsch-Gordan equation (III, § 5)

$$\mathfrak{D}_j \times \mathfrak{D}_{j'} = \mathfrak{D}_{j+j'} + \mathfrak{D}_{j+j'-1} + \cdots + \mathfrak{D}_{|j-j'|} \qquad (1.3)$$

then tells us: *If the two parts are in the simple states with inner quantum numbers* j, j' *then the whole has each of the simple states with inner quantum number*

$$J = j + j', \quad j + j' - 1, \cdots, |j - j'| \qquad (1.4)$$

associated with it, each exactly once. To include the signature we must add: *If the parts have as signatures the values* δ, δ' ($\delta = \pm 1$), *the signature of the whole has the value* $\delta\,\delta'$.

Compare the results which we have obtained with the corresponding results in classical mechanics. In both the moment of momentum is constant in time and the moment of momentum of the whole is equal to the sum of the moments of momentum of the two parts. Denoting the magnitude of the moment of momentum in classical theory by j, we have, in agreement with (1.4),

$$|j - j'| \leqq J \leqq j + j',$$

for the resultant of two vectors of magnitudes j, j' is a vector whose magnitude J lies within these limits. *Quantum mechanics deviates from classical mechanics in the following three respects:*

1. *In quantum mechanics the square of the moment of momentum is* $j(j+1)$, *in classical mechanics it is* j^2;

2. *Here* j *can assume only the discrete values* $0, \frac{1}{2}, 1, \frac{3}{2}, \cdots$, *there it may have any non-negative value ;*

3. *Here the* J *obtained on compounding two partial systems can assume only those values between* $|j - j'|$, $j + j'$ *which differ from them by an integer, there it can assume any value between these limits.*

Already before the rise of the new quantum mechanics a semi-empirical description of the regularities observed in spectra had been given with the aid of a vector model consisting of the vectorial moments of momentum of the individual electrons and of the atom as a whole ; the observations, assisted by the older quantum mechanics, had already led to these three modifications of classical theory.[1]

The reader will perhaps have wondered why we consider only the virtual rotations of space and not the translations, which must also be taken into account in order to arrive at a complete description of the homogeneity of space. The reason for this is that in studying atoms or ions we treat only the electrons as particles, taking the nucleus as a fixed centre of force situated in the origin. That this is at least approximately correct is due to the fact that the mass of the nucleus is many times the mass of the electrons. Space is thereby transformed from a homogeneous into a centred space ; such a procedure naturally allows us to consider only atoms or ions, which have a single nucleus. Diatomic molecules are accordingly described with the aid of the 1-parameter group of rotations about the axis joining the two nuclei, and not by the full 3-parameter group of rotations of space—to this we must add reflection in the plane which bisects the axis perpendicularly in case the two nuclei are physically equivalent.[2] If we are dealing with three or more fixed nuclei the symmetry either disappears entirely or is reduced to at most a finite group of rotations.[3]

§ 2. Simple States and Term Analysis. Examples

To each characteristic value E' of the energy H there belongs a definite sub-space $\mathfrak{R}'$ of $\mathfrak{R}$, the sub-space of quantum states with energy level E' ; it consists of all states $\mathfrak{x}$ which are transformed into $E' \cdot \mathfrak{x}$ by the operator H and is accordingly the characteristic space $\mathfrak{R}(E')$ associated with the characteristic value E' of H. Since the energy is a scalar, the considerations applied in the preceding paragraph to the total space $\mathfrak{R}$ can also be applied to $\mathfrak{R}'$: $\mathfrak{R}'$ is invariant under the operators induced in system-space by the rotation group and is consequently the carrier of a certain representation of this group, which can be reduced into its irreducible constituents. If the energy levels are of at most finite multiplicity we are faced with the problem of reducing only representations of finite degree. Accordingly $\mathfrak{R}$ is decomposed into the "simple spaces" $\mathfrak{R}_j$ associated with the rotation group in such a way that not only the square of the angular momentum and the signature have definite values in $\mathfrak{R}_j$, but also the energy has a sharply defined value E_j. This energy level E_j is necessarily $(2j + 1)$-fold degenerate; we speak of *an accidental degeneracy* when the energy levels of different simple sub-spaces $\mathfrak{R}_j$ are equal. I, M_z, M^2 and H are all simultaneously in diagonal form; that this is possible is due to the fact that these four operators all commute among themselves. *In this way the reduction into simple states can be employed in term analysis:* each energy level E_j possesses an inner quantum number j which gives the term the natural multiplicity $2j + 1$.

On subjecting the atom to a perturbing field which destroys its natural spherical symmetry this $(2j + 1)$-fold term is broken up into $2j + 1$ terms. Let the perturbation, i.e. its Hamiltonian function W, possess axial symmetry about the z-axis; if E_j possesses no accidental degeneracy, then in accordance with the theory of perturbations the perturbed energy levels are given to a first approximation by the portion of the Hermitian operator W in which $\mathfrak{R}_j$ intersects itself:

$$x(m) \to \sum W(m, m')\, x(m') \quad (m' = j, j - 1, \cdots, -j).$$

The rotation about the z-axis with meridian angle ϕ transforms $x(m)$ into $e(-m\phi) \cdot x(m)$, and in virtue of the symmetry assumed for W this correspondence of $\mathfrak{R}_j$ on itself must also be represented by

$$e(-m\phi) \cdot x(m) = \sum W(m, m') \cdot e(-m'\phi)\, x(m'),$$

or

$$W(m, m')\, e[(m - m')\phi] = W(m, m').$$

But this means that all elements $W(m, m')$ except those in the main diagonal vanish, whence

$$E_j + W(m, m) \tag{2.1}$$

are the $2j + 1$ perturbed terms. The quantum number m, which is capable of assuming the values $j, j - 1, \cdots, -j$, thus serves to label these components. Perhaps the most important axially symmetric perturbation is that due to a homogeneous magnetic field in the direction of the z-axis (*Zeeman effect*); because of this m is called the ***magnetic quantum number***. The inner quantum number j of a term can be determined spectroscopically by counting the number of terms appearing in the *Zeeman* effect. *Sommerfeld* first concluded, from the spectroscopic data, that j as well as m must be allowed to assume half-integral values. If we consider the *Zeeman* effect to be described by the analogue of the classical formula II, (12.5) then

$$W = \frac{eh}{2\mu c}(\mathfrak{H}\mathfrak{M}) = ho M_z, \quad o = \frac{e|\mathfrak{H}|}{2\mu c}, \tag{2.2}$$

and W is rigorously in diagonal form:

$$W(m, m) = hom. \tag{2.3}$$

Our analysis shows that the breaking up of energy levels due to an axially symmetric perturbation parallels the reduction of an irreducible representation of the rotation group $\mathfrak{d}_3$ when this is restricted to the group $\mathfrak{d}_2$ of rotations about the z-axis: by this $\mathfrak{D}_j$ is reduced into the $2j + 1$ one-dimensional representations which we have previously denoted by $\mathfrak{D}^{(m)}$:

$$x(m) \rightarrow e(-m\phi) \cdot x(m).$$

If two kinematically independent parts, which are in the simple states $\mathfrak{R}_j$, $\mathfrak{R}'_{j'}$, are compounded together, the state of the composite system is in the $(2j + 1)(2j' + 1)$-dimensional product space $\mathfrak{R}_{jj'} = \mathfrak{R}_j \times \mathfrak{R}'_{j'}$. If the parts have the energies E_j, $E'_{j'}$ then the whole has the energy $E_j + E'_{j'}$, assuming no interaction between the parts. Introducing a weak interaction between the two partial systems and assuming that there is no accidental degeneracy, i.e. assuming that all the remaining energy levels of the unperturbed system are different from $E_{jj'}$, it suffices, to a first approximation, to consider the section $\langle H \rangle$ of the energy operator H in which $\mathfrak{R}_{jj'}$ intersects itself; it is an Hermitian correspondence of $\mathfrak{R}_{jj'}$ on itself. We can apply the considerations, which were applied above to the total system-space $\mathfrak{R} \times \mathfrak{R}'$, to each of these $\mathfrak{R}_{jj'}$: $\mathfrak{R}_{jj'}$ is to be decomposed into sub-spaces belonging to numerically distinct characteristic values of $\langle H \rangle$. The rotation group induces a certain representation in each of these sub-spaces, and this can be further decomposed into its irreducible constituents. The result is that $\mathfrak{R}_j \times \mathfrak{R}'_{j'}$ is, in accordance with the Clebsch-Gordan series, reduced into the simple spaces $\mathfrak{R}_J$, $J = j + j'$, $j + j' - 1, \cdots, |j - j'|$, in such a way that in each of them the energy $\langle H\rangle$ has a definite value E_J. Different E_J can only "accidentally" have the same numerical value. Consequently the term $E_{jj'}$ is broken up by the perturbation into terms E_J in exactly the same way as the representation $\mathfrak{D}_j \times \mathfrak{D}_{j'}$ is reduced into the irreducible representations $\mathfrak{D}_J$. But this is only correct to the approximation characteristic of perturbation theory. As we have seen above, an inner quantum number J can be rigorously ascribed to a term E; in the approximation with which we have been dealing here there is associated with it in addition the inner quantum numbers j, j' of the parts, in the last analysis of the electrons themselves: the energy level E arises from a definite term $E_{jj'}$ of the unperturbed system by interaction of the two parts. Such an association is rigorously possible for "simple states," but the rules based on it lead only indirectly and approximately to an analysis of the terms.[4]

Examples

If we take the Schrödinger scalar wave theory to be valid for a single electron, then a simple quantum state of the electron in the field of the nucleus is characterized by the principal quantum number n and the azimuthal quantum number l (we here use the word "azimuthal" instead of "inner"). Such a term is $(2l+1)$-fold degenerate, and we assume there is no further accidental degeneration. The moment of momentum is represented by the operator $\mathfrak{L}$ taken over from classical theory; the square of its absolute magnitude is $l(l+1)$ and the signature has the value $(-1)^l$. If f electrons come together to form an atom we obtain a term, neglecting interaction between the electrons,

$$E(n_1 l_1) + E(n_2 l_2) + \cdots + E(n_f l_f) \tag{2.4}$$

of multiplicity $(2l_1+1)\cdots(2l_f+1)$. The quantum numbers n and l refer to the individual electrons. The interaction causes a separation which parallels the complete reduction, obtained with the aid of the Clebsch-Gordan series, of

$$\mathfrak{D}_{l_1} \times \mathfrak{D}_{l_2} \times \cdots \times \mathfrak{D}_{l_f} \tag{2.5}$$

into its irreducible constituents $\mathfrak{D}_L$ with total azimuthal quantum number L. Each such term is associated with the quantum numbers

$$(n_1 l_1, n_2 l_2, \cdots, n_f l_f; L). \tag{2.6}$$

If $f \geqq 3$ certain $\mathfrak{D}_L$ appear more than once in (2.5), and we may therefore have several $(2L+1)$-fold terms associated with the same set (2.6); these must then be distinguished from each other by some further index. The square of the total moment of momentum is $L(L+1)$ and the signature $(-1)^{l_1+l_2+\cdots+l_f}$. In spectroscopy it is usual to characterize the values $l = 0, 1, 2, 3, 4, \cdots$ by the small Latin letters $s, p, d, f, \cdots$ and the values $L = 0, 1, 2, 3, \cdots$ by the corresponding capitals $S, P, D, F, \cdots$.

We cannot expect the scalar wave theory to be correct, but must be prepared to describe the state of the wave field in terms of a quantity ψ with several, say a, components $(\psi_1, \psi_2, \cdots, \psi_a)$, i.e. by a covariant quantity of a definite kind $\mathfrak{A}$. Each component is a function of the spatial co-ordinates xyz; the components will depend on the choice of the Cartesian co-ordinate system in such a way that on going over to a new co-ordinate system by the rotation s the components will undergo among themselves that transformation $A(s)$ which corresponds to s in the representation $\mathfrak{A}$. Again, consider $\mathfrak{d}_3$ replaced by $\mathfrak{u}_2$ as the fundamental group. The general component $\psi_\alpha(xyz)$ of the "vector" ψ has two indices, the index α running from 1 to a and the index (xyz) running through all the points of space. Let $\mathfrak{R}_t$ be the vector space of functions $\psi(xyz)$ and $\mathfrak{R}_a$ the a-dimensional vector space; the state space of a single electron is then $\mathfrak{R}_a \times \mathfrak{R}_t$. Under the influence of the rotation s which sends xyz into $x'y'z'$ the state ψ goes over into the state ψ' defined by the equation

$$\psi'_\alpha(x'y'z') = \sum_\beta a_{\alpha\beta}\psi_\beta(xyz), \quad \| a_{\alpha\beta} \| = A(s);$$

the representation induced in system-space is accordingly $\mathfrak{R} = \mathfrak{A} \times \mathfrak{C}$. The moment of momentum $\mathfrak{M}$ of the electron consists of two parts:

$$\mathfrak{M} = (\mathfrak{S} \times \mathbf{1}) + (\mathbf{1} \times \mathfrak{L}), \tag{2.7}$$

the first of which refers to the a-dimensional "spin space" $\mathfrak{R}_a$, the second to the "translation space" $\mathfrak{R}_t$. $(\mathbf{1} \times L_x)$, or simply L_x, is the operator $\frac{1}{i}\left(y\frac{\partial}{\partial z} - z\frac{\partial}{\partial y}\right)$ which acts on each of the a components in the same way; it affects only the index (xyz), leaving the index α unaltered. $\frac{1}{i}S_x$ is the unitary transformation corresponding to the infinitesimal rotation about the x-axis in the representation $\mathfrak{A}$; $(S_x \times \mathbf{1})$, or simply S_x, consequently affects only the index α and leaves (xyz) unchanged. Only the part $\mathfrak{L}$ appears in classical mechanics; we call it the ***orbital moment of momentum***, and the remaining part $\mathfrak{S}$ the ***spin moment of momentum***, or simply the ***spin***. Its appearance is unavoidable so long as the wave quantity ψ is not simply a scalar or a set of scalars. Each of the two parts satisfies separately the commutation rules III, (15.7), but in general only the total angular momentum satisfies the law of conservation. If the quantity ψ is of a simple kind, i.e. if $\mathfrak{A}$ is an irreducible representation $\mathfrak{D}_s$, then $a = 2s + 1$ and the spin $\mathfrak{S}$ is equal to the moment of momentum $\mathfrak{M}_s$ associated with the representation $\mathfrak{D}_s$.

Since the Schrödinger theory has proved itself at least approximately correct, one should assume that to a first approximation each of the components ψ_α satisfies the Schrödinger scalar wave equation. So long as we consider this approximation, the a components have only the effect of multiplying the multiplicity of each energy level by a. But in reality the correct differential equations must contain a term, the "*spin perturbation*," which introduces a coupling between the various components ψ_α. The electron can thus be considered *in abstracto* as a composite system, consisting of the ***electron translation*** with system-space $\mathfrak{R}_t$ and the ***electron spin*** with system-space $\mathfrak{R}_a$; the spin perturbation is the weak interaction between these two. Because of this the method of composition can here be applied. Let $\mathfrak{A} = \mathfrak{D}_s$. Decompose the translation space $\mathfrak{R}_t$ into the $(2l + 1)$-dimensional sub-spaces $\mathfrak{R}(nl)$; the corresponding energy term $E(nl)$ with ***azimuthal quantum number*** l has, on neglecting the spin perturbation, the multiplicity $a(2l + 1)$ and its characteristic space is the space $\mathfrak{R}_a \times \mathfrak{R}(nl)$ of the same dimensionality. On taking the first order spin perturbation into account this term is separated into the terms E_j with inner quantum number j and multiplicity $(2j + 1)$ in a manner paralleling the decomposition of the representation $\mathfrak{D}_s \times \mathfrak{D}_l$ into its irreducible constituents:

$$\mathfrak{D}_s \times \mathfrak{D}_l = \Sigma \mathfrak{D}_j, \quad j = s + l, s + l - 1, \cdots, |l - s|, \qquad (2.8)$$

with the aid of the Clebsch-Gordan series. Care must be taken to differentiate sharply between the azimuthal and inner quantum numbers l and j. The latter is capable of assuming the values given in (2.8); whenever $l \geqq s$ the number of different terms in such a "*multiplet*" is $2s + 1$. L^2 is *approximately* equal to the constant $l(l + 1)$, S^2 is *approximately* equal to the constant $s(s + 1)$, and M^2 is *rigorously* constant and *exactly* equal to $j(j + 1)$. We can thus speak of the azimuthal quantum number of an actual energy term only to within the approximation characteristic of perturbation theory. It is well to set forth these considerations beforehand and to approach the spectroscopic data, as we shall in § 4, with them well in mind.

§ 3. Selection and Intensity Rules

We return to the consideration of our system as a whole, without resolving it into its individual electrons, and again denote the total inner quantum number by j. Let A be any physical quantity of the system, and let it be represented by the Hermitian form A; we write that portion of this form in which $\mathfrak{R}_j$ intersects $\mathfrak{R}'_{j'}$ in the form

$$\sum a(mm')\bar{x}(m)x'(m'), \tag{3.1}$$

where the indices m, m' run through the values

$$m = j,\ j-1,\ \cdots,\ -j\,;\ m' = j',\ j'-1,\ \cdots,\ -j'. \tag{3.2}$$

If the quantity A is a scalar, the operator A commutes with the operators $U(s)$ induced in system-space by the rotations s. On decomposition into these irreducible sub-spaces $\mathfrak{R}_j$, $\mathfrak{R}'_{j'}$ it follows from the fundamental theorem III, (10.5), of the theory of representations that *the section* (3.1) *of A corresponding to the transition $\mathfrak{R}_j \to \mathfrak{R}'_{j'}$ is zero if $j' \neq j$ and a multiple of the* $(2j+1)$*-dimensional unit form*

$$\sum_m \bar{x}(m)\, x'(m),$$

if $j' = j$.

An analogous situation exists for the group $\mathfrak{d}_2$ of rotations about the z-axis. With respect to it the total system space decomposes into 1-dimensional invariant sub-spaces $\mathfrak{R}^{(m)}$ in which the rotation with angle ϕ induces the representations $\mathfrak{D}^{(m)}: x(m) \to e(-m\phi)\, x(m)$. If we only assume that the physical quantity A possesses axial symmetry about the z-axis it follows that the coefficient $a(mm')$ is necessarily zero when the magnetic quantum numbers m and m' of the initial and final states are different.

We now consider a vectorial quantity $\mathfrak{q}$ with the three components q_x, q_y, q_z instead of the scalar quantity A. This is of particular importance because such a quantity, i.e. the electric dipole moment $\mathfrak{q}$ of the atom, determines the interaction between the atom and radiation—to that approximation in which the linear dimensions of the atom may be neglected in comparison with the wave-length of the emitted light. If the degeneracy of the energy level E_j is destroyed by an external axially symmetric perturbation, e.g. a homogeneous magnetic field in the direction of the z-axis, then the spectral line caused by the transition $\mathfrak{R}_j \to \mathfrak{R}'_{j'}$ from the term E_j to $E'_{j'}$ is broken up into the lines associated with all possible transitions

$(\mathfrak{R}_j, m) \to (\mathfrak{R}'_{j'}, m')$. On calculating the part of the Hermitian form representing the electric dipole moment in which the sub-space $\mathfrak{R}_j$ intersects $\mathfrak{R}'_{j'}$:

$$\Sigma \mathfrak{q}(mm')\bar{x}(m)x'(m'), \tag{3.3}$$

the ratios of the squares $|\mathfrak{q}(mm')|^2$ of the absolute values of its coefficients *determine the relative intensities of these* $(2j + 1)(2j'+1)$ *lines.* Since q_z is axially symmetric about the z-axis $q_z(mm') = 0$ unless $m' = m$; we thus have the selection rule

$$q_z : m \to m \tag{3.4}$$

for the z-component of the electric moment. On performing the rotation with angle ϕ about the z-axis $x(m)$, $q_x + iq_y$, $q_x - iq_y$ are multiplied by $e(-m\phi)$, $e(\phi)$, $e(-\phi)$ respectively. Since $\bar{x}(m)x'(m')$ is therefore multiplied by $e[(m - m')\phi]$ we obtain the selection rules

$$q_x + iq_y : \quad m \to m - 1, \quad q_x - iq_y : \quad m \to m + 1 \tag{3.4'}$$

for the x- and y-components of $\mathfrak{q}$. *Only the transitions*

$$m \quad \to \quad m - 1, \quad m, \quad m + 1 \tag{3.5}$$

of the magnetic quantum number are allowed; the first and the last generate two waves which are circularly polarized in the xy-plane in opposite directions, and the remaining transition $m \to m$ *generates a wave which is linearly polarized in the z-direction.* If the equation (2.3) holds for Zeeman effect, the wave number of the component $m \to m'$ is displaced by an amount $o(m - m')$ from its unperturbed value. Thus in "*normal Zeeman effect*" we obtain instead of $(2j + 1)(2j' + 1)$ components only three, whose polarization is as described above and whose wave numbers are displaced by the amounts 0, $\pm o$. That the resolution of the two terms E_j, $E'_{j'}$, is almost entirely hidden is due to the fact that the factor of proportionality ho in (2.3) has the same value for both terms. Fortunately most of the cases actually observed show "*anomalous Zeeman effect,*" in which the resolution of the terms can be seen clearly; in order to explain it we must change the expression (2.2) for the perturbation due to the magnetic field. But the above selection rule for the magnetic quantum number, which has been obtained from fundamental principles of group theory, is valid in all cases.

The selection rule for the inner quantum number j is obtained in an analogous manner. The three components q_x, q_y, q_z of $\mathfrak{q}$ suffer the transformation s among themselves when the $x(m)$, $x'(m')$ are subjected to the transformations corresponding to s in the representations $\mathfrak{D}_j$, $\mathfrak{D}_{j'}$ respectively. Or, if we wish to express it in terms of $\mathfrak{u}_2$ instead of $\mathfrak{d}_3$, s is that transformation which is associated with the element σ of $\mathfrak{u}_2$ in the representation $\mathfrak{D}_1$. This is, of course, merely an expression of the fact that $\mathfrak{q}$ is a vector. Now, in accordance with the terminology introduced in III, § 14, (3.3) is a vectorial quantity in the representation space of $\bar{\mathfrak{D}}_j \times \mathfrak{D}_{j'}$, and we are interested in determining how many linearly independent quantities of this kind there are. Their number is given by the number of times $\mathfrak{D}_1$ is contained in $\bar{\mathfrak{D}}_j \times \mathfrak{D}_{j'}$ or $\mathfrak{D}_j \times \mathfrak{D}_{j'}$ as an irreducible constituent. But in accordance with (1.3) $\mathfrak{D}_1$ occurs in $\mathfrak{D}_j \times \mathfrak{D}_{j'}$ *exactly once* if

$$j' = j - 1 \quad \text{or} \quad j \quad \text{or} \quad j + 1$$

and otherwise not at all, and we must further exclude the case $j = 0$, $j' = 0$. *We thus obtain the selection rule*

$$j \to j - 1, \quad j, \quad j + 1 \tag{3.6}$$

with the proviso that $0 \to 0$ *does not occur.* Since there exists but *one* linearly independent vectorial quantity in the representation space of $\mathfrak{D}_j \times \mathfrak{D}_{j'}$ in the cases in which the selection rule is satisfied, the components of $\mathfrak{q}(m, m')$ are determined by purely group-theoretic considerations to within a constant factor of proportionality.

In order to calculate the vectorial quantity (3.3) for $j' = j - 1$ we proceed as follows. Let ξ, η; ξ', η' be two arbitrary points on the unit sphere which transform cogrediently under $\mathfrak{u}$. $\bar{\xi}\xi' + \bar{\eta}\eta'$ is then the fundamental invariant, and the three forms which are obtained from

$$\frac{1}{k!}(\bar{\xi}\xi' + \bar{\eta}\eta')^k \tag{3.7}$$

by multiplication with

$$-\bar{\xi}^2, \quad \bar{\eta}^2, \quad \bar{\xi}\bar{\eta} \tag{3.8}$$

transform in the same way as the $(x + iy)$-, $(x - iy)$-, z-components of a vector, respectively. They are linear in the monomials $\bar{\xi}^r\bar{\eta}^s$ of degree $k + 2 = 2j$ and in the monomials $\xi'^{r'}\eta'^{s'}$ of degree $k = 2j'$. Introducing

$$x(m) = \frac{\xi^r\eta^s}{\sqrt{r!\,s!}}, \quad x'(m') = \frac{\xi'^{r'}\eta'^{s'}}{\sqrt{r'!\,s'!}}$$

$(2j = r + s = k + 2, \quad 2m = r - s; \quad 2j' = r' + s' = k,$
$2m' = r' - s')$

as co-ordinates in the representation spaces of $\mathfrak{D}_j$, $\mathfrak{D}_{j'}$ we find that the three forms above are of the type (3.3) with $j' = j - 1$. For example, we obtain for the $(x + iy)$-component

$$-\sum_{(r-2)+s=k} \frac{(\bar{\xi}\xi')^{r-2}(\bar{\eta}\eta')^s}{(r-2)!\,s!} \cdot \bar{\xi}^2 = -\sum \frac{\bar{\xi}^r\bar{\eta}^s}{\sqrt{r!\,s!}} \frac{\xi'^{r-2}\eta'^s}{\sqrt{(r-2)!\,s!}} \sqrt{r(r-1)}$$

$$= -\sum_m \sqrt{(j+m)(j+m-1)}\,\bar{x}(m)x'(m-1).$$

In agreement with the selection rule $m \to m - 1$ there occur here only those terms for which $m' = m - 1$. Calculating the $(x - iy)$- and z-components in the same way, we find for the transition

$\underline{j \to j' = j - 1:}$

$$\begin{aligned} (q_x + iq_y)(m, m-1) &= -\sqrt{(j+m)(j+m-1)}, \\ (q_x - iq_y)(m, m+1) &= \sqrt{(j-m)(j-m-1)}, \\ q_z(m, m) &= \sqrt{(j+m)(j-m)}. \end{aligned} \tag{3.9}$$

In order to calculate the components for the transition $j = j'$ we must replace the factors (3.8) by

$$2\eta'\bar{\xi}, \quad 2\xi'\bar{\eta}, \quad \xi'\bar{\xi} - \eta'\bar{\eta}$$

which also transform like the $(x + iy)$-, $(x - iy)$- and z-components of a vector. Finally, for the transition $j' = j + 1$ we must replace (3.8) by η'^2, $-\xi'^2$, $\xi'\eta'$. Since the angular momentum $\mathfrak{M}$ is a vector, the formulæ for the transition $j \to j$ must naturally agree with those already obtained for $\mathfrak{M}$ [III (15.9)], and since $\mathfrak{q}$ is Hermitian the formulæ for the transition $j \to j + 1$ must agree with those obtained by taking the Hermitian conjugate of the components for the transition $j \to j - 1$.

$\underline{j \to j' = j.}$

$$\begin{aligned} (q_x + iq_y)(m, m-1) &= \sqrt{(j+m)(j-m+1)}, \\ (q_x - iq_y)(m, m+1) &= \sqrt{(j-m)(j+m+1)}, \\ q_z(m, m) &= m. \end{aligned} \tag{3.9}$$

$\underline{j \to j' = j + 1.}$

$$\begin{aligned} (q_x + iq_y)(m, m-1) &= \sqrt{(j-m+1)(j-m+2)}, \\ (q_x - iq_y)(m, m+1) &= -\sqrt{(j+m+1)(j+m+2)}, \\ q_z(m, m) &= \sqrt{(j+m+1)(j-m+1)}. \end{aligned} \tag{3.9}$$

In each of these three sets of formula the right-hand sides are determinate only to within a common factor of proportionality which is independent of m, but which can be completely determined only by integrating the wave equation of the dynamic model of the atom, and not by the theory of groups alone. The coefficients which do not occur explicitly in the above formulæ are all null. *The squares of the absolute values of these coefficients yield the (rational!) intensity ratios of the components into which a line is split by the perturbation.*

Already before the rise of the new quantum mechanics the intensity formulæ (3.9) for the components of a line emitted under the influence of a magnetic field were obtained from the observational data under the guidance of the correspondence principle.[5] In the new quantum mechanics they are, as we have seen, a consequence of the most general principles, and we would find ourselves in serious difficulties if they were incorrect. Nevertheless it is to be remembered that they can be invalid (1) if the spherical symmetry of the system is destroyed by external perturbing fields, or (2) if for short wave-lengths the interaction between matter and radiation is no longer determined primarily by the electric dipole moment.

Since the dipole moment is a proper vector, as the components q_x, q_y, q_z go over into $-q_x, -q_y, -q_z$ on reflection i in the origin, the representation $\mathfrak{D}_1$ induced on them by $\mathfrak{u}_2'$ has as signature -1. If the signatures of $\mathfrak{R}_j, \mathfrak{R}'_{j'}$ are δ, δ', then under the influence of the reflection i (3.3) is multiplied by the factor $\delta\delta'$. The coefficients $\mathfrak{q}(mm')$ must accordingly all vanish unless $\delta\delta' = -1$; *the selection rule for the signature is*

$$\delta \rightarrow -\delta.$$

If the individual electrons are governed by the scalar wave theory the total azimuthal quantum number L of the atom can jump only to $L-1$, L or $L+1$, while *the sum of the azimuthal quantum numbers of the individual electrons $l_1 + l_2 + \cdots + l_f$ can change only by an odd integer (Laporte's rule).* In the case of a single electron, $f = 1$, only the transitions $l \rightarrow l \pm 1$ are consistent with these rules; this result has already been obtained in II, § 5, from the theory of spherical harmonics.

The formulæ (3.9) allow us to solve a problem which we shall here, for the sake of future application, introduce from the physical standpoint. A partial system in the simple state $\mathfrak{R}_j$ is compounded with a second in the simple state $\mathfrak{R}'_{j'}$ to form a single system. In $\mathfrak{R}_{jj'} = \mathfrak{R}_j \times \mathfrak{R}'_{j'}$, $\mathfrak{u}_2$ induces the representation $\mathfrak{D} = \mathfrak{D}_j \times \mathfrak{D}_{j'}$; let the corresponding moment of momentum be $\mathfrak{M}$. On adapting the normal co-ordinate system in $\mathfrak{R}_{jj'}$ to the complete reduction of $\mathfrak{D}$ into its irreducible constituents $\mathfrak{D}_J$, $\mathfrak{M}$ is broken up into square sub-matrices $\mathfrak{M}_J$ of length $2J + 1$, arranged along the principal diagonal, corresponding to the decomposition of $\mathfrak{R}_{jj'}$ into sub-spaces $\mathfrak{R}_J$. But the same is not true of the moment of momentum $\mathfrak{M}_j \times \mathbf{1}$ of the first partial system, and we wish to determine the portion of this matrix in which $\mathfrak{R}_J$ intersects itself. That is, in physical language, we wish to determine the temporal mean value $\langle \mathfrak{M}_j \rangle$ of the moment of momentum of the first system in the state defined by the quantum numbers j, j' ; J of the two parts and the whole. We assume that the interaction between the two parts resolves the energy level $E_{jj'}$ into *distinct* levels E_J on applying the theory of perturbations. Since $\mathfrak{M}_j$ is a vector we know, from the same considerations as we applied to the electric dipole moment above, that the portion of it corresponding to the transition $J \rightarrow J$ must be a multiple of $\mathfrak{M}_J$:

$$\langle \mathfrak{M}_j \times \mathbf{1} \rangle_J = \kappa_J \cdot \mathfrak{M}_J. \tag{3.10}$$

In order to evaluate the proportionality factor κ we construct the scalar product of the matrices $(\mathfrak{M}_j \times \mathbf{1})$ and $\mathfrak{M}$; since

$$\mathfrak{M} = (\mathfrak{M}_j \times \mathbf{1}) + (\mathbf{1} \times \mathfrak{M}_{j'})$$

these two matrices commute and we have

$$(\mathbf{1} \times \mathfrak{M}_{j'})^2 = \mathfrak{M}^2 + (\mathfrak{M}_j \times \mathbf{1})^2 - 2\mathfrak{M}(\mathfrak{M}_j \times \mathbf{1})$$

or

$$2\mathfrak{M}(\mathfrak{M}_j \times \mathbf{1}) = j(j+1) - j'(j'+1) + \mathfrak{M}^2, \tag{3.11}$$

for since in the original co-ordinate system $(\mathfrak{M}_j \times \mathbf{1})^2$ was $j(j + \mathbf{1})$ times the unit matrix, it remains the same in the new co-ordinates. And, on the other hand, $\mathfrak{M}(\mathfrak{M}_j \times \mathbf{1})$ is equal to $\kappa_J \cdot J(J + 1)$ times the unit matrix in the sub-space $\mathfrak{R}_J$, as follows from (3.10). Hence from (3.11)

$$2\kappa_J J(J+1) = j(j+1) - j'(j'+1) + J(J+1),$$

$$\kappa_J = \frac{1}{2} + \frac{j(j+1) - j'(j'+1)}{2J(J+1)}. \tag{3.12}$$

§ 4. The Spinning Electron, Multiplet Structure and Anomalous Zeeman Effect

We have hitherto ignored the fact that the terms of the *alkali spectra*, characterized by the two quantum numbers n, l, are in reality not simple. Each of these terms—with the exception of the s terms $l = 0$—actually consists of a fine *doublet*. By § 2 the (n, l) term should be resolved into $2l + 1$ components in a magnetic field; instead we find that one of the doublet terms breaks up into $2l$ components and the other into $2l + 2$. We should accordingly ascribe to them the inner quantum numbers $j = l - \frac{1}{2}$, $j = l + \frac{1}{2}$, respectively.

Our general considerations immediately give us a hint as to how this discrepancy is to be explained. *The quantity* ψ *describing the wave field is not a scalar, but is instead a covariant quantity of the kind* $\mathfrak{D}_{\frac{1}{2}}$, *having two components* (ψ_1, ψ_2). This is the theory of doublet phenomena as developed by *W. Pauli.*[6] It seems indeed easy to arrive at this conclusion after the preparation of the preceding paragraphs, but historically this systematic foundation was developed only after *Pauli's* discovery. It is quite immaterial whether we associate the matrix $+\mathbf{1}$ or the matrix $-\mathbf{1}$ with the element ι in the representation $\mathfrak{D}_{\frac{1}{2}}$ of $\mathfrak{u}_2'$. Taking the first of these alternatives, the signature has the value $(-1)^l$ in the quantum state (nlj); hence *Laporte's rule remains rigorously correct on taking the spin into account.* We have as further rigorous selection rules those concerning the total inner and the total magnetic quantum numbers. In the representation $\mathfrak{D}_{\frac{1}{2}}$ the transformation σ itself corresponds to the element σ of $\mathfrak{u}_2$, and by III, (15.6), the spin moment of momentum is $\frac{1}{2}\mathfrak{S}$, where $\mathfrak{S}$ is the vector already defined with components

$$S_x = \begin{Vmatrix} 0 & 1 \\ 1 & 0 \end{Vmatrix}, \quad S_y = \begin{Vmatrix} 0 & -i \\ i & 0 \end{Vmatrix}, \quad S_z = \begin{Vmatrix} 1 & 0 \\ 0 & -1 \end{Vmatrix}.$$

We shall not as yet attempt to find the specific effect of the spin perturbation on the wave equation. This was done originally by picturing the electron as a small material sphere, the rotation of which gave rise to the spin; the additional moment of momentum required by spectroscopic observations was first introduced in this way by *Goudsmit and Uhlenbeck.*[7] Since S_z is capable of assuming only the values ± 1 it appears as if the spin axis can only be quantized along the positive or negative z-axis; we need not go into the false conclusions this assertion can lead to on interpreting it literally. The spin perturbation must appear in going over from classical to relativistic mechanics. The terms of the hydrogen atom, calculated in accordance with the scalar non-relativistic wave mechanics, depend only on the principal quantum number n, but the theory of relativity introduces a correction which causes the terms corresponding to the various values of l to split apart and form the so-called *fine structure.* We should therefore expect the same scheme of terms in hydrogen as in the alkalies, but observation shows that the doublet separation of an l term into two terms with $j = l \pm \frac{1}{2}$ is just such that two terms with the same j, but with different $l = j \pm \frac{1}{2}$, exactly coincide. Hence the spin perturbation in hydrogen agrees quantitatively with the separation caused by the relativity correction.

The alkali doublets show anomalous Zeeman effect. Other elements, such as alkaline earth metals, have (in addition to triplets) a system of singlet terms, and singlet terms always show normal Zeeman effect in a magnetic field. It therefore seems probable that the anomalies in Zeeman effect are closely connected with the spin. The magnetic separation of an alkali term is quite independent of the principal quantum number n; all the terms of a series behave in the same way. A term (l, j) splits up into $2j + 1$ equi-distant components, characterized by the magnetic quantum number m, but their separation is hog instead of ho, where g is a rational function of l and j (the "Landé g-factor"). *The energy value of the component m is therefore displaced by an amount*

$$hog \cdot m \quad (m = j, j-1, \cdots, -j) \tag{4.1}$$

from its unperturbed value. The empirical formula for the factor g, which is due to Landé, is

$$g = \frac{2j+1}{2l+1}. \tag{4.2}$$

This formula holds for *weak magnetic fields*, in which the separation is of a smaller order of magnitude than the doublet separation. If $l = 0$, $j = \frac{1}{2}$, we have in particular $g = 2$.

This latter fact gives a hint toward the solution of the puzzle: If the total moment of momentum consisted only of the spin ($\mathfrak{L} = 0$), its magnetic effect would be twice as great as if it consisted of $\mathfrak{L}$ alone. We therefore assume that *the magnetic effect of the spin $\frac{1}{2}\mathfrak{S}$ is twice as great as that of the orbital angular momentum $\mathfrak{L}$; the perturbation due to an external magnetic field $\mathfrak{H}$ is therefore to be taken as*

$$W = \frac{eh}{2\mu c}(\mathfrak{H}, \mathfrak{L} + \mathfrak{S}) = \frac{eh}{2\mu c}(\mathfrak{H}, \mathfrak{M} + \tfrac{1}{2}\mathfrak{S}). \tag{4.3}$$

The spin offers an explanation of why the beam in the Stern-Gerlach experiment is separated into two parts. The valence electron of the univalent silver atom is, in the normal state, in an s-orbit ($l = 0$); hence $j = \frac{1}{2}$ and m can assume only the values $\pm \frac{1}{2}$. Although the component of the mechanical moment of momentum in the direction of the magnetic field can have only the values $\pm \frac{h}{2}$, the experiment shows that the value of the magnetic moment of the atom is a whole Bohr magneton, and not the half of one; but we now see that since the mechanical moment of momentum consists only of spin it *should* give rise to twice the expected magnetic moment. The connection between magnetic moment and mechanical moment of momentum is even more apparent in the *magneto-mechanical effect:* the demagnetization of a vertically suspended bar of weak iron must result in giving to it an angular momentum. The ratio between the change in the magnetic moment and the moment of momentum was expected to be $\frac{e}{2\mu c}$, but the experiment, which was performed only on ferro-magnetic bodies, yielded twice this value. The anomalous magnetic behaviour of the spin also accounts for this result, if we assume that the mechanical moment of momentum in ferro-magnetic substances is due entirely to the electron spin.[8]

Does this hypothesis also explain the general Landé formula (4.2)? This is answered by the formula (3.12) obtained toward the end of § 3, in which j, j', J must be taken as $\frac{1}{2}$, l, j in order that it apply to the composition of electron spin and electron translation. We find that in the state (lj) the temporal mean value of the spin $\frac{1}{2}\mathfrak{S}$ is equal to $\mathfrak{M}$ multiplied by the factor

$$g - 1 = \frac{1}{2} + \frac{\frac{3}{4} - l(l+1)}{2j(j+1)}$$

or

$$g - 1 = \pm \frac{1}{2l+1} \quad \text{for} \quad j = l \pm \frac{1}{2}. \tag{4.4}$$

Hence by (4.3)

$$\langle W \rangle = \frac{eh}{2\mu c} \cdot g \cdot (\mathfrak{H}\mathfrak{M}) = ho g \cdot M_z.$$

So long as the magnetic separation is small compared with the spin perturbation the Zeeman separation of the term (lj) is determined primarily by $\langle W\rangle$; (4.4) then leads, in fact, to equation (4.2), in agreement with the empirical data.

If the atom consists of several, say f, electrons, the situation then arising can be understood with the aid of the general rule of composition. If the electrons are in quantum states with inner quantum numbers j_r and energy levels $E(j_r)$, $(r = 1, 2, \cdots, f)$, then on neglecting the interaction between the electrons the total system has a $(2j_1 + 1) \cdots (2j_f + 1)$-fold energy level $E(j_1) + \cdots + E(j_f)$. If this level coincides with none of the other levels it is resolved by a small perturbation into terms with total inner quantum numbers J in a manner corresponding to that in which the product

$$\mathfrak{D}_{j_1} \times \mathfrak{D}_{j_2} \times \cdots \times \mathfrak{D}_{j_f} = \Sigma \mathfrak{D}_J \tag{4.5}$$

is reduced into its irreducible constituents $\mathfrak{D}_J$ (*Clebsch-Gordan* series). Obviously in order that this (jj) coupling lead to an adequate description the mutual interactions between the electrons must be small compared with the spin perturbation.

The situation usually met is, however, the opposite of that contemplated above: *the normal term order* corresponds to the Russell-Saunders or (sl) coupling. Neglecting for the moment the interaction between the electrons as well as the spin perturbation, we are led to a $2^f(2l_1 + 1) \cdots (2l_f + 1)$-fold energy level (2.4) in whose characteristic space the rotation group induces the representation

$$\mathfrak{D}_{\frac{1}{2}}^f \times (\mathfrak{D}_{l_1} \times \mathfrak{D}_{l_2} \times \cdots \times \mathfrak{D}_{l_f}). \tag{4.6}$$

Due to the interaction between the electron translations the second factor is reduced in a manner analogous to (4.5); a single term with azimuthal quantum number L has now the multiplicity $2^f(2L + 1)$. We next reduce

$$\mathfrak{D}_{\frac{1}{2}}^f = \Sigma \mathfrak{D}_s, \tag{4.7}$$

and finally, as the last step, we carry out the reduction

$$\mathfrak{D}_s \times \mathfrak{D}_L = \Sigma \mathfrak{D}_J, \quad (J = L + s, L + s - 1, \cdots, |L - s|), \tag{4.8}$$

associated with the coupling between the spin and the orbital moment of momentum. The terms which result from this last reduction form together a ***multiplet***. Each multiplet is therefore associated with a definite *azimuthal quantum number* L and a *spin quantum number* s; the individual members of the multiplet are distinguished by the *inner quantum number* J. We call $2s + 1$ the *multiplicity*, although the number of terms in the multiplet is only actually equal to this when $L \geqq s$, as by (4.8) their number is less if $L < s$. The 2^f-dimensional representation $\mathfrak{D}_{\frac{1}{2}}^f$ is even or odd according as f is even or odd. The reduction (4.7) into irreducible constituents accordingly yields only integral values for s when f is even and only half-integral values when f is odd: *The term multiplicities alternate regularly between even and odd as we run through the atomic table in the order of increasing atomic number* (H even, He odd, Li even, Be odd, etc: "*alternation law*"). For $f = 2$ we have, for example,

$$\mathfrak{D}_{\frac{1}{2}}^2 = \mathfrak{D}_0 + \mathfrak{D}_1.$$

It is empirically found that *the bivalent alkaline earth metals have in fact a singlet and a triplet system of terms.* But in the triplet system the S terms, for which $L = 0$, are simple; only the $P, D, \cdots$, terms have the actual multiplicity 3.

Instead of considering all the electrons at once as in (4.6) we can build up the atom by successively adding one electron after another. On adding a next electron, say the fth, to an atom or an ion A^+, a multiplet of A^+ characterized by azimuthal quantum number L and spin s breaks up into all those multiplets contained in the representation $(\mathfrak{D}_s \times \mathfrak{D}_{\frac{1}{2}}) \times (\mathfrak{D}_L \times \mathfrak{D}_l)$, where $l_f = l$ is the azimuthal quantum number of the electron added. Since

$$\mathfrak{D}_s \times \mathfrak{D}_{\frac{1}{2}} = \mathfrak{D}_{s+\frac{1}{2}} + \mathfrak{D}_{s-\frac{1}{2}},$$
$$\mathfrak{D}_L \times \mathfrak{D}_l = \Sigma \mathfrak{D}_{L^*}, \quad L^* = L + l, \ L + l - 1, \cdots, \ |L - l|,$$

this results in multiplets (s^*, L^*), one for each of the pairs

$$s^* = s \pm \frac{1}{2}, \quad L^* = L + l, \ L + l - 1, \cdots, \ |L - l| \tag{4.9}$$

("*branching rule*"). The alternation law is again contained in the first of the above equations. It is to be noted, however, that the *Pauli exclusion principle* for equivalent orbits, which will be discussed in part C of this chapter, materially restricts the array of multiplets allowed by this rule.[9]

Again applying (3.12) to the composition of spin and orbital moment of momentum, we find that *the $2J + 1$ components into which a J term of a multiplet (s, L) is split in a weak magnetic field are displaced from the unperturbed positions by the amounts*

$$hog \cdot m \quad (m = J, J - 1, \cdots, -J) \tag{4.10}$$

where the separation factor g is given by

$$g = 1 + \frac{J(J + 1) - L(L + 1) + s(s + 1)}{2J(J + 1)}. \tag{4.11}$$

This is exactly the formula which was derived empirically by *Landé*; we here see the importance of the fact that the square of the absolute value of the moment of momentum $\mathfrak{M}$ (or $\mathfrak{L}$ or $\mathfrak{S}$) is calculated from the quantum number J (or L or s) by $J(J+1)$, etc., instead of J^2, etc., as in the older quantum mechanics.

When the magnetic field increases to such an extent that the magnetic separation becomes comparable with the separation between the terms of the multiplet we must handle both the perturbation to which the multiplet separation is due and the magnetic perturbation together. In order to express the smallness of the term in the Hamiltonian function to which this former perturbation is due, we introduce a factor ρ which will appear in the same way as the factor o in the magnetic term; the case of a weak magnetic field may then be expressed by saying that o is small in comparison with ρ. We can consider o and ρ as variables which increase gradually from 0 to their actual values and follow the dependence of the separation on their ratio. We therefore write the perturbation term in the Hamiltonian function in the form

$$W = \rho W' + oW''.$$

Since the decomposition (4.8) need not for present purposes be expressed in terms of its ultimate constituents, the individual electrons, we may here denote the azimuthal and inner quantum numbers by l and j. Let the representation spaces of $\mathfrak{D}_s$, $\mathfrak{D}_l$ be $\mathfrak{r}_s$, $\mathfrak{R}_l$ with co-ordinates $\xi(m_s)$, $x(m_l)$ respectively. Denote the moments of momentum $\mathfrak{M}_s$, $\mathfrak{M}_l$ of these two representations by $\mathfrak{s}$, $\mathfrak{L}$ respectively; if the magnetic field has as its direction the z-axis, then

$$W'' = h(L_z + 2s_z). \tag{4.12}$$

The co-ordinate system is again to be so chosen that the rotations about the z-axis appear in reduced form; to such a rotation of angle ϕ corresponds the transformation

$$\xi(m_s) \to e(-m_s\phi)\cdot\xi(m_s), \quad x(m_l) \to e(-m_l\phi)\cdot x(m_l);$$

the range of the quantum numbers m_s and m_l is given by

$$m_s = s, s-1, \cdots, -s; \ m_l = l, l-1, \cdots, -l. \tag{4.13}$$

The variables of $\mathfrak{r}_s \times \mathfrak{R}_l$ then behave like the $(2s+1)(2l+1)$ products

$$\xi(m_s)\cdot x(m_l) \tag{4.14}$$

and are multiplied, under the influence of a rotation ϕ about the z-axis, by $e(-m\phi)$, where

$$m = m_s + m_l.$$

We now reduce $\mathfrak{D}_s \times \mathfrak{D}_l$ into its irreducible constituents $\mathfrak{D}_j$. Let the co-ordinates of the $(2j + 1)$-dimensional irreducible sub-space of $\mathfrak{r}_s \times \mathfrak{R}_l$, in which the representation $\mathfrak{D}_j$ takes place, be denoted by

$$x(j\,;\ m) \quad (m = j, j-1, \cdots, -j).$$

m is the magnetic quantum number, i.e. under the influence of the rotation ϕ about the z-axis $x(j\,;\,m)$ is multiplied by $e(-m\phi)$. The co-ordinate transformation which leads to the complete reduction of $\mathfrak{D}_s \times \mathfrak{D}_l$ into its constituents $\mathfrak{D}_j$ is obviously of such a kind that $x(j\,;\ m)$ is a linear combination of those of the variables (4.14) for which $m_s + m_l$ has the value m.

If the unperturbed system possesses no accidental degeneration the separation is determined by that part of the matrix (4.12) in which the sub-space $\mathfrak{r}_s \times \mathfrak{R}_l$ of $\mathfrak{R}^f$ intersects itself. We must therefore solve a secular equation G of degree $(2s+1)(2l+1)$; but the problem is materially simplified by the fact that the perturbation term possesses rotational symmetry about the z-axis, as the only non-vanishing elements of the matrix W are those for which $m \rightarrow m$. The one secular equation G is consequently broken up into $2(l+s)+1$ secular equations G_m corresponding to the possible values

$$m = l+s, l+s-1, \cdots, -(l+s)$$

of m. The degree of G_m is given by the number of possible partitions of m into two summands $m_s + m_l$ which run through the ranges (4.13). In the case of a single electron, $f = 1$, we have only equations of the first and second degrees, and the calculation can therefore be carried through completely for this case.[10]

The roots of the secular equation G_m are the displacements of the energy terms due to the perturbation. Since the trace of a matrix is an invariant, the sum of the term displacements which are associated with a definite value m of the magnetic quantum number (the roots of the secular equation G_m) is equal to the sum of the terms in the principal diagonal of this portion of W, i.e. to

$$\sum_{(m_s + m_l = m)} W(m_s m_l,\ m_s m_l).$$

It is therefore a homogeneous linear function of ρ and o (" *sum rule* "). We obtain the part due to the magnetic field by putting $\rho = 0$; by (4.12) this is

$$oW''(m_s m_l,\ m_s m_l) = ho(m_l + 2m_s).$$

On the other hand, the formula (4.10), (4.11) determine the term displacements in the case in which o is small in comparison with ρ. In consequence of the sum rule these two results must agree. l and s being fixed once and for all, we denote the Landé g-factor (4.11) by $g(j)$, and we then have

$$\Sigma(m_l + 2m_s) = m \cdot \Sigma g(j).$$

The sum on the left is extended over all partitions of $m = m_l + m_s$ for given m, and that on the right over all values of j which are consistent with the conditions

$$j = |m|, |m| + 1, \cdots; \ j = l + s, l + s - 1, \cdots, |l - s|.$$

$g(j)$ can in fact be determined from this equation. For $m = l + s$ both sums reduce to a single term; we then have

$$l + 2s = (l + s) \cdot g(l + s).$$

For $m = l + s - 1$ there are two possibilities for (m_s, m_l) and two for j: $m_l = l$, $m_s = s - 1$ or $m_l = l - 1$, $m_s = s$; $j = l + s$ or $l + s - 1$. Consequently we must have

$$2l + 4s - 3 = (l + s - 1)\{g(l + s) + g(l + s - 1)\}.$$

In this way we obtain recursion formulæ for the successive calculation of $g(l + s)$, $g(l + s - 1)$, $\cdot \cdot \cdot$. The reader can readily verify that the result of the first few steps agrees with (4.11).

It is to be noted that in following the terms from a weak to a strong magnetic field they cannot cross each other, considered as functions of the monotonic increasing parameter $o : \rho$; the "singular elements" of a unitary group, i.e. those elements for which two or more characteristic values coincide, constitute a manifold of *three*, and not simply *one*, fewer dimensions.[11]

B. The Lorentz Group

§ 5. Relativistically Invariant Equations of Motion of an Electron

We have as yet obtained no specific expression for the spin perturbation; that for the magnetic effect due to an external field was set up with the aid of the experimental facts. It is clear that we can arrive at a satisfactory theory of the electron only when we are able to express its fundamental laws of motion in a form which is invariant under Lorentz transformations, as required by the restricted theory of relativity. The solution of this problem is due to *Dirac*.[12] We saw in III, § 8, how the 2-dimensional representation $\mathfrak{D}_{\frac{1}{2}}$ of the rotation group, which, following *Pauli*, characterizes the covariant quantity $\psi = (\psi_1, \psi_2)$ describing the wave field, can be extended to the group of positive Lorentz transformations. ψ_1, ψ_2 play the same rôle as the variables ξ, η introduced in connection with $\mathfrak{D}_{\frac{1}{2}}$.

Following *de Broglie* we took as the wave equation of a particle of mass m in field-free space

$$\left(\frac{\partial^2}{\partial x^2} + \frac{\partial^2}{\partial y^2} + \frac{\partial^2}{\partial z^2} - \frac{1}{c^2}\frac{\partial^2}{\partial t^2}\right)\psi = m_0^2\,\psi \quad \left(m_0 = \frac{cm}{h}\right). \tag{5.1}$$

But this equation is not in agreement with the general scheme of quantum mechanics, which requires that only first order derivatives with respect to the time appear. The formulation of a relativistically invariant differential equation satisfying this requirement is, as *Dirac* discovered, made possible by the transition from the scalar wave function ψ to one with two components. We seek to derive these dynamical equations from a Hamiltonian principle.

Let

$$x_0 = ct, \quad x_1 = x, \quad x_2 = y, \quad x_3 = z$$

constitute a normal co-ordinate system in our 4-dimensional space-time. If the quantity ω is of the same kind as ψ, the quantities $\tilde{\psi}\, S_\alpha\, \omega$ behave, in accordance with III, (8.16), like the four components of a 4-vector; the S_α are the matrices defined in III, (8.15). Hence in particular

$$\sum_{\beta=0}^{3} \tilde{\psi}\, S_\alpha \frac{\partial\psi}{\partial x_\beta} dx_\beta$$

are the components ds_α of an infinitesimal vector; we are here dealing with a linear correspondence which is independent of the co-ordinate system employed and which sends the vector dx over into ds. Its trace

$$\sum_\alpha \tilde{\psi} S_\alpha \frac{\partial\psi}{\partial x_\alpha} \tag{5.2}$$

is consequently a scalar and its integral (multiplied by $1/i$)

$$M = \frac{1}{i}\int \sum_\alpha \tilde{\psi}\, S_\alpha \frac{\partial\psi}{\partial x_\alpha} \cdot dx \quad (dx = dx_0\, dx_1\, dx_2\, dx_3), \tag{5.3}$$

extended over any finite portion of the world, is a quantity which is independent of the co-ordinate system.*

* The letter M used for the *material* part of the action is not to be confused with the moment of momentum.

Although M may not be real, it is *practically real* in the sense that $M - \overline{M}$ is the integral of a complete divergence. For since the S_α are Hermitian matrices,

$$\overline{M} = -\frac{1}{i}\int \sum_\alpha \frac{\partial \tilde{\psi}}{\partial x_\alpha} S_\alpha \psi \cdot dx$$

and $M - \overline{M}$ is in fact the integral of

$$\frac{1}{i}\sum_\alpha \frac{\partial(\tilde{\psi}\, S_\alpha\, \psi)}{\partial x_\alpha}.$$

In using M as an action we are not interested in M itself, but only in its variations δM caused by arbitrary infinitesimal variations $\delta\psi$ of $\psi = (\psi_1, \psi_2)$ which vanish outside of a given finite portion of the world (the integral is then extended over the entire world or, what amounts to the same, over this finite portion). The circumstances mentioned above guarantee that δM is real; on writing it in the form

$$\delta M = \int (\delta\tilde{\psi} \cdot \omega + \tilde{\omega} \cdot \delta\psi) dx$$

we find on comparison with (5.3) that

$$\omega = \frac{1}{i}\sum_\alpha S_\alpha \frac{\partial\psi}{\partial x_\alpha}.$$

We thus arrive at the first order differential operator

$$\nabla = \sum_\alpha S_\alpha \frac{\partial}{\partial x_\alpha}. \tag{5.4}$$

From the invariance of (5.2) it follows that this operator transforms $\psi = (\psi_1, \psi_2)$ into a quantity $\psi' = (\psi'_1, \psi'_2)$ which transforms contragrediently to $\bar{\psi} = (\bar{\psi}_1, \bar{\psi}_2)$ under the influence of an arbitrary positive Lorentz transformation. If we wish to guarantee that M is real, we may replace the original definition by

$$M = \frac{1}{2i}\int \sum_\alpha \left(\tilde{\psi} S_\alpha \frac{\partial\psi}{\partial x_\alpha} - \frac{\partial\tilde{\psi}}{\partial x_\alpha} S_\alpha\, \psi \right) \cdot dx. \tag{5.5}$$

In III, § 8, we found it necessary to introduce quantities ψ'_1, ψ'_2 which transform contragrediently to $\bar{\psi}_1, \bar{\psi}_2$ in order to be able to extend the restricted Lorentz group to the complete group. And just as ∇ applied to ψ generates a quantity of the kind ψ', in the same way the " conjugate " operator

$$\nabla' = \sum_\alpha S'_\alpha \frac{\partial}{\partial x_\alpha}$$

transforms ψ' into a quantity of the kind ψ. $\nabla'\nabla$ is, as is readily verified, the operator

$$\frac{\partial^2}{\partial x_0^2} - \left(\frac{\partial^2}{\partial x_1^2} + \frac{\partial^2}{\partial x_2^2} + \frac{\partial^2}{\partial x_3^2}\right).$$

Consequently equation (5.1) for ψ_1, ψ_2 can be written in the form

$$\left.\begin{aligned} \frac{1}{i}\nabla\psi + m_0\psi' = 0, \\ \frac{1}{i}\nabla'\psi' + m_0\psi = 0 \end{aligned}\right\} \qquad (5.6)$$

on introducing an auxiliary pair of components ψ'. From now on we denote the column of the *four* components ψ_1, ψ_2; ψ_1', ψ_2' by ψ and employ S_α as the symbol for the transformations of these four components as in the latter part of Chapter III; with this understanding the differential equations (5.6) arise from an action integral which is composed additively of the quantity M, (5.3), and the invariant [cf. III, (8.19)]

$$M' = m_0 \int \tilde{\psi} T \psi \cdot dx.$$

M and M' are also invariant with respect to interchange of right and left, and under the spatial reflection i in the origin.

In accordance with the general scheme of quantum mechanics the differential equations for ψ should, as already remarked, contain only the first derivative of ψ with respect to time; the additional requirement that it be relativistically invariant then leads to the conclusion that it can also contain only first derivatives with respect to the spatial co-ordinates. We have here been able to satisfy these requirements without altering the actual content of de Broglie's equation (for the components ψ_1, ψ_2); the equations thus obtained are to be taken as the equations for a free particle. This formal transition to first order equations will become physically significant only when we pass to the derivation of the equations of motion in an electromagnetic field with the aid of the principle of gauge invariance developed in II, § 12. According to it, if $-\phi_0$ is the scalar and ϕ_1, ϕ_2, ϕ_3 the vector potential, we must replace

$$\frac{1}{i}\frac{\partial}{\partial x_\alpha} \quad \text{by} \quad \frac{1}{i}\frac{\partial}{\partial x_\alpha} + \frac{e}{hc}\phi_\alpha. \qquad (5.7)$$

It will be found convenient in the following to introduce the quantities f_α obtained by multiplying the potentials ϕ_α by the factor $\frac{e}{hc}$. Then in

$$M = \frac{1}{i}\int \tilde{\psi} \cdot \nabla\psi \cdot dx \tag{5.8}$$

the operator ∇ is defined by

$$\frac{1}{i}\nabla = \sum_\alpha S_\alpha\left(\frac{1}{i}\frac{\partial}{\partial x_\alpha} + f_\alpha\right). \tag{5.9}$$

Because of this gauge invariance the quantities M, M' are unchanged on replacing simultaneously

$$\psi \text{ by } e^{i\lambda}\psi \text{ and } f_\alpha \text{ by } f_\alpha - \frac{\partial \lambda}{\partial x_\alpha}, \tag{5.10}$$

where λ is an arbitrary function of position in space-time. Now take λ to be an infinitesimal function which vanishes outside a certain finite portion of the world; then δM and $\delta M'$ must automatically vanish for the variations

$$\delta\psi = i\lambda \cdot \psi, \quad \delta f_\alpha = -\frac{\partial \lambda}{\partial x_\alpha}. \tag{5.11}$$

The complete expression

$$\delta(M + M') = \int[(\delta\tilde{\psi} \cdot \omega + \tilde{\omega} \cdot \delta\psi) + \sum_\alpha s^\alpha \, \delta f_\alpha] dx$$

for the variation automatically tells us that under the assumption that the laws of matter (5.6) are satisfied, i.e. that $\omega = 0$,

$$\delta(M + M') = \int \sum_\alpha s^\alpha \, \delta f_\alpha \cdot dx.$$

Hence we have as a consequence of the laws of matter

$$-\int \sum_\alpha s^\alpha \cdot \frac{\partial \lambda}{\partial x_\alpha} \cdot dx = \int \lambda \cdot \sum_\alpha \frac{\partial s^\alpha}{\partial x_\alpha} dx = 0,$$

i.e. the continuity equation

$$\sum_\alpha \frac{\partial s^\alpha}{\partial x_\alpha} = 0. \tag{5.12}$$

A glance at the explicit expression for M shows that

$$s^\alpha = \tilde{\psi} S_\alpha \psi; \tag{5.13}$$

these are the quantities which formed the starting-point for the theory of the transformations of ψ as developed in III, § 8, and we already know that they form the components of a 4-vector which is independent of the particular space-time co-ordinates employed. The time component

$$s^0 = \tilde{\psi}\psi = (\bar{\psi}_1\psi_1 + \bar{\psi}_2\psi_2) + (\bar{\psi}'_1\psi'_1 + \bar{\psi}'_2\psi'_2) \qquad (5.14)$$

is the *probability density* and hence $c\mathfrak{s} = c(s^1, s^2, s^3)$ is what may be called the *probability current:* in order to obtain the number of particles which will on the average pass through a surface element do in time unit, multiply the total number of particles present into the product of the area do and the normal component of the vector $c\mathfrak{s}$. On integrating the equation (5.12) over a volume V we find that the increase in the mean number of particles in V per unit time is equal to the mean number of particles entering V through the surface in unit time. *In contrast to the provisional scalar theory, the Dirac theory leads in a most natural way to expressions for the probability density, as well as the probability current, which depend on* ψ *alone.*

On integrating

$$\int s^0 \, dx_1 \, dx_2 \, dx_3$$

over the whole of space we find that the integral is independent of time—and, in accordance with the statistical interpretation of ψ, is to be so normalized that its value is 1. Consequently, in the dynamical law

$$\frac{1}{i}\frac{d\psi}{dt} + H\psi = 0$$

the energy H/h is a Hermitian operator, as should be. *We shall from now on take h as the unit of action, with corresponding units for linear and angular momentum.* The result of this is that the quantity h disappears completely from the laws of quantum mechanics. With the usual abbreviation, $p_\alpha = \frac{1}{i}\frac{\partial}{\partial x_\alpha}$,

$$\frac{1}{c}H = f_0 + \sum_{r=1}^{3} S_r(p_r + f_r) + m_0 T. \qquad (5.15)$$

The influence of the electro-magnetic field on the matter is taken care of by (5.9), but, on the other hand, the matter generates the electro-magnetic field in accordance with Maxwell's equations. In order to express this explicitly we must add to $M + M'$ the Maxwellian action

$$F = \frac{1}{2}\int \{(f_{23}^2 + f_{31}^2 + f_{12}^2) - (f_{10}^2 + f_{20}^2 + f_{30}^2)\} dx \qquad (5.16)$$

of the electro-magnetic field, where the

$$f_{\alpha\beta} = \frac{\partial f_\beta}{\partial x_\alpha} - \frac{\partial f_\alpha}{\partial x_\beta}$$

are the field strengths—which are unaffected by the change of gauge (5.10). F is obtained from

$$\frac{1}{2}\iint(\mathfrak{H}^2 - \mathfrak{E}^2)dV\,dt \tag{5.17}$$

by multiplication with $c\left(\frac{e}{hc}\right)^2 = \frac{e^2}{ch^2}$. (5.17) is the action in Heaviside units, which are best adapted to the electro-magnetic field theory. Since we have taken h as the unit of action, the total action of our system, consisting of matter plus field, is

$$W = M + M' + \frac{1}{\alpha}F \quad \left(\alpha = \frac{e^2}{ch}\right). \tag{5.18}$$

For reasons which will be apparent later the real number $\alpha/4\pi$ is called the ***fine structure constant.*** Whereas the variation of the ψ in the Hamiltonian integral $\int W \cdot dx$ yields the *equations of matter*, variation of f_α leads to the *equations of the electro-magnetic field* with

$$-e \cdot s^\alpha = -e \cdot \tilde{\psi} S_\alpha \psi \tag{5.19}$$

appearing as the 4-vector of *charge and current density.* The only constants occurring in the field equations are the two combinations

$$m_0 = \frac{cm}{h}, \quad \alpha = \frac{e^2}{ch} \tag{5.20}$$

of fundamental atomic constants; the first is a reciprocal length and the second a pure number.

Schrödinger, in his fundamental papers on wave mechanics, thought he could explain the quantum behaviour of matter and radiation "classically" by setting up a closed system of field equations such as we have obtained above. In particular, he held that the charge of the electron was actually "smeared" over the whole of space with the density $-e \cdot s^0$. But there can be no doubt at the present time that the field equations are not to be interpreted in this classical manner; they must rather be interpreted in accordance with the statistical view-point developed in Chapter II. The expression (5.14) for the density then guarantees the *atomistic structure of electricity.* To show this we first remark that the charge in a volume V is represented by $-e$ times the Hermitian form

$$\iiint\limits_{(V)} \tilde{\psi}\psi dx_1 dx_2 dx_3.$$

But this is an "idempotent" form with respect to the "vector" ψ; its characteristic values are 1, 0 and the corresponding characteristic functions are those quantities ψ which vanish outside or inside V, respectively. The charge contained in V is accordingly capable of assuming only the values $-e$ and 0, i.e. according to whether the electron is found in V or not. In order to guarantee the atomicity of electricity the electric charge density must equal $-e$ times the probability density. But if we base our theory on the de Broglie wave equation, modified by introducing the electro-magnetic potentials in accordance with the rule (5.7), we find as the expression for the charge density one involving the temporal derivative $\frac{\partial\psi}{\partial t}$ in addition to ψ; this expression has nothing to do with the probability density and is not even an idempotent form. According to *Dirac* this is the most conclusive argument for the stand that the differential equations for the motion of an electron in an electro-magnetic field must contain only first order derivatives with respect to the time.[13] Since it is not possible to obtain such an equation with a scalar wave function which satisfies at the same time the requirement of relativistic invariance, *the spin appears as a phenomenon necessitated by the theory of relativity.*

The theorem of the conservation of electricity (5.12) follows, as we have seen, from the equations of matter, but it is at the same time a consequence of the electro-magnetic equations. The fact that (5.12) is a consequence of both sets of field laws means that these sets are not independent, i.e. that there exists an identity between them. The true ground for this identity is to be found in the gauge invariance, for it is equivalent to the assertion that δW vanishes identically when ψ and f_α are subjected to variations of the form (5.11). We have

$$\delta W = \int\{(\delta\tilde{\psi}\cdot\omega + \tilde{\omega}\cdot\delta\psi) + \sum_\alpha L^\alpha\,\delta f_\alpha\}dx,$$

where $\omega = 0$ are the equations of matter and $L^\alpha = 0$ the Maxwellian equations. On substituting the variations from (5.11) and integrating the last term in the integral by parts,

$$\frac{1}{i}(\tilde{\psi}\,\omega - \tilde{\omega}\,\psi) + \sum_\alpha \frac{\partial L^\alpha}{\partial x_\alpha} = 0.$$

Because of the arbitrariness of the gauge the number of independent equations must be one less than the number of unknown functions ψ and f_α.

§ 6. Energy and Momentum. Remarks on the Interchange of Past and Future

I. Energy and Momentum.

The complete field equations are explicitly

$$\left.\begin{aligned} &\sum_\alpha S_\alpha\Big(\frac{1}{i}\frac{\partial}{\partial x_\alpha} + f_\alpha\Big)\psi + m_0 \cdot T\psi = 0\,; \\ &\operatorname{div}\mathfrak{E} + \rho = 0, \quad \frac{\partial\mathfrak{E}}{\partial x_0} - \operatorname{curl}\mathfrak{H} = \mathfrak{s}. \end{aligned}\right\} \tag{6.1}$$

Where $\mathfrak{E}$ and $\mathfrak{H}$ are the electric and magnetic field strengths :

$$E_1 = \frac{1}{\alpha}\Big(\frac{\partial f_0}{\partial x_1} - \frac{\partial f_1}{\partial x_0}\Big), \cdots; \quad H_1 = \frac{1}{\alpha}\Big(\frac{\partial f_3}{\partial x_2} - \frac{\partial f_2}{\partial x_3}\Big), \cdots; \tag{6.2}$$

ρ is the charge density $\tilde{\psi}\psi$, and the components $s_1, \cdots$ of the current $\mathfrak{s}$ are given by

$$s_1 = \tilde{\psi} S_1 \psi, \cdots. \tag{6.3}$$

In addition to the differential law

$$\frac{\partial\rho}{\partial x_0} + \operatorname{div}\mathfrak{s} = 0, \tag{6.4}$$

expressing *the conservation of electricity*, we have a vector conservation law governing *energy and momentum.* A completely satisfactory expression for the tensor representing density and flux of energy and momentum is only to be obtained along the lines employed in the general theory of relativity. Here we give only the result for the density of energy $-c \cdot t_0^0$ and momentum (t_1^0, t_2^0, t_3^0), and in doing so we separate the material from the electro-magnetic part. We have for the part referring to matter

$$\left.\begin{aligned} -t_0^0 &= \frac{1}{2i}\sum_{p=1}^{3}\Big\{\tilde{\psi}S_p\Big(\frac{\partial}{\partial x_p} + if_p\Big)\psi - \Big(\frac{\partial}{\partial x_p} - if_p\Big)\tilde{\psi}\cdot S_p\psi\Big\} \\ &\qquad\qquad + m_0\tilde{\psi}T\psi\,; \\ t_1^0 &= \frac{1}{2i}\Big(\tilde{\psi}\frac{\partial\psi}{\partial x_1} - \frac{\partial\tilde{\psi}}{\partial x_1}\psi\Big) + \frac{1}{4}\Big(\frac{\partial s_3'}{\partial x_2} - \frac{\partial s_2'}{\partial x_3}\Big), \cdots. \end{aligned}\right\} \tag{6.5}$$

We have here introduced, in addition to S_p, the operator S_p' $(p = 1, 2, 3)$ which acts on all four components of ψ ; whereas the former subjects ψ_1, ψ_2 to the 2-dimensional transformation S_p [III, (8.15)] and ψ'_1, ψ'_2 to $-S_p$, the latter exercises the same 2-dimensional transformation S_p on both pairs of components. Correspondingly

$$s'_p = \tilde{\psi} S'_p \psi.$$

The density of energy and momentum due to the electro-magnetic field is given by the familiar Maxwellian expressions

$$\left.\begin{aligned} -t_0^0 &= \frac{\alpha}{2}\{(E_1^2 + +) + (H_1^2 + +)\}; \\ t_1^0 &= \alpha(E_2 H_3 - E_3 H_2), \cdots. \end{aligned}\right\} \tag{6.6}$$

We find the *conservation laws*

$$\sum_{\alpha=0}^{3} \frac{\partial t_0^\alpha}{\partial x_\alpha} = 0; \quad \sum_{\alpha=0}^{3} \frac{\partial t_1^\alpha}{\partial x_\alpha} = 0, \cdots \tag{6.7}$$

as consequences of the field equations. Furthermore, the tensor t is symmetric—not identically, but in consequence of the field equations; in this sense we have

$$t_0^p + t_p^0 = 0 \; (p = 1, 2, 3); \quad t_p^q = t_q^p \; (p, q = 1, 2, 3). \tag{6.8}$$

On combining these with (6.7) we obtain the divergence conditions

$$\sum_{\alpha=0}^{3} \frac{\partial(x_2 t_3^\alpha - x_3 t_2^\alpha)}{\partial x_\alpha} = 0, \cdots; \tag{6.9}$$

$$\sum_{\alpha=0}^{3} \frac{\partial(x_0 t_1^\alpha + x_1 t_0^\alpha)}{\partial x_\alpha} = 0, \cdots. \tag{6.10}$$

These results can all be verified directly, but their deeper significance can be understood only by going over to the *general theory of relativity* as mentioned above. Just as the theorem of the conservation of electricity follows from the gauge invariance of the equations, the theorems for the conservation of energy and momentum follow from the circumstance that the action integral, formulated as in the general theory of relativity, *is invariant under arbitrary* (infinitesimal) *transformations of co-ordinates.* In this general relativistic formulation we need further to erect a normal set of co-ordinate axes at each point P of space-time, consisting of four mutually perpendicular directions at P (" orthogonal ennuple "), in order to fix the metric at P and to be able to describe the wave quantity ψ in terms of its components; all permissible orthogonal ennuples at P are obtainable from each other by local Lorentz transformations which leave P invariant. But the rotations of these *local ennuples* can be performed in the various points P *quite independently*—the quantities at various points are not bound to each other as in the special theory of relativity. The symmetry of the energy-momentum tensor can be traced back to the invariance with respect to such rotations. One can in fact take it as a general rule that every invariance property of the kind met in general relativity, involving an arbitrary function, gives rise to a differential conservation theorem. In particular, gauge invariance is only to be understood from this standpoint. It follows from the transformation laws for ψ that its four components ψ_ρ relative to the local ennuple are determined only to within a common factor $e^{i\lambda}$ of proportionality, the exponent λ of which depends arbitrarily on position in space-time; in consequence of this it is necessary, in order to obtain a unique covariant differential for ψ, to set up a linear form $\sum_\alpha f_\alpha dx_\alpha$ which is coupled with the gauge factor contained in ψ in the manner required by the principle of gauge invariance.[14]

We obtain the *integral conservation laws* from the differential ones by integration. We set up the integral

$$\int t_\alpha^0 \, dV = J_\alpha \quad (dV = dx_1 \, dx_2 \, dx_3)$$

over a section $x_0 = \text{const.}$ of space-time and find that it is independent of x_0. $-cJ_0 = H$ is the *energy* and (J_1, J_2, J_3) the *linear momentum*. The material part is, on a simple integration by parts,

$$-J_0 = \int \tilde{\psi} \left\{ \sum_{p=1}^{3} S_p \left(\frac{1}{i} \frac{\partial}{\partial x_p} + f_p \right) + m_0 T \right\} \psi \, dV;$$

$$J_1 = \int \tilde{\psi} \cdot \frac{1}{i} \frac{\partial \psi}{\partial x_1} dV, \cdots.$$

These are Hermitian forms in the "vector" ψ. They again lead us to associate the operators $\frac{1}{i}\left(\frac{\partial}{\partial x_1}, \frac{\partial}{\partial x_2}, \frac{\partial}{\partial x_3}\right)$ with the components (J_1, J_2, J_3) of linear momentum, i.e. to the assumptions with which we, following *de Broglie* and *Schrödinger*, began. For the energy we obtain (on dividing by c) the operator

$$\frac{1}{c} H = \sum_{p=1}^{3} S_p \left(\frac{1}{i} \frac{\partial}{\partial x_p} + f_p \right) + m_0 T,$$

without the additive term f_0 as in (5.15); the differential equations of matter are therefore

$$\left(\frac{1}{i} \frac{\partial}{\partial x_0} + f_0 \right) \psi + \frac{1}{c} H \psi = 0.$$

Moreover, we must not forget that to the part due to matter we must yet add that due to the electro-magnetic field.

The quantities

$$M_1 = \int(x_2 t_3^0 - x_3 t_2^0)dV, \cdot \cdot \cdot, \tag{6.11}$$

which are by (6.9) also constant, are the components of the *moment of momentum.* We find from (6.5) that the part due to matter is

$$M_1 = \int\tilde{\psi}\left\{\frac{1}{i}\left(x_2\frac{\partial}{\partial x_3} - x_3\frac{\partial}{\partial x_2}\right) + \frac{1}{2}S_1'\right\}\psi dV, \cdot \cdot \cdot.$$

In agreement with our earlier assumptions we here obtain the operator which is composed of the sum of the x_1-component $\frac{1}{i}\left(x_2\frac{\partial}{\partial x_3} - x_3\frac{\partial}{\partial x_2}\right)$ of the orbital moment of momentum and the spin moment of momentum $\frac{1}{2}S_1'$. The vector

$$\frac{1}{2}\mathfrak{S}' = \frac{1}{2}(S_1', S_2', S_3')$$

is actually the spin, for in accordance with the law of transformation of both ψ pairs (ψ_1, ψ_2), (ψ_1', ψ_2') of components suffer the same transformation σ as in the Pauli theory of the spin under the influence of the transformation σ (spatial rotation) of $\mathfrak{u}_2$.

On integrating equations (6.10) over the spatial section $x_0 =$ const. we obtain

$$J_1 = -\frac{d}{dx_0}\int x_1 t_0^0 dV, \cdot \cdot \cdot,$$

which we may consider as the law of inertia of energy. The integral may be written $J_0 \cdot \xi_1 = -\frac{H}{c} \cdot \xi_1$, where ξ_1, ξ_2, ξ_3 are the co-ordinates of the "centre of energy"; the equations are then

$$J_1 = \frac{H}{c^2} \cdot \frac{d\xi_1}{dt}, \cdot \cdot \cdot.$$

We thus obtain the familiar mechanical law: *Momentum is equal to mass times velocity*, where the velocity is to be taken as that of the centre of energy and the mass as $1/c^2$ times the energy content of the field. Nevertheless it is advisable *not* to divide by H in defining the centre of energy, as the energy density $-t_0^0$ is here no longer positive-definite, and we cannot be certain that the energy content H will turn out to be positive.

Our theory is a classical field theory, the quantum features entering only in the statistical interpretation. With this interpretation the field laws are concerned with *a single electron.* At the present stage of our development we can deal only with the additional quantities due to the electro-magnetic field by assuming a given external field affecting the motion of the particle, without the particle reacting on the field; we must then surrender our Maxwellian equations. The true laws governing the interaction between electrons and quanta will only be obtained, in analogy with II, § 13, on subjecting the system of field equations to the process of quantization, just as was done by Heisenberg for any system of classical mechanical differential equations.

The fact that we are led back to our original assumptions concerning the operators representing position and momentum is due to the particular expressions we have chosen for the action, from which the field equations were obtained; indeed, it depends entirely on the part M. These original postulates of quantum theory are accordingly of less interest from the standpoint of general principles than we at first believed. But, on the other hand, this connection seems to indicate that M cannot be replaced in its rôle as representing the action due to matter. M is also responsible for the fact that the charge and probability densities agree, which is unconditionally required as a guarantee of the atomistic structure of electric charge. These connections with the most fundamental physical observations thus require that the action be composed additively of M and further terms which are invariant not only under change of gauge (5.10) as is M, but also on replacing ψ by $e^{i\lambda} \cdot \psi$ and f_α by $f_\alpha - \frac{\partial \mu}{\partial x_\alpha}$, where λ and μ are two independent arbitrary functions in space-time. M' and the Maxwellian action F are in fact of this kind. Further relativistic invariant scalars satisfying these conditions are readily found—indeed it is not difficult to set up the most general action possible with the quantities at our disposal. But we have yet to be convinced by physical observation that the three quantities M, M', F here employed do not suffice.

11. Electric and Magnetic Spin Perturbations.

In order to be able to compare Dirac's theory with the facts, we eliminate ψ_1', ψ_2' in the same way as we did in the absence of the electro-magnetic field. We obtain the equation

$$-\nabla'\nabla\psi = m_0^2\psi$$

with the new definition (5.9) of ∇ and ∇'. The substitutions S_α in two variables satisfied the equations

$$S_0S_1 = S_1S_0 = S_1\,; \quad S_2S_3 = -S_3S_2 = iS_1\,;$$

and consequently those denoted by the same letters but operating on all four variables obey

$$S_0S_1 = S_1S_0 = S_1\,; \quad S_2S_3 = -S_3S_2 = iS_1'.$$

$\nabla'\nabla$ contains terms of the following four types ;

$$(1) \qquad \Big(\frac{\partial}{\partial x_0} + if_0\Big)\Big(\frac{\partial}{\partial x_0} + if_0\Big),$$

$$(2) \qquad -\Big(\frac{\partial}{\partial x_1} + if_1\Big)\Big(\frac{\partial}{\partial x_1} + if_1\Big),$$

$$(3) \qquad S_1\Big\{\Big(\frac{\partial}{\partial x_0} + if_0\Big)\Big(\frac{\partial}{\partial x_1} + if_1\Big) - \Big(\frac{\partial}{\partial x_1} + if_1\Big)\Big(\frac{\partial}{\partial x_0} + if_0\Big)\Big\},$$

$$(4) \quad -iS_1'\Big\{\Big(\frac{\partial}{\partial x_2} + if_2\Big)\Big(\frac{\partial}{\partial x_3} + if_3\Big) - \Big(\frac{\partial}{\partial x_3} + if_3\Big)\Big(\frac{\partial}{\partial x_2} + if_2\Big)\Big\}.$$

We collect together terms of types (1) and (2) to form the "regular term" in which the components of ψ are not coupled with each other :

$$\Big(\frac{\partial^2}{\partial x_0^2} - \frac{\partial^2}{\partial x_1^2} - -\Big) + \sum_\alpha f_\alpha f^\alpha + \frac{2}{i}\sum_\alpha f^\alpha \frac{\partial}{\partial x_\alpha}.$$

[The transition from lower to upper indices, i.e. from "covariant" to "contravariant" components, is performed in accordance with the equations $f^0 = -f_0$, $f^p = f_p (p = 1, 2, 3)$.] The irregular term consists of the electric part

$$iS_1\Big(\frac{\partial f_1}{\partial x_0} - \frac{\partial f_0}{\partial x_1}\Big) + + = \frac{1}{i}(S_1 f_{10} + +)$$

and the magnetic part

$$S_1'\Big(\frac{\partial f_3}{\partial x_2} - \frac{\partial f_2}{\partial x_3}\Big) + + = (S_1' f_{23} + +).$$

These become, on multiplying by the factor h and expressing the electric and magnetic field strengths $\mathfrak{E}$ and $\mathfrak{H}$ in the usual units,

$$\frac{e}{ic}(\mathfrak{S}\mathfrak{E}), \quad \frac{e}{c}(\mathfrak{S}'\mathfrak{H}).$$

We have already (II, § 12) calculated the regular term for a homogeneous magnetic field and found it to be $\frac{e}{c}(\mathfrak{H}\mathfrak{L})$. On adding the regular and irregular terms we obtain, on neglecting the squares f_α^2 of the potentials,

$$\frac{e}{c}(\mathfrak{H}, \mathfrak{L} + \mathfrak{S}').$$

This contains the fact, which was already derived in § 4 from spectroscopic data, that to the spin $\frac{1}{2}\mathfrak{S}'$, twice as great a magnetic moment is to be ascribed as to the same amount of orbital moment of momentum; we have now obtained a convincing theoretical foundation for this procedure. The laws governing the interaction of a general inhomogeneous magnetic field with orbital and spin momenta emphasize still more emphatically the essential difference between $\mathfrak{L}$ and $\mathfrak{S}'$. The irregular electric term, calculated for the central-symmetric field originating in the nucleus, is the spin perturbation.

The description of the electron given earlier, according to which it was a composite structure composed of two kinematically independent parts—the electron translation, with an ∞-dimensional system-space, and the electron spin, with a 2-dimensional system space—is, in view of the *Dirac* theory, no longer quite appropriate. But the classification of spectra given there is none the less valid here, for it depends only on the fact that to the group of rotations of physical space corresponds the representation $\mathfrak{D}_{\frac{1}{2}} \times \mathfrak{E}$ in the total system-space.

From the field equations (6.1) as they are to be understood for the present, i.e. as the laws of motion of an electron in an external electro-magnetic field, *dispersion phenomena* can be (approximately) calculated; they tell us how the motion of the electron in the normal or other quantum states is affected by the incident light wave. From the perturbed ψ we then determine the scattered light with the aid of Maxwell's equations; to this class of phenomena belong in particular the *Compton* and *Smekal-Raman effects*.[15] Spontaneous emission can be handled similarly if we take the considerations of II, § 13, as justifying the following procedure: The polarization and intensity of light emitted by the quantum jump $n \rightarrow n'$ of the atom is to be calculated by integrating Maxwell's equations, where the expressions $\tilde{\psi}\psi$, $\tilde{\psi}\mathfrak{S}\psi$ for charge and current density are to be understood as $\tilde{\psi}^{(n)}\psi^{(n')}$, $\tilde{\psi}^{(n)}\mathfrak{S}\psi^{(n')}$, $\psi^{(n)}$ being the characteristic function of the atom in the nth quantum state.

III. Interchange of Past and Future.

The action is so constructed that it is invariant under *interchange of right and left;* the corresponding substitution is

$$\left.\begin{array}{ll} \left.\begin{array}{ll} x_0 \to x_0, & x_p \to -x_p; \\ f_0 \to f_0, & f_p \to -f_p; \end{array}\right\} (p = 1, 2, 3) & \\ \psi_1 \to \psi_1', \quad \psi_2 \to \psi_2'; & \psi_1' \to \psi_1, \quad \psi_2' \to \psi_2. \end{array}\right\} \tag{6.12}$$

Does a corresponding result hold for the interchange of past and future? The foundations of the theory lead to the hope that it will be able to take account of the essential difference between the two time directions, so obvious in Nature. But *Dirac* has remarked that M, M' go over into $-M$, $-M'$ under the influence of the substitution

$$\left.\begin{array}{l} x_\alpha \to -x_\alpha, \quad f_\alpha \to -f_\alpha \quad (\alpha = 0, 1, 2, 3); \\ \psi_1 \to \psi_1, \quad \psi_2 \to \psi_2; \quad \psi_1' \to -\psi_1', \quad \psi_2' \to -\psi_2'. \end{array}\right\} \tag{6.13}$$

Hence when, in dealing with the motion of an electron in an external electro-magnetic field, we obtain a solution ψ which contains the time in the factor $e^{-i\nu t}$, this substitution will lead us to a new solution which contains the time in the factor $e^{i\nu t}$; or, more precisely, a solution of the problem obtained by changing f into $-f$. But this can be done by retaining the same external field with potentials ϕ and replacing e by $-e$. We denote such a particle, whose mass is the same as that of the electron but whose charge is e instead of $-e$, as a "positive electron"; it is not observed in Nature! It follows from what has been said above that the energy levels of such a particle are $-h\nu$, where $h\nu$ are those of the negative electron. Disregarding this difference in sign, the two particles behave the same. *The electron will possess, in addition to its positive energy levels, negative ones as well,* the latter arising from the positive energy levels of the positive electron on changing signs as above. Obviously something is wrong here; we should be able to get rid of these negative energy levels of the electron. But that seems impossible, for under the influence of the radiation field transitions should occur between the positive and negative terms. That we have twice as many terms as we should is obviously related to the fact that our quantity ψ has *four* instead of *two* components (satisfying first order differential equations). The solution of this difficulty would seem to lie in the direction of interpreting our four differential equations as including the proton in addition to the electron.

The substitution (6.13) transforms the terms M, M' of the action into $-M$, $-M'$, but leaves the Maxwellian term F unaltered. Our field equations as a whole, i.e. when we also take into account the reaction of the particle on the radiation field, are consequently *not* invariant under this substitution. However, there does exist a substitution which reverses the direction of time and which at the same time leaves all terms in the action invariant. We mentioned in III, § 8 that the expression (5.13) formed from a ψ with two components takes on the sign $\delta_\alpha : \delta_0 = 1$, $\delta_p = -1 (p = 1, 2, 3)$ on going over from ψ_1, ψ_2 to $\bar{\psi}_2, -\bar{\psi}_1$. Hence if ω is a quantity which transforms in the same way as ψ then

$$\tilde{\psi} S_\alpha \omega \rightarrow \delta_\alpha \cdot \tilde{\omega} S_\alpha \psi;$$

on applying this to $\omega = \sum_\beta \frac{\partial \psi}{\partial x_\beta} \partial x_\beta$ we find that

$$\tilde{\psi} S_\alpha \frac{\partial \psi}{\partial x_\beta} \rightarrow \delta_\alpha \cdot \frac{\partial \tilde{\psi}}{\partial x_\beta} S_\alpha \psi.$$

Hence if we make in addition the substitution

$$x_0 \rightarrow -x_0, \quad x_p \rightarrow x_p \quad (p = 1, 2, 3)$$

then

$$\sum_{\alpha=0}^{3} \tilde{\psi} S_\alpha \frac{\partial \psi}{\partial x_\alpha} \rightarrow -\sum_{\alpha=0}^{3} \frac{\partial \tilde{\psi}}{\partial x_\alpha} S_\alpha \psi$$

and consequently M, formula (5.5), remains invariant. In the presence of an electro-magnetic field its components must change signs in accordance with

$$f_0 \rightarrow f_0, \quad f_p \rightarrow -f_p \quad (p = 1, 2, 3).$$

We have thus found that M, M' and F all remain invariant under the substitution

$$\left.\begin{array}{l} \left.\begin{array}{l} x_0 \rightarrow -x_0, \quad x_p \rightarrow x_p; \\ f_0 \rightarrow f_0, \quad f_p \rightarrow -f_p; \end{array}\right\} (p = 1, 2, 3) \\ \psi_1 \rightarrow \bar{\psi}_2, \quad \psi_2 \rightarrow -\bar{\psi}_1; \quad \psi_1' \rightarrow \bar{\psi}_2', \quad \psi_2' \rightarrow -\bar{\psi}_1' \end{array}\right\} \tag{6.14}$$

This shows that the past and the future enter into our field theory in precisely the same manner—in spite of the fact that the sign in the exponent of the time factor $e^{-i\nu t}$ of a solution of the quantum problem is unchanged by the substitution (6.14). We must of course suspend judgment as to whether the laws governing interaction between photons and electrons allow us to distinguish between these two directions in time until we have carried through the quantization (§ 12).

§ 7. Electron in Spherically Symmetric Field

We now proceed to the discussion of the behaviour of an electron in a spherically symmetric electrostatic field in *Dirac's* theory.

I. Dirac's Conservation Theorem.

From the definitions follow immediately the commutation rules:

$$S_pT = -TS_p, \quad S'_pT = TS'_p \quad (p = 1, 2, 3).$$

We need further the results

$$S'_1S'_1 = 1, \quad S'_2S'_3 = -S'_2S'_3 = iS'_1$$

and the commutation rules

$$L_1p_1 - p_1L_1 = 0, \quad (\mathfrak{p}\mathfrak{L}) = p_1L_1 + p_2L_2 + p_3L_3 = 0,$$
$$L_1p_2 - p_2L_1 = ip_3, \quad L_2p_1 - p_1L_2 = -ip_3,$$

for the components of linear and angular momenta $\mathfrak{p} = (p_1, p_2, p_3)$ and $\mathfrak{L} = (L_1, L_2, L_3)$.

In a spherically symmetric electrostatic field $f_1 = f_2 = f_3 = 0$ and $f_0 = \Phi$ is a function only of the distance r from the centre. With the aid of the formula given above it is easily shown that

$$M_1 = L_1 + \frac{1}{2}S'_1$$

commutes with Φ, T, $(\mathfrak{S}'\mathfrak{p})$ and consequently with each term in the expression

$$\frac{1}{c}H = \Phi + (\mathfrak{S}\mathfrak{p}) + m_0T \tag{7.1}$$

for the energy H. Indeed, this conservation law for the total moment of momentum $\mathfrak{M} = \mathfrak{L} + \frac{1}{2}\mathfrak{S}'$ was already known to us from general considerations. We further find that $(\mathfrak{S}'\mathfrak{L})$ commutes with Φ and T, but that

$$(\mathfrak{S}'\mathfrak{L})(\mathfrak{S}'\mathfrak{p}) + (\mathfrak{S}'\mathfrak{p})(\mathfrak{S}'\mathfrak{L}) = -2(\mathfrak{S}'\mathfrak{p})$$

or

$$(\mathfrak{S}'\mathfrak{p})\{(\mathfrak{S}'\mathfrak{L}) + 1\} + \{(\mathfrak{S}'\mathfrak{L}) + 1\}(\mathfrak{S}'\mathfrak{p}) = 0.$$

Hence $(\mathfrak{S}'\mathfrak{L}) + 1$ anti-commutes with $(\mathfrak{S}'\mathfrak{p})$ and therefore also with $(\mathfrak{S}\mathfrak{p})$; its commutation properties with respect to the three terms of (7.1) are therefore the same as those of T. Hence on setting

$$\boxed{(\mathfrak{S}'\mathfrak{L}) + 1 = kT,} \tag{7.2}$$

k is a scalar which commutes with the energy H (where by scalar we mean invariant under the group of rotations of space). Consequently we can decompose the system-space of the electron into irreducible sub-spaces $\mathfrak{R}_k$ associated with the rotation group, in such a way that the quantity k, which we call the ***auxiliary quantum number,*** as well as the energy H, possesses a definite value in each of the sub-spaces. Now

$$\begin{aligned}(\mathfrak{S}'\mathfrak{L})^2 &= \{L_1^2 + + \} + \{S_2'S_3'(L_2L_3 - L_3L_2) + + \} \\ &= \mathfrak{L}^2 - (S'_1L_1 + +) = \mathfrak{L}^2 - (\mathfrak{S}'\mathfrak{L})\end{aligned}$$

and consequently

$$\{(\mathfrak{S}'\mathfrak{L}) + 1\}^2 = \mathfrak{L}^2 + (\mathfrak{S}'\mathfrak{L}) + 1 = \left(\mathfrak{L} + \frac{1}{2}\mathfrak{S}'\right)^2 + \frac{1}{4} = \mathfrak{M}^2 + \frac{1}{4}$$

$$\mathfrak{M}^2 = k^2 - \frac{1}{4}.$$

This agrees with

$$\mathfrak{M}^2 = j(j+1) = \left(j + \frac{1}{2}\right)^2 - \frac{1}{4} \tag{7.3}$$

when we put

$$j = |k| - \frac{1}{2}, \quad |k| = j + \frac{1}{2}. \tag{7.4}$$

Accordingly, the auxiliary quantum number k is a non-vanishing integer. The conservation theorem (7.2) goes beyond (7.3) in giving us in addition the sign of k. For a given half-integral j the two values $k = \pm\left(j + \frac{1}{2}\right)$ are both possible; they must correspond to the two possibilities $l = j \pm \frac{1}{2}$ of our previous notation. The *single* quantum number k replaces the *two* l, j.

II. The Differential Equation for the Determination of the Characteristic Values.

Since the field is spherically symmetric, it suffices to carry through the calculation for the point $x = 0$, $y = 0$, $z = r$. At this point

$$L_1 = -\frac{z}{i}\frac{\partial}{\partial y}, \quad L_2 = \frac{z}{i}\frac{\partial}{\partial x}, \quad L_3 = 0$$

and the *Dirac* conservation law (7.2) becomes

$$\left.\begin{aligned} -z\Big(\frac{\partial}{\partial x} - i\frac{\partial}{\partial y}\Big)\psi_2 &= k\psi_1' - \psi_1 \\ z\Big(\frac{\partial}{\partial x} + i\frac{\partial}{\partial y}\Big)\psi_1 &= k\psi_2' - \psi_2 \end{aligned}\right\} \qquad (7.5)$$

together with the equations obtained from these by interchanging the two pairs ψ_1, ψ_2 and ψ_1', ψ_2' of components. The differential equation (6.1) for the characteristic vector ψ, which contains the time only in the factor $e^{-i\nu t}$, has as its four components the two

$$\left.\begin{aligned} U\psi_1 + i\Big(\frac{\partial}{\partial x} - i\frac{\partial}{\partial y}\Big)\psi_2 + i\frac{\partial\psi_1}{\partial z} - m_0\psi_1' &= 0 \\ U\psi_1' - i\Big(\frac{\partial}{\partial x} - i\frac{\partial}{\partial y}\Big)\psi_2' - i\frac{\partial\psi_1'}{\partial z} - m_0\psi_1 &= 0 \end{aligned}\right\} \qquad (7.6)$$

and two others of analogous structure ; we have here written

$$E = \frac{\nu}{c}, \quad E - \Phi = U.$$

The derivatives with respect to x and y which appear in (7.6) can be eliminated with the aid of (7.5) ; the resulting equations are

$$\left.\begin{aligned} \Big[U + i\Big(\frac{1}{r} + \frac{d}{dr}\Big)\Big]f - \Big(m_0 + \frac{ik}{r}\Big)g &= 0 \\ \Big[U - i\Big(\frac{1}{r} + \frac{d}{dr}\Big)\Big]g + \Big(-m_0 + \frac{ik}{r}\Big)f &= 0 \end{aligned}\right\} \qquad (7.7)$$

where

$$f = \psi_1(0, 0, r), \quad g = \psi_1'(0, 0, r).$$

The remaining two equations are obtained by writing (ψ_2', ψ_2) in place of (ψ_1, ψ_1'). At an arbitrary point $P = P(x, y, z)$ the first and third components of ψ satisfy the equations (7.7) in a *rotated* co-ordinate system whose positive z-axis passes through P. We shall find it convenient to introduce rf and rg as variables in place of f and g, as

$$\frac{1}{r}\frac{d(rf)}{dr} = \Big(\frac{1}{r} + \frac{d}{dr}\Big)f.$$

If we wish to avoid the explicit appearance of i in the equations, we may write

$$rf = v + iw, \quad rg = v - iw$$

and obtain, finally, the fundamental equations

$$\boxed{\begin{aligned} Uv - \frac{dw}{dr} - m_0 v - \frac{k}{r} w &= 0, \\ Uw + \frac{dv}{dr} + m_0 w - \frac{k}{r} v &= 0. \end{aligned}} \tag{7.8}$$

III. Spherical Harmonics with Spin.

Let $f(r)$, $g(r)$ be a solution of equations (7.7); then in the rotated co-ordinate system

$$\psi_1 = f \cdot \rho, \; \psi_1' = g \cdot \rho; \; \psi_2' = f \cdot \tau, \; \psi_2 = g \cdot \tau$$

where the factors ρ, τ are constants independent of r. On returning to the original co-ordinate system each of the pairs ψ_1, ψ_2; ψ_1', ψ_2' undergoes the transformation σ associated with the rotation s. Consequently

$$\begin{array}{l|l} \psi_1 = f\rho_1 + g\tau_1 & \psi_1' = g\rho_1 + f\tau_1 \\ \psi_2 = f\rho_2 + g\tau_2 & \psi_2' = g\rho_2 + f\tau_2 \end{array} \tag{7.9}$$

in which f and g depend only on r, and the factors ρ, τ only on direction, i.e. on the spherical co-ordinates θ, ϕ introduced by setting

$$x + iy = r \sin\theta \, e^{i\phi}, \quad z = r\cos\theta;$$

the coefficients in (7.9) must further satisfy the conditions

$$\begin{aligned} \rho_1(1 - \cos\theta) - \rho_2 \sin\theta \, e^{-i\phi} &= 0, \\ \tau_1(1 + \cos\theta) + \tau_2 \sin\theta \, e^{-i\phi} &= 0. \end{aligned} \tag{7.10}$$

On substituting the expression for $\mathfrak{L}$ in polar co-ordinates [II, (4.10)] into the *Dirac* conservation law, we are led, with the aid of (7.9) and (7.10), to the differential equations

$$\begin{aligned} \sin\theta \frac{\partial\rho_1}{\partial\theta} + i\frac{\partial\rho_1}{\partial\phi} + k(1 + \cos\theta)\tau_1 &= 0, \\ \sin\theta \frac{\partial\tau_1}{\partial\theta} - i\frac{\partial\tau_1}{\partial\phi} - k(1 - \cos\theta)\rho_1 &= 0. \end{aligned} \tag{7.11}$$

We have thereby accomplished the *transformation of the Dirac wave equation into polar co-ordinates.* (7.9) corresponds to the substitution $\psi = f(r) Y_l$ of the scalar theory; in place of the single factor f depending only on the distance r we have here the pair f, g and in place of the surface harmonic Y_l depending only on the direction we have the matrix

$$\left\| \begin{matrix} \rho_1 & \tau_1 \\ \rho_2 & \tau_2 \end{matrix} \right\|.$$

The equations (7.11), together with the conditions (7.10), define the "*surface harmonics with spin of order k*"; they are quite independent of the potential Φ. The characteristic values E of the equations (7.7) or (7.8) are the energy levels associated with quantum number k.

As in the theory of the ordinary spherical harmonics, we here again seek out those spherical harmonics with spin which contain the meridian angle only in the multiplicative factor $e^{im\phi}$:

$$\rho_1 = e^{im\phi}(\sin\theta)^{-m}\cdot P, \quad \tau_1 = e^{im\phi}(\sin\theta)^{-m}\cdot Q. \tag{7.12}$$

Substituting these expressions in (7.11) and taking $z = \cos\theta$ as the independent variable, we find

$$\boxed{\begin{aligned}(1-z)\frac{dP}{dz} &= -mP + kQ,\\ (1+z)\frac{dQ}{dz} &= mQ - kP.\end{aligned}} \tag{7.13}$$

We denote the solutions P, Q of these equations which lead to non-singular functions ρ, τ on the sphere more precisely by $P_k^{(m)}$, $Q_k^{(m)}$. It suffices to consider the case $k > 0$, for $(-P, Q)$ is a solution of the equations obtained by changing k into $-k$:

$$P_{-k}^{(m)}(z) = -P_k^{(m)}(z), \quad Q_{-k}^{(m)}(z) = Q_k^{(m)}(z). \tag{7.14}$$

Furthermore,

$$\frac{dP^{(m)}}{dz} = P^{(m-1)}, \quad \frac{dQ^{(m)}}{dz} = Q^{(m-1)},$$

for the derivatives of $P^{(m)}$, $Q^{(m)}$ satisfy the differential equations (7.13) with $m - 1$ in place of m. For $m = -k$, $P = 1$, $Q = -1$ is a solution which satisfies all continuity requirements on the sphere, since the multiplicative factor

$$(\sin\theta)^{-m}e^{im\phi} = (x - iy)^{-m}$$

is finite for negative m. Consequently we find *polynomial solutions* of (7.13), the degrees of which are $0, 1, \cdots, 2k-1$ corresponding to the values $m = -k, -k+1, \cdots, k-1$. The solution for $m = k - 1$ is

$$P(z) = (1-z)^{k-1}(1+z)^k, \quad Q(z) = (1+z)^{k-1}(1-z)^k.$$

We thus finally obtain the following explicit expressions for the spherical harmonics with spin:

$$P_k^{(m)}(z) = \frac{d^p}{dz^p}\{(1-z)^{k-1}(1+z)^k\}, \quad Q_k^{(m)}(z) = \frac{d^p}{dz^p}\{(1+z)^{k-1}(1-z)^k\} \tag{7.15}$$

where $p = k - 1 - m$. They behave very much like the ordinary spherical harmonics. The following equations are also of importance:

$$P_k^{(m)}(-z) = (-1)^p \cdot Q_k^{(m)}(z), \quad Q_k^{(m)}(-z) = (-1)^p \cdot P_k^{(m)}(z) \qquad (7.16)$$

§ 8. Selection Rules. Fine Structure

I. Selection Rules.

In a solution ψ defined by (7.9), (7.12) ψ_1, like ρ_1 and τ_1, contains ϕ only in the factor $e^{im\phi}$ and ψ_2, like ρ_2 and τ_2, only in the factor $e^{i(m+1)\phi}$; correspondingly for ψ_1', ψ_2'. Hence

$$M_z \psi_1 = \frac{1}{i}\frac{\partial \psi_1}{\partial \phi} + \frac{1}{2}\psi_1 = \left(m + \frac{1}{2}\right)\psi_1,$$

$$M_z \psi_2 = \frac{1}{i}\frac{\partial \psi_2}{\partial \phi} - \frac{1}{2}\psi_2 = \left(m + \frac{1}{2}\right)\psi_2.$$

The z-component of the moment of momentum in the state (k, m) is accordingly $m + \frac{1}{2}$. This change in the meaning of the quantum number m is to be carefully noted: $m + \frac{1}{2}$ runs through the values

$$k - \frac{1}{2},\ k - \frac{3}{2},\ \ldots,\ -k + \frac{1}{2} = j,\ j - 1,\ \cdots,\ -j.$$

as it should.

In order to obtain the *selection rules* for the possible transitions $(k, m) \to (k', m')$ and to obtain the corresponding *intensities* we must calculate the matrix which represents the energy of interaction between the atom and radiation in terms of the co-ordinate system determined by the characteristic functions $\psi^{(n)}$ defining the quantum states n of the atom. Proceeding as in II, § 13, we see from (5.15) that this matrix is

$$e\sum_{p=1}^{3} S_p \phi_p\ .$$

The vector $e c \mathfrak{S}$ here plays the same rôle as $\mathfrak{q}$ there. The intensities are essentially determined by the elements $\mathfrak{S}(nn')$, the three components of which are

$$S_p(nn') = \int \tilde{\psi}^{(n)} S_p \psi^{(n')} dV.$$

The selection rules are merely consequences of the fact that $\mathfrak{S}$ is a vector. We first obtain the old result for m and j from considerations involving the proper rotations of space. The rule for j asserts that the auxiliary quantum number k may go over into

$$\pm (k - 1), \quad \pm k, \quad \pm (k + 1). \tag{8.1}$$

To the reflection i corresponds the interchange T of the two pairs (ψ_1, ψ_2), (ψ_1', ψ_2'). In polar co-ordinates this reflection consists in the transition from (θ, ϕ) to $(\pi - \theta, \pi + \phi)$; $z = \cos\theta$ is thereby transformed into $-z$ and the factor $e^{im\phi}$ takes on the sign $(-1)^m$. In accordance with (7.15) and the expressions for ρ_1, τ_1; ρ_2, τ_2 this results in an interchange of ρ_1, τ_1 with possible change of sign, as represented by the substitution

$$(-1)^{p+m} \begin{Vmatrix} 0 & 1 \\ 1 & 0 \end{Vmatrix} = (-1)^{k-1} \begin{Vmatrix} 0 & 1 \\ 1 & 0 \end{Vmatrix},$$

and the same for ρ_2, τ_2. By (7.9) we therefore have for ψ with auxiliary quantum number k:

$$T\psi(-x, -y, -z) = (-1)^{k-1}\psi(x, y, z).$$

The sub-space $\mathfrak{R}_k$ thus has the signature $\delta = (-1)^{k-1}$; this result was derived under the assumption $k > 0$. On replacing k by $-k$ and applying (7.14) we find in place of (7.16):

$$P^{(m)}_{-k}(-z) = (-1)^{p+1}Q^{(m)}_{-k}(z), \quad Q^{(m)}_{-k}(-z) = (-1)^{p+1}P^{(m)}_{-k}(z).$$

The signature corresponding to auxiliary quantum number $-k$ $(k > 0)$ is accordingly $(-1)^k$. On setting

$$\left.\begin{aligned} l = -k \text{ when } k \text{ is negative } \left(j = -k - \frac{1}{2} = l - \frac{1}{2}\right), \\ l = k - 1 \text{ when } k \text{ is positive } \left(j = k - \frac{1}{2} = l + \frac{1}{2}\right), \end{aligned}\right\} \tag{8.2}$$

both possibilities are included under $\delta = (-1)^l$, or we could also write $\delta = \operatorname{sgn} k \cdot (-1)^{k-1}$. The only coefficients occurring in a proper vector are those corresponding to transitions in which the signature is reversed. Our selection rule (8.1) for k is thus narrowed down to

$$k \rightarrow k - 1, \; -k, \; k + 1. \tag{8.3}$$

The following table gives the value of the auxiliary quantum number k associated with each possible combination of l and j:

j \ l	0	1	2	3	4 · · ·
$j = l - \frac{1}{2}$		−1	−2	−3	−4 · · ·
$j = l + \frac{1}{2}$	1	2	3	4	5 · · ·

II. Transition to the limit $c \to \infty$.

In order to return from relativistic to ordinary mechanics we must pass to the limit $c \to \infty$. Before applying this to equations (7.8) we must replace U, v by $m_0 + \frac{U}{c}$, cv; we then have, on neglecting $\frac{U}{c^2}$ in comparison with $\frac{2m}{h}$,

$$Uv = \left(\frac{d}{dr} + \frac{k}{r}\right)w,$$

$$2mw = -h\left(\frac{d}{dr} - \frac{k}{r}\right)v;$$

on eliminating w we obtain

$$Uv = -\frac{h}{2m}\left(\frac{d}{dr} + \frac{k}{r}\right)\left(\frac{d}{dr} - \frac{k}{r}\right)v$$

or

$$\frac{h}{2m}\left(\frac{d^2}{dr^2} - \frac{k(k-1)}{r^2}\right)v + Uv = 0.$$

On introducing l by (8.2) we have in both cases $k(k-1)=l(l+1)$. Hence in the limit terms with the same l, and therefore those with auxiliary quantum numbers k and $-k-1$, coincide with that one associated with azimuthal quantum number l in the scalar theory of Chapter II. *The doublet found in alkali spectra —and in general the multiplet structure of spectral lines—is accordingly explained as a relativistic phenomenon.*

III. H, He^+, · · ·.

In a Coulomb field with nuclear charge Ze we have

$$-\Phi = \frac{Z\alpha}{4\pi r},$$

employing Heaviside units, which are better adapted to a field theory. In the following calculations we shall denote the multiple $\frac{Z\alpha}{4\pi}$ of the fine-structure constant $\frac{\alpha}{4\pi}$ simply by α itself, and we shall set $m_0 c = \nu_0$. In order to integrate equations (7.8) we first perform the substitution

$$v = e^{-\beta r} \cdot F, \quad w = e^{-\beta r} \cdot G,$$

where β is a positive constant. Our equations are then

$$\begin{aligned} \left(\frac{d}{dr} + \frac{k}{r}\right)G - \frac{\alpha}{r}F &= \left(\frac{\nu}{c} - \frac{\nu_0}{c}\right)F + \beta G, \\ \left(\frac{d}{dr} - \frac{k}{r}\right)F + \frac{\alpha}{r}G &= \beta F - \left(\frac{\nu}{c} + \frac{\nu_0}{c}\right)G. \end{aligned} \tag{8.4}$$

Our method will lead to a solution if we choose the constant β in such a way that the determinant of the linear combinations of F and G on the right vanishes :

$$\left(\frac{\nu}{c}\right)^2 - \left(\frac{\nu_0}{c}\right)^2 + \beta^2 = 0, \quad c\beta = \sqrt{\nu_0^{\,2} - \nu^2}. \tag{8.5}$$

We now seek a power series solution

$$F = \Sigma a_\mu r^\mu, \quad G = \Sigma b_\mu r^\mu,$$

where the exponent μ begins with an initial value μ_0 and runs through the values μ_0, $\mu_0 + 1$, $\mu_0 + 2$, $\cdots$. On substituting these in (8.4) we obtain the recursion formulæ

$$\begin{aligned} (\mu + k)b_\mu - \alpha\, a_\mu &= \left(\frac{\nu}{c} - \frac{\nu_0}{c}\right)a_{\mu-1} + \beta\, b_{\mu-1}, \\ \alpha\, b_\mu + (\mu - k)a_\mu &= \beta\, a_{\mu-1} - \left(\frac{\nu}{c} + \frac{\nu_0}{c}\right)b_{\mu-1}. \end{aligned} \tag{8.6}$$

The initial exponent $\mu = \mu_0$ is determined by the fact that the determinant of the coefficients of a_μ, b_μ on the left must vanish for this value of the index :

$$\mu^2 - k^2 + \alpha^2 = 0\,; \quad \mu_0 = \sqrt{k^2 - \alpha^2}.$$

Because of the manner in which β was determined in (8.5) there exists a linear relation, with coefficients $\frac{\nu}{c} + \frac{\nu_0}{c}$, β between the right-hand sides of (8.6) which is satisfied identically in $a_{\mu-1}$, $b_{\mu-1}$. Hence for all μ

$$\left(\frac{\nu}{c} + \frac{\nu_0}{c}\right)[(\mu + k)b_\mu - \alpha\, a_\mu] + \beta[\alpha\, b_\mu + (\mu - k)a_\mu] = 0$$

or

$$b_\mu\left[\left(\frac{\nu}{c} + \frac{\nu_0}{c}\right)(\mu + k) + \alpha\beta\right] + a_\mu\left[\beta(\mu - k) - \left(\frac{\nu}{c} + \frac{\nu_0}{c}\right)\alpha\right] = 0. \tag{8.7}$$

The power series will *break off* with the term with exponent μ if on replacing $a_{\mu-1}$, $b_{\mu-1}$ by a_μ, b_μ the right-hand side of (8.6) is made to vanish. The condition for this is that

$$\beta\, b_\mu + \left(\frac{\nu}{c} - \frac{\nu_0}{c}\right)a_\mu = 0\,; \tag{8.8}$$

it will be satisfied in virtue of (8.7) if the determinant of the coefficients in these two equations vanishes :

$$\left(\frac{\nu}{c} - \frac{\nu_0}{c}\right)\left[\left(\frac{\nu}{c} + \frac{\nu_0}{c}\right)(\mu + k) + \alpha\beta\right] - \beta\left[\beta(\mu - k) - \left(\frac{\nu}{c} + \frac{\nu_0}{c}\right)\alpha\right] = 0$$

or by (8.5)

$$\alpha \cdot \frac{\nu}{c} - \beta \mu = 0, \quad \frac{\nu}{c\beta} = \frac{\mu}{\alpha}.$$

Since the exponent μ with which the series break off must be of the form $\mu_0 + n$, where n is a positive integer, we obtain the *fine structure formula*

$$\boxed{\frac{\nu}{\sqrt{\nu_0^2 - \nu^2}} = \frac{E}{\sqrt{m_0^2 - E^2}} = \frac{1}{\alpha}(n + \sqrt{k^2 - \alpha^2})}. \tag{8.9}$$

The solution ψ of our differential equations, for the characteristic values $\nu = cE$ defined by (8.9), is of the form

$$e^{-\beta r} \cdot r^{\mu_0} \cdot (\text{polynomial of degree } n \text{ in } r)$$

and satisfies the condition that the spatial integral of $|\psi|^2$ converge in the neighbourhood of the singular points $r = 0, \infty$. *These E consequently constitute the discrete term spectrum of an ion with nuclear charge Ze and having but one electron outside the nucleus.* If we neglect the small constant α in comparison with k, E depends only on $n + |k|$. This fine structure formula further tells us that the two terms with auxiliary quantum numbers k and $-k$, or the two terms with the same j and for which $l = j \pm \frac{1}{2}$, exactly coincide. That this is in fact found to be the case has already been mentioned in § 4. Equation (8.9) has had a remarkable history. It was first derived on the basis of the older quantum theory by *Sommerfeld* and, at about the same time, verified by the experiments of *Paschen*; it was perhaps the greatest triumph of that theory, next to *Bohr's* explanation of the Balmer series and his calculation of the *Rydberg* number from universal atomic constants. The new quantum theory at first destroyed this beautiful agreement, as in its scalar form it led to (8.9) with the half-integral quantum number j in place of the integral $|k|$. Sommerfeld's original formula was only completely re-established with the advent of the *Dirac* theory here discussed. The quantum number k, which was used in the older quantum mechanics in place of l and which may assume the value 0, has also re-appeared and is now supplied with a sign. But on the other hand, the number of components in the fine structure is now greater than in Sommerfeld's theory, as in addition to the transitions $k \to k-1$, $k+1$ we may now also have $k \to -k$; this addition is also in agreement with experiment.

Our conclusion that (8.8) was to be satisfied in virtue of equation (8.7) for the unknowns a_μ, b_μ, assuming that the determinant of the two equations vanished, fails when both coefficients of equation (8.6) are zero :

$$\frac{\nu + \nu_0}{c\beta} = - \frac{\alpha}{\mu + k} = \frac{\mu - k}{\alpha}.$$

It follows from this that then $\mu = \sqrt{k^2 - \alpha^2}$, or $n = 0$, and that $\mu + k < 0$, or $k < 0$. *There actually exist no terms* $n = 0$, $k = -1, -2, \cdots$. For the coefficients a_μ, b_μ of the beginning term in the corresponding solution, which is at the same time the end term, would by (8.6), (8.8) necessarily satisfy the equations

$$(\mu + k)b_\mu - \alpha a_\mu = 0, \;\; \alpha b_\mu + (\mu - k)a_\mu = 0, \;\; \beta\, b_\mu + \frac{\nu - \nu_0}{c} a_\mu = 0$$

or

$$\left(-\frac{b_\mu}{a_\mu} = \right) - \frac{\alpha}{\mu + k} = \frac{\mu - k}{\alpha} = \frac{\nu - \nu_0}{c\beta},$$

and this is impossible because of the condition $|\nu| < \nu_0$.[16]

In accordance with the foregoing we may describe *the normal state of the hydrogen atom;* $n = 0$, $k = 1$ $(l = 0)$, as follows. We take the quantum number m, which may assume either of the values 0, -1, to be 0. Let $a = 0{\cdot}532$ Å. be the radius of the first Bohr orbit and $\alpha = 7{\cdot}29 \cdot 10^{-3}$ the fine-structure constant. ψ_1, ψ_2; ψ_1', ψ_2' are obtained by multiplying the radial function

$$\lambda(r) = e^{-r/a} \cdot r^{\sqrt{1-\alpha^2} - 1}$$

with the factors

$$\begin{array}{ll|l} (1 + \sqrt{1 - \alpha^2}) + i\alpha \cos\theta, & i\alpha \sin\theta\, e^{i\phi} & \psi_1, \psi_2 \\ (1 + \sqrt{1 - \alpha^2}) - i\alpha \cos\theta, & -i\alpha \sin\theta\, e^{i\phi} & \psi_1', \psi_2'. \end{array}$$

We find from these expressions that the probability density $\tilde{\psi}\psi$ is distributed spherical-symmetrically in accordance with the law

$$\rho = [\lambda(r)]^2.$$

The normalization is here not chosen in such a way that the integral of ρ over all space is unity ; it is actually

$$4\pi\left(\frac{a}{2}\right)^{1 + 2\sqrt{1-\alpha^2}} \cdot \Gamma(1 + 2\sqrt{1 - \alpha^2}).$$

We have already seen that in a certain sense the probability density multiplied by $-e$ represents the distribution of charge in the atom. Considering the probability current as determining the convection of this continuous charge distribution ρ, we find that it represents a circulation about the z-axis with velocity $\alpha c \sin \theta$ (αc is the velocity of the electron in the first Bohr orbit on the older theory). On giving the axis of rotation all possible directions ψ runs through the 2-parameter family of characteristic solutions for which $n = 0$, $k = 1$; we may take as a basis for this family of solutions the above ($m = 0$) and that for which $m = -1$, representing a circulation in the opposite direction.

C. The Permutation Group

§ 9. Resonance between Equivalent Individuals

The Hermitian forms Q, which represent in system-space all possible physical quantities of a given system, constitute a totality Σ within which addition and multiplication is defined. If Σ were reducible we could choose our co-ordinate system in system-space in such a way that all Q would be simultaneously completely reduced; these individual parts into which the whole would be divisible would then each constitute solutions of the quantum problem which were merely accidentally joined together to form the given solution. In accordance with the fundamental Aristotelian postulate of "*nihil frustra*" Nature could hardly be expected to indulge in such a superfluous luxury. Hence we propose the thesis that *Σ is an irreducible system.* On introducing as fundamental quantities the canonical variables as in II, § 11, this assumption contains the *requirement that it be impossible to choose co-ordinates in system-space in such a way that the $2f$ matrices $q_1, \cdots, q_f$; $p_1, \cdots, p_f$ are simultaneously completely reduced. This postulate is to be added to the Heisenberg commutation rules as an essential supplement.*

In accordance with Burnside's theorem [III, § 10], which we carry over without scruple from spaces with a finite number of dimensions to those with infinitely many, the irreducibility postulate allows us to assert that there can exist no linear homogeneous relation $\mathrm{tr}(AQ) = 0$ between the components of Q which is satisfied for all Q. Since in the domain of the Q's not only is multiplication possible—as presupposed in Burnside's theorem—but also addition, we arrive at the conclusion that *all* Hermitian matrices in system-space are contained in Σ. It is perhaps desirable to express our requirement directly in the form: *any* Hermitian form represents a physical quantity of the system. In accordance with II, § 7 there is associated with each statistical ensemble a positive definite Hermitian form A in such a way that tr(AQ) is the expectation of the quantity represented by Q. Burnside's theorem asserts that the equation

$$\text{tr}\,(AQ) = \text{tr}\,(A'Q)$$

can be satisfied for all Q only if $A = A'$, or *it is impossible to distinguish between the two statistical aggregates represented by the positive definite Hermitian forms only if* $A = A'$. In particular it follows from this that the states represented by two rays in system space are physically different if the two rays are distinct; this was to be expected, or even required, from the outset. These consequences show the naturalness and cogency of the irreducibility postulate, from which it can conversely be deduced.

The states of physical entities I which are fully equivalent, as, for example, the electrons in an atom, are to be represented by vectors $\mathfrak{x} = (x_i)$ or rays in the *same* system-space $\mathfrak{R}$. If two such individuals unite to form a single physical system I^2 the vectors of the corresponding system-space $\mathfrak{R} \times \mathfrak{R} = \mathfrak{R}^2$ are, in accordance with the general rule of $\times$-multiplication, the tensors (x_{ik}) of order two. But, by III, § 5, $\mathfrak{R}^2$ is reducible into two independent sub-spaces $\{\mathfrak{R}^2\}$ and $[\mathfrak{R}^2]$, the space of anti-symmetric and the space of symmetric tensors of 2nd order. Physical quantities Q of I^2 have only an objective physical significance if they depend *symmetrically* on the two individuals. This requirement is expressed in terms of the elements of the Hermitian form

$$Q = \Sigma q_{ik,\,i'k'}\, \bar{x}_{ik}\, x_{i'k'}$$

by the symmetry condition

$$q_{ki,\,k'i'} = q_{ik,\,i'k'} \tag{9.1}$$

On reducing (x_{ik}) into its anti-symmetric and its symmetric parts,

$$x_{ik} = x\{ik\} + x(ik) \tag{9.2}$$

Q is reduced, in virtue of (9.1), into two Hermitian forms in $x\{ik\}$ and $x(ik)$ respectively. For on substituting (9.2) into Q we obtain four terms: those in which $\{\mathfrak{R}^2\}$, $[\mathfrak{R}^2]$ intersect themselves, and the two in which $\{\mathfrak{R}^2\}$ intersects $[\mathfrak{R}^2]$ or conversely. These last two then vanish, for if we interchange the dummy indices i with k, i' with k' in

$$[Q] = \Sigma q_{ik,\,i'k'}\, \bar{x}\{ik\}\, x(i'k')$$

and then replace

$$q_{ki,\,k'i'},\ x\{ki\},\ x(k'i') \text{ by } q_{ik,\,i'k'},\ -x\{ik\},\ x(i'k')$$

we find $[Q] = -[Q]$, or $[Q] = 0$. The totality of Hermitian forms Q which represent the quantities of I^2 depending symmetrically on the two individuals is therefore not irreducible; it can be reduced in accordance with the decomposition

$$\mathfrak{R}^2 = \{\mathfrak{R}^2\} + [\mathfrak{R}^2] \tag{9.3}$$

of the space $\mathfrak{R}^2$.

In particular, every possible *interaction between the two individuals* depends symmetrically on them, even when other physical elements, such as a radiation field, are also involved. Hence if I^2 is at any time in a state contained in one of the sub-spaces $\{\mathfrak{R}^2\}$ or $[\mathfrak{R}^2]$ it is for all time impossible to get it out of this sub-space *by any influence whatsoever.* Again, we expect Nature to make use of but one of these sub-spaces, but the irreducibility postulate offers us no clue as to which one she has decided on.

Take as co-ordinates in the system space $\mathfrak{R}$ of the individual I the principal axes $\mathfrak{e}_i$ of the energy associated with the characteristic numbers E_i. Disregarding the interaction between the two individuals for the moment, the system I^2 has as energy levels $E_i + E_k$ with characteristic vectors $\mathfrak{e}_i \times \mathfrak{e}_k = \mathfrak{e}_{ik}$; each characteristic number of the type $E_1 + E_2$ appears twice, and the corresponding characteristic space is spanned by the vectors $\mathfrak{e}_{12}$ and $\mathfrak{e}_{21}$. On introducing the interaction as a small perturbation the two states $\mathfrak{e}_{12}$ and $\mathfrak{e}_{21}$ are in resonance with each other. Denoting the components of the total Hamiltonian function by $H(ik, i'k')$, the transformation of the sub-matrix

$$\left\| \begin{matrix} H(1\,2,\ 1\,2) & H(1\,2,\ 2\,1) \\ H(2\,1,\ 1\,2) & H(2\,1,\ 2\,1) \end{matrix} \right\|$$

to principal axes, as required by perturbation theory, can in the present case be performed in a manner which is universally valid; we need only to replace the fundamental vectors $\mathfrak{e}_{12}$, $\mathfrak{e}_{21}$ by

$$\frac{1}{\sqrt{2}}(\mathfrak{e}_{12} - \mathfrak{e}_{21}), \quad \frac{1}{\sqrt{2}}(\mathfrak{e}_{12} + \mathfrak{e}_{21}). \tag{9.4}$$

Denoting $H(1\,2,\ 1\,2) = H(2\,1,\ 2\,1)$ by $h\nu$ and the numbers $H(1\,2,\ 2\,1) = H(2\,1,\ 1\,2)$, which must be real in virtue of the condition $H(1\,2,\ 2\,1) = \bar{H}(2\,1,\ 1\,2)$ of Hermitian symmetry, by $h\alpha$, the resonance equations become

$$\frac{1}{i}\frac{dx_{12}}{dt} + (\nu x_{12} + \alpha x_{21}) = 0,$$

$$\frac{1}{i}\frac{dx_{21}}{dt} + (\alpha x_{12} + \nu x_{21}) = 0$$

from which it follows that

$$\frac{d(x_{12}-x_{21})}{dt} = -i(\nu-\alpha)(x_{12}-x_{21}),$$

$$\frac{d(x_{12}+x_{21})}{dt} = -i(\nu+\alpha)(x_{12}+x_{21}).$$

Taking as initial conditions $x_{12}=1$, $x_{21}=0$ for $t=0$, we find

$$x_{12}-x_{21}=e^{-i(\nu-\alpha)t}, \quad x_{12}+x_{21}=e^{-i(\nu+\alpha)t}; \tag{9.5}$$
$$|x_{12}|^2=\cos^2\alpha t, \quad |x_{21}|^2=\sin^2\alpha t.$$

We see from this how the two states $\mathfrak{e}_{12}$, $\mathfrak{e}_{21}$ alternate back and forth with the beat period $\frac{2\pi}{\alpha}$, whereas the components (9.5) along the axes (9.4) have always the same constant absolute magnitudes.

The only characteristic numbers associated with the system space $\{\mathfrak{R}^2\}$ are those of the type E_1+E_2, each of which appears exactly once, but the sub-space $[\mathfrak{R}^2]$ has simple characteristic numbers of the type $2E_1$ in addition to these. Hence if Nature decides in favour of $\{\mathfrak{R}^2\}$ both individuals can never be simultaneously in the same quantum state with energy E_1—assuming this energy level for the individual system is non-degenerate. That E_1+E_2 occurs only once in $\{\mathfrak{R}^2\}$ and only once in $[\mathfrak{R}^2]$ means: the possibility that one of the identical twins Mike and Ike is in the quantum state E_1 and the other in the quantum state E_2 does not include two differentiable cases which are permuted on permuting Mike and Ike; it is impossible for either of these individuals to retain his identity so that one of them will always be able to say " I'm Mike " and the other " I'm Ike." Even in principle one cannot demand an alibi of an electron! In this way the Leibnizian principle of *coincidentia indiscernibilium* holds in quantum mechanics.[17]

On passing from 2 to f equivalent individuals I it is not so easy to reduce the representation $(\mathfrak{c})^f$ of the complete linear or of the unitary group in system-space $\mathfrak{R}$ into its irreducible constituents; we shall go into this matter in the last chapter. Nevertheless we know from III, § 5, that the anti-symmetric and the symmetric tensors of order f with components

$$x\{k_1k_2\cdots k_f\}, \quad x(k_1k_2\cdots k_f),$$

respectively, each yield such an irreducible representation. A physical quantity Q of the total system I^f which depends symmetrically on all f individuals will be represented by an Hermitian operator Q, the coefficients $q(k_1k_2\cdots k_f;\, k_1'k_2\cdots k_f')$

of which are unchanged on subjecting $k_1k_2 \cdots k_f$ and $k'_1k'_2 \cdots k'_f$ simultaneously to the same permutation. It is evident that such an operator always sends an anti-symmetric tensor $x\{k_1k_2 \cdots k_f\}$ into an anti-symmetric tensor x' :

$$x'\{k_1k_2 \cdots k_f\} = \sum_{k'} q(k_1k_2 \cdots k_f\,;\, k'_1k'_2 \cdots k'_f)x\{k'_1k'_2 \cdots k'_f\}.$$

Hence the sub-space $\{\mathfrak{R}^f\}$ of anti-symmetric tensors is reduced out of the system space $\mathfrak{R}^f$ of I^f, determined in accordance with the general rule of $\times$-multiplication, in such a way that if I^f is ever in the system space $\{\mathfrak{R}^f\}$ it remains there forever, regardless of what influences may act upon it. The sub-space $[\mathfrak{R}^f]$ of all symmetric tensors $x(k)$ of order f can similarly be separated out of $\mathfrak{R}^f$. The energy level $E_1 + E_2 + \cdots + E_f$, which is $f!$-fold degenerate in $\mathfrak{R}^f$, appears in $\{\mathfrak{R}^f\}$ as a simple level. Only characteristic numbers of this type appear in $\{\mathfrak{R}^f\}$, but the characteristic numbers of $[\mathfrak{R}^f]$ are all numbers which can be obtained by summation of f distinct or non-distinct energies E.

If the system space is n-dimensional, $\{\mathfrak{R}^f\}$ is only possible if $f \leqq n$. *If E is an n-fold energy level of the individual I* then the quantum states with energy E constitute an n-dimensional sub-space $\mathfrak{R}(E)$. If it should happen that only $\{\mathfrak{R}^f\}$ is realized in Nature, then in view of the foregoing it would be *impossible to have more than n individuals of the system I^f in the quantum state E.*

The reduction of $\mathfrak{R}^f$ to $\{\mathfrak{R}^f\}$ or $[\mathfrak{R}^f]$ involves relationships which frustrate any attempt at description in terms of our old intuitive pictures with their orbits and billiard-ball electrons. But the difficulty enters already with the general composition rule, according to which the manifold of possible pure states of a system composed of two parts is much greater than the manifold of combinations in which each of the partial systems is itself in a pure state.

§ 10. The Pauli Exclusion Principle and the Structure of the Periodic Table

One of the most fundamental facts of Nature, *the ordering of the chemical elements in the periodic table,* can be understood only with the help of these considerations. We go from one atom to the following, which we denote by A, in two steps: the first is preparatory and consists in increasing the charge on the nucleus by 1, and the second and final step consists in adding an electron to the ion A^+ so obtained. To obtain the normal state of A this additional electron must be bound as tigntly as possible, i.e. the energy of the total system A must be a minimum. If we disregard the mutual perturbations of the electrons for a moment, although they may be very considerable, we might expect to find every electron in an unexcited atom in the lowest energy level, i.e. with principal quantum number $n = 1$. But instead we find the following: The 1 electron of H and the 2 electrons of He are in the 1*s* orbit, i.e. they are in the quantum state $n = 1$, $l = 0$. But the next 2 electrons, which are added in going over to Li, Be, are in a 2*s* orbit, and the additional 6, the addition of each of which gives rise to one of the elements from B to Ne, enter the 2*p* orbit. Then follow Na, Mg, each with a new electron in the 3*s* state, the elements from Al to A, the additional electrons entering the 3*p* orbit, etc. These facts are readily seen on writing the wave number of the lowest *S* term in the form $-R/n^2_*$; in H, He, Li the "effective quantum number" n_* has the values 1·00. 0·74, 1·59. That n_* sinks on going from H to He is understandable in view of the "screening" effect of the original electron on the new one. We should expect that if the next electron also went into the orbit $n = 1$ the corresponding value of n_* would be something like 0·59, but we find instead a number which is greater than this by unity. The same occurs on going from Be to B or from Mg to Al; the normal states of these atoms are formed by the valence electrons entering 2*p* or 3*p* orbits because the 2*s* or 3*s* orbits are already "occupied," and if the valence electron is raised to an *s* state by excitation, it can only be raised to one for which $n \geqq 3$ or $n \geqq 4$.* Obviously the essential features of the regularities expressed in the periodic table depend on this mysterious *numerus clausus* for the various states with principal quantum numbers $n = 1, 2, \cdots$ and on the fact that in consequence of this the electrons in the atom are added on in definite layers or "shells." Stated more precisely, in an *ns* orbit ($n = 1, 2, \cdots$) there is room for but 2 electrons, in an *np* orbit ($n = 2, 3, \cdots$) for but 6; in general the situation is described by *Stoner's rule: there can be at most* $2(2l + 1)$ *electrons in a state with quantum numbers* n, l.

On taking into account the duplicity caused by the spin we see that this number is exactly the dimensionality of the sub-space $\mathfrak{R}(nl)$ in the system-space of a single electron. Neglecting the spin perturbation, which is indeed much smaller than the

* The physical significance of the "true principal quantum number" n is contained in these considerations: we think of the term in the Hamiltonian function which represents the energy of interaction between the various electrons as multiplied by a numerical factor λ and let λ decrease steadily from 1 to 0; this virtual adiabatic process sends each electron into a definite hydrogenic orbit with a principal quantum number n, the "true quantum number" of the electron.

mutual perturbations of the electrons, the energy level associated with this sub-space is $2(2l + 1)$-fold degenerate. This degeneracy can be removed by the introduction of the spin perturbation and a weak magnetic field; the energy level is then broken up into $2(2l + 1)$ simple components distinguished by the quantum numbers

$$j = l \pm \frac{1}{2}, \quad m = j, j - 1, \cdots, -j.$$

Stoner's rule led *Pauli* to postulate the exclusion of equivalent orbits: *it is impossible for two electrons in an atom to be simultaneously in the same quantum state* (n, l, j, m). This shows that $\mathfrak{R}^f$ is obviously not the system space of the physical system I^f in which f electrons revolve about a fixed nucleus, but that the reduction to $\{\mathfrak{R}^f\}$ takes place: *Nature has decided in favour of the reduction to the space of anti-symmetric tensors, at least in the case of electrons.* In view of the considerations advanced in the previous paragraph this principle leads conversely to Stoner's rule.[18]

If the formation of one atom from the preceding one were an entirely regular process the occupation of the various states would take place in accordance with the following table, the lower row of which indicates the number of electrons captured, on going from atom to atom, by the orbit immediately above:

$$\frac{1s;\ 2s,\ 2p;\ 3s,\ 3p,\ 3d;\ 4s,\ 4p,\ 4d,\ 4f;\ \cdots}{2;\ 2+6;\ 2+6+10;\ 2+6+10+14;\ \cdots}.$$

This would indeed be the case if we could increase the charge on the nuclei by some large fixed amount, for the mutual perturbations of the electrons could thus be made arbitrarily small in comparison with the Coulomb attraction of the nucleus. But even a rough calculation shows that these perturbations are actually too considerable not to lead to displacements in the above table, i.e. to changes in the order in which the various shells are filled. For example, after the $3p$ shell is filled, which is accomplished with A, the next 2 electrons go into $4s$ states to form K, Ca, and only then do we find electrons entering the $3d$ orbits to form Sc, Ti, $\cdots$. For details consult the books by *Hund*, *Pauling and Goudsmit* or *Ruark and Urey* mentioned in the Introduction.

It is not the purpose of this book to report on the extensive empirical data of spectroscopy, nor to show how the two main principles required to lead beyond the general scheme of quantum mechanics to the interpretation of spectra were wrested from this material; I here refer to the introduction of the inner quantum number j in addition to the azimuthal l, or the spinning electron, on the one hand, and to the reduction of $\mathfrak{R}^f$ to $\{\mathfrak{R}^f\}$ by means of the *Pauli* exclusion principle on the other. *Millikan* begins his report to the American Philosophical Society on "Recent Developments in Spectroscopy" [*Proc. Am. Phil. Soc.* **66,** p. 211 (1927)], with the words: "Never in the history of science has a subject sprung so suddenly from a state of complete obscurity and unintelligibility to a condition of full illumination and predictability as has the field of spectroscopy since the year **1913.**" The theory of groups offers the appropriate mathematical tool for the description of the order thus won.

The lines of the optical spectrum are caused by quantum jumps of the electrons which are most loosely bound. In the alkalies Li, Na, K, $\cdots$ the one involved is accordingly in the state $2s$, $3s$, $4s$, $\cdots$. We also understand why their cores Li^+, Na^+, K^+, $\cdots$ are spherically symmetric, and therefore why their spectra may be approximately calculated in terms of the motion of an electron in a spherically symmetric field; the real reason behind this is the following. That an electron has the quantum numbers n, l means that its state is in a sub-space $\mathfrak{R}_l$ of $\lambda = 2(2l + 1)$ dimensions. The sub-space $\{\mathfrak{R}_l \times \mathfrak{R}_l \times \cdots \times \mathfrak{R}_l\}$ with λ factors, as obtained by the anti-symmetric reduction of $\mathfrak{R}_l^\lambda$, is 1-*dimensional* and the rotation group induces in it the 1-dimensional identical representation; i.e. *a shell consisting of λ electrons in the state n, l acts spherical-symmetrically; its presence does not increase the manifold of terms.* Hence the "closedness" of those elements with which a shell is completed; the rare gases, which precede the alkalies, are elements of this kind. But we should also expect Cu, Ag, Au to have alkali-like spectra, as they contain but a single electron in the s state, while all the others are bound more tightly in a "closed" configuration with an external field which is spherically symmetric. The valence of the elements must obviously find its explanation in these terms; indeed, it gave the clues which originally led to the discovery of the periodic table. But only in recent times have we been able to call on the assistance of spectra, interpreted and arranged with the aid of atomic theory by *Bohr* and others, and they have verified the principal features of the table, while modifying, supplementing and improving its details.

The consequences of the Pauli principle for the term analysis of atomic spectra will be discussed in detail in Chapter V, particularly in § 15. We here mention briefly the results for the case of 2-electron spectra $f = 2$.

Just as the alkalies may be treated as if they were but 1-electron atoms, in dealing with the alkaline earth metals we need only take into account the two most loosely bound electrons which occupy an s orbit outside a spherically symmetric closed shell. As before, we obtain *one* singlet and *one* triplet term

$$(nl, n'l' ; L)$$

whose total azimuthal quantum number L assumes the values

$$L = l + l', l + l' - 1, \cdots, |l - l'|$$

assuming that the two quantum states (nl), $(n'l')$ of the individual electrons are distinct. The only difference is that now such a term appears only once, whereas before it appeared twice, corresponding to a permutation of the electrons. The situation is, however, more complicated if $(nl) = (n'l')$. *The only singlet terms*

$$(nl ; nl ; L)$$

which actually occur are those with even $L = 0, 2, \cdots, 2l$ *and the only triplet terms are those with odd* $L = 1, 3, \cdots, 2l - 1$. This rule is thoroughly in accord with the empirical data.

The best-known lines of the spectra are those arising from transitions in which only one electron is not in the normal state and is jumping between higher energy levels. Hence if one of the two electrons (not saying which!) is in the normal state $n' = n_0$, $l' = 0$ ($n_0 = 1, 2, 3, 4, \cdots$ for He, Be, Mg, Ca, $\cdots$) we have $L = l$ and the two quantum numbers (n, l) suffice to determine the singlets or triplets. The lowest S term ($L = 0$) of the singlet system has the principal quantum number $n = n_0$, but there is no such term in the triplet system; it begins with $n = n_0 + 1$. We find that the lowest S term in such a triplet system (which is, as we know, simple), e.g. in the spectrum of Mg, actually does lie in the neighbourhood of the *second lowest* S term of the singlet system instead of the lowest.

§ 11. The Problem of Several Bodies and the Quantization of the Wave Equation

In this paragraph we depart from our usual terminology and denote the number of individuals by n instead of f. We first consider more fully the reduction of $\mathfrak{R}^n$ to $[\mathfrak{R}^n]$, for we shall find that although it does not apply to electrons, it does to photons. Let $H = \|H_{\alpha\beta}\|$ be the Hamiltonian function of an individual. The variables $\psi(n_1, n_2, \cdots)$ of the unitary space $[\mathfrak{R}^n]$ behave like the monomials

$$\frac{x_1^{n_1} x_2^{n_2} \cdots}{\sqrt{n_1!\, n_2! \cdots}} \quad (n_1 + n_2 + \cdots = n), \tag{11.1}$$

of degree n which are formed from the components x_α of an arbitrary vector in $\mathfrak{R}$; we denote this monomial (11.1), without the denominator, by $\phi(n_1, n_2, \cdots)$. We shall have occasion to use the differentiation formula

$$d(x_1^{n_1} x_2^{n_2} \cdots) = (n_1 x_1^{n_1-1} x_2^{n_2} \cdots dx_1) + (n_2 x_1^{n_1} x_2^{n_2-1} \cdots dx_2) + \cdots.$$

In the absence of interaction between the individuals we obtain from

$$\frac{1}{i}\frac{dx_\alpha}{dt} + \sum_\beta H_{\alpha\beta} x_\beta = 0 \tag{11.2}$$

the equation

$$\begin{aligned} -\frac{1}{i}\dot{\phi}(n_1, n_2, \cdots) &= n_1 \sum_\beta H_{1\beta}\phi(n_1 - 1, n_2, \cdots, n_\beta + 1, \cdots) \\ &\quad + n_2 \sum_\beta H_{2\beta}\phi(n_1, n_2 - 1, \cdots, n_\beta + 1, \cdots) \\ &\quad + \cdots\cdots\cdots\cdots\cdots\cdots. \end{aligned}$$

In the sum on the right $\phi(n_1 - 1, n_2, \cdots, n_\beta + 1, \cdots)$ is to be interpreted as $\phi(n_1, n_2, \cdots)$ for $\beta = 1$; similarly for the term with $\beta = 2$, etc. We can also write this equation

$$\begin{aligned} -\frac{1}{i}\dot{\phi}(n_1, n_2, \cdots) &= \sum_\alpha n_\alpha H_{\alpha\alpha} \cdot \phi(n_1, n_2, \cdots) \\ &\quad + \sum_{\alpha \neq \beta} n_\alpha H_{\alpha\beta} \cdot \phi(\cdots, n_\alpha - 1, \cdots, n_\beta + 1, \cdots). \end{aligned}$$

On introducing the binomial coefficients in accordance with (11.1) we obtain as the equations of motion

$$\begin{aligned} -\frac{1}{i}\frac{d\psi(n_1, n_2, \cdots)}{dt} &= \sum_\alpha n_\alpha H_{\alpha\alpha} \cdot (n_1, n_2, \cdots) \\ &\quad + \sum_{\alpha \neq \beta} \sqrt{n_\alpha(n_\beta + 1)} H_{\alpha\beta} \cdot \psi(\cdots, n_\alpha - 1, \cdots, n_\beta + 1, \cdots). \end{aligned} \tag{11.3}$$

These equations are of the form

$$\frac{1}{i}\frac{d\psi}{dt} + \mathsf{H}\psi = 0, \quad \mathsf{H} = \sum_{\alpha, \beta} H_{\alpha\beta}\eta_{\alpha\beta} \tag{11.4}$$

where the matrices $\eta_{\alpha\beta}$ are defined by

$$\eta_{\alpha\alpha}(n_1, n_2, \cdots; n_1', n_2', \cdots) = \begin{cases} n_\alpha \text{ if } n_1' = n_1, n_2' = n_2, \cdots \\ 0 \text{ otherwise} \end{cases} \tag{11.5}$$

and for $\alpha \neq \beta$

$$\eta_{\alpha\beta}(n_1, n_2, \cdots; n_1', n_2', \cdots) = \begin{cases} \sqrt{n_\alpha(n_\beta + 1)} \\ 0 \end{cases} \tag{11.5'}$$

where the first alternative holds when all $n' = n$ with the exception of $n_\alpha' = n_\alpha - 1$, $n_\beta' = n_\beta + 1$ and the second in all other cases. **H** is, as it should be, an Hermitian matrix. If H is in diagonal form the fundamental vectors forming our co-ordinate system are the quantum states of the various individuals; $|\psi(n_1, n_2, \cdots)|^2$ is then the probability that there are simultaneously n_1 individuals in the first quantum state, n_2 in the second, etc. On reduction from $\mathfrak{R}^n$ to $[\mathfrak{R}^n]$ it becomes impossible to identify the individuals as Mike, Ike, $\cdots$ and we therefore may not ask for the probability that Mike is in the αth state, Ike is in the βth, $\cdots$. If we have in addition to H a perturbation εW affecting the individuals (and symmetric with respect to these individuals), then equation (11.3) governs the change of the probabilities $|\psi(n_1, n_2, \cdots)|^2$ in time.

The Hamiltonian function **H** reminds us of the one which we obtained in Chapter II, § 13 by quantizing Maxwell's equations; there the individuals were photons. Maxwell's equations are to be considered as the quantum-theoretical wave equations of an individual photon. If we replace the photon by an individual whose state (x_α) varies in accordance with equation (11.2) we are led to a new way of treating the problem of several bodies, which we call the "method of second quantization" in contrast to the "method of composition" or "$\times$-multiplication" developed in Chapter II, § 10. In this we consider (11.2) as the classical equations of motion of a physical system whose canonical variables are the real and imaginary parts q_α, p_α of x_α, and as such subject them to the process of quantization.[19] We here tie on to the development given in Chapter II, § 11. Introduce the complex quantities

$$x_\alpha = \frac{1}{\sqrt{2}}(q_\alpha + ip_\alpha), \quad \bar{x}_\alpha = \frac{1}{\sqrt{2}}(q_\alpha - ip_\alpha)$$

into the Hamiltonian function H as independent variables in place of q_α, p_α; the Hamiltonian equations are then

$$\frac{dx_\alpha}{dt} = -i\frac{\partial \mathrm{H}}{\partial \bar{x}_\alpha}, \quad \frac{d\bar{x}_\alpha}{dt} = i\frac{\partial \mathrm{H}}{\partial x_\alpha}. \tag{11.6}$$

In order that (11.2) may be considered as the classical equations of motion of a system with infinitely many degrees of freedom, in accordance with our programme, they must be of the form (11.6). But this is in fact the case; the Hamiltonian function is then

$$\mathrm{H} = \sum_{\alpha,\beta} H_{\alpha\beta}\bar{x}_\alpha x_\beta.$$

In quantizing x_α, $\bar{x}_\alpha$ are to be replaced by Hermitian conjugate matrices $\boldsymbol{x}_\alpha$, $\tilde{\boldsymbol{x}}_\alpha$ which satisfy the following commutation rules:

$$\left.\begin{aligned} \boldsymbol{x}_\alpha \boldsymbol{x}_\beta - \boldsymbol{x}_\beta \boldsymbol{x}_\alpha = 0, \quad \tilde{\boldsymbol{x}}_\alpha \tilde{\boldsymbol{x}}_\beta - \tilde{\boldsymbol{x}}_\beta \tilde{\boldsymbol{x}}_\alpha = 0, \\ \boldsymbol{x}_\alpha \tilde{\boldsymbol{x}}_\beta - \tilde{\boldsymbol{x}}_\beta \boldsymbol{x}_\alpha = \delta_{\alpha\beta} = \begin{cases} 1 \ (\alpha = \beta) \\ 0 \ (\alpha \neq \beta). \end{cases} \end{aligned}\right\} \tag{11.7}$$

The Hamiltonian function H then becomes the matrix

$$\mathsf{H} = \sum_{\alpha,\beta} H_{\alpha\beta}\, \tilde{\boldsymbol{x}}_\alpha \boldsymbol{x}_\beta\,; \tag{11.8}$$

if H is in diagonal form then

$$\mathsf{H} = \sum_{\alpha} E_\alpha \tilde{\boldsymbol{x}}_\alpha \boldsymbol{x}_\alpha.$$

We are here dealing with an infinite set of oscillators, the individual members of which are distinguished by the index α; the energy of the αth is given in terms of the complex co-ordinates x_α, $\bar{x}_\alpha$ by $E_\alpha \bar{x}_\alpha x_\alpha$.

The quantum theory of a single oscillator as developed in II, § 3 gives us as the irreducible solution of

$$\boldsymbol{x}\tilde{\boldsymbol{x}} - \tilde{\boldsymbol{x}}\boldsymbol{x} = \mathbf{1},$$

where $\boldsymbol{x}$, $\tilde{\boldsymbol{x}}$ are two Hermitian conjugate matrices normalized in such a way that the energy $\tilde{\boldsymbol{x}}\boldsymbol{x}$ is in diagonal form, the matrices

$$x(n, n+1) = \sqrt{n+1}, \ \tilde{x}(n, n-1) = \sqrt{n}\,; \ \tilde{x}x(n, n) = n,$$

all other components vanishing; the quantum number n assumes the values 0, 1, 2, · · ·. From this we obtain the solution of (11.7) by composition:

$$x_\alpha(n_1, n_2, \cdots;\ n_1', n_2', \cdots) = \begin{cases} \sqrt{n_\alpha + 1}, & \text{if all } n' = n \text{ except } n_\alpha' = n_\alpha + 1, \\ 0 & \text{otherwise}; \end{cases}$$

$$\tilde{x}_\alpha(n_1, n_2, \cdots;\ n_1', n_2, \cdots) = \begin{cases} \sqrt{n_\alpha} & \text{if all } n' = n \text{ except } n_\alpha = n_\alpha - 1, \\ 0 & \text{otherwise}. \end{cases}$$

The products $\tilde{\boldsymbol{x}}_\alpha \boldsymbol{x}_\alpha$ are of course in diagonal form; $\tilde{\boldsymbol{x}}_\alpha \boldsymbol{x}_\beta$ is the matrix $\eta_{\alpha\beta}$ introduced above, and (11.8) coincides with (11.4):

the method of second quantization leads to the same result as the method of composition supplemented by the " symmetric reduction " of $\mathfrak{R}^n$ *to* $[\mathfrak{R}^n]$. But now the number

$$n_1 + n_2 + \cdots = n$$

of individuals is not prescribed; however **H** is reduced into sub-matrices in accordance with the various values of n, for all components $\mathsf{H}(n_1 n_2 \cdots ; n_1' n_2' \cdots)$ for which $n_1' + n_2' + \cdots \neq n_1 + n_2 + \cdots$ vanish. The total number of photons is not conserved, and to this extent Maxwell's equations do not fit completely into the quantum-theoretical picture—unless we wish to consider " non-existence " as a particular quantum state of the photon.

The method of composition remains applicable in the presence of interaction between the individuals, provided it is an instantaneous action at a distance determined by the simultaneous values of the canonical variables of the various individuals. But it breaks down when, as in the theory of relativity, account is taken of the finite velocity of propagation, which led to the introduction of continuous fields in the classical theories. The difficulty arises from the fact that the wave function ψ must contain the *one* time t as argument in addition to the spatial co-ordinates of each particle, whereas the theory of relativity requires that the proper time of each particle appear as argument in ψ as well as the spatial co-ordinates. The method of second quantization shows its superiority in dealing with such problems.

As we have seen, the method of second quantization in accordance with Heisenberg's commutation rules is equivalent to a reduction of the system space $\mathfrak{R}^n$ to $[\mathfrak{R}^n]$. Since we have seen in II, § 13 that this leads to the correct laws of radiation phenomena, we must conclude that the behaviour of photons corresponds to this reduction. But in the case of electrons the reduction is to the space $\{\mathfrak{R}^n\}$, and we must now investigate to what kind of quantization this corresponds.[20] The vectors of the unitary space $\{\mathfrak{R}^n\}$ are the anti-symmetric tensors with components

$$x\{\alpha_1, \alpha_2, \cdots, \alpha_n\} \sim |x_{\alpha_1}, x_{\alpha_2}, \cdots, x_{\alpha_n}| \tag{11.9}$$

in the space $\mathfrak{R}$, where the one row in the determinant stands for the n rows formed in the same manner from n vectors $\mathfrak{x} = \mathfrak{x}^{(1)}, \mathfrak{x}^{(2)}, \cdots, \mathfrak{x}^{(n)}$ of $\mathfrak{R}$. We can obtain the totality of linearly independent components by restricting the indices by the condition

$$\alpha_1 < \alpha_2 < \cdots < \alpha_n. \tag{11.10}$$

We now denote (11.9) by $\psi(n_1, n_2, \cdots)$, where $n_\alpha = 1$ or 0 according to whether α appears in the set of indices $\alpha_1, \alpha_2, \cdots, \alpha_n$ or not; these quantum numbers n_α may thus only assume one of two values. On replacing $\alpha_1 = \alpha$ in (11.9) by an index $\beta \neq \alpha$, (11.9) vanishes if β is equal to one of the remaining indices $\alpha_2, \cdots, \alpha_n$; if β is different from $\alpha_2, \cdots, \alpha_n$ it becomes

$$x\{\beta\alpha_2 \cdots \alpha_n\} = \pm\psi(n_1, \cdots, n_\alpha - 1, \cdots, n_\beta + 1, \cdots),$$

the sign ± 1 being $(-1)^r$ where r is the number of indices in the set $\alpha_2, \cdots, \alpha_n$ lying between α and β:

$$r = \sum_\lambda n_\lambda$$

where the sum is extended over all indices λ between α and β. We again obtain equations of the form (11.4); (11.5) is then valid as it stands but (11.5′) is to be replaced by

$$\eta_{\alpha\beta}(n_1, n_2, \cdots; n_1', n_2', \cdots) = \pm 1 \text{ or } 0,$$

where the first alternative applies to the case in which all $n' = n$ except $n_\alpha = 1$, $n_\alpha' = 0$; $n_\beta = 0$, $n_\beta' = 1$, the sign being again determined in accordance with the above rule. On writing a matrix $\|a(nn')\|$ in the form

$$\begin{Vmatrix} a(0\ 0) & a(0\ 1) \\ a(1\ 0) & a(1\ 1) \end{Vmatrix}$$

and introducing the abbreviations

$$\begin{Vmatrix} 1 & 0 \\ 0 & 1 \end{Vmatrix} = \mathbf{1}, \quad \begin{Vmatrix} 1 & 0 \\ 0 & -1 \end{Vmatrix} = \mathbf{1}',$$

we may write

$$\eta_{\alpha\alpha} = \mathbf{1}\times\mathbf{1}\times \cdots \times \begin{Vmatrix} 0 & 0 \\ 0 & 1 \end{Vmatrix} \times\mathbf{1}\times\mathbf{1}\times \cdots,$$

$$\eta_{\alpha\beta} = \mathbf{1}\times\mathbf{1}\times \cdots \times \begin{Vmatrix} 0 & 0 \\ 1 & 0 \end{Vmatrix} \times\mathbf{1}'\times \cdots \times\mathbf{1}'\times \begin{Vmatrix} 0 & 1 \\ 0 & 0 \end{Vmatrix} \times\mathbf{1}\times \cdots \quad (\alpha \neq \beta),$$

where the matrix that is written explicitly in the first equation is in the αth place and those in the second in the αth and β^{th} places respectively. We must now attempt to write these matrices in the form $\tilde{\boldsymbol{x}}_\alpha \boldsymbol{x}_\beta$; this can in fact be accomplished by taking

$$\left.\begin{aligned} \boldsymbol{x}_\alpha &= \mathbf{1}' \times \mathbf{1}' \times \cdots \times \mathbf{1}' \times \begin{Vmatrix} 0 & 1 \\ 0 & 0 \end{Vmatrix} \times \mathbf{1} \times \mathbf{1} \times \cdots, \\ \tilde{\boldsymbol{x}}_\alpha &= \mathbf{1}' \times \mathbf{1}' \times \cdots \times \mathbf{1}' \times \begin{Vmatrix} 0 & 0 \\ 1 & 0 \end{Vmatrix} \times \mathbf{1} \times \mathbf{1} \times \cdots, \end{aligned}\right\} \quad (11.11)$$

the small explicit matrices being in the αth place. $\boldsymbol{x}_\alpha$, $\tilde{\boldsymbol{x}}_\alpha$ are Hermitian conjugates, and $\mathbf{H}$ can now be written in terms of them in the desired form (11.8). Instead of the commutation rules (11.7) we now have

$$\boldsymbol{x}_\alpha \boldsymbol{x}_\beta + \boldsymbol{x}_\beta \boldsymbol{x}_\alpha = 0, \tilde{\boldsymbol{x}}_\alpha \tilde{\boldsymbol{x}}_\beta + \tilde{\boldsymbol{x}}_\beta \tilde{\boldsymbol{x}}_\alpha = 0, \boldsymbol{x}_\alpha \tilde{\boldsymbol{x}}_\beta + \tilde{\boldsymbol{x}}_\beta \boldsymbol{x}_\alpha = \delta_{\alpha\beta}. \quad (11.12)$$

(11.1) is the irreducible solution of these equations by a pair of Hermitian conjugate matrices $\boldsymbol{x}_\alpha$, $\tilde{\boldsymbol{x}}_\alpha$ which are so normalized that $\tilde{\boldsymbol{x}}_\alpha \boldsymbol{x}_\alpha$ is a diagonal matrix.

In order to show that the equations (11.4) for the vector ψ in system-space yield the Hamiltonian equations (11.6) for the forms

$$\boldsymbol{x}_\alpha = \sum_{n,n'} x_\alpha(n; n') \bar{\psi}(n) \psi(n') \quad \text{and} \quad \tilde{\boldsymbol{x}}_\alpha,$$

we must prove that the formula

$$\boldsymbol{x}_\alpha \mathbf{H} - \mathbf{H} \boldsymbol{x}_\alpha = \frac{\partial \mathbf{H}}{\partial \tilde{\boldsymbol{x}}_\alpha}$$

employed in II, § 11, holds here as well. We find that it does not hold for an *arbitrary* polynomial $\mathbf{H}$ in $\boldsymbol{x}_\alpha$, $\tilde{\boldsymbol{x}}_\alpha$, but that it does for even polynomials in general and so in particular for the Hermitian form (11.8). For we have, for example,

$$\boldsymbol{x}_1 \tilde{\boldsymbol{x}}_\alpha \boldsymbol{x}_\beta = \delta_{1\alpha} \boldsymbol{x}_\beta - \tilde{\boldsymbol{x}}_\alpha \boldsymbol{x}_1 \boldsymbol{x}_\beta = \delta_{1\alpha} \boldsymbol{x}_\beta + \tilde{\boldsymbol{x}}_\alpha \boldsymbol{x}_\beta \boldsymbol{x}_1,$$

whence

$$\boldsymbol{x}_1 \cdot \tilde{\boldsymbol{x}}_\alpha \boldsymbol{x}_\beta - \tilde{\boldsymbol{x}}_\alpha \boldsymbol{x}_\beta \cdot \boldsymbol{x}_1 = \delta_{1\alpha} \boldsymbol{x}_\beta, \quad \boldsymbol{x}_1 \mathbf{H} - \mathbf{H} \boldsymbol{x}_1 = \sum_\beta H_{1\beta} \boldsymbol{x}_\beta.$$

On introducing real quantities, i.e. Hermitian forms, $\boldsymbol{p}_\alpha$, $\boldsymbol{q}_\alpha$ by

$$\boldsymbol{x}_\alpha = \frac{1}{2}(\boldsymbol{q}_\alpha + i\boldsymbol{p}_\alpha), \quad \tilde{\boldsymbol{x}}_\alpha = \frac{1}{2}(\boldsymbol{q}_\alpha - i\boldsymbol{p}_\alpha)$$

and denoting the set $\boldsymbol{p}_1, \boldsymbol{q}_2;\ \boldsymbol{p}_2, \boldsymbol{q}_2;\ \cdots$ straight through by $\boldsymbol{p}_1, \boldsymbol{p}_2, \boldsymbol{p}_3, \boldsymbol{p}_4, \cdots$ we obtain the relations

$$\boldsymbol{p}_\alpha^2 = 1, \quad \boldsymbol{p}_\alpha \boldsymbol{p}_\beta + \boldsymbol{p}_\beta \boldsymbol{p}_\alpha = 0 \quad (\alpha \neq \beta) \qquad (11.13)$$

The $\boldsymbol{p}_\alpha$ are not only Hermitian but unitary as well, as can be seen from the first of these equations or directly. Here again we meet the matrices

$$\left\| \begin{matrix} 0 & 1 \\ 1 & 0 \end{matrix} \right\|, \quad \left\| \begin{matrix} 0 & -i \\ i & 0 \end{matrix} \right\|$$

which occurred in connection with the spinning electron.

We have thus discovered the correct way to quantize the field equations defining electron waves and matter waves. Here again we find, as in the case of the spinning electron, that quantum kinematics is not to be restricted by the assumption of Heisenberg's specialized commutation rules.

§ 12. Quantization of the Maxwell-Dirac Field Equations [21]

The field laws arise from a Hamiltonian principle which is analogous to the Hamiltonian principle of classical mechanics. This latter is expressed in terms of a Lagrangian function L which depends on the positional co-ordinates q_i and their derivatives $\dot{q}_i$ with respect to time, and asserts that the first variation of

$$\int L(q_i, \dot{q}_i)dt \tag{12.1}$$

vanishes when the q_i are assigned arbitrary infinitesimal increments δq_i which vanish outside a certain finite time interval. This principal yields, on integration by parts, the differential equations

$$\frac{dp_i}{dt} + L_i = 0 \quad \text{with} \quad p_i = -\frac{\partial L}{\partial \dot{q}_i}, \; L_i = \frac{\partial L}{\partial q_i}. \tag{12.2}$$

Defining

$$H = L + \sum_i \dot{q}_i p_i$$

and noting that

$$\delta L = \sum_i L_i \delta q_i - \sum_i p_i \delta \dot{q}_i$$

we obtain for the differential of H the expression

$$\delta H = \sum_i L_i \delta q_i + \sum_i \dot{q}_i \delta p_i.$$

Expressing H as a function of the q_i and the generalized momenta p_i associated with them, we have

$$\frac{\partial H}{\partial q_i} = L_i, \quad \frac{\partial H}{\partial p_i} = \dot{q}_i$$

and by (12.2) these are just the Hamiltonian canonical equations

$$\frac{dq_i}{dt} = \frac{\partial H}{\partial p_i}, \; \frac{dp_i}{dt} = -\frac{\partial H}{\partial q_i}.$$

In quantum theory the q_i, p_i are operators satisfying Heisenberg's commutation rules.

This reasoning can be carried over without difficulty to the case of a continuum, as appears in field theories. On replacing for the moment the 3-dimensional space by the 1-dimensional interval $0 \leqq x \leqq 1$ described by the co-ordinate x and assuming, for the sake of simplicity, that only *one* state function $q = q(x, t)$ is involved, the integral (12.1) is then to be replaced by

$$\int\int_0^1 L(q, \dot{q})\, dx\, dt.$$

Naturally L may depend on the spatial derivative $\frac{\partial q}{\partial x}$, or even higher derivatives, in addition to q. The continuous variable x takes the place of the index i and the Lagrangian function, in the sense of (12.1), is now the integral $\int_0^1 L(q, \dot{q})dx$ with respect to the spatial variable instead of L itself. We first replace the continuum by a discrete set of equidistant points defined by $\Delta x = \frac{i}{n}$ $(i = 0, 1, \cdot \cdot \cdot, n - 1)$. The differential quotients with respect to x are naturally to be replaced by difference quotients with the difference $\Delta x = 1/n$, and the integrals become sums. In accordance with the outline above we must now set

$$p_i = -\frac{\partial L(q, \dot{q})}{\partial \dot{q}} \cdot \Delta x,$$

calculated at the point $x = i/n$. For the continuum we have analogously to set

$$p = -\frac{\partial L(q, \dot{q})}{\partial \dot{q}},$$

and H is to be defined by

$$H = L + \int_0^1 \dot{q}\, p\, dx.$$

The *commutation rules* which are satisfied by q, p in quantum mechanics cause some trouble. As long as we employ the discrete set of points in place of the continuum they are

$$q(x)\, p(x') - p(x')\, q(x) = \frac{\sqrt{-1}}{\Delta x} \cdot \delta_{xx'}$$

where x, x' run independently through the set i/n and $\delta_{xx'}$ is 1 or 0 according as x' coincides with x or not. For fixed x'

$$\frac{1}{\Delta x} \cdot \delta_{xx'} = \delta(x - x')$$

is a function of x which vanishes for all values of the argument other than x' and is there so large that the sum $\sum_x \delta(x - x') \cdot \Delta x$ has the value 1. In dealing with the continuum we therefore introduce with *Dirac* a function $\delta(x - x')$ which vanishes at all points $x \neq x'$ and is so large at the point x' that its integral has the value 1 (cf. I, § 7). Of course there exists no such function, but it can be "arbitrarily closely approximated" by a function which vanishes everywhere except in a very small interval about x' and assumes very large values within this interval. Only in this sense can we perform the passage to the limit $\Delta x = 0$ and write the commutation rules symbolically in the form

$$q(x)\, p(x') - p(x')\, q(x) = i\, \delta(x - x'). \tag{12.3}$$

A good illustration of the mathematical interpretation of this pathological function $\delta(x - x')$ arises in the theory of orthogonal sets of functions $\phi_i(x)$, for with its aid the completeness condition may be formulated

$$\sum_i \bar{\phi}_i(x)\, \phi_i(x') = \delta(x - x').$$

This is literally correct as long as x only runs through a discrete set of points, but the rigorous mathematical formulation for the case of a continuum is given by

$$\lim_{n\to\infty} \int_0^1\!\!\int_0^1 \sum_{i=1}^{n} \bar{\phi}_i(x)\, \phi_i(x') \cdot u(x)\, v(x')\, dx\, dx' = \int_0^1 \bar{u}(x)\, v(x)\, dx$$

where $u(x)$, $v(x)$ are any two continuous functions in the interval $(0, 1)$. Hence from the more rigorous standpoint (12.3) must be replaced by the equation

$$\int_0^1\!\!\int_0^1 \bar{u}(x)\{q(x)\, p(x') - p(x')\, q(x)\}v(x')\, dx\, dx' = i\int_0^1 \bar{u}(x)\, v(x)\, dx$$

containing two arbitrary functions $u(x)$, $v(x)$; furthermore, it is to be noted that the p, q in the brackets are first to be replaced by approximations $p^{(n)}$, $q^{(n)}$—e.g. by the nth partial sum of their expansion in terms of orthogonal functions—and the passage to the limit $n \to \infty$ is to take place *after*, not before, the integration. This interpretation offers a sound mathematical method of dealing with the relation (12.3). It is to be emphasized that (12.3) refers to two points of space x, x' at the same moment t, i.e. in a section of the world in which $t =$ const.; the arguments of q and p are to be written more precisely as (x, t), (x', t) respectively.

On applying this general scheme to the action

$$W = M + M' + \frac{1}{\alpha} F, \tag{5.18}$$

from which the field equations for the electron and for the electromagnetic field are obtained, we find ourselves faced with a difficulty arising from the fact that the Lagrangian function does not contain the time derivative of the scalar potential f_0, for the generalized momentum associated with f_0 then vanishes identically and cannot possibly satisfy a commutation relation such as (12.3). We avoid this difficulty for the moment by utilizing the principle of gauge invariance to remove f_0 from the expression of the Lagrangian function by setting it equal to 0; this device has already been employed in II, § 13. The set of independent functions describing the state is then

$$\psi = (\psi_1, \psi_2, \psi_3, \psi_4), \quad \mathfrak{f} = (f_1, f_2, f_3),$$

where we have written ψ_3, ψ_4 in place of ψ_1', ψ_2'. The momenta associated with these quantities are then found to be: $i\bar{\psi}_\rho$ with ψ_ρ and $-E_p$ with f_p. The commutation rules which are to be applied in quantizing the field equations are accordingly

$$\bar{\psi}_\rho(P)\psi_\sigma(P') + \psi_\sigma(P')\bar{\psi}_\rho(P) = \delta_{\rho\sigma}\cdot\delta(P-P') \quad [\rho, \sigma = 1, 2, 3, 4], \tag{12.4'}$$

$$f_p(P)E_q(P') - E_q(P')f_p(P) = \frac{1}{i}\delta_{pq}\cdot\delta(P-P') \quad [p, q = 1, 2, 3], \tag{12.4''}$$

where P and P' are any two points of the *same* spatial section $t =$ const. We have here taken account of the fact that the quantities ψ describing matter are not to satisfy Heisenberg's commutation rules, but are instead to satisfy those obtained by replacing the minus sign which occurs in them by a plus sign. These rules must be supplemented by the assertion that the ψ_ρ satisfy in addition the equations

$$\psi_\rho(P)\psi_\sigma(P') + \psi_\sigma(P')\psi_\rho(P) = 0, \tag{12.5}$$

and the same for $\bar{\psi}_\rho$; that the f_p at any two points P, P' are commutative and the same for the E_p; and finally that the material quantities ψ, $\bar{\psi}$ on the one hand and the electromagnetic quantities f_p, E_p on the other are kinematically independent, and that every quantity of the first kind at a point P commutes with every quantity of the second kind at any point P' (in the same section $t =$ const. of the world).

As in II, § 13, we again consider the whole system enclosed in an insulated and perfectly reflecting cavity which is at rest.

In order to describe the electro-magnetic potentials we make use of a complete orthogonal set of solutions $\mathfrak{f}$ of

$$\Delta\mathfrak{f} + \nu^2\mathfrak{f} = 0 \tag{12.6}$$

in the cavity, which satisfy the conditions

$$\operatorname{div} \mathfrak{f} = 0, \quad \mathfrak{f} \text{ normal}$$

at the walls. The construction of such a system is readily obtained from the Gauss divergence theorem

$$\int(\operatorname{curl} \mathfrak{f} \cdot \operatorname{curl} \mathfrak{g} + \operatorname{div} \mathfrak{f} \cdot \operatorname{div} \mathfrak{g} + \mathfrak{f} \cdot \Delta\mathfrak{g})dV$$

$$= \int([\mathfrak{f}, \operatorname{curl} \mathfrak{g}]_n + \mathfrak{f}_n \operatorname{div} \mathfrak{g})\, do \quad (n \text{ denoting normal component})$$

for the vector $[\mathfrak{f}, \operatorname{curl} \mathfrak{g}] + \mathfrak{f} \operatorname{div} \mathfrak{g}$, $\mathfrak{f}$ and $\mathfrak{g}$ being two arbitrary vector fields.[22] We first determine the scalar functions $\phi = \phi_\lambda$ which satisfy the equation $\Delta\phi + \lambda^2\phi = 0$ and vanish on the walls, and from them construct the vector fields $\mathfrak{f}_\lambda = \operatorname{grad} \phi_\lambda$; these vectors $\mathfrak{f}_\lambda$ automatically satisfy the conditions above, are of course mutually orthogonal and can be normalized in accordance with the equation

$$\int(\mathfrak{f}_\lambda \cdot \mathfrak{f}_{\lambda'})dV = \delta_{\lambda\lambda'}\Big[= \lambda^2\int\phi_\lambda\phi_{\lambda'}dV\Big].$$

We also determine a complete normal orthogonal system $\mathfrak{f}_\nu$ of solutions of (12.6) which are normal to the walls but which satisfy the condition $\operatorname{div} \mathfrak{f}_\nu = 0$ everywhere, not only at the walls. The $\mathfrak{f}_\lambda$ are then orthogonal to these $\mathfrak{f}_\nu$ and they constitute together a complete orthogonal system for vector fields in the cavity. We may consequently write

$$\left.\begin{aligned} \mathfrak{f} &= \sum_\nu q_\nu \mathfrak{f}_\nu + \sum_\lambda p_\lambda \mathfrak{f}_\lambda \\ -\mathfrak{E} &= \sum_\nu p_\nu \mathfrak{f}_\nu - \sum_\lambda q_\lambda \mathfrak{f}_\lambda \end{aligned}\right\} \tag{12.7}$$

in the section $t = \text{const.}$ The $\mathfrak{f}_\lambda$, $\mathfrak{f}_\nu$ are vectorial functions of position in space and have as values ordinary numbers, whereas the p, q are scalar quantum mechanical matrices which are independent of position and which satisfy the commutation rules

$$q_\nu p_\nu - p_\nu q_\nu = i, \quad q_\lambda p_\lambda - p_\lambda q_\lambda = i;$$

all q commute among themselves and all p among themselves, and any p commutes with any q whose index is not the same. [These rules are perhaps most readily obtained by solving (12.7) for the " Fourier coefficients " p, q in terms of integrals of scalar products of $\mathfrak{f}$, $\mathfrak{E}$ with $\mathfrak{f}_\lambda$, $\mathfrak{f}_\nu$ and applying the commutation rules (12.4).] The energy

$$\frac{1}{2}\int\left\{\alpha\,\mathfrak{E}^2 + \frac{1}{\alpha}(\text{curl}\,\mathfrak{f})^2\right\}dV$$

of the electro-magnetic field becomes

$$\sum_\nu \frac{1}{2}\left(\alpha\,p_\nu^2 + \frac{\nu^2}{\alpha}q_\nu^2\right) + \frac{\alpha}{2}\sum_\lambda q_\lambda^2.$$

We already know the solution of the commutation rules which reduces this expression for the energy to diagonal form. The individual components of the vector on which the p, q operate are distinguished by means of the quantum numbers N_ν, corresponding to the ν, and the values of the continuous variables q_λ, corresponding to the λ. On setting $q_\nu = \sqrt{\alpha}\cdot Q_\nu$, Q_ν is an operator which affects only the index N_ν in accordance with the equations

$$Q_\nu(N_\nu, N_\nu - 1) = \sqrt{\frac{N_\nu}{2\nu}},\quad Q_\nu(N_\nu, N_\nu + 1) = \sqrt{\frac{N_\nu + 1}{2\nu}}\,;$$

all other components, corresponding to transitions $N_\nu \to N_\nu'$ in which N_ν' is neither $N_\nu \pm 1$, vanish. N_ν assumes the integral values 0, 1, 2, $\cdots$ and can be considered as the number of photons of the kind ν. The momentum p_λ associated with the continuous variable q_λ is, following *Schrödinger*, represented by the operator $\frac{1}{i}\frac{\partial}{\partial q_\lambda}$. The electro-magnetic energy is then in diagonal form and, on neglecting the (infinite!) null-point energy, multiplies the vector component $(N_\nu\,;\; q_\lambda)$ with

$$\sum_\nu \nu N_\nu + \frac{\alpha}{2}\sum_\lambda q_\lambda^2. \tag{12.8}$$

We thus see how it happens that the electro-static part, which is described by the continuous variable q_λ, is separated off from the part due to the radiation, described by the discrete N_ν giving the number of photons of kind ν.

The ψ appear in the part of the energy due to matter only in combinations of the form $\bar{\psi}_\rho\,\psi_\sigma$. Consequently it will be found advantageous in dealing with electrons to apply the method of composition followed by anti-symmetric reduction; we have shown in the preceding section that this procedure is equivalent to quantizing in accordance with the rules (12.4′). Since the electro-magnetic quantities commute with the ψ_ρ, $\bar{\psi}_\rho$ they may here be considered as ordinary numbers. The quantized wave equations then refer to a " vector " $\mathfrak{z}$ with components

$$z_{\rho_1 \ldots \rho_n}(P_1 \cdots P_n;\ N_\nu;\ q_\lambda),$$

where $P_1, \cdots, P_n$ are the positions of the n electrons and $\rho_1, \cdots, \rho_n$ are their spin variables, each of which runs through the four values 1, 2, 3, 4. We write $z_{\rho_1 \ldots \rho_n}$ as a column consisting of 4^n terms; this z is anti-symmetric with respect to a permutation affecting the P_r and ρ_r alike. $\mathfrak{S}^{(r)} = (S_1^{(r)}, S_2^{(r)}, S_3^{(r)})$ is the spin vector (S_1, S_2, S_3) operating only on the rth index ρ_r, $T^{(r)}$ is similarly the operation on the rth index ρ_r which interchanges ψ_1, ψ_2 with ψ_3, ψ_4, and $\text{grad}^{(r)}$ is the gradient with respect to P_r. The part of the Hermitian energy operator $-\mathcal{J}_0$ in the equation

$$\frac{1}{i}\frac{d\mathfrak{z}}{dx_0} - \mathcal{J}_0\mathfrak{z} = 0 \qquad (x_0 = ct,\ H = -c\mathcal{J}_0)$$

which depends only on matter is

$$\sum_{r=1}^{n}\left(\mathfrak{S}^{(r)},\ \frac{1}{i}\,\text{grad}^{(r)} + \sqrt{\alpha}\sum_{\nu} Q\cdot \mathfrak{f}_\nu(P_r) + \frac{1}{i}\sum_{\lambda}\text{grad}\,\phi_\lambda(P_r)\cdot\frac{\partial}{\partial q_\lambda}\right) + m_0\sum_{r=1}^{n} T^{(r)} \qquad (12.9)$$

and to this must be added the electro-magnetic part (12.8).

Since we have throughout taken the scalar potential $f_0 = 0$ we have lost the equation

$$\text{div}\,\mathfrak{E} + \rho = 0 \qquad (12.10)$$

arising from the variation of f_0. This equation contains no derivatives with respect to time, and consequently represents a condition on the state of the field at a moment $t =$ const.; we must naturally take it into account. On substituting the value of $\mathfrak{E}$ from (12.7) we obtain

$$\sum_{\lambda} q_\lambda \Delta\phi_\lambda + \rho = 0$$

and on multiplying with ϕ_λ and integrating over the space under consideration

$$q_\lambda - \int \rho\,\phi_\lambda\,dV = 0.$$

From the standpoint of quantum mechanics the left-hand side of this equation is an operator D_λ, and the meaning of the equation $D_\lambda = 0$ is that only those vectors $\mathfrak{z}$ which satisfy the equation $D_\lambda \mathfrak{z} = 0$ are to be allowed. D_λ also consists of an electrical part q_λ and a material part

$$\int \rho \phi_\lambda dV = \int (\bar{\psi}_1 \psi_1 + \bar{\psi}_2 \psi_2 + \bar{\psi}_3 \psi_3 + \bar{\psi}_4 \psi_4) \phi_\lambda dV.$$

The operator D_λ which is to be applied to $\mathfrak{z}$ is accordingly

$$D_\lambda = q_\lambda - \sum_{r=1}^{n} \phi_\lambda(P_r).$$

The equations $D_\lambda \mathfrak{z} = 0$ then assert that all components $z(P_r;\ N_\nu;\ q_\lambda)$ of $\mathfrak{z}$ vanish except those for which $q_\lambda = \sum_{r=1}^{n} \phi_\lambda(P_r)$; we may therefore write the non-vanishing components as

$$\psi(P_r;\ N_\nu) = z[P_r;\ N_\nu;\ \sum_{r=1}^{n} \phi_\lambda(P_r)].$$

But then

$$\text{grad}^{(r)} \psi = \text{grad}^{(r)} z + \sum_\lambda \text{grad}\, \phi_\lambda(P_r) \cdot \frac{\partial z}{\partial q_\lambda}$$

is exactly the combination which appears in (12.9). $\sum_\lambda q_\lambda^2$ is now given by

$$\sum_{r,s=1}^{n} \sum_\lambda \phi_\lambda(P_r)\, \phi_\lambda(P_s) = \sum_{r,s=1}^{n} G(P_r, P_s)$$

where

$$G(P, P') = \sum_\lambda \phi_\lambda(P)\, \phi_\lambda(P')$$

is the ordinary Green's function for the cavity. We consequently obtain the quantum equation

$$\frac{1}{i} \frac{d\psi}{dx_0} - \mathfrak{J}_0 \psi = 0$$

for ψ, in which the operator

$$-\mathfrak{J}_0 = \sum_{r=1}^{n} \left\{ \frac{1}{i} (\mathfrak{S}^{(r)}, \text{grad}^{(r)}) + m_0 T^{(r)} \right\} + \frac{\alpha}{2} \sum_{r,s=1}^{n} G(P_r, P_s)$$
$$+ \sum_\nu \nu N_\nu + \sqrt{\alpha} \sum_\nu \sum_{r=1}^{n} \{ (\mathfrak{S}^{(r)}, \mathfrak{f}_\nu(P_r)) \cdot Q_\nu \}. \qquad (12.11)$$

In *Dirac's* theory

$$\frac{1}{i}(S, \text{grad}) + m_0 T$$

is the energy operator for a single free particle. $\alpha G(P, P')$ is the classical potential due to the electro-static repulsion between two electrons situated at P and P'. The next term represents the sum of the energies ν of the photons in the various frequency states ν, and finally the last term represents the interaction between photons and electrons by emission and absorption. The meaning of each of the terms from which the energy operator (12.11) is constructed is thus apparent. The quantum theory had previously dealt with fields, such as that which binds the electron in hydrogen to the nucleus, in a manner entirely different from that with which it treated the field of the emitted radiation; the first was calculated classically and purely electro-statically as an action at a distance described by the Coulomb potential, whereas the second was broken up into discrete photons with the aid of *Bohr's* frequency condition. We have now obtained a theoretical justification for this procedure which led to good agreement with experiment.

Our expression shares with classical electro-dynamics the disadvantage that it contains the term $G(P_r, P_r)$ representing the infinitely large reaction of the rth electron with itself, for as we allow P' to approach P, $G(P, P')$ becomes infinite like the reciprocal of the distance $\overline{PP'}$. We should therefore replace $G(P, P)$ by the finite $\Gamma(P, P)$ where

$$\Gamma(P, P') = G(P, P') - \frac{1}{4\pi \cdot \overline{PP'}},$$

for this amounts to dropping an infinitely large additive constant from $\mathfrak{J}_0$. $\Gamma(P, P)$ represents the effect on an electron at P of the field obtained by reflecting the field of P in the walls of the cavity. (12.11) shows explicitly how the various terms of $\mathfrak{J}_0$ depend on the value of the fine-structure constant α; on developing the solution in powers of α we are faced again and again with infinitely large terms of the same kind as $G(P_r, P_r)$. The operator $\mathfrak{J}_0$ contains singularities which, at the present stage, frustrate all attempts to carry through the theory. We may indeed conclude with *P. Jordan* that the problem of the *existence* of the electron is solved, but that that of its *constitution* has as yet eluded us. Our equations further suffer from the fundamental disadvantage of the Dirac theory that the individual spin variables ρ_r assume 4 instead of 2 different values.[23]

There is, of course, nothing to prevent us from quantizing the matter waves in a manner analogous to that applied to electro-magnetic waves. We should then develop our quantities describing the material field in a series of characteristic functions $\psi = \psi^{(\mu)}$ (with four components) of the Dirac equation

$$\left\{\frac{1}{i}(\mathfrak{S}, \text{grad}) + m_0 T\right\}\psi + \mu\psi = 0 \qquad (12.12)$$

which constitute, on imposing appropriate boundary conditions, a complete orthogonal system. The general component z of the vector $\mathfrak{z}$, on which the energy $-c\mathcal{J}_0$ operates, will then depend on the quantum number n_μ, which corresponds to the characteristic values μ and which may assume only the values 0 and 1, and in addition on the numbers N_ν of photons of the various frequencies ν and on the continuous variables q_λ. But then the operators D_λ, which commute among themselves and with $\mathcal{J}_0$, are not in diagonal form, and the elimination of q_λ cannot be accomplished as in the above method.

Instead of introducing a cavity as in the above we may employ a rectangular parallelepipedon with the "boundary condition" that all functions are to be periodic functions whose periods are the lengths of the sides of the parallelepipedon. We can then introduce *running* instead of *standing* waves as characteristic functions for the electro-magnetic field; this gives rise to a better agreement with the physical picture in which a photon corresponds to a homogeneous plane wave. The *energy* and the *momenta* are then also in diagonal form if we neglect the interaction between matter and light. Equation (12.10) then causes some difficulty, as its right-hand side 0 must be replaced by the constant mean value of the charge throughout the entire space in order that a periodic solution be possible. On taking account of protons in the theory this will automatically correct itself, as the total charge will then be 0.

The dynamical law allows only those quantum jumps of the particles in which *one* n_μ falls from 1 to 0 and another $n_{\mu'}$ jumps at the same time from 0 to 1. Consequently the total number of particles $\sum_\mu n_\mu$, and therefore the charge, remains fixed; hence that portion of the dynamical laws in which the total number is a given finite n is separated off from the remaining portion and intercombinations between the two do not arise. *Dirac* has proposed to interpret the presence or the absence of a *proton* in the state of positive energy μ as the absence or the presence, respectively, of an electron in the corresponding negative energy state $-\mu$; our laws will then include protons as well as electrons.[21] Remembering that the numbers $n_\mu = 0$, 1 were at first introduced merely as an arbitrary index indicating the rows of a matrix, there is nothing to prevent us from replacing the numbers $n_{-\mu}$ for negative $-\mu$ by $n_\mu^- = 1 - n_{-\mu}$, keeping $n_\mu^+ = n_\mu$ for positive μ. The theorem of the conservation of charge is then

$$\Sigma n_\mu^+ - \Sigma n_\mu^- = \text{const.} \quad (\mu > 0).$$

But we thereby alter the content, as well as the notation, of the theory; we are now interested in that part of the dynamical equations in which only a finite number of n_μ with positive μ are different from 0 and *only a finite number of n_μ with negative μ are different from* 1! The quantum jump of an electron between positive and negative energy levels, which was so undesirable in the Dirac theory as formulated in the previous section, now appears as a process in which an electron and a proton are simultaneously destroyed and as the inverse process. The assumption of such an occurrence, for which our terrestrial experiments offer no justification, has long been entertained in atrophysics, as it seems otherwise extremely difficult to explain the source of the energy emitted by stars.

However attractive this idea may seem at first, it is certainly impossible to hold without introducing other profound modifications to square our theory with the observed facts. Indeed, according to it the mass of a proton should be the same as the mass of an electron; furthermore, no matter how the action is chosen (so long as it is invariant under interchange of right and left), this hypothesis leads to the essential equivalence of positive and negative electricity under all circumstances—even on taking the interaction between matter and radiation rigorously into account.

Having now quantized the field equations, we must return to the question of how the constituents M, M', F of the action behave under the substitutions (6.12), (6.13), (6.14). The first two substitutions, which we may call (*a*) and (*b*), have exactly the same effect as before. But the third substitution (*c*), which sends the components of ψ over into the components of $\bar{\psi}$ or their negative, now affects M and M' differently, for ψ and $\bar{\psi}$ are no longer commutative with respect to multiplication—they are, in fact, almost anti-commutative. From this it is found that M, M', F behave under (*c*) in exactly the same way as they do under (*b*), i.e. they are multiplied by the signs $-$, $-$, $+$ respectively. *Hence past and future play essentially different rôles in the quantized field equations;* we find no substitution which leaves these equations unchanged while reversing the direction of time. It seems to me that we have thereby reached an extraordinarily important goal of physics. We can now obtain the substitution

$$\left.\begin{array}{l} f_\alpha \to -f_\alpha \quad (\alpha = 0, 1, 2, 3) \\ \psi_1 \to -\bar{\psi}_4, \psi_2 \to \bar{\psi}_3, \psi_3 \to \bar{\psi}_2, \psi_4 \to -\bar{\psi}_1 \end{array}\right\}$$

on combining (*a*), (*b*) and (*c*); this substitution neither affects the co-ordinates nor disturbs the quantized wave equations. In view of Dirac's theory of the proton this means that positive and negative electricity have essentially the same properties in the sense that the laws governing them are invariant under a certain substitution which interchanges the quantum numbers of the electrons with those of the protons. The dissimilarity of the two kinds of electricity thus seems to hide a secret of Nature which lies yet deeper than the dissimilarity of past and future.

§ 13. The Energy and Momentum Laws of Quantum Physics. Relativistic Invariance

In quantizing the wave equations the spatial and temporal variables were treated so differently that the relativistic invariance of the resulting laws might seem to be open to serious doubt. But a thorough investigation due to *Heisenberg and Pauli* reassures us on this point.[25] We carry through these considerations on our action principle—but in such a way that the general validity of the argument may be readily seen. At the same time this offers an opportunity to discuss the meaning of the quantization more thoroughly than we have done hitherto.

1. The Energy and Momentum Laws of Quantum Physics.

We begin with the $4 + 3 + 3$ operators ψ_ρ, f_p, E_p which are functions in 3-dimensional space satisfying the commutation rules (12.4) and the supplementary rules there set forth. There exists one, and in the sense of equivalence only one, irreducible solution of these conditions. From it we obtain the energy density t_0^0 defined by (6.5), (6.6) and integrate it over all of space:

$$\mathfrak{J}_0 = \int t_0^0 \, dV. \tag{13.1}$$

We next construct the "commutator"

$$\delta\Phi = [\mathfrak{J}_0, \Phi] = \frac{1}{i}(\mathfrak{J}_0 \Phi - \Phi \mathfrak{J}_0)$$

of an arbitrary operator Φ with $\mathfrak{F}_0$. Consider the result of this for the particular operators $\Phi = \psi_\rho, f_p, E_p$; it should be possible to evaluate these commutators using (12.4) and the supplementary rules alone ; if one of the quantities involved appears as a derivative with respect to a spatial co-ordinate it should be transformed by integrating (13.1) by parts—or by deducing commutation rules for it from (12.4) in terms of appropriately defined derivates of the δ function. If $\frac{\partial}{\partial x_0}$ is that process involving only differentiations with respect to the spatial variables, but which coincide with the derivative with respect to time in virtue of the Maxwell-Dirac field equations, we find

$$\delta\psi_\rho = \left(\frac{\partial}{\partial x_0} + if_0\right)\psi_\rho ; \quad \delta f_p = \frac{\partial f_p}{\partial x_0} - \frac{\partial f_0}{\partial x_p} = -\,\alpha E_p ;$$

$$\delta E_p = \frac{\partial E_p}{\partial x_0}. \quad (13.2)$$

We now drop the normalization $f_0 = 0$. It follows from these equations that $\delta\Phi$ for any *gauge invariant* operator Φ coincides with its time derivative as defined in terms of its spatial derivatives by means of the field laws. We may therefore replace the Maxwell-Dirac field equations by the quantum mechanical dynamical law

$$\frac{1}{i}\frac{d\mathfrak{z}}{dx_0} = \mathfrak{F}_0\,\mathfrak{z} \quad (13.3)$$

$\mathfrak{z}$ represents the probability state of the physical system (pure state !) at the time x_0 ; it is a vector of that vector-space in which our operations take place. The fundamental concepts here involved are contained in the general programme of quantum mechanics as set forth in II, § 7. The " density of electricity at the point P " is, for example, represented by the operator $\rho = \bar{\psi}_1\psi_1 + + +$ which is independent of time. The changes in the probability distribution for this physical quantity in course of time are due to the changes in the state $\mathfrak{z}$ and not to changes in ρ itself ; the rule for the calculation of this probability distribution from ρ and $\mathfrak{z}$ is given in the general programme referred to above. The same remarks apply to any gauge invariant quantity Φ. However, it is more desirable to consider the " density of electricity " (without specifying either time or position) as a fixed physical quantity represented by a definite operator ρ, and to ascribe the variations in its probability distribution in time and space to changes in the probability state $\mathfrak{z}$ *considered as a function of the spatial co-ordinates*

x_1, x_2, x_3 *in addition to the time* x_0. We should then expect to find four equations

$$\frac{1}{i}\frac{\partial \mathfrak{z}}{\partial x_\alpha} = \mathcal{J}_\alpha \mathfrak{z} \quad (\alpha = 0, 1, 2, 3) \tag{13.4}$$

in place of the one (13.3) in which the operators

$$\mathcal{J}_\alpha = \int t^0_\alpha \, dV$$

are those representing energy and momentum. Only now that we have formulated the general scheme of quantum physics in a manner which is symmetric with respect to the spatial and temporal co-ordinates, as required by the theory of relativity, can we consider it as complete. In order to determine the mean value of a quantity such as the electric density ρ we must assign to the spatial co-ordinates x_1, x_2, x_3, on which the operator ρ depends, any definite values x^0_p (e.g. 0). The spatial components of equation (13.4) tell us that the replacement of (x^0_p) by a neighbouring point $(x^0_p + dx_p)$ amounts to the same thing as subjecting the normal co-ordinate system in system space, to which the vectors $\mathfrak{z}$ are referred, to the infinitesimal rotation

$$i(\mathcal{J}_1 \, dx_1 + \mathcal{J}_2 \, dx_2 + \mathcal{J}_3 \, dx_3).$$

We must not forget that the equation (13.3) is not equivalent to the complete set of field equations, for we have omitted the one

$$\sigma(P) \equiv \operatorname{div} \mathfrak{E} + \rho = 0$$

which does not involve differentiation with respect to time. We must therefore restrict ourselves to vectors $\mathfrak{z}$ which satisfy all the equations

$$\sigma(P)\mathfrak{z} = 0. \tag{13.5}$$

These equations define a linear sub-space $\mathfrak{R}_\sigma$ of the original system-space $\mathfrak{R}$. The operators $\sigma(P)$, $\sigma(P')$ associated with any two points P, P' of space are commutative:

$$\sigma(P)\,\sigma(P') - \sigma(P')\,\sigma(P) = 0.$$

It is of prime importance that $\sigma(P)$ commute with $\mathcal{J}_0$, i.e. that

$$\delta\sigma \equiv \frac{1}{i}(\mathcal{J}_0\sigma - \sigma\mathcal{J}_0) = 0\,;$$

that this is the case follows from the fact that the equation $\dfrac{\partial \sigma}{\partial x_0} = 0$ is a consequence of the remaining field equations in the classical field theory, and consequently—independently of our field equations—we may conclude that the gauge invariant operator σ satisfies the equation $\delta\sigma = 0$. This commutativity of $\sigma(P)$ and $\mathfrak{F}_0$ guarantees that the infinitesimal rotation $i\mathfrak{F}_0 dx_0$ of system-space during the time interval dx_0 does not carry the vector $\mathfrak{z}$ lying in the sub-space $\mathfrak{R}_\sigma$ out of $\mathfrak{R}_\sigma$.

Continuing our programme, we now set

$$\mathfrak{F}_1 = \int t_1^0 \, dV$$

and investigate the " commutator "

$$\delta\Phi = [\mathfrak{F}_1, \Phi]$$

of an operator Φ with $\mathfrak{F}_1$; we shall denote this commutator by δ_1 whenever confusion might arise between it and the commutator $\delta = \delta_0$ with $\mathfrak{F}_0$. We find the equations *

$$\left.\begin{aligned} \delta\psi_\rho &= \left(\frac{\partial}{\partial x_1} + if_1\right)\psi_\rho\,;\ \delta f_1 = 0,\ \delta f_2 = H_3 \equiv \frac{\partial f_2}{\partial x_1} - \frac{\partial f_1}{\partial x_2}, \cdots; \\ \delta E_1 &= -\left(\frac{\partial E_2}{\partial x_2} + \frac{\partial E_3}{\partial x_3} + \rho\right) = \frac{\partial E_1}{\partial x_1} - \sigma,\ \delta E_2 = \frac{\partial E_2}{\partial x_1}, \cdots. \end{aligned}\right\} \tag{13.6}$$

From this it follows that for any *gauge invariant* quantity Φ we have $\delta\Phi = \dfrac{\partial\Phi}{\partial x_1}$ *on taking the equation* $\sigma = 0$ *into account.* Hence the way in which gauge invariant quantities depend on the spatial co-ordinates can in fact be described as we predicted: the operators representing them are constant, but the vector $\mathfrak{z}$ representing the probability state varies in space in accordance with the equations (13.4) for $\alpha = 1, 2, 3$.

That the four equations (13.4) are consistent also follows from these considerations. In the first place we have

$$\delta_1\sigma = 0 \quad \text{or} \quad \sigma(P)\mathfrak{F}_1 - \mathfrak{F}_1\sigma(P) = 0$$

in the entire space $\mathfrak{R}$; this follows from (13.6). In the classical field theory the differential conservation theorem

$$\frac{\partial t_1^0}{\partial x_0} + \left(\frac{\partial t_1^1}{\partial x_1} + \frac{\partial t_1^2}{\partial x_2} + \frac{\partial t_1^3}{\partial x_3}\right) = 0.$$

is a consequence of the field equations. Since t_1^0 is a gauge invariant, it follows that after the quantization the operators satisfy the relation

$$\delta_0 t_1^0 + \left(\frac{\partial t_1^1}{\partial x_1} + \frac{\partial t_1^2}{\partial x_2} + \frac{\partial t_1^3}{\partial x_3}\right) = 0$$

* In contrast with (6.2) we now employ the letter $\mathfrak{H}$, without the factor $1/\alpha$, as an abbreviation for curl $\mathfrak{f}$.

in the space $\mathfrak{R}_\sigma$ defined by (13.5). Integrating over the space $x_0 =$ const. we obtain

$$\delta_0 \int t_1^0 dV = 0 \text{ or } \mathcal{J}_0 \mathcal{J}_1 - \mathcal{J}_1 \mathcal{J}_0 = 0. \tag{13.7}$$

[The equation which takes the place of (13.7) for the entire space $\mathfrak{R}$ is

$$\mathcal{J}_0 \mathcal{J}_1 - \mathcal{J}_1 \mathcal{J}_0 = \int \sigma E_1 dV.]$$

Furthermore,

$$\frac{\partial t_3^0}{\partial x_2} = \delta_2 t_3^0$$

in $\mathfrak{R}_\sigma$, and on integrating this over space we find

$$\delta_2 \int t_3^0 dV = 0 \text{ or } \mathcal{J}_2 \mathcal{J}_3 - \mathcal{J}_3 \mathcal{J}_2 = 0.$$

We thus see that the operators $\mathcal{J}_\alpha$ are commutative in $\mathfrak{R}_\sigma$, and consequently equations (13.4) possess one and only one solution $\mathfrak{z}$ when the initial value of $\mathfrak{z}$ (i.e. at the origin of the space-time co-ordinate system) is a given vector in $\mathfrak{R}_\sigma$.

II. Relativistic Invariance.

On transforming from the normal co-ordinate system x_α in space-time to another x'_α by means of a Lorentz transformation

$$\Lambda: \; x'_\alpha = \sum_{\beta=0}^{3} o_{\alpha\beta} x_\beta$$

the solution of the equations

$$\frac{1}{i}\frac{\partial \mathfrak{z}}{\partial x'_\alpha} = \mathcal{J}_\alpha \mathfrak{z} \tag{13.4'}$$

is, as we shall show, obtained from the solution of (13.4) by means of a unitary transformation U induced in system-space by Λ. That is, there exists a unitary transformation U such that

$$\frac{1}{i} d(U\mathfrak{z}) = (\sum_\alpha \mathcal{J}_\alpha dx'_\alpha)(U\mathfrak{z})$$

is satisfied in virtue of (13.4) :

$$U \cdot \sum_\alpha \mathcal{J}_\alpha dx_\alpha = \sum_\beta \mathcal{J}_\beta dx'_\beta \cdot U$$

or

$$U\mathcal{J}_\alpha = \sum_\beta o_{\beta\alpha} \mathcal{J}_\beta \cdot U. \tag{13.8}$$

We could also say that (13.4′) have the same solution $\mathfrak{z}$ as (13.4) but that the normal co-ordinate system employed in system-space has undergone the unitary rotation U, for the vector $U\mathfrak{z}$ has the same components with respect to the new co-ordinate system as $\mathfrak{z}$ had with respect to the old. We are only able to give the transformation U explicitly for infinitesimal Λ:

$$\|o_{\alpha\beta}\| = \mathbf{1} + \|\delta o_{\alpha\beta}\|; \quad U = 1 + \frac{1}{i}\delta M.$$

The equations (13.8) which are to be verified are then

$$\sum_{\beta} \mathfrak{J}_\beta \, \delta o_{\beta\alpha} = [\delta M, \mathfrak{J}_\alpha].$$

In particular, the operators in system-space which correspond to infinitesimal rotations in physical space are, as we have long known, those representing moment of momentum; that δM corresponding to the infinitesimal rotation D_x:

$$\delta x_0 = 0, \quad \delta x_1 = 0, \quad \delta x_2 = -x_3, \quad \delta x_3 = x_2 \tag{13.9}$$

about the x_1-axis is the x_1-component of moment of momentum:

$$(M_1 =)M_{23} = \int (x_2 t_3^0 - x_3 t_2^0) dV. \tag{13.10}$$

The infinitesimal Lorentz transformations which actually represent a re-partitioning of the world into a new space and a new time are dealt with in exactly the same manner; it will suffice to consider as typical of such transformations

$$\delta x_0 = x_1, \quad \delta x_1 = x_0, \quad \delta x_2 = 0, \quad \delta x_3 = 0.$$

The δM associated with this transformation is

$$M_{10} = \int x_1 t_0^0 \, dV + \int x_0 t_1^0 \, dV;$$

the second term, which vanishes for $x_0 = 0$, can be omitted, for we have already shown that $\mathfrak{J}_1$ commutes with all $\mathfrak{J}_\alpha$. This term does not fit into the present scheme, in which all the operators are functions of x_1, x_2, x_3 alone. Our problem is thus reduced to showing that in $\mathfrak{R}_\sigma$

$$\left.\begin{aligned} [M_{23}, \mathfrak{J}_\alpha] &= \quad 0, \qquad 0, \quad \mathfrak{J}_3, \; -\mathfrak{J}_2 \\ [M_{10}, \mathfrak{J}_\alpha] &= -\mathfrak{J}_1, \; -\mathfrak{J}_0, \quad 0, \qquad 0 \end{aligned}\right\} \text{for } \alpha = 0, 1, 2, 3. \qquad \begin{aligned}(13.11)\\(13.12)\end{aligned}$$

Furthermore, the invariance of equations (13.5) which define the sub-space $\mathfrak{R}_\sigma$, will be proved by showing that the equations

$$[M_{23}, \sigma] = 0, \quad [M_{10}, \sigma] = 0 \tag{13.13}$$

hold in the entire space $\mathfrak{R}$.

In order to prove (13.11) we make use of the identities

$$\int \frac{\partial(x_2\, t_3^0 - x_3\, t_2^0)}{\partial x_\alpha} dV = 0 \quad [\alpha = 1,\, 2,\, 3].$$

Introducing the Kronecker δ_{ik}, the integrand may be written

$$(\delta_{\alpha 2}\, t_3^0 - \delta_{\alpha 3}\, t_2^0) + \left(x_2 \frac{\partial t_3^0}{\partial x_\alpha} - x_3 \frac{\partial t_2^0}{\partial x_\alpha}\right).$$

In consequence of $\sigma = 0$ and since $t = t_2^0,\, t_3^0$ are gauge invariants the operations

$$\frac{\partial t}{\partial x_\alpha} \text{ may be replaced by } \delta_\alpha t = [\mathcal{J}_\alpha,\, t],$$

whence

$$(\delta_{\alpha 2}\, \mathcal{J}_3 - \delta_{\alpha 3}\, \mathcal{J}_2) + \delta_\alpha \int (x_2\, t_3^0 - x_3\, t_2^0) dV = 0$$

or

$$\delta_\alpha M_{23} = [\mathcal{J}_\alpha,\, M_{23}] = \delta_{\alpha 3}\, \mathcal{J}_2 - \delta_{\alpha 2}\, \mathcal{J}_3 \quad [\alpha = 1,\, 2,\, 3].$$

In the classical field theory the conservation law

$$\sum_{\alpha=0}^{3} \frac{\partial(x_2\, t_3^\alpha - x_3\, t_2^\alpha)}{\partial x_\alpha} = 0$$

is a consequence of the field equations, whence on quantizing

$$\delta_0(x_2\, t_3^0 - x_3\, t_2^0) + \sum_{\alpha=1}^{3} \frac{\partial(x_2\, t_3^\alpha - x_3\, t_2^\alpha)}{\partial x_\alpha} = 0$$

holds identically in $\mathfrak{R}_\sigma$. Integrating over the whole of physical space we obtain

$$\delta_0 M_{23} = [\mathcal{J}_0,\, M_{23}] = 0\,;$$

equations (13.11), i.e.

$$[M_{23},\, \mathcal{J}_\alpha] = \delta_{\alpha 2}\, \mathcal{J}_3 - \delta_{\alpha 3}\, \mathcal{J}_2 \quad [\alpha = 0,\, 1,\, 2,\, 3],$$

are thus completely verified.

The relations (13.12) are obtained in an analogous manner from

$$\int \frac{\partial(x_1\, t_0^0)}{\partial x_\alpha} dV = 0 \quad [\text{for } \alpha = 1,\, 2,\, 3]$$

and from the equation

$$\int \{\delta_0(x_1\, t_0^0) + t_1^0\} dV = 0$$

which parallels the conservation theorem

$$\int \frac{\partial(x_1 t_0^0 + x_0 t_1^0)}{\partial x_0} dV = 0$$

of the classical field theory.

We should expect the operator functions expressed by the ψ_ρ, f_p, E_p, depending on the spatial co-ordinates, to be invariant if we associate with an infinitesimal rotation of the spatial co-ordinate system an appropriate linear transformation of the components ψ_ρ among themselves and of the vector components f_p, E_p, and at the same time subject the normal co-ordinate system in system space $\mathfrak{R}_\sigma$ to the corresponding unitary transformation. In formulæ: We expect the process

$$\delta\Phi = [M_{23}, \Phi]$$

to yield the equations

$$\delta\psi = \delta'\psi - \frac{1}{2i} S_1' \psi,$$

$$\delta f_p = \delta' f_p + (\delta_{p2} f_3 - \delta_{p3} f_2),$$
$$\delta E_p = \delta' E_p + (\delta_{p2} E_3 - \delta_{p3} E_2),$$

where we have written

$$\delta'\Phi = x_2 \frac{\partial\Phi}{\partial x_3} - x_3 \frac{\partial\Phi}{\partial x_2}.$$

But we find by direct calculation that

$$\delta\psi = \delta'\psi + i(x_2 f_3 - x_3 f_2)\psi - \frac{1}{2i} S_1' \psi,$$

$$\delta f_1 = x_2 H_2 + x_3 H_3, \; \delta f_2 = -x_2 H_1, \; \delta f_3 = -x_3 H_1,$$
$$\delta E_p = \delta' E_p + \delta_{p2}(E_3 + x_3 \sigma) - \delta_{p3}(E_2 + x_2 \sigma).$$

We first observe that these equations yield

$$\delta\sigma = [M_{23}, \sigma] = 0$$

independently of the condition $\sigma = 0$. On introducing the condition $\sigma = 0$ we find from these equations that *gauge invariant* quantities Φ exhibit the expected behaviour. The second of the equations (13.13) can be obtained by an analogous computation.

D. Quantum Kinematics

§ 14. Quantum Kinematics as an Abelian Group of Rotations

If we consider the operators ip, iq as infinitesimal unitary rotations of the ray field in system space, then Heisenberg's commutation rules [II, (11.4)] assert that these rotations are commutative; consequently they generate a $2f$-parameter Abelian group, where f is the number of degrees of freedom. Let us therefore investigate the properties of Abelian groups of unitary rotations in the ray field of n-dimensional space! On introducing a gauge as in III, § 16, to each such " rotation " there corresponds a transformation of vector space with matrix A and between any two matrices A, B there exists an equation of the form

$$AB = \varepsilon BA. \tag{14.1}$$

This equation is possible only if ε is an nth root of unity, for on evaluating the determinant of both sides we obtain $\varepsilon^n = 1$. From (14.1) we obtain by mathematical induction

$$\left.\begin{aligned} A^k B &= \varepsilon^k B A^k, \\ AB^l &= \varepsilon^l B^l A, \end{aligned}\right\} \tag{14.2}$$

for $k, l = 1, 2, 3, \cdots$. On combining these two equations by applying the second to A^k and B instead of A and B we find the general rule

$$A^k B^l = \varepsilon^{kl} B^l A^k. \tag{14.3}$$

Taking $k = n$ in (14.2) we are led to the equation $A^n B = BA^n$; if the Abelian rotation group is irreducible Schur's fundamental lemma allows us to conclude that since A^n commutes with all elements B of the group it must be a multiple of the unit matrix: $A^n \simeq \mathbf{1}$. *The order of any element of an irreducible Abelian rotation group in n dimensions is consequently a factor of n.*

An f-parameter continuous rotation group is generated by an f-dimensional linear family $\mathfrak{g}$ of infinitesimal unitary correspondences

$$\sigma_1 C_1 + \sigma_2 C_2 + \cdots + \sigma_f C_f \tag{14.4}$$

in terms of a basis formed by any f independent elements $C_1, C_2, \cdots, C_f$ of the family. The numerical parameters $\sigma_1, \sigma_2, \cdots, \sigma_f$ may assume all real values. Setting $\sigma_i = \alpha_i \, d\tau$

and reiterating the infinitesimal transformation (14.4), we find that at " time " τ the resulting transformation is

$$U(\sigma_1, \sigma_2, \cdots, \sigma_f) = e^{\sigma_1 C_1 + \sigma_2 C_2 + \cdots + \sigma_f C_f}, \tag{14.5}$$

where we have replaced $\alpha_i\tau$ by σ_i. U runs through the entire group, which is now expressed in terms of the parameters σ. If the group of unitary transformations of the *vector* space is Abelian the C_ν must satisfy the conditions

$$C_\mu C_\nu - C_\nu C_\mu = 0. \tag{14.6}$$

From this it then follows that all the elements (14.5) of the group are mutually commutative, for if $AB - BA = 0$ we have, as in the domain of ordinary numbers,

$$e^A \cdot e^B = e^{A+B}.$$

The parameters σ in (14.5) are added on composition :

$$U(\sigma_1, \cdots, \sigma_f)U(\sigma_1', \cdots, \sigma_f') = U(\sigma_1 + \sigma_1', \cdots, \sigma_f + \sigma_f').$$

If, however, only the rotations of the *ray space* are commutative, we find in place of (14.6) conditions of the form

$$C_\mu C_\nu - C_\nu C_\mu = ic_{\mu\nu}\mathbf{1},$$

where the $c_{\mu\nu}$ constitute an anti-symmetric system of real numbers. The commutator of the infinitesimal transformations with matrices

$$A = \sigma_1 C_1 + \cdots + \sigma_f C_f, \quad B = \tau_1 C_1 + \cdots + \tau_f C_f$$

is

$$AB - BA = i\sum_{\mu,\nu} c_{\mu\nu}\sigma_\mu\tau_\nu \cdot \mathbf{1}.$$

We shall refer to the anti-symmetric form

$$\sum_{\mu,\nu} c_{\mu\nu}\sigma_\mu\tau_\nu = h(\sigma, \tau)$$

as the *commutator form ;* it is invariant under change of basis. On writing $\mathbf{1} + \frac{A}{m}$, $\mathbf{1} + \frac{B}{m}$ in (14.3) in place of A, B and allowing $k = l = m \to \infty$, we find that the commutator of any two elements $U(\sigma_1, \sigma_2, \cdots, \sigma_f) = U(\sigma)$ and $U(\tau)$ of the group is

$$U(\sigma)U(\tau)U^{-1}(\sigma)U^{-1}(\tau) = e[h(\sigma, \tau)] \cdot \mathbf{1}. \tag{14.7}$$

If the rotation group is *irreducible* a fixed $U(\sigma)$ can only commute with *all* $U(\tau)$ if it is a multiple of the unit matrix, i.e. if all its parameters σ vanish. From this we conclude that the commutator form is non-degenerate, i.e. that it cannot vanish identically in τ_i for a fixed set of values σ_i, unless all $\sigma_i = 0$—this amounts to the same as the condition $|c_{ik}| \neq 0$. Such a form exists only if the number f of variables is *even*, in which case it can, by appropriate choice of the basis (i.e. by transforming the variables σ_i and τ_i cogrediently under an appropriate transformation), be reduced to the canonical form in which the matrix $\|c_{ik}\|$ is decomposed into 2-rowed sub-matrices

$$\begin{Vmatrix} 0 & 1 \\ -1 & 0 \end{Vmatrix}$$

arranged along the principal diagonal.* It is then desirable to write $2f$ in place of f and to denote the "*canonical basis*" so obtained by

$$iP_\nu,\ iQ_\nu \quad (\nu = 1, 2, \cdots, f)$$

and the corresponding parameters by σ_ν, τ_ν. The factor i has been introduced in order to express the results in terms of Hermitian operators P_ν, Q_ν. The basic elements then satisfy the commutation rules

$$i(P_\nu Q_\nu - Q_\nu P_\nu) = \mathbf{1}, \quad i(P_\mu Q_\nu - Q_\nu P_\mu) = 0$$

for $\mu \neq \nu$ and

$$P_\mu P_\nu - P_\nu P_\mu = 0, \quad Q_\mu Q_\nu - Q_\nu Q_\mu = 0$$

for all μ, ν. The elements

$$U(\sigma) = e(\sigma_1 P_1 + \sigma_2 P_2 + \cdots + \sigma_f P_f) \quad [e(x) = e^{ix}]$$

then constitute an f-parameter Abelian group of unitary (vector) correspondences, as do also the

$$V(\tau) = e(\tau_1 Q_1 + \tau_2 Q_2 + \cdots + \tau_f Q_f).$$

But the commutator of elements $U(\sigma)$, $V(\tau)$ belonging to these two sets, respectively, is

$$U(\sigma)V(\tau)U^{-1}(\sigma)V^{-1}(\tau) = e(\sigma_1\tau_1 + \cdots + \sigma_f\tau_f) \cdot \mathbf{1}.$$

We have now carried our development to a point where we can profitably return to the considerations of II, § 11. In the case of a system with *one* degree of freedom in classical mechanics any physical quantity associated with the system is expressed mathematically as a function $f(p, q)$ of the canonical variables p, q. In making the transition to quantum mechanics we had previously restricted ourselves to polynomials in p, q. But the Fourier representation

$$f(p, q) = \iint_{-\infty}^{+\infty} e(\sigma p + \tau q)\, \xi(\sigma, \tau)\, d\sigma\, d\tau \tag{14.8}$$

* See Appendix 3.

of a function f is applicable to a much larger class of functions; this integral need not be interpreted literally, the essential point being that it represents a linear combination of the simple functions $e(\sigma p + \tau q)$. On considering ip, iq as infinitesimal unitary correspondences in ray space which are commutative in accordance with the relation

$$i(pq - qp) = \mathbf{1}, \tag{14.9}$$

$e(\sigma p + \tau q)$ runs through the group generated by them. If we now consider $\xi(\sigma, \tau)$ as the components of an element in the resulting group algebra, then (14.8) is its group matrix in the representation obtained by associating with (σ, τ) the unitary transformation $e(\sigma p + \tau q)$. This group matrix is Hermitian if the element is real, i.e. if

$$\bar{\xi}(\sigma, \tau) = \xi(-\sigma, -\tau).$$

A quantity f is consequently carried over from classical to quantum mechanics in accordance with the rule: *replace p and q in the Fourier development* (14.8) *of f by the Hermitian operators representing them in quantum mechanics.* In particular, the derivatives of f are represented by

$$f_p = i\iint_{-\infty}^{+\infty} e(\sigma p + \tau q) \cdot \sigma\,\xi(\sigma, \tau)\, d\sigma\, d\tau,$$

$$f_q = i\iint_{-\infty}^{+\infty} e(\sigma p + \tau q) \cdot \tau\,\xi(\sigma, \tau)\, d\sigma\, d\tau.$$

On letting $U(\tau)$ in (14.7) again in infinitesimal we find, with the aid of the commutation rules (14.9), that

$$p \cdot e(\sigma p + \tau q) - e(\sigma p + \tau q) \cdot p = \tau \cdot e(\sigma p + \tau q),$$
$$q \cdot e(\sigma p + \tau q) - e(\sigma p + \tau q) \cdot q = -\sigma \cdot e(\sigma p + \tau q).$$

We therefore have in general

$$if_p = q \cdot f - f \cdot q, \quad -if_q = p \cdot f - f \cdot p$$

as required in order that the Hamiltonian equations

$$\frac{dq}{dt} = H_p, \quad \frac{dp}{dt} = -H_q$$

be equivalent to the quantum-theoretical equations of motion for the vectors of system space.

We have thus found a very natural interpretation of quantum kinematics as described by the commutation rules. *The kinematical structure of a physical system is expressed by an irreducible Abelian group of unitary ray rotations in system space. The real elements of the algebra of this group are the physical quantities of the system; the representation of the abstract group by rotations of system space associates with each such quantity a definite Hermitian form which "represents" it.* If the group is *continuous* this procedure automatically leads to *Heisenberg's* formulation; in particular, we have seen how the *pairs of canonical variables* then result from the requirement of irreducibility, whence the number of parameters in such an irreducible Abelian group must be *even*.[26]

If one of the canonical co-ordinates, say q, is a cyclical co-ordinate with period 2π, then all quantities of the physical system are represented by periodic functions with period 2π. Consequently the only values assumed by the parameter τ associated with q in (14.8) are multiples of 2π and the integral is to be replaced by a sum. In such a case we are no longer dealing with a continuous group, but with a mixed (continuous-discrete) group.

Our general principle allows for the possibility that the Abelian rotation group is entirely discontinuous, or that it may even be a finite group. Thus we have discussed in III, § 16, a group of order 4 and an irreducible ray representation $\mathfrak{B}$ of it in 2 dimensions. That such groups actually occur in Nature is shown by the fact that the group we have just mentioned characterizes the kinematics of the electron spin discussed in § 4. It can be readily shown that $\mathfrak{B}$ is the only irreducible representation of this group, and that it is in fact the only irreducible 2-dimensional group of unitary rotations in ray space. These results emphasize the remarkable nature of this simplest case. The quantization of the problem of several electrons discussed in § 11 also falls within our general scheme. In dealing with it we are interested in that Abelian group whose basic elements p_α $(\alpha = 1, 2, \cdots, 2f)$ are all of order 2; such a group consists of the totality of the 4^f different elements

$$p_1^{n_1} p_2^{n_2} \cdots p_{2f}^{n_{2f}} \quad (n_\alpha = 1 \text{ or } 0).$$

The gauge can be so chosen that the corresponding unitary matrices $\boldsymbol{p}_\alpha$ in the irreducible ray representation in 2^f dimensions satisfy the equations

$$\boldsymbol{p}_\alpha^2 = 1, \quad \boldsymbol{p}_\beta \boldsymbol{p}_\alpha = -\boldsymbol{p}_\alpha \boldsymbol{p}_\beta \quad (\alpha \neq \beta). \tag{14.10}$$

The kinematics of the spinning electron is described by the simplest case $f = 1$ of this representation.

Because of these results I feel certain that the general scheme of quantum kinematics formulated above is correct. But the field of discrete groups offers many possibilities which we have not as yet been able to realize in Nature; perhaps these holes will be filled by applications to nuclear physics. However, it seems more probable that the scheme of quantum kinematics will share the fate of the general scheme of quantum mechanics: to be submerged in the concrete physical laws of the only existing physical structure, the actual world.

§ 15. Derivation of the Wave Equation from the Commutation Rules

We now show by actual construction that there exists but *one* irreducible ray representation (excluding the identity) of a 2-parameter continuous Abelian group: namely, that one which leads to the wave equation.

We obtain our 2-parameter continuous group as the limiting case of a finite group with 2 basic elements; our proof is rigorous only insofar as the validity of this limiting process is admitted. Let A, B be two commutative rotations of an n-dimensional unitary space. On introducing the gauge we have an equation between their matrices:

$$AB = \varepsilon BA, \tag{14.1}$$

in which, as we know already, ε is an nth root of unity. The system consisting of the two matrices A, B shall be irreducible. Let their commutator, the number ε, be a *primitive* m^{th} root of unity, i.e. ε^m is the lowest power of ε which is equal to 1; m is then a factor of n. The orders of the rotations A, B are also factors of n: $A^n \simeq \mathbf{1}$, $B^n \simeq \mathbf{1}$, so the gauge may be chosen in such a way that $A^n = \mathbf{1}$, $B^n = \mathbf{1}$. Let B be reduced to diagonal form by an appropriate choice of our normal co-ordinate system; the elements b_i in the main diagonal are then all n^{th} roots of unity. Equation (14.1) then yields the following conditions on the elements of $A = \|a_{ik}\|$:

$$\frac{b_k}{b_i} a_{ik} = \varepsilon a_{ik}. \tag{15.1}$$

We divide the indices i and the corresponding variables x_i into classes in accordance with the rule that i and k belong to the same class if the quotient b_i/b_k is an mth root of unity, i.e. a power of ε. That this process really results in such a division into classes is shown by the fact that if b_i/b_k and b_k/b_l are powers of ε, then b_i/b_l is also. By (15.1) $a_{ik} = 0$ if i and k belong to different classes; hence the matrix A is reduced in accordance with the division of the indices into classes. But in view of the assumption that the system A, B was irreducible there can therefore exist but *one* such class.

Having established this result, we now proceed to a finer division into classes: i and k shall now be considered as belonging to the same class if $b_i = b_k$. We arbitrarily choose as the first of these classes that one for which $b_i = b$ and let the second consist of those for which $b_i = \varepsilon b$, the third with $b_i = \varepsilon^2 b, \cdots$, the mth with $b_i = \varepsilon^{m-1} b$; this exhausts the set, for the $(m+1)^{\text{st}}$

class $b_i = \varepsilon^m b$ coincides with the first. Let the variables be arranged and numbered in this order. It then follows from equation (15.1) that all sub-matrices (i, k) of the matrix A are empty, i.e. $a_{ik} = 0$, unless their row index i and their column index k belong to successive classes. The matrix A then has the form indicated in Fig. 3, in which all elements in the non-shaded portions are zero (and we have taken $m = 4$). The shaded portions are occupied by the sub-matrices $A^{(1)}$, $A^{(2)}$, $\cdots$, $A^{(m)}$. Since A is unitary the sum of the squares of the absolute values of the elements of a row or column is 1; the

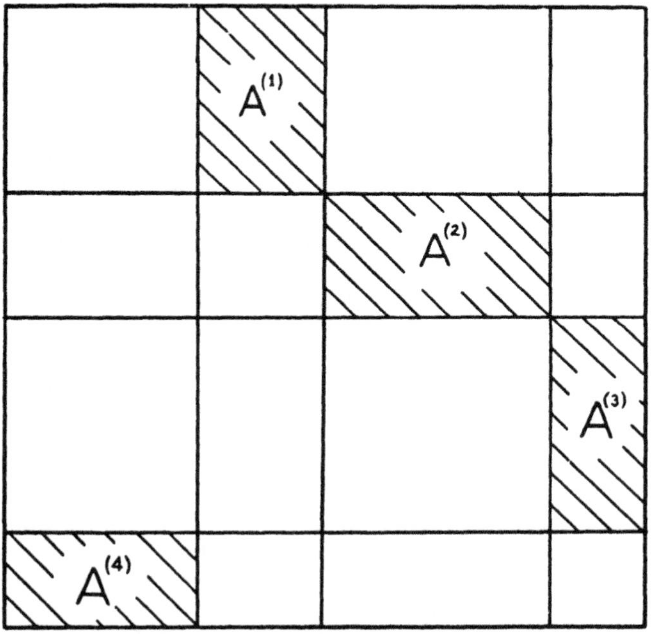

FIG. 3.

same must therefore also hold for the rows and columns of each of the sub-matrices. The sum of the absolute values of the squares of *all* elements in $A^{(1)}$ must then be equal, on the one hand, to the number of rows and, on the other, to the number of columns; the rectangle $A^{(1)}$ is consequently a square, and the number of indices in the second class is equal to the number in the first class, say d. By the same argument we see that *the number of individuals in each of the m classes is d, and hence $n = md$.* The figure is to be corrected accordingly; each of the shaded matrices is now unitary. On subjecting the variables of the first class to the unitary transformation with matrix $A^{(1)*}$ the sub-matrix $A^{(1)}$ is reduced to the d-dimensional unit matrix. This normal form is undisturbed by a unitary transformation affecting the variables of the first set and the variables of the second set alike; we can therefore reduce the second sub-matrix to a multiple of the d-dimensional unit matrix, and so on through the $(m-1)^{\text{st}}$. The normal form so obtained is unchanged on subjecting the variables of each class to the same d-dimensional unitary transformation; we may therefore choose as this last transformation one which reduces $A^{(m)}$ to diagonal form. But the matrix A is then decomposed into d-sub-matrices, as can be seen by renumbering the variables, taking first the first members in each set, then the second, etc. The irreducibility assumption then tells us that there can be but one member in each set: $d = 1$, $n = m$. Our matrices are now in the normal form:

$$A = \begin{Vmatrix} 0 & 1 & & & & \\ & 0 & 1 & & & \\ & & 0 & 1 & & \\ \cdot & \cdot & \cdot & \cdot & \cdot & \cdot \\ a & 0 & 0 & 0 & \cdots & 0 \end{Vmatrix}, \quad B = \begin{Vmatrix} \varepsilon^r & & & & \\ & \varepsilon^{r+1} & & & \\ & & \varepsilon^{r+2} & & \\ \cdot & \cdot & \cdot & \cdot & \cdot \\ & & & & \varepsilon^{n+r-1} \end{Vmatrix};$$

all elements not explicitly indicated are zero. The exponents in B are n successive integers and ε is a *primitive* n^{th} root of unity. Finally, the equation $A^n = \mathbf{1}$ yields $a = 1$. We number the variables from r on and take indices which are congruent mod. n as equal; the two correspondences are then

$$A: \ x'_k = x_{k+1}, \quad B: \ x'_k = \varepsilon^k x_k.$$

On reiteration we find

$$A^s: \ x'_k = x_{k+s}, \quad B^t: \ x'_k = \varepsilon^{kt} x_k. \tag{15.2}$$

The transition to continuous groups is now accomplished by passing to the limit $n \to \infty$. Let the basis iP, iQ of the continuous 2-parameter Abelian rotation group be normalized in accordance with (14.9). We identify the matrix A of the above considerations with the infinitesimal $e(\xi P)$ and B with $e(\eta Q)$ where ξ and η are real infinitesimal constants. Then $e(\sigma P) = A^s$, $e(\tau Q) = B^t$ when in the limit $s\xi \to \sigma$, $t\eta \to \tau$. ε is now $e(\xi\eta)$ and $\varepsilon^{kt} = e(\xi k \tau)$. $e(\tau Q)$ represents the physical quantity $e^{\tau q}$; the values which it may assume are given by $e^{i\tau\xi k}$, where τ is real and k runs through all integral values. In other words, the quantity q may assume the values $k\xi$; *q may assume all real numbers from $-\infty$ to $+\infty$.* (Of course k is to be considered mod.n and $k\xi$ mod.$n\xi$, but $n\xi$ is a multiple of 2π η

and may consequently be infinite in the limit.) We therefore write q in place of $k\xi$, where q is understood to be a variable which runs through the possible values of the physical quantity q, and $\sqrt{\xi} \cdot \psi(q)$ in place of x_k. $\psi(q)$ is an arbitrary function, whose values are complex numbers, which satisfies the normalizing condition

$$\int |\psi(q)|^2 dq = 1.$$

On passing to the limit in the second equation of (15.2) we find that the quantity $e^{i\tau q}$ is represented by the linear operator

$$\psi(q) \rightarrow e^{i\tau q} \cdot \psi(q).$$

Similarly we find from the first equation of (15.2) that

$$\psi(q) \rightarrow \psi(q + \sigma)$$

is the operator representing $e^{i\sigma p}$. On returning from finite to infinitesimal unitary transformations we find

$$q : \delta\psi(q) = q \cdot \psi(q), \quad p : \delta\psi(q) = \frac{1}{i}\frac{d\psi}{dq}. \tag{15.3}$$

We have thus finally justified the assumption from which we started in Chapter II.

The extension of these results to systems with several degrees of freedom causes no trouble. *The kinematics of a system which is expressed by a continuous Abelian group of rotations is consequently determined uniquely by the number f of degrees of freedom.* The postulate of irreducibility allows us to conclude that the particular operators (15.3) of the Schrödinger theory are a necessary consequence of Heisenberg's commutation rules.[27]

P. Jordan and E. Wigner[28] have given a very elegant group-theoretic proof that there exists but one irreducible matrix solution of equations (14.10), i.e. that one of degree 2^f there mentioned and given in greater detail at the end of §11.

CHAPTER V

THE SYMMETRIC PERMUTATION GROUP AND THE ALGEBRA OF SYMMETRIC TRANSFORMATIONS

A. GENERAL THEORY

§ 1. The Group Induced in Tensor Space and the Algebra of Symmetric Transformations

THE principal problem we propose to solve in this chapter is the group-theoretic classification of line spectra of an atom consisting of an arbitrary number, say f, of electrons, taking into account the reduction of the space $\mathfrak{R}^f$ to $\{\mathfrak{R}^f\}$, as required by the Pauli exclusion principle, and the spinning electron. For this it is necessary to consider in detail the representations of the symmetric group, i.e. the group π_f of all $f!$ permutations of f things. These are most intimately related to the representations of the group $\mathfrak{u}$ of all unitary transformations or the group $\mathfrak{c}$ of all homogeneous linear transformations of a space $\mathfrak{R}_n$. This connection has already been touched upon in Chapter III, § 5: the substratum of a representation of $\mathfrak{c}$ or $\mathfrak{u}$ consists of the linear manifold of all tensors of order f in $\mathfrak{R}_n$ which satisfy certain symmetry conditions, and the symmetry properties of a tensor are expressed by linear relations between it and the tensors obtained from it by the $f!$ permutations.

A tensor F of order f in the n-dimensional vector space $\mathfrak{R} = \mathfrak{R}_n$ is defined by its n^f components or, as we prefer to say, " coefficients " $F(i_1 i_2 \cdots i_f)$; each of the indices i runs from 1 to n. Tensors can be added and multiplied by arbitrary numbers ; hence the totality of such tensors F constitute a linear " vector space " $\mathfrak{R}^f$ of n^f dimensions. Further, F can be subjected to an arbitrary permutation s of its f indices, which can be thought of as a permutation of the f numbers $1, 2, \cdots, f$ attached to the indices i in the general component above ; if s is the permutation

$$1 \rightarrow 1', \ 2 \rightarrow 2', \ \cdots, \ f \rightarrow f'$$

then the tensor $\boldsymbol{s}F$ obtained by applying s to F is, by definition, that tensor whose coefficients are

$$\boldsymbol{s}F(i_1 i_2 \cdots i_f) = F(i_{1'} i_{2'} \cdots i_{f'}). \tag{1.1}$$

It follows from this definition that for any two permutations s and t

$$\boldsymbol{t}(\boldsymbol{s}F) = (\boldsymbol{ts})F.$$

A linear correspondence $F \to F'$:

$$F'(i_1 \cdots i_f) = \sum_{(k)} a(i_1 \cdots i_f;\ k_1 \cdots k_f) F(k_1 \cdots k_f) \tag{1.2}$$

is said to be ***symmetric*** if the coefficient

$$a(i_1 \cdots i_f;\ k_1 \cdots k_f)$$

is unaltered on subjecting the sub-indices $1, 2, \cdots, f$ of both the indices i and k to the *same* arbitrary permutation s. The processes of addition, multiplication by a number and permutation, in the sense defined above, applied to tensors are invariant under symmetric linear transformations; and conversely, any transformation of tensor space under which these processes are invariant is linear and symmetric. The totality of symmetric correspondences constitutes an algebra Σ: if A and B are elements of Σ then $A + B$, AB and cA (c an arbitrary number) are also. The problem with which we shall concern ourselves is the reduction of $\mathfrak{R}^f$ into linear sub-spaces $\mathfrak{P}$ which are invariant with respect to Σ, i.e. with respect to all symmetric linear transformations. Wherever in the following we employ the terms invariant, irreducible, etc., in referring to the tensor space $\mathfrak{R}^f$, they are to be interpreted with respect to the algebra Σ.

We give a brief résumé of our terminology. We are dealing with a vector space $\mathfrak{R}$ and a system Σ of linear correspondences

$$\mathfrak{x} \to \mathfrak{x}' = A\mathfrak{x}$$

of $\mathfrak{R}$ on itself; we may often prefer to use the term "linear projection" instead of "linear correspondence (operator)" in order to bring out the fact that the correspondence need not be one-to-one. A (linear) sub-space $\mathfrak{P}$ of $\mathfrak{R}$ is *invariant* if an arbitrary projection A of the system Σ sends every vector $\mathfrak{x}$ of $\mathfrak{P}$ over into a vector of $\mathfrak{P}$; $\mathfrak{P}$ is *irreducible* if it contains no invariant sub-space other than itself and the space 0 consisting only of the vector 0. We shall always understand by a *complete reduction* $\mathfrak{P} = \mathfrak{P}_1 + \mathfrak{P}_2$ of the invariant sub-space $\mathfrak{P}$ a complete reduction into two linearly independent *invariant* sub-spaces $\mathfrak{P}_1$, $\mathfrak{P}_2$, even when this is not explicitly stated. A linear projection $\mathfrak{x} \to \mathfrak{x}'$ of the invariant sub-space $\mathfrak{P}$ on the invariant sub-space $\mathfrak{P}'$ is *similar* if two vectors $\mathfrak{x}$ and $\mathfrak{y}$ of $\mathfrak{P}$ which are related by a correspondence A of the system : $\mathfrak{y} = A\mathfrak{x}$, are always projected into two vectors $\mathfrak{x}'$ and $\mathfrak{y}'$ of $\mathfrak{P}'$ which are related by the *same* A : $\mathfrak{y}' = A\mathfrak{x}'$. $\mathfrak{P}$ and $\mathfrak{P}'$ are similar or *equivalent:* $\mathfrak{P}' \sim \mathfrak{P}$ if a one-to-one linear and similar correspondence can be set up between $\mathfrak{P}$ and $\mathfrak{P}'$. In particular, these concepts are to be applied to the case in which the vector space is the tensor space $\mathfrak{R}^f = \mathfrak{R}^f_n$ of n^f dimensions and Σ is the totality of symmetric transformations.

In quantum theory the state of a system consisting of f equivalent individuals (electrons) with a system-space $\mathfrak{R}$ is described by a tensor of order f in $\mathfrak{R}$. The energy necessarily depends on each of the f individuals in exactly the same way ; hence the Hermitian operator which represents the energy is necessarily symmetric in our sense. The fundamental dynamical law therefore allows us to conclude that an invariant sub-space $\mathfrak{P}$ of $\mathfrak{R}^f$ has the property that if the tensor describing the state of the system is at any time in $\mathfrak{P}$ no influence whatever can drive it out. A complete reduction of $\mathfrak{R}^f$ into invariant sub-spaces $\mathfrak{P}$ implies a corresponding reduction of the operator representing the energy ; hence the term spectrum is reduced into classes of terms belonging to the various $\mathfrak{P}$, such that the members of one class can under no conditions combine with the members of another. Naturally this division into non-combining classes is to be carried as far as possible. But this problem is exactly the one proposed above—the only difference being that we are here only concerned with the totality $\Sigma^{(h)}$ of symmetric *Hermitian* operators. However, this restriction is quite irrelevant, for any symmetric operator can be written in the form $A = A_1 + iA_2$ where

$$A_1 = \frac{1}{2}(A + \tilde{A}), \quad A_2 = \frac{1}{2i}(A - \tilde{A}),$$

are both Hermitian.

On going over to a new co-ordinate system in the fundamental vector space $\mathfrak{R}$ by means of a non-singular transformation

$$x'_i = \sum_{k=1}^{n} a(ik)x_k \tag{1.3}$$

the coefficients of a tensor F are transformed in accordance with

$$F'(i_1 i_2 \cdots i_f) = \sum_{(k)} a(i_1k_1)a(i_2k_2) \cdots a(i_fk_f) \cdot F(k_1k_2 \cdots k_f) \tag{1.4}$$

The transformation (1.3) in vector space induces the *symmetric* transformation (1.4) in tensor space. These induced transformations, which we shall call "*special symmetric transformations*," constitute a group Σ_0 which is isomorphic with the complete linear group $\mathfrak{c} = \mathfrak{c}_n$; this representation of $\mathfrak{c}$ was previously denoted by $(\mathfrak{c})^f$. The *group* Σ_0 is contained in the *algebra* Σ. Hence a sub-space $\mathfrak{P}$ of $\mathfrak{R}^f$ which is invariant under the algebra Σ is *a fortiori* invariant under the group Σ_0. That the converse of this result is also valid is not so self-evident. Nevertheless for all questions involving only linearity Σ_0 can be replaced by the more extended Σ, for Σ is what we might call an *enveloping algebra for the group* Σ_0; by this we mean that any symmetric transformations can be expressed as a linear combination of appropriately chosen special symmetric transformations.[1] To show this we prove the theorem:

A homogeneous linear relation

$$\sum_{(i;\,k)} c(i_1 \cdots i_f;\ k_1 \cdots k_f)\, x(i_1 \cdots i_f;\ k_1 \cdots k_f) = 0 \tag{1.5}$$

is satisfied identically by all symmetric transformations

$$\|x(i_1 \cdots i_f;\ k_1 \cdots k_f)\|,$$

if it is satisfied by all special symmetric transformations, i.e. if the equation

$$\sum_{(i;\,k)} c(i_1 \cdots i_f;\ k_1 \cdots k_f) x(i_1 k_1) \cdots x(i_f k_f) = 0 \tag{1.6}$$

is satisfied for all values of the n^2 variables $x(ik)$ for which the determinant $|x(ik)| \neq 0$.

Proof. Denoting the pair (ik) of indices by j and calling the $n^2 = m$ values of j simply $1, 2, \cdots, m$, the left-hand side of (1.6) is a homogeneous polynomial of order f in the m variables $x(ik) = x_j$:

$$\phi(x_1 x_2 \cdots x_m) = \sum_{(f)} b(f_1, f_2, \cdots, f_m) x_1^{f_1} x_2^{f_2} \cdots x_m^{f_m}$$

where $f_1 + f_2 + \cdots + f_m = f$ and $b(f_1, f_2, \cdots, f_m)$ is $\dfrac{f!}{f_1! f_2! \cdots f_m!}$ times that coefficient $c(j_1 j_2 \cdots j_f)$ whose indices contain $j = 1$ f_1 times, $j = 2$ f_2 times, etc. On denoting that variable $x(j_1 j_2 \cdots j_f)$ in which the indices $j = 1, 2, \cdots, m$ occur $f_1, f_2 \cdots, f_m$ times by $y(f_1, f_2, \cdots, f_m)$ the left-hand side of equation (1.5) becomes

$$\sum_{(f)} b(f_1, f_2, \cdots, f_m) y(f_1, f_2, \cdots, f_m).$$

The determinant of the $x(ik)$ is a certain polynomial $D(x_1 x_2 \cdots x_m)$ in the variables x_j. Our assertion is thus reduced to the well-known theorem of algebra: let $\phi(x)$, $D(x)$ be two polynomials in the variables $x_1\, x_2 \cdots x_m$, the second of which does not vanish algebraically, i.e. its coefficients do not all vanish. If $\phi(x)$ is zero for all values of the variables for which the value of $D(x) \neq 0$, then $\phi(x)$ vanishes algebraically.

This theorem is proved for a single variable x as follows. If $\phi(x)$ does not vanish algebraically it has a definite degree $p \geqq 0$; let q be the degree of $D(x)$. There are then at most $p + q$ values of the variable x for which $\phi(x)$ or $D(x)$ vanish; for any one of the remaining infinitude of possible values of x neither $\phi(x)$ nor $D(x)$ can vanish, contrary to assumption. The theorem is readily extended to polynomials in any number of variables by mathematical induction. The principal point is that the *analytical* vanishing of a polynomial for all values of the independent variables implies that it vanishes *algebraically*.

In quantum theory the vector space $\mathfrak{R}$ is unitary: the transition from one normal co-ordinate system to another such is accomplished by an arbitrary *unitary* transformation (1.3). The transformations thus induced constitute a sub-group $\Sigma_0^{(u)}$ of Σ_0 which is isomorphic to the unitary group $\mathfrak{u}_n$, i.e. the representation $(\mathfrak{u})^f$ of the unitary group. I assert that a sub-space $\mathfrak{P}$ of $\mathfrak{R}^f$ which is invariant and irreducible with respect to Σ remains irreducible not only under the group Σ_0, but under the more restricted group $\Sigma_0^{(u)}$ as well. To prove this we must show that the identity (1.5) holds even when we assume only that (1.6) is true for those values of the variables $x(ik)$ with unitary matrix.

One of the most natural proofs of the above theorem concerning the formal vanishing of a form ϕ of order f depends on the process of "polarization": we assign arbitrary infinitesimal increments dx_j to the values of the variables x_j; the identical vanishing of ϕ then allows us to conclude that the differential

$$\sum_j \frac{\partial \phi}{\partial x_j} dx_j$$

vanishes for arbitrary values of x_j and dx_j. This procedure also leads us to the desired conclusion in the case under consideration. Denoting by Φ the matrix obtained by transposing rows and columns in $\left\| \frac{\partial \phi}{\partial x(ik)} \right\|$, we have

$$\mathrm{tr}\,(\Phi dX) = 0$$

where X, $X + dX$ are two arbitrary neighbouring unitary matrices. In order that this be the case we must have

$$dX = iX \cdot \delta X$$

where δX is an arbitrary Hermitian matrix: the "rotation" $X + dX$ is obtained by following up the rotation X with the infinitesimal rotation $1 + i \cdot \delta X$. But the equation

$$\operatorname{tr}(\Phi X \cdot \delta X) = 0$$

implies the vanishing of ΦX. This is seen immediately from the fact that a linear form

$$\Sigma a_{ik} y_{ik}$$

in the variables $y_{ik} = \delta x(ik)$ vanishes identically if it vanishes for all values satisfying the condition $y_{ki} = \bar{y}_{ik}$; indeed, any matrix $Y = \|y_{ik}\|$ can be written in the form $Y_1 + iY_2$ where Y_1 and Y_2 are Hermitian. On multiplying the right-hand side of $\Phi X = 0$ by X^{-1} we find $\Phi = 0$: all derivatives $\dfrac{\partial \phi}{\partial x(ik)}$ vanish in the same sense as ϕ itself, i.e. for arbitrary $x(ik)$ whose matrix is unitary. But these derivatives are forms of order $f - 1$; the truth of our assertion above is thus proved by mathematical induction.

Every invariant sub-space $\mathfrak{P}$ of $\mathfrak{R}^f$ is the representation space of representations of the groups $\mathfrak{c}$ and $\mathfrak{u}$ which are contained in $(\mathfrak{c})^f$ and $(\mathfrak{u})^f$ respectively. Hence the above results prove that if $\mathfrak{P}$ is irreducible these representations are also.

§ 2. Symmetry Classes of Tensors

One of the most natural methods of obtaining invariant manifolds of tensors F consists in subjecting F to linear symmetry conditions of the form

$$\sum_s a(s) \cdot \boldsymbol{s}F = 0. \tag{2.1}$$

This suggests introducing the ***symmetry operator***

$$\boldsymbol{a} = \sum_s a(s) \cdot \boldsymbol{s}. \tag{2.2}$$

Such operators can be added and multiplied with arbitrary numbers, and two operators $\boldsymbol{a}$, $\boldsymbol{b}$ can be applied successively with the same result as the symmetry operator $\boldsymbol{c} = \boldsymbol{ba}$ defined by

$$c(s) = \sum_{tt'=s} b(t)\, a(t'). \tag{2.3}$$

In other words, we are here led in a most natural way to the algebra ρ of the symmetric group $\pi = \pi_f$ of all permutations s. The elements of this algebra, which constitute an $f!$-dimensional linear space $\mathfrak{r}$, appear as operators which can be applied to tensors of order f. We may call the numbers $a(s)$ appearing in (2.2) the *components* of the element $\boldsymbol{a}$. In particular, $\boldsymbol{a}$ is an Hermitian operator in the tensor space $\mathfrak{R}^f$ if it is a *real* element, i.e. if it coincides with its Hermitian conjugate $\tilde{\boldsymbol{a}}$ defined by the equation

$$\tilde{a}(s) = \bar{a}(s^{-1}). \tag{2.4}$$

Hence these real symmetry operators represent physical quantities of the physical system consisting of f equivalent individuals, whose total system space is $\mathfrak{R}^f$; quantities of this kind are unknown in classical physics and cannot be pictured in terms of the usual spatial and temporal models.[2]

(2.1) or

$$\sum_s a(s)x(s) = 0$$

is a linear condition which is imposed on the element $\boldsymbol{x} = \boldsymbol{F}$ defined by $x(s) = \boldsymbol{s}F$. A **symmetry class** is defined by one or more equations of this kind; we are thus led to the definition:

Each linear sub-space $\mathfrak{p}$ *of* $\mathfrak{r}$ *determines a symmetry class* $\mathfrak{P}$ *of tensors.* F *belongs to* $\mathfrak{P}$ *when the corresponding symmetry quantity or element* $\boldsymbol{F}$ *is in* $\mathfrak{p}$. It will be found convenient to denote the process by which $\mathfrak{P}$ is generated from $\mathfrak{p}$ by a symbol; we write $\mathfrak{P} = \sharp\mathfrak{p}$.

If the reader finds it difficult to operate with elements $\boldsymbol{F}$ whose components $\boldsymbol{s}F$ are tensors rather than numbers he may replace the tensor by the totality of its coefficients $F(i_1 i_2 \cdots i_f)$ and $\boldsymbol{F}$ by the elements

$$\boldsymbol{x} = \boldsymbol{F}(i_1 i_2 \cdots i_f)$$

associated with each definite set of indices $(i_1 i_2 \cdots i_f)$; this $\boldsymbol{x}$ is defined by the equation

$$x(s) = \boldsymbol{s}F(i_1 i_2 \cdots i_f).$$

The requirement that $\boldsymbol{F}$ belong to $\mathfrak{p}$ means that $\boldsymbol{F}(i_1 i_2 \cdots i_f)$ belongs to $\mathfrak{p}$ *for all the* n^f *possible combinations of the indices* i. That the symmetry class $\mathfrak{P} = \sharp\mathfrak{p}$ is invariant with respect to all symmetric transformations (1.2) is due to the fact that (1.2) implies the corresponding equation for the *elements* $\boldsymbol{F}$, $\boldsymbol{F}'$. $\boldsymbol{F}'(i_1 i_2 \cdots i_f)$ is a linear combination of the elements $\boldsymbol{F}(k_1 k_2 \cdots k_f)$ associated with the various combinations $(k_1 k_2 \cdots k_f)$ of indices k.

If $\boldsymbol{F}$ belongs to $\mathfrak{p}$ then $\boldsymbol{a} \cdot \boldsymbol{F}$ does also, where $\boldsymbol{a}$ is any element whatever of the algebra. To show this we note that the s-component of

$$\boldsymbol{H}(i_1 \cdots i_f) = \boldsymbol{a} \cdot \boldsymbol{F}(i_1 \cdots i_f)$$

is given by

$$\sum_r a(r^{-1}) \cdot \boldsymbol{rs}F(i_1 \cdots i_f) = \sum_r a(r^{-1}) \cdot \boldsymbol{s}F(k_1 \cdots k_f)$$

where the $k_1, \cdots, k_f$ are obtained from $i_1, \cdots, i_f$ by the permutation r. Hence $\boldsymbol{H}\ (i_1, \cdots, i_f)$ is a linear combination of those $\boldsymbol{F}(k_1 \cdots k_f)$ whose indices k are obtained by a permutation of the indices i.

The principal question now is whether every invariant sub-space $\mathfrak{P}$ can be generated from a $\mathfrak{p}$ by the process $\sharp$, and further, whether or to what extent this generating $\mathfrak{p}$ is uniquely determined by $\mathfrak{P}$. The answer is perhaps best expressed with the aid of the inverse process $\natural$ which generates a $\mathfrak{p}$ from the given $\mathfrak{P}$. The following geometrical analogy may be useful in enabling the reader to understand the situation with which we are dealing. Let the points $\boldsymbol{x}$ of a plane with a fixed centre correspond to the elements of the algebra ρ and the line segments F going out from the origin correspond to the tensors. On contracting the entire plane, leaving the centre invariant, in the fixed ratio $\tau \quad (0 \leqq \tau \leqq 1)$ the point $\boldsymbol{x}$ goes into the point $\tau\boldsymbol{x}$ and the segment F into the segment τF; this contraction of segments shall be the analogue of the symmetrical transformations of tensors. $\mathfrak{P}$ will now denote an "invariant" set of segments, i.e. a set such that if it contains the segment F it also contains all the contracted segments τF. Just as we associated the symmetry elements $\boldsymbol{F}(i_1 \cdots i_f)$ with the tensor F we now associate with the segment F the continuum of points $\boldsymbol{F}(\tau)$ of F; $\boldsymbol{F}(\tau)$ is the end point of the segment τF. Let $\mathfrak{p}$ be any set of points; the segment F will then be included in the set $\mathfrak{P} = \sharp\mathfrak{p}$ if and only if all its points $\boldsymbol{F}(\tau)$ are in $\mathfrak{p}$. Obviously the only segment sets $\mathfrak{P}$ which can be obtained in this way are those which are invariant, and all such invariant sets can be so obtained. Only the "core" $\mathfrak{p}_0$ of the point set $\mathfrak{p}$ is essential to this construction; $\mathfrak{p}_0$ consists only of those points $\boldsymbol{x}$ such that $\tau\boldsymbol{x}$ belongs to $\mathfrak{p}$ for all τ (in the interval $0 \leqq \tau \leqq 1$). $\mathfrak{p}_0$ is invariant in the sense that with $\boldsymbol{x}$ all $\tau\boldsymbol{x}$ belong to $\mathfrak{p}_0$. That only the core $\mathfrak{p}_0$ is essential means that our construction generates the same segment set $\mathfrak{P}$ from two point sets $\mathfrak{p}, \mathfrak{p}'$ if these latter have the same core; hence we can restrict ourselves *ab initio* to the consideration of invariant point sets $\mathfrak{p} = \mathfrak{p}_0$. It is extraordinarily easy to find the point set $\mathfrak{p}$ which generates a given segment set $\mathfrak{P}$: we include in $\mathfrak{p}$ those and only those points lying on the segments of $\mathfrak{P}$, and this $\mathfrak{p}$ is automatically invariant.

If the reader will think through this geometrical illustration, which we have formulated here in such a pedantic manner, he will have no trouble in understanding the analogous situation for tensors and symmetry elements. A linear sub-space $\mathfrak{p}$ of $\mathfrak{r}$ is to be called invariant if all elements $\boldsymbol{ax}$ are in $\mathfrak{p}$, where $\boldsymbol{x}$ is an arbitrary element of $\mathfrak{p}$ and $\boldsymbol{a}$ is any element whatever.* Hence such a $\mathfrak{p}$ is invariant under the totality of correspondences of the form

$$(a): \boldsymbol{x} \rightarrow \boldsymbol{x}' = \boldsymbol{ax} \tag{2.5}$$

On associating this correspondence (a) of $\mathfrak{r}$ on itself with the element $\boldsymbol{a}$ we obviously obtain a representation of the algebra ρ (and therefore of the group π_f); it is called the ***regular representation.*** ($\mathfrak{r}$ appears here twice: once as the representation space and again as the algebra ρ represented in this space; the first will be expressed by the German letter $\mathfrak{r}$, the second by the Greek ρ. We are here doing the same thing as in III, § 2, where we obtained a realization of the group $\mathfrak{g}$ by associating with the element a of $\mathfrak{g}$ the correspondence $s \rightarrow s' = as$ of the group manifold on itself.) This regular representation supplies us with material from which we can construct all—and hence in particular the inequivalent irreducible—representations of the algebra ρ. When we use the terms invariant, irreducible, etc., in $\mathfrak{r}$ they will always refer to the algebra of all correspondences (a) of $\mathfrak{r}$ on itself, which is simply isomorphic with the algebra ρ of all symmetry elements $\boldsymbol{a}$. $\mathfrak{p}$ being an invariant sub-space of $\mathfrak{r}$, we shall always refer to the representation induced in $\mathfrak{p}$ by the regular representation simply as the *regular representation in* $\mathfrak{p}$; it associates with each element $\boldsymbol{a}$ the correspondence (2.5) *of* $\mathfrak{p}$ *on itself.* The equation $\boldsymbol{x}' = \boldsymbol{ax}$ is, in terms of components,

$$x'(s) = \sum_r a(r^{-1})x(rs).$$

Let $\boldsymbol{x}$ be an arbitrary element of $\mathfrak{p}$; the requirement that $\mathfrak{p}$ be invariant allows us to conclude that the element $\boldsymbol{x}'$ defined by $x'(s) = x(rs)$ is also in $\mathfrak{p}$, where r is any fixed permutation.

Let $\mathfrak{p}$ be an arbitrary sub-space of $\mathfrak{r}$; we say that $\boldsymbol{x}$ belongs to the *core* $\mathfrak{p}_0$ of $\mathfrak{p}$ if and only if all quantities of the form $\boldsymbol{ax}$ belong to $\mathfrak{p}$; this $\mathfrak{p}_0$ is invariant. We thus have the theorem that two linear sub-spaces $\mathfrak{p}$, $\mathfrak{p}'$ generate the same symmetry class $\mathfrak{P} = \sharp\mathfrak{p} = \sharp\mathfrak{p}'$ of tensors if they have the same core. We may therefore restrict ourselves *ab initio* to the consideration of *invariant* sub-spaces $\mathfrak{p}$.

* This "invariant sub-space" is not the same as an "invariant sub-algebra" as defined in Chap. III, § 13; to conform with our previous nomenclature it should be called a "left-invariant sub-algebra."

It is possible that certain relations (2.1) will be satisfied by *all* tensors. Let $\mathfrak{r}_0$ denote the smallest sub-space of r which contains the elements $\boldsymbol{F}(i_1 i_2 \cdots i_f)$ associated with *all* tensors F and all values of the indices $(i_1 i_2 \cdots i_f)$. Then $\mathfrak{p}$ generates the same $\mathfrak{P} = \sharp\mathfrak{p}$ as the intersection of $\mathfrak{p}$ with $\mathfrak{r}_0$; it is therefore natural to restrict ourselves further to the consideration of *invariant sub-spaces* $\mathfrak{p}$ *of* $\mathfrak{r}_0$. These remarks are not applicable if the dimensionality $n \geqq f$, for certainly the $f!$ coefficients

$$\boldsymbol{s}F(1, 2, \cdots, f) = F(1', 2', \cdots, f')$$

of the arbitrary tensor F are independent. But the situation is different in case $n < f$: for example, let $\delta_s = \pm 1$ according as s is an even or an odd permutation; then

$$\sum \delta_s \cdot \boldsymbol{s}F$$

is an anti-symmetric tensor and must therefore vanish in case the dimensionality n is less than the order f.

We can at most hope that conversely $\mathfrak{p}$ is uniquely determined by $\mathfrak{P}$ if we restrict ourselves to *invariant sub-spaces* $\mathfrak{p}$ *which are contained in* $\mathfrak{r}_0$. In order to prove that this is indeed the case we attempt to find the inverse process which leads from $\mathfrak{P}$ to $\mathfrak{p}$, following the programme outlined by the geometrical analogy considered above. In case $n \geqq f$ this is readily done as follows: if F is any tensor in $\mathfrak{P}$ we let the element $\boldsymbol{x} = \boldsymbol{F}(1, 2, \cdots, f)$ in $\mathfrak{r}$ correspond to it; $\mathfrak{p}$ consists of all the elements $\boldsymbol{x}$ so obtained. But in order to obtain a method which is also applicable to the case $n < f$ we must alter the procedure. *We understand by* $\mathfrak{p} = \natural\mathfrak{P}$ *the smallest linear manifold containing the totality of elements* $\boldsymbol{F}(i_1, i_2, \cdots, i_f)$ *associated with all possible tensors* F *of* $\mathfrak{P}$ *and all possible combinations of indices* $(i_1 i_2 \cdots i_f)$. If the tensors E_α constitute a basis for $\mathfrak{P}$, $\mathfrak{p}$ consists of all elements of the form

$$\boldsymbol{x} = \sum_{\alpha}\sum_{(i)} c_\alpha(i_1 \cdots i_f) \cdot \boldsymbol{E}_\alpha(i_1 \cdots i_f) \tag{2.6}$$

That such a $\mathfrak{p}$ is invariant has already been shown above, for if $\boldsymbol{x} = \boldsymbol{F}(i_1 i_2 \cdots i_f)$ the element $\boldsymbol{x}'$ defined by $x'(s) = x(rs)$ is equal to $\boldsymbol{F}(k_1 k_2 \cdots k_f)$ where $k_1 k_2 \cdots k_f$ are obtained from $i_1 i_2 \cdots i_f$ by the fixed permutation r.

We now denote the $\mathfrak{r}_0$ introduced above by $\natural\mathfrak{R}^f$; it coincides with the entire space $\mathfrak{r}$ when $n \geqq f$. Let the symbol $\prec$ denote "is contained in"; the following results then follow immediately from the definitions: If $\mathfrak{p}$ is a linear sub-space of $\mathfrak{r}$ and $\mathfrak{P} = \sharp\mathfrak{p}$, then $\natural\mathfrak{P} \prec \mathfrak{p}$. If $\mathfrak{P}$ is any linear sub-space of $\mathfrak{R}^f$ and $\mathfrak{p} = \natural\mathfrak{P}$, then conversely $\mathfrak{P} \prec \sharp\mathfrak{p}$. We can at most expect that the symbol $\subseteq$ can be replaced by $=$ if in the first theorem $\mathfrak{p}$ is an invariant sub-space of $\mathfrak{r}_0$ and in the second if $\mathfrak{P}$ is an invariant sub-space of $\mathfrak{R}^f$. That these converse theorems are in fact true under these limitations will be proved in § 4.

§ 3. Invariant Sub-spaces in Group Space

We are in need of a fundamental theorem concerning the algebra of a group as a preparation for carrying through the investigation proposed above; we here prove this theorem for a general finite group. However, we do not alter the notation, so here π denotes any finite group of order h.

Theorem (3.1). *If* $\mathfrak{p}$ *is an invariant sub-space of* $\mathfrak{r}$ *there exists an element* $\boldsymbol{e}$ *of the group algebra having the following two properties:* (1) *every element of the form* $\boldsymbol{xe}$ *belongs to* $\mathfrak{p}$, (2) *every element* $\boldsymbol{x}$ *of* $\mathfrak{p}$ *satisfies the equation* $\boldsymbol{xe} = \boldsymbol{x}$.

In particular (1) implies that $\boldsymbol{e} = \mathbf{1}\boldsymbol{e}$ itself belongs to $\mathfrak{p}$, and hence by (2) $\boldsymbol{ee} = \boldsymbol{e}$; $\boldsymbol{e}$ is ***idempotent***.[3] It is a "***generating unit***" of $\mathfrak{p}$ in the sense that $\mathfrak{p}$ consists of all elements of the form $\boldsymbol{xe}$.

Proof. Let $\boldsymbol{e}_1, \boldsymbol{e}_2, \cdots, \boldsymbol{e}_h$ be a co-ordinate system in the vector space $\mathfrak{r}$ which is adapted to the g-dimensional sub-space $\mathfrak{p}$ in such a way that $\mathfrak{p}$ is the linear set defined by $\boldsymbol{e}_1, \boldsymbol{e}_2, \cdots, \boldsymbol{e}_g$. The parallel projection which transforms

$$\boldsymbol{x} = x_1\boldsymbol{e}_1 + \cdots + x_h\boldsymbol{e}_h \text{ into } \boldsymbol{x}' = x_1\boldsymbol{e}_1 + \cdots + x_g\boldsymbol{e}_g$$

has the two properties (1) it projects every $\boldsymbol{x}$ into an $\boldsymbol{x}'$ lying in $\mathfrak{p}$, and (2) within $\mathfrak{p}$ it is the identity. In the original co-ordinate system defined by the simple elements $\boldsymbol{s}$ of the algebra this projection is given by

$$x'(s) = \sum_t d(s, t)x(t),$$

where the matrix $d(s, t)$ is necessarily of the form

$$d(s, t) = e_1(s)\breve{e}_1(t) + \cdots + e_g(s)\breve{e}_g(t)$$

and the $\breve{e}_i(s)$ are defined by

$$\sum_s \breve{e}_i(s)e_k(s) = \delta_{ik} \quad (i, k = 1, 2, \cdots, g).$$

The fact that $\mathfrak{p}$ is invariant implies that if $\boldsymbol{x}$ is in $\mathfrak{p}$ then the element $\boldsymbol{x}_r$ defined by $x_r(s) = x(rs)$ is also in $\mathfrak{p}$. Consequently the projection with the matrix $d(rs, rt)$ has the same two properties (1) and (2), where r is any fixed permutation (i.e. element of the group π) whatever. Hence the assertions also hold for the correspondence with the matrix

$$e(s, t) = \frac{1}{h} \sum_r d(rs, rt) \tag{3.2}$$

obtained by summing over all elements r of the group. This matrix satisfies the equation

$$e(rs, rt) = e(s, t),$$

whence $e(s, t)$ depends only on the combination $t^{-1}s$: $e(s, t) = e(t^{-1}s)$. The linear projection

$$x'(s) = \sum_t e(s, t)\, x(t)$$

may therefore be written briefly $\boldsymbol{x}' = \boldsymbol{xe}$, which proves the validity of the theorem.

Let the invariant sub-space $\mathfrak{p}$ be completely reduced into two invariant sub-spaces: $\mathfrak{p} = \mathfrak{p}_1 + \mathfrak{p}_2$, and let $\boldsymbol{e}$ be the generating unit of $\mathfrak{p}$. Any element in $\mathfrak{p}$ can be written as the sum of its components in $\mathfrak{p}_1$ and $\mathfrak{p}_2$; hence in particular $\boldsymbol{e} = \boldsymbol{e}_1 + \boldsymbol{e}_2$. From this it follows that for an arbitrary element $\boldsymbol{x}$ of $\mathfrak{p}$

$$\boldsymbol{x} = \boldsymbol{xe} = \boldsymbol{xe}_1 + \boldsymbol{xe}_2.$$

But since $\boldsymbol{x}_1 = \boldsymbol{xe}_1$ is in $\mathfrak{p}_1$ and $\boldsymbol{x}_2 = \boldsymbol{xe}_2$ is in $\mathfrak{p}_2$, $\boldsymbol{x}_1$ and $\boldsymbol{x}_2$ are the (unique) components of $\boldsymbol{x}$ in $\mathfrak{p}_1$ and $\mathfrak{p}_2$. These two components for the element $\boldsymbol{e}_1$ are obviously $\boldsymbol{e}_1$ and 0, whence

$$\boldsymbol{e}_1\boldsymbol{e}_1 = \boldsymbol{e}_1, \quad \boldsymbol{e}_1\boldsymbol{e}_2 = 0;$$

similarly

$$\boldsymbol{e}_2\boldsymbol{e}_1 = 0, \quad \boldsymbol{e}_2\boldsymbol{e}_2 = \boldsymbol{e}_2.$$

Hence $\boldsymbol{e}_1$, $\boldsymbol{e}_2$ are the generating idempotent units of $\mathfrak{p}_1$, $\mathfrak{p}_2$ respectively; they are "***independent***" in the sense of the equations

$$\boldsymbol{e}_1\boldsymbol{e}_2 = 0, \quad \boldsymbol{e}_2\boldsymbol{e}_1 = 0.$$

On completely reducing $\mathfrak{p}$ into any number of components: $\mathfrak{p} = \sum_i \mathfrak{p}_i$, the generating unit $\boldsymbol{e}$ of $\mathfrak{p}$ is decomposed into

$$\boldsymbol{e} = \sum_i \boldsymbol{e}_i,$$

the components of which satisfy the analogous equations

$$\boldsymbol{e}_i\boldsymbol{e}_k = 0 \quad (i \neq k), \quad \boldsymbol{e}_i\boldsymbol{e}_i = \boldsymbol{e}_i.$$

The existence of the generating unit offers a means of obtaining a new and simpler proof of the fact that reducibility implies complete reducibility:

Theorem (3.3). *If* $\mathfrak{p}$, $\mathfrak{p}_1$ *are invariant and* $\mathfrak{p}_1 \prec \mathfrak{p}$, *then* $\mathfrak{p}$ *can be reduced into* $\mathfrak{p}_1 \dotplus \mathfrak{p}_2$ *in such a way that* $\mathfrak{p}_2$ *is also invariant.*

Proof. Let $\boldsymbol{e}_1$ be the generating unit of $\mathfrak{p}_1$. We decompose every element of $\mathfrak{p}$ in accordance with the equation

$$\boldsymbol{x} = \boldsymbol{x}\boldsymbol{e}_1 + (\boldsymbol{x} - \boldsymbol{x}\boldsymbol{e}_1). \tag{3.4}$$

The first component $\boldsymbol{x}_1 = \boldsymbol{x}\boldsymbol{e}_1$ lies in $\mathfrak{p}_1$, and the second

$$\boldsymbol{x}_2 = \boldsymbol{x} - \boldsymbol{x}\boldsymbol{e}_1$$

runs through a certain linear sub-space $\mathfrak{p}_2$ of $\mathfrak{p}$ when $\boldsymbol{x}$ runs through all elements of $\mathfrak{p}$. This sub-space $\mathfrak{p}_2$ is also invariant, for

$$\boldsymbol{a}\boldsymbol{x}_2 = \boldsymbol{a}\boldsymbol{x} - (\boldsymbol{a}\boldsymbol{x})\boldsymbol{e}_1$$

as $\boldsymbol{a}\boldsymbol{x}$ is in $\mathfrak{p}$ if $\boldsymbol{x}$ is. The elements $\boldsymbol{x}_1$, $\boldsymbol{x}_2$ of $\mathfrak{p}_1$, $\mathfrak{p}_2$ respectively satisfy the equations

$$\boldsymbol{x}_1\boldsymbol{e}_1 = \boldsymbol{x}_1, \quad \boldsymbol{x}_2\boldsymbol{e}_1 = 0.$$

From this it follows that the sum of an element $\boldsymbol{y}_1$ of $\mathfrak{p}_1$ and an element $\boldsymbol{x}_2$ of $\mathfrak{p}_2$ cannot vanish unless both $\boldsymbol{y}_1$ and $\boldsymbol{x}_2$ also vanish; hence $\mathfrak{p}_1$ and $\mathfrak{p}_2$ are independent. To prove this we merely note that on multiplying $\boldsymbol{y}_1 + \boldsymbol{x}_2 = 0$ by $\boldsymbol{e}_1$ we find $\boldsymbol{y}_1\boldsymbol{e}_1 = \boldsymbol{y}_1 = 0$. Equation (3.4) represents the reduction of any element of $\mathfrak{p}$ into its components in $\mathfrak{p}_1$ and $\mathfrak{p}_2$.

Any idempotent element $\boldsymbol{e}$ generates an invariant sub-space $\mathfrak{p}_e$ consisting of elements of the form $\boldsymbol{x}\boldsymbol{e}$. If $\boldsymbol{e}_1$, $\boldsymbol{e}_2$ are two independent idempotent elements ($\boldsymbol{e}_1\boldsymbol{e}_2 = 0$, $\boldsymbol{e}_2\boldsymbol{e}_1 = 0$) then the sub-spaces $\mathfrak{p}_1$, $\mathfrak{p}_2$ which they generate are independent, and the idempotent element $\boldsymbol{e} = \boldsymbol{e}_1 + \boldsymbol{e}_2$ generates $\mathfrak{p} = \mathfrak{p}_1 + \mathfrak{p}_2$. An idempotent element $\boldsymbol{e}$ is said to be ***primitive*** if it can only be expressed as the sum of two idempotent elements $\boldsymbol{e}_1 + \boldsymbol{e}_2$ if one of the summands is 0 (and the other $\boldsymbol{e}$). *In order that* $\mathfrak{p}_e$ *be irreducible it is necessary and sufficient that* $\boldsymbol{e}$ *be primitive.*

Obviously any idempotent element $\boldsymbol{e}$, in particular the modulus **1** of the algebra, can be reduced into the sum of independent primitive idempotent elements. For if we have a reduction into independent non-vanishing idempotent elements

$$\boldsymbol{e} = \boldsymbol{e}_1 + \boldsymbol{e}_2 + \cdots + \boldsymbol{e}_m$$

and if, for example, $\boldsymbol{e}_1$ is not primitive, it can be further reduced to the sum of two independent non-vanishing idempotent elements $\boldsymbol{e}_1' + \boldsymbol{e}_1''$; in this way we obtain a complete reduction of $\boldsymbol{e}$ into $m + 1$ independent terms, for we have, for example,

$$\boldsymbol{e}_1'\boldsymbol{e}_2 = \boldsymbol{e}_1'\boldsymbol{e}_1\boldsymbol{e}_2 = 0; \quad \text{similarly } \boldsymbol{e}_2\boldsymbol{e}_1' = 0.$$

This process must certainly cease after at most h steps. Our analysis allows us to assert that we thus obtain a complete reduction of $\mathfrak{p}_e$ into independent *irreducible* sub-spaces.

We have seen that the theorem concerning the complete reducibility is a consequence of the existence of a generating unit. But the converse is also true: If $\mathfrak{p}$ appears as a summand in a complete reduction $\mathfrak{r} = \mathfrak{p} + \mathfrak{p}'$ of our given algebra $\mathfrak{r}$, then it possesses a generating unit. We need only to specialize the considerations developed above by applying them to the modulus $\mathbf{1}$ of $\mathfrak{r}$; $\mathbf{1}$ can be completely reduced into the two components $\boldsymbol{e} + \boldsymbol{e}'$ lying in $\mathfrak{p}$ and $\mathfrak{p}'$, and the generating units of $\mathfrak{p}$ and $\mathfrak{p}'$ are $\boldsymbol{e}$ and $\boldsymbol{e}'$ respectively.

The mathematician will find it worthy of note that all these considerations are still applicable when the algebra is defined over any field whatever. Instead of dealing with the continuum of real or complex numbers, as in analysis, we may in abstract algebra operate in an arbitrary *field*, i.e. a domain of elements, called numbers, in which the two fundamental operations of addition and multiplication and their inverses, subtraction and division, are defined in accordance with the formal laws of ordinary arithmetic. Our development depended only on these rules of operation—with a slight restriction. There are fields in which a definite integer, say h, times any number of the field yields zero; we may say that *h annihilates*. Such "modular" fields must be excluded, for we wish to retain the possibility of finding a number such that its product with h is any given number. When our reasoning involves no more restrictive assumptions concerning the number field, we are operating in a relatively elementary theoretical domain. However, such theorems as the "fundamental theorem" III, (10.5), and that of Burnside-Frobenius-Schur, which depend on the fundamental theorem of algebra, belong to a deeper layer. These theorems hold only in "***algebraically closed***" number fields, in which any algebraic equation (with coefficients in the field) is soluble. Finally such concepts as "Hermitian," "unitary," etc., involve the transition from a number to its conjugate complex and have no place in general abstract fields. Our earlier proof of the theorem of complete reducibility was obtained with the aid of such tools foreign to the general concept of a field.

Theorem (3.5). *A similarity projection $\boldsymbol{x} \rightarrow \boldsymbol{x}'$ of the invariant sub-space $\mathfrak{p}$ on the invariant sub-space $\mathfrak{p}'$ is necessarily expressed by an equation of the form $\boldsymbol{x}' = \boldsymbol{x}\boldsymbol{b}$.* (In particular, when $\mathfrak{p}$ and $\mathfrak{p}'$ are equivalent this theorem is applicable to the one-to-one similarity correspondence $\mathfrak{p} \leftrightarrows \mathfrak{p}'$.)

Proof. Let the given similarity correspondence send the

generating unit $\boldsymbol{e}$ of $\mathfrak{p}$ over into $\boldsymbol{b}$. In virtue of the similarity $\boldsymbol{xe}$ then goes over into $\boldsymbol{x}' = \boldsymbol{xb}$, where $\boldsymbol{x}$ is any element in $\mathfrak{p}$; but for such an element $\boldsymbol{xe} = \boldsymbol{x}$.

Additional remark. The projection sends $\boldsymbol{e}$ into $\boldsymbol{eb}$; hence $\boldsymbol{eb} = \boldsymbol{b}$. On the other hand, if $\boldsymbol{e}'$ is the generating element of $\mathfrak{p}'$, then since $\boldsymbol{b}$ is in $\mathfrak{p}'$ we have $\boldsymbol{be}' = \boldsymbol{b}$:

$$\boldsymbol{b} = \boldsymbol{eb} = \boldsymbol{be}' = \boldsymbol{ebe}'.$$

We express this result, i.e. that $\boldsymbol{b}$ is of the form $\boldsymbol{exe}'$, by saying ***b** has the character* $(\boldsymbol{e}, \boldsymbol{e}')$. Our considerations show that such a projection can always be expressed in terms of a unique element $\boldsymbol{b}$ of character $(\boldsymbol{e}, \boldsymbol{e}')$.

If we are operating in the field of complex numbers, with which the investigations of analysis (e.g. the theory of functions) deal and in which we are exclusively interested in quantum theory, we may supplement the theorem (3.1) concerning the existence of a generating unit $\boldsymbol{e}$ in an invariant sub-space $\mathfrak{p}$ by the following:

The generating unit may be so chosen that it is real; it is then determined uniquely by $\mathfrak{p}$.

To prove this we choose as the basis $\boldsymbol{e}_1, \boldsymbol{e}_2, \ldots, \boldsymbol{e}_g$ of $\mathfrak{p}$ a unitary-orthogonal system of vectors; then

$$\sum_s \bar{e}_i(s)\, e_k(s) = \delta_{ik} \qquad (i, k, = 1, 2, \ldots, g).$$

In constructing $d(s, t)$, which we now denote by $e(s, t)$, we may therefore choose $\check{e}_i = \bar{e}_i$:

$$e(s, t) = \sum_{i=1}^{g} e_i(s)\, \bar{e}_i(t). \tag{3.6}$$

I assert that the equation

$$e(rs, rt) = e(s, t) \tag{3.7}$$

is automatically satisfied—it is no longer necessary to take its mean value as in (3.2). The element $\boldsymbol{e}$ defined by $e(t^{-1}s) = e(s, t)$ is then the real generating unit of $\mathfrak{p}$.

In order to establish the validity of (3.7) it is only necessary to note that $e(s, t)$ is independent of the particular unitary basis $\boldsymbol{e}_1, \boldsymbol{e}_2, \ldots, \boldsymbol{e}_g$ chosen; for on going over to a new unitary basis $\boldsymbol{e}_1', \boldsymbol{e}_2', \ldots, \boldsymbol{e}_g'$ by a unitary transformation U the bilinear form (3.6) remains invariant. Now in particular the equation

$$e_i'(s) = e_i(rs),$$

in which r is a fixed element of the group, defines a transition to a new unitary basis.

To prove that this real generating unit $\boldsymbol{e}$ of $\mathfrak{p}$ is unique, assume there exists a second, $\boldsymbol{e}'$; then all elements $\boldsymbol{x}$ of $\mathfrak{p}$ satisfy the equations

$$\boldsymbol{xe} = \boldsymbol{x}, \quad \boldsymbol{xe}' = \boldsymbol{x}.$$

On applying the first equation for $\boldsymbol{x} = \boldsymbol{e}'$ and the second for $\boldsymbol{x} = \boldsymbol{e}$ we have

$$\boldsymbol{e}'\boldsymbol{e} = \boldsymbol{e}', \quad \boldsymbol{ee}' = \boldsymbol{e}.$$

But since $\boldsymbol{e}$ and $\boldsymbol{e}'$ are both real, the first of these results yields, on going over to the Hermitian conjugates,

$$\boldsymbol{e}\boldsymbol{e}' = \boldsymbol{e}',$$

and from this and the previous result we conclude that $\boldsymbol{e}' = \boldsymbol{e}$.

Under these conditions the content of theorem (3.3) can be extended and its proof simplified. If $\boldsymbol{e}$, $\boldsymbol{e}_1$ are the real generating units of $\mathfrak{p}$, $\mathfrak{p}_1$ respectively, then since $\boldsymbol{e}_1$ is in $\mathfrak{p}$ $\boldsymbol{e}_1\boldsymbol{e} = \boldsymbol{e}_1$, and on going over to the Hermitian conjugates we find $\boldsymbol{e}\boldsymbol{e}_1 = \boldsymbol{e}_1$. Hence the idempotent element $\boldsymbol{e}_2$ introduced by $\boldsymbol{e} = \boldsymbol{e}_1 + \boldsymbol{e}_2$ is real and independent of $\boldsymbol{e}_1$; $= \mathfrak{p}_1 + \mathfrak{p}_2$ is thus completely reduced into $\mathfrak{p}_1$ and an invariant sub-space $\mathfrak{p}_2$ which is *unitary-orthogonal* to $\mathfrak{p}_1$ and which has as its real generating unit $\boldsymbol{e}_2$.

§ 4. Invariant Sub-spaces in Tensor Space

We now return to the investigation of tensors of order f, the totality of which constitutes the space $\mathfrak{R}^f$. Let π again be the group of all permutations of f things and $\mathfrak{r}$ ($= \rho$) the corresponding group space (algebra). Let $\boldsymbol{a}$ be a symmetry quantity, i.e. an element of the algebra ρ, with components $a(s)$; the element $\hat{\boldsymbol{a}}$ is then defined by

$$\hat{a}(s) = a(s^{-1}) \tag{4.1}$$

The relation

$$F' = \boldsymbol{a}F,$$

which asserts that the tensor F' is obtained from F by the operator $\boldsymbol{a}$, is equivalent to the equation

$$\boldsymbol{F}' = \boldsymbol{F} \cdot \hat{\boldsymbol{a}}$$

between the corresponding elements $\boldsymbol{F}$ and $\boldsymbol{F}'$ of the algebra ρ. For

$$\boldsymbol{s}F' = \sum_t \hat{a}(t^{-1}) \cdot \boldsymbol{st}F$$

is in fact obtained from

$$F' = \sum_t a(t) \cdot \boldsymbol{t}F = \sum_t \hat{a}(t^{-1}) \cdot \boldsymbol{t}F$$

by operating on it with the permutation s.

In the following considerations, which are concerned with symmetry classes of tensors, $\mathfrak{p}$ (with or without index) always denotes an invariant sub-space of $\mathfrak{r}$, $\hat{\boldsymbol{e}}$ the generating unit of $\mathfrak{p}$ and $\mathfrak{P}$ the corresponding $\sharp\mathfrak{p}$. We may then say that $\boldsymbol{e}$ is the *generating idempotent operator of the symmetry class* $\mathfrak{P}$ in the following sense:

(1) $\boldsymbol{e}F$ lies in $\mathfrak{P}$, F being any tensor whatever;

(2) if F is in $\mathfrak{P}$ it is reproduced by the operator $\boldsymbol{e}$: $\boldsymbol{e}F = F$.

In this way we obtain a *constructive definition of the symmetry class* $\mathfrak{P}$ *as the totality of all tensors of the form* $\boldsymbol{e}F$. This definition is considerably simpler than the original one in terms of $\mathfrak{p}$, for it depends on a single element $\boldsymbol{e}$ instead of a *manifold* $\mathfrak{p}$. If, for example, we are dealing with the class $\mathfrak{P}$ of all completely symmetric tensors

$$\frac{1}{f!}\sum_s \boldsymbol{s}.$$

is such an operator; the corresponding operator for the class of all anti-symmetric tensors is the alternating sum

$$\frac{1}{f!}\sum_s \delta_s \boldsymbol{s}.$$

Theorem (4.2). *If* $\mathfrak{p}' \prec \mathfrak{p}$ *or* $\mathfrak{p} = \mathfrak{p}_1 + \mathfrak{p}_2$, *we have* $\mathfrak{P}' \prec \mathfrak{P}$, $\mathfrak{P} = \mathfrak{P}_1 + \mathfrak{P}_2$ *respectively.*

We need to prove only the latter part of this theorem, i.e. for the case of complete reduction. The generating unit $\hat{\boldsymbol{e}} = \hat{\boldsymbol{e}}_1 + \hat{\boldsymbol{e}}_2$ of $\mathfrak{p}$ has as components $\hat{\boldsymbol{e}}_1$, $\hat{\boldsymbol{e}}_2$ in $\mathfrak{p}_1$, $\mathfrak{p}_2$ the generating units of $\mathfrak{p}_1$, $\mathfrak{p}_2$ respectively. The formula

$$\boldsymbol{e}F = \boldsymbol{e}_1 F + \boldsymbol{e}_2 F$$

defines the corresponding complete reduction of $\mathfrak{P}$ into the independent invariant sub-spaces $\mathfrak{P}_1$, $\mathfrak{P}_2$.

Theorem (4.3). *If* $\mathfrak{p}_1 \sim \mathfrak{p}_2$ *then* $\mathfrak{P}_1 \sim \mathfrak{P}_2$.

The similarity correspondence $\boldsymbol{x}_1 \to \boldsymbol{x}_2$ of $\mathfrak{p}_1$ on $\mathfrak{p}_2$ is, by theorem (3.5), of the form

$$\boldsymbol{x}_2 = \boldsymbol{x}_1 \hat{\boldsymbol{b}}, \quad \boldsymbol{x}_1 = \boldsymbol{x}_2 \hat{\boldsymbol{b}}'.$$

Hence

$$F_2 = \boldsymbol{b} F_1, \quad F_1 = \boldsymbol{b}' F_2$$

define a one-to-one similar correspondence of $\mathfrak{P}_1$ on $\mathfrak{P}_2$ and its inverse.

Theorem (4.4). *If* $\mathfrak{p} \prec \mathfrak{r}_0$ *then* $\mathfrak{p} = \natural\mathfrak{P}$.

The only non-trivial part of this first converse theorem which remains to be proved is that $\mathfrak{p} \prec \natural\mathfrak{P}$. All tensors of the form $F_\alpha = \boldsymbol{e} E_\alpha$ are in $\mathfrak{P}$, where (E_α) is a basis for the entire tensor space $\mathfrak{R}^f$; hence all elements of the form

$$\boldsymbol{y} = \sum_{\alpha, i} c_\alpha(i_1 \cdots i_f) \cdot \boldsymbol{F}_\alpha(i_1 \cdots i_f)$$

are in $\natural\mathfrak{P}$. On introducing

$$\boldsymbol{x} = \sum_{\alpha, i} c_\alpha(i_1 \cdots i_f) \cdot \boldsymbol{E}_\alpha(i_1 \cdots i_f)$$

we have $\boldsymbol{y} = \boldsymbol{x}\hat{\boldsymbol{e}}$. On recalling the definition of $\mathfrak{r}_0 = \natural\mathfrak{R}^f$ we see that $\boldsymbol{x}\hat{\boldsymbol{e}}$ belongs to $\natural\mathfrak{P}$ if $\boldsymbol{x}$ lies in $\mathfrak{r}_0$. But in virtue of the assumption that $\mathfrak{p} \prec \mathfrak{r}_0$ this is automatically satisfied if $\boldsymbol{x}$ is an arbitrary element of $\mathfrak{p}$; but then $\boldsymbol{x}\hat{\boldsymbol{e}} = \boldsymbol{x}$. Hence every element $\boldsymbol{x}$ of $\mathfrak{p}$ is contained in $\natural\mathfrak{P}$.

In order to formulate the converse of these theorems let $\mathfrak{P}$ (with or without index) now denote an arbitrary invariant sub-space of $\mathfrak{R}^f$ and $\mathfrak{p}$ the corresponding $\natural\mathfrak{P}$.

Theorem (4.5). *If* $\mathfrak{P}' \prec \mathfrak{P}$ *or* $\mathfrak{P} = \mathfrak{P}_1 + \mathfrak{P}_2$, *then* $\mathfrak{p}' \prec \mathfrak{p}$, $\mathfrak{p} = \mathfrak{p}_1 + \mathfrak{p}_2$, *respectively.*

Theorem (4.6). *If* $\mathfrak{P} \sim \mathfrak{P}'$ *then* $\mathfrak{p} \sim \mathfrak{p}'$.

Theorem (4.7). $\mathfrak{P} = \sharp\mathfrak{p}$.

The last theorem is by far the most important of all; it asserts that *every* $\mathfrak{P}$ *is a symmetry class of tensors.* It is desirable to prove it first, i.e. to prove that $\sharp\mathfrak{p} \prec \mathfrak{P}$. Let $\hat{\boldsymbol{e}}$ again denote the generating unit of $\mathfrak{p}$; $\sharp\mathfrak{p}$ then consists of all tensors of the form $F' = \boldsymbol{e}F$. Since the element $\hat{\boldsymbol{e}}$ belongs to $\mathfrak{p}$ it is necessarily of the form

$$\hat{e}(s) = e(s^{-1}) = \sum_{\alpha, k} e_\alpha(k_1 \cdots k_f) \cdot \boldsymbol{s}E_\alpha(k_1 \cdots k_f), \tag{4.8}$$

where the tensors E_α constitute a basis for the space $\mathfrak{P}$. Now the trivial equation

$$\sum_i \boldsymbol{s}c(i_1 \cdots i_f) \cdot \boldsymbol{s}F(i_1 \cdots i_f) = \sum_i c(i_1 \cdots i_f) \cdot F(i_1 \cdots i_f)$$

shows, on replacing $\boldsymbol{s}c$ by c, that

$$\sum_i c(i_1 \cdots i_f) \cdot \boldsymbol{s}F(i_1 \cdots i_f) = \sum_i \boldsymbol{s}^{-1}c(i_1 \cdots i_f) \cdot F(i_1 \cdots i_f).$$

Hence we may replace (4.8) by

$$e(s) = \sum_{\alpha, k} \boldsymbol{s}e_\alpha(k_1 \cdots k_f) \cdot E_\alpha(k_1 \cdots k_f)$$

and the coefficients of F' are then given by

$$F'(i_1 \cdots i_f) = \sum_{\alpha, k} c_\alpha(i_1 \cdots i_f;\ k_1 \cdots k_f) E_\alpha(k_1 \cdots k_f)$$

where

$$c_\alpha(i_1 \cdots i_f;\ k_1 \cdots k_f) = \sum_s \boldsymbol{s}F(i_1 \cdots i_f) \cdot \boldsymbol{s}e_\alpha(k_1 \cdots k_f).$$

Because of the summation over all elements s of the group π this transformation with coefficients c_α is symmetric; hence the assumption that the sub-space $\mathfrak{P}$ is invariant allows us to conclude that F' lies in $\mathfrak{P}$ if the E_α do. But this establishes our theorem.

The theorem can also be proved directly, without calling on the theorems of § 3, in the following way. That F is in

$\sharp\mathfrak{p}$ means that $\boldsymbol{F}(i_1 i_2 \cdots i_f)$ is in $\mathfrak{p}$ and is consequently of the form (2.6):

$$\boldsymbol{F}(i_1 \cdots i_f) = \sum_{\alpha, k} b_\alpha(i_1 \cdots i_f;\ k_1 \cdots k_f) \cdot \boldsymbol{E}_\alpha(k_1 \cdots k_f).$$

The E_α constitute a basis of $\mathfrak{P}$. Writing down the s-component of this equation and replacing the indices $i_{1'}, \cdots, i_{f'}$ by $i_1 \cdots, i_f$ we find the equation

$$F(i_1 \cdots i_f) = \sum_{\alpha, k} \boldsymbol{s}^{-1} b_\alpha(i_1 \cdots i_f;\ k_1 \cdots k_f) \cdot E_\alpha(k_1 \cdots k_f)$$

for the components of F. Since this holds for every permutation s^{-1} we may sum over the elements of the group and obtain

$$F(i_1 \cdots i_f) = \sum_{\alpha, k} c_\alpha(i_1 \cdots i_f;\ k_1 \cdots k_f) \cdot E_\alpha(k_1 \cdots k_f),$$

where the coefficients

$$c_\alpha(i_1 \cdots i_f;\ k_1 \cdots k_f) = \frac{1}{f!} \sum_s \boldsymbol{s} b_\alpha(i_1 \cdots i_f;\ k_1 \cdots k_f)$$

are symmetric. Hence since the E_α belong to the invariant sub-space $\mathfrak{P}$ and F is obtained from them by a symmetric transformation, F also belongs to $\mathfrak{P}$.

The only part of theorem (4.5) which is not self-evident is the assertion that $\mathfrak{p}_1$, $\mathfrak{p}_2$ are independent. By theorem (4.7) we have the relations

$$\sharp\mathfrak{p}^* \prec \sharp\mathfrak{p}_1 \prec \mathfrak{P}_1, \quad \sharp\mathfrak{p}^* \prec \mathfrak{P}_2$$

for the (invariant) intersection $\mathfrak{p}^*$ of $\mathfrak{p}_1$ and $\mathfrak{p}_2$. But since $\mathfrak{P}_1$, $\mathfrak{P}_2$ are independent it follows that $\sharp\mathfrak{p}^*$, and therefore $\mathfrak{p}^*$, is empty.

Theorem (4.5) shows the $\mathfrak{P}$ associated with an *irreducible* $\mathfrak{p}$ is also *irreducible*. Hence it follows, in particular, that *the manifold of symmetric and the manifold of anti-symmetric tensors are irreducible and invariant*, not only with respect to the algebra of symmetric transformations, but also with respect to the transformations induced in tensor space by the affine or unitary groups of transformations in the vector space $\mathfrak{R}$. Applying this to the 2-dimensional vector space, we see that the representations $\mathfrak{C}_f$ of $\mathfrak{c} = \mathfrak{c}_2$ or $\mathfrak{u}$ constructed in III, § 5, are irreducible.

In order to prove (4.6) we must first examine the nature of $\mathfrak{r}_0$ (for $n < f$) in some detail. We call the component $a(\mathbf{I})$ of an element $\boldsymbol{a}$ of the algebra the ***trace*** of $\boldsymbol{a}$. Hence the trace of the product $\boldsymbol{ab}$, which we call the ***scalar product*** $\mathrm{tr}(\boldsymbol{ab})$ of $\boldsymbol{a}$ and $\boldsymbol{b}$, is

$$\mathrm{tr}(\boldsymbol{ab}) = \sum a(s) b(s^{-1}).$$

The trace of $\boldsymbol{a}$ is then $\mathrm{tr}(\boldsymbol{a}\mathbf{l}) = \mathrm{tr}(\mathbf{l}\boldsymbol{a}) = \mathrm{tr}(\boldsymbol{a})$. The scalar product is obviously symmetric in $\boldsymbol{a}$ and $\boldsymbol{b}$, and the symmetric bilinear form $\mathrm{tr}(\boldsymbol{ab})$ is non-degenerate, i.e. $\boldsymbol{a} = 0$ is the only element for which the equation $\mathrm{tr}(\boldsymbol{ax}) = 0$ is satisfied identically in $\boldsymbol{x}$.

Auxiliary theorem (4.9). *$\mathfrak{r}_0$ is a left- as well as right-invariant sub-algebra of $\mathfrak{r}$. $\mathrm{tr}(\boldsymbol{ab})$ is non-degenerate within $\mathfrak{r}_0$, i.e. the only element $\boldsymbol{a}$ of $\mathfrak{r}_0$ whose scalar product with every element $\boldsymbol{x}$ of $\mathfrak{r}_0$ vanishes is $\boldsymbol{a} = 0$.*

The first part of this theorem is almost self-evident. For if $\boldsymbol{x} = \boldsymbol{F}(i_1 \cdots i_f)$, the element $\boldsymbol{x}'$ defined by $x'(s) = x(sr)$ is $\boldsymbol{F}'(i_1 \cdots i_f)$ where $F' = \boldsymbol{r}F$.

Let $\boldsymbol{i}$ be the generating unit of $\mathfrak{r}_0$, $\boldsymbol{a}$ an element of $\mathfrak{r}_0$ and $\boldsymbol{x}$ an arbitrary element. Then since $\mathfrak{r}_0$ is right-invariant $\boldsymbol{ax}$ is also in $\mathfrak{r}_0$, whence

$$\boldsymbol{ax} = \boldsymbol{ax} \cdot \boldsymbol{i}, \quad \mathrm{tr}(\boldsymbol{ax}) = \mathrm{tr}(\boldsymbol{a} \cdot \boldsymbol{xi}).$$

Now $\boldsymbol{xi}$ is in $\mathfrak{r}_0$; hence if the scalar product of $\boldsymbol{a}$ with every element $\boldsymbol{xi}$ of $\mathfrak{r}_0$ vanishes then $\mathrm{tr}(\boldsymbol{ax}) = 0$ without restriction on $\boldsymbol{x}$. It therefore follows that $\boldsymbol{a} = 0$, as asserted.

Proof of theorem (4.6). Let E_α be a basis for $\mathfrak{P}$, and let the similarity correspondence of $\mathfrak{P}$ on $\mathfrak{P}'$ send E_α into the basis E'_α for $\mathfrak{P}'$. Let $c_\alpha(i_1 \cdots i_f)$ be a given system of coefficients and write

$$\boldsymbol{c} = \sum_{\alpha, i} c_\alpha(i_1 \cdots i_f) \cdot \boldsymbol{E}_\alpha(i_1 \cdots i_f) \tag{4.10}$$

$$\boldsymbol{c}' = \sum_{\alpha, i} c_\alpha(i_1 \cdots i_f) \cdot \boldsymbol{E}'_\alpha(i_1 \cdots i_f).$$

The desired similarity correspondence between $\mathfrak{p}$ and $\mathfrak{p}'$ is naturally to be defined by $\boldsymbol{c} \to \boldsymbol{c}'$. However, this is only possible provided two systems of coefficients $c_\alpha(i_1 \cdots i_f)$ which define the same $\boldsymbol{c}$ also define the same $\boldsymbol{c}'$; or a system of coefficients which causes $\boldsymbol{c}$ to vanish must also cause $\boldsymbol{c}'$ to vanish.

We first remark that if a tensor F satisfies the equation

$$G \equiv \sum_s c(s^{-1}) \cdot \boldsymbol{s}F = 0$$

then also

$$G' \equiv \sum_s c'(s^{-1}) \cdot \boldsymbol{s}F = 0.$$

By (4.10)

$$c(s^{-1}) = \sum_{\alpha, k} \boldsymbol{s}\, c_\alpha(k_1 \cdots k_f) \cdot E_\alpha(k_1 \cdots k_f),$$

whence

$$G(i_1 \cdots i_f) = \sum_\alpha \sum_k c_\alpha(i_1 \cdots i_f;\ k_1 \cdots k_f) E_\alpha(k_1 \cdots k_f)$$

where

$$c_\alpha(i_1 \cdots i_f;\ k_1 \cdots k_f) = \sum_s s F(i_1 \cdots i_f) \cdot s c_\alpha(k_1 \cdots k_f).$$

These c_α define a symmetric transformation. Hence the given similarity transformation $\mathfrak{P} \to \mathfrak{P}'$, which sends E_α into E_α', sends G into G'. This proves our assertion that the vanishing of G implies the vanishing of G'.

If $\boldsymbol{c} = 0$ we then have

$$\sum_s c'(s^{-1}) \cdot \boldsymbol{s}F(i_1 \cdots i_f) = \mathrm{tr}[\boldsymbol{c}' \cdot \mathrm{F}(i_1 \cdots i_f)] = 0$$

for all tensors F and all combinations of indices $i_1 \cdots i_f$, or $\mathrm{tr}(\boldsymbol{c}'\boldsymbol{x}) = 0$ for all elements $\boldsymbol{x}$ of $\mathfrak{r}_0$. Hence by the auxiliary theorem (4.9) $\boldsymbol{c}' = 0$.

The result of our investigations is that there exists a one-to-one correspondence between the invariant sub-spaces $\mathfrak{p}$ *of* $\mathfrak{r}_0$ *and the invariant sub-spaces* $\mathfrak{P}$ *of* $\mathfrak{R}^f$. *This correspondence is as close as possible ; irreducibility, complete reduction, equivalence and inequivalence on the one hand imply the same on the other.* In particular, we emphasize the further consequence :

Theorem (4.11). *Every invariant sub-space* $\mathfrak{P}$ *of* $\mathfrak{R}^f$, *in particular* $\mathfrak{R}^f$ *itself, can be completely reduced into irreducible invariant sub-spaces.*

I hope that our elementary methods have made this correspondence quite apparent.

It is evident *a priori* that we can completely reduce the modulus $\mathbf{1}$ of the algebra ρ into a sum $\boldsymbol{e}_1 + \boldsymbol{e}_2 + \cdots + \boldsymbol{e}_m$ of independent primitive idempotent elements. The formula

$$F = \boldsymbol{e}_1 F + \boldsymbol{e}_2 F + \cdots + \boldsymbol{e}_m F$$

then gives the complete reduction of $\mathfrak{R}^f$ into independent invariant sub-spaces $\mathfrak{P}_1, \mathfrak{P}_2, \cdots, \mathfrak{P}_m$, each of which is generated by one of the idempotent operators $\boldsymbol{e}$. ($\mathfrak{P}_1$ consists of all tensors of the form $\boldsymbol{e}_1 F$.) From this point of view we might consider as the only non-trivial result of our investigation the assertion that the $\mathfrak{P}$ generated by a *primitive* $\boldsymbol{e}$ is *irreducible* (with respect to the algebra Σ of all symmetric transformations). Physically this means that the class of terms corresponding to such a $\mathfrak{P}$ cannot be further divided into parts which cannot under any conditions interact with each other. If in spite of this there does exist such a decomposition it is *accidental*—i.e. attributable to the special dynamical situation in the case in question.

§ 5. Fields and Algebras

We here interrupt our development in order to present an axiomatic treatment of the two fundamental concepts *field* and *algebra*; our investigation has revealed the importance of these concepts for quantum theory. The physicist who is not particularly interested in such a treatment may well omit these sections.

A ***field*** is a domain of elements, called ***numbers,*** within which the two operations of *addition* and *multiplication* are defined and which associate with any two numbers α, β of the field certain unique numbers $\alpha + \beta$, $\alpha\beta$ respectively. Addition obeys the commutative and associative laws

$$\alpha + \beta = \beta + \alpha, \quad (\alpha + \beta) + \gamma = \alpha + (\beta + \gamma)$$

and has a unique inverse, *subtraction.* From this follows the existence of a unique number o (zero) with the property $\alpha + o = o + \alpha = \alpha$ for all α. Further, associated with each number α is a number $-\alpha$, its negative, such that $\alpha + (-\alpha) = o$.

We require that multiplication obey the associative law

$$(\alpha\beta)\gamma = \alpha(\beta\gamma)$$

and the distributive laws

$$(\alpha + \beta)\gamma = (\alpha\gamma) + (\beta\gamma), \quad \alpha(\beta + \gamma) = (\alpha\beta) + (\alpha\gamma)$$

with respect to addition. From the distributive law follow the relations

$$\alpha o = o\alpha = o.$$

Multiplication need not be commutative; in case it is we speak of a ***commutative field.*** Further, division by any number other than o shall be possible and shall lead to a unique quotient, i.e. each of the equations

$$\alpha\xi = \beta, \quad \eta\alpha = \beta$$

have for given $\alpha \neq o$ and given β one and only one solution ξ, η respectively. From this it follows that the product $\alpha\beta$ of two numbers can only be o if one of the two factors is o. As a further consequence, there exists a number ε, "one" or "unity," with the property that

$$\alpha\varepsilon = \varepsilon\alpha = \alpha$$

for all α. We explicitly assume that not all numbers equal o; then in particular $\varepsilon \neq o$. Every number $\alpha \neq o$ possesses a unique reciprocal α^{-1} with the property $\alpha\alpha^{-1} = \alpha^{-1}\alpha = \varepsilon$.

We must introduce in addition to the numbers of our field the ordinary numerical symbols 1, 2, 3, $\cdots$. Their interpretation as multipliers is given by the equations

$$1\alpha = \alpha, \ 2\alpha = \alpha + \alpha, \ 3\alpha = (2\alpha) + \alpha, \ \cdots,$$

in general

$$(n + 1)\alpha = (n\alpha) + \alpha.$$

In particular we can construct the series

$$1\varepsilon, \ 2\varepsilon, \ \cdots, \ n\varepsilon, \ \cdots \tag{5.1}$$

of multiples of ε. We then have two possibilities. (1) All the numbers of this set may differ from ε; then they are all different, and we can conclude with the aid of the equation

$$n\beta = n\varepsilon \cdot \beta$$

and the division axiom that for a given number α there exists one and only one number $\beta = \frac{\alpha}{n}$ which satisfies the equation $n\beta = \alpha$; we can then introduce ordinary rational numbers as multipliers. (2) The second possibility is that one of the multiples in (5.1) is equal to ε itself; let the least multiple of this kind be $p\varepsilon$. Then the numbers of the series (5.1) repeat in cycles of length p. p must be a prime number, for if p were the product of two integers m, n smaller than p we would then have

$$o = p\varepsilon = m\varepsilon \cdot n\varepsilon,$$

but by assumption neither $m\varepsilon$ nor $n\varepsilon$ are o, for $p\varepsilon$ is the lowest multiple of this kind, and this is contrary to the division axiom. In this case we are dealing with *a finite field of* **modulus** *p*.[4]

In order not to lose ourselves in too broad generalities we now take as our number domain a *commutative field* and define a ***linear associative algebra*** *of finite order* over this field. By *number* we mean the elements of the field, and denote its zero o and its unit ε by 0 and 1; by ***element*** we mean an *element of the algebra*. We denote the former by small Greek and the latter by small Latin letters. An algebra is characterized by three fundamental operations: *addition* of two elements, $a + b$; *multiplication* of an element *by a number*, γa; *multiplication* of two elements, ab. The first and second of these operations obey the familiar axioms of vector calculus (I, § 1), which we set forth here again for the sake of completeness.

Addition is commutative and associative and has a unique

inverse, subtraction. It then follows that there exists a null-element o. Multiplication by a number obeys the laws

$$1a = a, \quad \alpha(\beta c) = (\alpha\beta)c,$$
$$(\alpha + \beta)c = (\alpha c) + (\beta c), \quad \alpha(b + c) = (\alpha\beta) + (\alpha c).$$

The ***order*** h is introduced by the dimensionality axiom: every $h + 1$ elements of the algebra are linearly dependent, the coefficients in the equations expressing the dependence being numbers of the field, but there exist h linearly independent elements. A set of h such elements $e_1, e_2, \cdots, e_h$, called "*basal units*," form a basis for the algebra in the sense that any element a can be expressed in one and only one way in the form

$$a = \alpha_1 e_1 + \alpha_2 e_2 + \cdots + \alpha_h e_h$$

and can be replaced by the set $(\alpha_1, \alpha_2, \cdots, \alpha_h)$ of h numerical components.

Multiplication of elements among themselves obeys the distributive laws

$$(a + b)c = (ac) + (bc), \quad c(a + b) = (ca) + (cb)$$

for both factors and the associative laws

$$\gamma a \cdot b = \gamma(ab), \quad b \cdot \gamma a = \gamma(ba),$$
$$(ab)c = a(bc)$$

We neither assume that multiplication is commutative nor that it possesses a unique inverse, division. But we do assume that the algebra possesses a "one," the *modulus* (or *principal unit*), i.e. an element e with the property $ae = ea = a$ for all elements a. We shall usually not hesitate to denote the zero and one of the elements of the algebra by 0 and 1.

If we assume the possibility of division the algebra reduces to a (in general *non-commutative*) *field* or ***division algebra*** of finite order h over the given field.

§ 6. Representations of Algebras

For the sake of the printer and in order to give the text a more peaceful appearance we no longer emphasize the elements of our algebra by expressing them in boldface type. This applies in particular to the elements of the algebra ρ of "symmetry quantities"—which we may often denote by this latter expression in case of possible confusion with the elements of the underlying group. We still employ this means of distinguishing between the tensor F and the symmetry element $\mathbf{F}$ or when we wish to consider an element as an operator acting on a tensor.

We start with an algebra ρ of finite order h, the elements of which constitute an h-dimensional vector space $\mathfrak{r}$, and associate with the element a of ρ the correspondence

$$(a): \quad x \rightarrow x' = ax$$

of $\mathfrak{r}$ on itself. We consider the algebra (ρ) of transformations (a), which is simply isomorphic with the algebra ρ, as fundamental for the vector space $\mathfrak{r}$, i.e. the term reducible, invariance, etc., as applied to sub-spaces of $\mathfrak{r}$ are with respect to the group of transformations (a). *We assume that* $\mathfrak{r}$ *can be completely reduced into irreducible sub-spaces* $\mathfrak{p}_1 + \mathfrak{p}_2 + \cdots$; each of these sub-spaces then contains an idempotent generating unit $e_1, e_2, \cdots$. We have already seen that this assumption is true for the algebra associated with any finite group—at least under the restriction that the field over which the algebra is defined does not have as modulus a prime number which is a factor of the order h of the group.

We discussed the *representations* of a group or of the corresponding algebra in Chapter III. We found that the irreducible representations are subject to certain important conditions which, surprisingly enough, limit their number and which, together with the as yet unproved "completeness theorem," lead to the reduction of the given algebra into independent simple matric algebras (III, § 13). That we were unable to prove the completeness theorem with the methods there employed was to be expected, for we assumed that the representations were *given* and examined their properties; we had no general process for the *construction* of representations of the given algebra. But we are now in possession of the materials for such a construction: the reduction of $\mathfrak{r}$ into irreducible sub-spaces $\mathfrak{p}_i$ reduces the regular representation into as many inequivalent irreducible representations of our algebra as there are inequivalent invariant sub-spaces $\mathfrak{p}_i$. We shall now carry out this construction process to the point of obtaining the reduction of our algebra into independent simple matric algebras; it will be desirable to derive the previous results again from this standpoint. A further difference between this investigation and that of Chapter III consists in the fact that we here refrain as long as possible from placing restrictive assumptions on the commutative field over which the algebra is defined; only at the end of the investigation do we discuss the advantages attributable to the fact that the continuum of complex numbers, the only field in which we are interested for the physical applications, is *algebraically closed.*

Theorem (6.1). *Every representation of the algebra ρ is completely reducible into irreducible representations. Each of these irreducible constituents is equivalent to the representation induced in some $\mathfrak{p}_i$ by the regular representation.*

(Hence the complete reducibility of the given algebra implies the complete reducibility of its representations. Further, every irreducible representation is contained in the regular representation, which therefore constitutes an appropriate starting point for obtaining all representations by the method of reduction).

Let $\mathfrak{H}$ be an n-dimensional representation, and let $\mathfrak{e}_1, \mathfrak{e}_2, \cdots, \mathfrak{e}_n$ be n fundamental vectors constituting a co-ordinate system in the representation space $\mathfrak{R}$ of $\mathfrak{H}$. If the element a of the algebra corresponds to the linear correspondence A in $\mathfrak{H}$, we interpret the equation

$$\mathfrak{x}' = a\mathfrak{x} \quad \text{as} \quad \mathfrak{x}' = A\mathfrak{x},$$

where $\mathfrak{x}'$, $\mathfrak{x}$ are vectors in $\mathfrak{R}$. If $\mathfrak{e}$ is a given fixed vector and x runs through all elements of one of the irreducible invariant sub-spaces $\mathfrak{p} = \mathfrak{p}_i$ of $\mathfrak{r}$ then, as we shall show immediately, $x\mathfrak{e}$ runs through a certain sub-space $\mathfrak{p}(\mathfrak{e})$ of $\mathfrak{R}$ which is invariant with respect to $\mathfrak{H}$. Indeed, the transformation A associated with an arbitrary element a sends $x\mathfrak{e}$ over into $(ax)\mathfrak{e}$, and if x is in $\mathfrak{p}$, ax is also. $\mathfrak{p}(\mathfrak{e})$ is either 0 or is similar to $\mathfrak{p}$ in the sense that different x generate different images $x\mathfrak{e}$, for those x of $\mathfrak{p}$ for which $x\mathfrak{e} = 0$ constitute an invariant sub-space $\mathfrak{p}'$ of $\mathfrak{p}$, and in virtue of the assumption that $\mathfrak{p}$ was irreducible $\mathfrak{p}'$ must either be 0 or $\mathfrak{p}$ itself. Hence if $\mathfrak{p}(\mathfrak{e}) \neq 0$ the representation induced in $\mathfrak{p}(\mathfrak{e})$ by $\mathfrak{H}$ is equivalent to the regular representation in $\mathfrak{p}$.

These considerations are to be supplemented by the following remark. If $\mathfrak{P}$ is any invariant sub-space of $\mathfrak{R}$ then $\mathfrak{p}(\mathfrak{e})$ is either independent of $\mathfrak{P}$ or is contained entirely in $\mathfrak{P}$, for those elements x of $\mathfrak{p}$ for which $x\mathfrak{e}$ lies in $\mathfrak{P}$ constitute an invariant sub-space of $\mathfrak{p}$, which is therefore necessarily either 0 or $\mathfrak{p}$ itself.

Now construct successively

$$\begin{array}{lll} \mathfrak{p}_1(\mathfrak{e}_1), & \mathfrak{p}_2(\mathfrak{e}_1), & \cdots, \\ \mathfrak{p}_1(\mathfrak{e}_2), & \mathfrak{p}_2(\mathfrak{e}_2), & \cdots, \\ \cdot\ \cdot\ \cdot & \cdot\ \cdot\ \cdot & \cdots, \\ \mathfrak{p}_1(\mathfrak{e}_n), & \mathfrak{p}_2(\mathfrak{e}_n), & \cdots; \end{array}$$

Each sub-space in this list is either entirely contained in the sum of the previous ones or is independent of this sum; on retaining only those sub-spaces for which this latter possibility is realized we obtain a reduction of $\mathfrak{R}$ into certain invariant sub-spaces $\mathfrak{p}_i(\mathfrak{e}_k)$. To prove this theorem we need only to note that the sum of the sub-spaces contained in the first row contains at least the vector $\mathfrak{e}_1$, that on adding to them the sum of those contained in the second row we obtain at least the vector $\mathfrak{e}_2$ in addition, etc.[5]

The theorem just proved is in particular applicable to the symmetric group π, and we now wish to establish the analogue for the algebra Σ of symmetric transformations in the space $\mathfrak{R}^f$ of tensors of order f. We already know that $\mathfrak{R}^f$ can be reduced into sub-spaces $\mathfrak{P}_i$ which are irreducible with respect to Σ (provided the number field over which Σ is defined does not have as modulus a prime $\leqq f$). Every transformation A of Σ is at the same time a transformation A_i of $\mathfrak{P}_i$ on itself and the correspondence $A \rightarrow A_i$ is naturally a representation of Σ, the " representation induced in $\mathfrak{P}_i$ by the algebra Σ." We wish to show that the representations of Σ are completely reducible into irreducible constituents, and that each of these constituents is equivalent to the representation induced in some $\mathfrak{P}_i$ by the algebra Σ. Naturally this does not follow immediately from theorem (6.1) ; in order to establish the connection between the two we must show that the complete reducibility of $\mathfrak{R}^f$ into irreducible invariant sub-spaces $\mathfrak{P}_i$ implies the same for the algebra Σ. We apply the notation and conventions given at the beginning of this section to the algebra Σ: (A) is the correspondence

$$S \rightarrow S' = AS$$

of the " vector space " Σ on itself, $A \rightarrow (A)$ the regular representation of Σ; the algebra of transformations (A), which is simply isomorphic with Σ, is taken as fundamental in the vector space Σ, i.e. the transformation group of Σ consists of the transformations (A).

Theorem (6.2). *Let Σ be an algebra of transformations in a vector space $\mathfrak{R}$, and let $\mathfrak{R}$ be completely reducible with respect to this system Σ of transformations into irreducible invariant sub-spaces $\mathfrak{P}_i$. Then Σ is itself completely reducible into irreducible invariant sub-spaces Π_j, and the representation induced by the regular representation in Π_j coincides with* (*more precisely, is equivalent to*) *the representation induced in one of the irreducible $\mathfrak{P}_i$ by the algebra Σ itself.*

This theorem holds without any restrictions on the field over which Σ is defined. Let Π be an irreducible invariant sub-space of Σ (consisting not merely of the transformation 0), and let $R \neq 0$ be a transformation of Π. There then exists a vector $\mathfrak{a}$ in $\mathfrak{R}$ such that $R\mathfrak{a} \neq 0$. Let $\mathfrak{a}$ be decomposed into its components $\mathfrak{a}_i$ in the various sub-spaces $\mathfrak{P}_i$; at least one of these components, say $\mathfrak{a}_i = \mathfrak{e}$, must be carried over into a vector $R\mathfrak{e} \neq 0$ by R. We now hold $\mathfrak{e}$ fixed and let S in $\mathfrak{s} = S\mathfrak{e}$ run through all transformations of Π; these $\mathfrak{s}$ then constitute an invariant sub-space $\Pi(\mathfrak{e})$ of $\mathfrak{P} = \mathfrak{P}_i$. The "typical reasoning" already applied in the proof of the previous theorem then allows us to conclude that:

(1) $\Pi(\mathfrak{e})$ is either 0 or $\mathfrak{P}$, as $\mathfrak{P}$ is irreducible; in this case it is necessarily $\mathfrak{P}$, for the vector $R\mathfrak{e} \neq 0$ belongs to $\Pi(\mathfrak{e})$.

(2) $S = 0$ is the only transformation in Π which sends $\mathfrak{e}$ over into 0, for those S of Π for which $S\mathfrak{e} = 0$ constitute an invariant sub-space of the irreducible sub-space Π. Hence $\mathfrak{s} = S\mathfrak{e}$ sets up a one-to-one correspondence between Π and $\mathfrak{P}$.

This correspondence is similar, for $S' = AS$ implies that the vectors $\mathfrak{s} = S\mathfrak{e}$, $\mathfrak{s}' = S'\mathfrak{e}$ satisfy the equation $\mathfrak{s}' = A\mathfrak{s}$. We have thus proved the second part of our theorem: the representation induced in Π by the regular representation coincides with the representation induced in $\mathfrak{P}$ by the algebra itself; briefly, Π is similar to some $\mathfrak{P}_i$.

Since $S\mathfrak{e}$ runs through the entire sub-space $\mathfrak{P}$ when S runs through Π there exists an E in Π such that $E\mathfrak{e} = \mathfrak{e}$; then $E^2\mathfrak{e} = \mathfrak{e}$. Since the transformations E and E^2 of Π both associate the same image with $\mathfrak{e}$ they are identical: E is idempotent. Hence Σ can be completely reduced into two independent sub-spaces $\Pi + \Sigma'$ in accordance with the formula

$$S = SE + (S - SE).$$

[Cf. the proof of Theorem (3.3).] Successive application of this procedure leads to the complete reduction of Σ into its constituents Π_j.

Having proved Theorem (6.2), we obtain from Theorem (6.1), under the same assumptions, the further theorem:

Theorem (6.3). *Every representation of Σ is completely reducible into irreducible representations. Every irreducible representation of Σ coincides with the representation $A \to A_i$ induced in some $\mathfrak{P}_i$ by the algebra Σ itself.*

Theorem (6.1) yields the further (rather uninteresting) fact that not only is every Π_j similar to some $\mathfrak{P}_i$, but also conversely every $\mathfrak{P}_i$ is similar to some Π_j.

As has already been indicated, all of these results are applicable to the algebra of symmetric transformations in tensor space $\mathfrak{R}^f$. But we have shown in § 1 that this algebra can be replaced by the group $(\mathfrak{c})^f$ induced in tensor space by the group $\mathfrak{c}$ of linear transformations

$$x_i' = \sum_{k=1}^{n} a(ik)x_k \quad [\det [a(ik)] \neq 0] \tag{1.3}$$

of n-dimensional vector space, i.e. by the representation $(\mathfrak{c})^f$ of $\mathfrak{c}$. We shall say that a representation of $\mathfrak{c}$ is of order f if the components of the matrix A, which corresponds to the element (1.3) of the group, are rational integral functions of the $a(ik)$ of order f. Our theorem then asserts :

Theorem (6.4). *Every* f^{th} *order representation of* $\mathfrak{c}$ *is completely reducible into irreducible representations, and every irreducible representation of order* f *of* $\mathfrak{c}$ *is contained in the representation* $(\mathfrak{c})^f$.

This theorem is still valid on restricting the affine group $\mathfrak{c}$ to its unitary sub-group $\mathfrak{u}$. (Naturally the concept " unitary " implies that we are then no longer dealing with an arbitrary field, but are operating in the field of all complex numbers.)

§ 7. Constructive Reduction of an Algebra into Simple Matric Algebras

We again assume that the algebra ρ of order h, which may at the same time be considered as a vector space $\mathfrak{r}$ of h dimensions, is completely reducible into irreducible invariant sub-spaces $\mathfrak{p}_i$. The generating units e_i of these irreducible $\mathfrak{p}_i$ are obtained by the corresponding reduction of the modulus ; we can then express an arbitrary element x of $\mathfrak{r}$ as the sum of its components in the various $\mathfrak{p}_i$:

$$1 = \sum_i e_i \quad (e_i \text{ in } \mathfrak{p}_i), \quad x = \sum_i x e_i. \tag{7.1}$$

If $\mathfrak{q}$ is a sub-space of $\mathfrak{r}$ we denote by $\mathfrak{q}a$ the totality of elements of the form xa where x runs through all elements of $\mathfrak{q}$; e, with or without index, is an idempotent element, usually primitive ; $\mathfrak{p} = \mathfrak{r}e$ the invariant sub-space generated by e ; $\mathfrak{h}$ the representation of ρ induced in $\mathfrak{p}$ by the regular representation.

We could consider in addition to the reduction (7.1) of $\mathfrak{r}$ into left-invariant sub-spaces the analogous reduction into right-invariant sub-spaces by means of the equation

$$x = \sum_i e_i x.$$

But the most complete separation into mutually independent components is obtained by carrying out both of these processes simultaneously :

$$x = \sum_{i,k} e_i x e_k = \sum_{i,k} x_{ik}. \tag{7.2}$$

The elements of the form $e_i x e_k$ are those of character (e_i, e_k), or briefly (ik). Let $\mathfrak{p}_{ik}$ be the sub-space consisting of all elements of this character. The various $\mathfrak{p}_{ik}$ are independent and the entire $\mathfrak{r}$ is reduced into the sum of the $\mathfrak{p}_{ik}$; the original left-invariant $\mathfrak{p}_k = \sum_i \mathfrak{p}_{ik}$. The important properties of $\mathfrak{p}_{ik}$ are given by the following:

Auxiliary Theorem (7.3). I. *If $\mathfrak{p}$, $\mathfrak{p}'$ are two inequivalent irreducible sub-spaces with generating units e, e', all elements of character (e, e') are $= 0$.*

II. *The elements of character (e, e) constitute a field or division algebra which is simply isomorphic with the system of similar projections of $\mathfrak{p}$ on itself.*

Proof. I. Let a be any element of character (e, e'). The transformation

$$[a] : x \rightarrow x' = xa \tag{7.4}$$

carries every element x of $\mathfrak{p}$ over into an element x' of $\mathfrak{p}'$ and defines a similar projection. Conversely, we know that any similar projection of $\mathfrak{p}$ on $\mathfrak{p}'$ is defined by an equation of this form, and that the generating element a of character (e, e') is uniquely determined by the projection. If $\mathfrak{p}$ and $\mathfrak{p}'$ are irreducible our "typical reasoning" leads us to the two usual alternatives: either the projection associates with every element x of $\mathfrak{p}$ the image $x' = 0$ or it defines a one-to-one correspondence of $\mathfrak{p}$ on $\mathfrak{p}'$. The equation $ea = a$ tells us that the first alternative is possible only if $a = 0$, and the second implies that $\mathfrak{p}$ and $\mathfrak{p}'$ are equivalent.

II. The above remarks are applicable to an element a of character (e, e) and the similarity projection of $\mathfrak{p}$ on itself which it generates. If $\mathfrak{p}$ is irreducible every such projection, except the one defined by $a = 0$, is one-to-one and consequently has an inverse. But the existence of an inverse is identical with the *possibility of division.* The isomorphism asserted in the theorem is apparent on reversing our usual procedure, and reading the resultant of two or more correspondences from left to right, for the resultant of the correspondences

$$x' = xa, \quad x'' = x'a'$$

is given by

$$x'' = x(aa').$$

We now proceed with the help of this auxiliary theorem as follows: Arrange the $\mathfrak{p}_i$ into classes of equivalent sub-spaces with generating units

$$e_1', \cdots, e_r'; \; e_1'', \cdots, e_s''; \; \cdots$$

and add together the generating units in each of these classes:

$$e'_1 + \cdots + e'_r = \varepsilon', \quad e''_1 + \cdots + e''_s = \varepsilon'', \cdots.$$

We then have

$$1 = \varepsilon' + \varepsilon'' + \cdots \tag{7.5}$$

$$\mathfrak{r} = \mathfrak{r}' + \mathfrak{r}'' + \cdots \tag{7.6}$$

where $\mathfrak{r}', \mathfrak{r}'', \cdots$ denote the inequivalent sub-spaces $\mathfrak{r}\varepsilon', \mathfrak{r}\varepsilon'', \cdots$ into which $\mathfrak{r}$ is reduced.

Part I of the auxiliary theorem above then tells us that, for example,

$$\varepsilon' x \varepsilon'' = 0.$$

Hence the product $a'a''$ of two elements belonging to different sub-spaces $\mathfrak{r}', \mathfrak{r}''$ is always 0, and the reduction

$$a = a' + a'' + \cdots = a\varepsilon' + a\varepsilon'' + \cdots$$

leads to the multiplication rule

$$ab = a'b' + a''b'' + \cdots.$$

From this it follows that $\mathfrak{r}'$ is both right- and left-invariant and *a fortiori* constitutes an algebra ρ' (" invariant sub-algebra "); ε' is the modulus of ρ'. *The given algebra is then the direct sum of the simple algebras* $\rho', \rho'', \cdots$, where the precise meaning of *direct sum* is defined by the following:

Let $\rho', \rho'', \cdots$ be algebras (defined over the same field), and consider as the elements of a new algebra ρ, *the* ***direct sum*** *of* $\rho', \rho'', \cdots$, all sets

$$a = (a', a'', \cdots)$$

consisting of an arbitrary element a' of ρ', an arbitrary a'' of $\rho'', \cdots$. The fundamental operations in ρ are defined by

$$(a', a'', \cdots) + (b', b'', \cdots) = (a' + b', a'' + b'', \cdots),$$
$$\lambda(a', a'', \cdots) = (\lambda a'; \lambda a'', \cdots),$$
$$(a', a'', \cdots)(b', b'', \cdots) = (a'b', a''b'', \cdots)$$

where λ is any number.

Note that the *central* of the algebra ρ obtained by direct summation is the direct sum of the centrals of the individual algebras $\rho', \rho'', \cdots$.

We investigate in detail *one* of these simple sub-algebras, say ρ', which we now denote simply by ρ; its modulus ε' may now be denoted by 1. On omitting the primes, the decomposition of 1 into *equivalent* primitive idempotent elements e_i is expressed by

$$1 = e_1 + e_2 + \cdots + e_r.$$

Every element a of ρ is reduced in accordance with the formula (double *Peirce* reduction)

$$a = \sum_{i,k=1}^{r} a_{ik} = \sum_{i,k} (e_i a e_k)$$

into components of characters (ik). The component c_{ik} of the product $c = ab$ is easily seen to be expressed in terms of the components a_{ik}, b_{ik} of a and b by the equation

$$c_{ik} = \sum_{j=1}^{r} a_{ij} b_{jk}.$$

We have thus already obtained the connection between our considerations and the matrix calculus.

The invariant sub-spaces $\mathfrak{p}_1, \mathfrak{p}_2, \cdots, \mathfrak{p}_r$ generated by the $e_1, e_2, \cdots, e_r$ are all equivalent. Let $\mathfrak{p}$ be any of these classes, e.g. $\mathfrak{p} = \mathfrak{p}_1$, and let Γ_i be any fixed one-to-one similarity correspondence of $\mathfrak{p}_i$ on $\mathfrak{p}$. In accordance with (7.4) any element

$$a = a_{ik} = e_i a e_k$$

of character (e_i, e_k) generates a similarity projection $[a]$ of $\mathfrak{p}_i$ on $\mathfrak{p}_k$; this projection can be written in the form

$$[a] = \Gamma_i \alpha \Gamma_k^{-1} \tag{7.7}$$

where α is a similarity projection of $\mathfrak{p}$ on itself. But by Part II of the auxiliary theorem proved above the similarity projections of $\mathfrak{p}$ on itself constitute a field (division algebra) Φ which is simply isomorphic with the set of elements of character (e, e). If Φ is of order v each of the r left-invariant sub-spaces

$$\mathfrak{p}_k = \sum_{i=1}^{r} \mathfrak{p}_{ik}$$

is of dimensionality $g = r \cdot v$. *The number of times r an irreducible representation occurs in the regular representation is accordingly a factor of the dimensionality g of the representation.*

Any element a can be reduced into its components a_{ik}, which may be any elements of the independent sub-spaces $\mathfrak{p}_{ik}$. In accordance with (7.7)

$$[a_{ik}] = \Gamma_i \alpha_{ik} \Gamma_k^{-1} \tag{7.8}$$

and a_{ik} may be replaced by the corresponding element α_{ik} of the field Φ. Since conversely any such element α_{ik} is by (7.8) associated with a similarity projection $[a_{ik}]$ of $\mathfrak{p}_i$ on $\mathfrak{p}_k$, and therefore with a definite element a_{ik} of character (ik), we obtain a one-to-one reciprocal correspondence between the totality of all elements a of the simple algebra ρ and the totality of matrices

$$\left\|\begin{matrix} \alpha_{11} & \alpha_{12} & \cdots & \alpha_{1r} \\ \alpha_{21} & \alpha_{22} & \cdots & \alpha_{2r} \\ \cdot & \cdot & \cdot & \cdot \\ \alpha_{r1} & \alpha_{r2} & \cdots & \alpha_{rr} \end{matrix}\right\| \tag{7.9}$$

of order r whose components α_{ik} are elements of the field Φ. The correspondence is such that to the three fundamental operations of the one (addition of elements, multiplication of an element by a number and multiplication of two elements) correspond to the same operations of the other. Note that in particular

$$[a_{ij}b_{jk}] = [a_{ij}][b_{jk}] = \Gamma_i\alpha_{ij}\Gamma_j^{-1} \cdot \Gamma_j\beta_{jk}\Gamma_k^{-1}$$
$$= \Gamma_i \cdot \alpha_{ij}\beta_{jk} \cdot \Gamma_k^{-1}.$$

We have thus proved :

Wedderburn's Theorem.[6] *Any of the simple algebras, whose direct sum constitutes the given algebra* ρ, *is simply isomorphic with a simple matric algebra in a certain field (division algebra)* Φ *defined over the field of the original algebra.*

(*Remark.* The invariant sub-space $\mathfrak{p}_k$ consists of all elements a such that the matrix $\|\alpha_{ik}\|$ has as its only non-vanishing column the kth. The element e_i is then described by that diagonal matrix all of whose components vanish except the one occupying the ith place, which is 1.)

It is readily seen that the central of the simple algebra ρ consists of those elements whose matrix (7.9) is of the form

$$\left\|\begin{matrix} \alpha & 0 & \cdots & 0 \\ 0 & \alpha & \cdots & 0 \\ \cdot & \cdot & \cdot & \cdot \\ 0 & 0 & \cdots & \alpha \end{matrix}\right\|$$

where α belongs to the central of the field Φ.

Our construction was divided into two steps. First $\mathfrak{r}$ was completely reduced into the sub-spaces $\mathfrak{r}'$, $\mathfrak{r}''$, $\cdots$ which are both right- and left-invariant and then these were further reduced into the left-invariant sub-spaces $\mathfrak{p}_i$. We must now return to the consideration of the first step. On multiplying $x\varepsilon'$ on the left by (7.5) we find

$$x\varepsilon' = \varepsilon' x\varepsilon',$$

and on multiplying $\varepsilon' x$ on the right by the same factor

$$\varepsilon' x = \varepsilon' x\varepsilon'.$$

Hence

$$x\varepsilon' = \varepsilon' x\,;$$

the $\varepsilon', \varepsilon'', \cdots$ *commute with all elements* and belong to the *central* of the algebra. The sub-spaces $\mathfrak{r}' = \rho'$, $\mathfrak{r}'', \cdots$ are both right- and left-invariant in the sense that neither the transformation $x' = xa$ nor $x' = ax$ leads out of them, and they are furthermore *irreducible* in this respect—indeed, it is for this reason we call them " simple." In order to show this we proceed as follows :

(7.10). If $\mathfrak{r}_0$ is a sub-space which is both right- and left-invariant then either e_i is contained in $\mathfrak{r}_0$ or $\mathfrak{r}_0 e_i = 0$. For $\mathfrak{r}_0 e_i$ is an invariant sub-space of the irreducible $\mathfrak{p}_i$ and is therefore either 0 or $\mathfrak{p}_i$ itself. In the second case we have

$$\mathfrak{p}_i = \mathfrak{r}_0 e_i \prec \mathfrak{r}_0$$

since $\mathfrak{r}_0$ is right-invariant ; hence e_i is contained in $\mathfrak{r}_0$.

(7.11). If e_i is in $\mathfrak{r}_0$ the same is true of any e which is equivalent to e_i. For the similarity projection $x' = xb$ of $\mathfrak{p}_i$ on $\mathfrak{p}$ associates e with some element a_i of $\mathfrak{p}_i$ by means of the equation $e = a_i b$, and since a_i is in $\mathfrak{r}_0$ e is also.

(7.12). If $\mathfrak{r}_0 \prec \mathfrak{r}'$ then since $\mathfrak{r}_0 = \sum_i \mathfrak{r}_0 e_i'$ not all the $\mathfrak{r}_0 e_i'$ can be empty, i.e. one of the e_i' must occur in $\mathfrak{r}_0$. But they must then all occur in $\mathfrak{r}_0$, hence also $\varepsilon' = \sum_i e_i'$, and consequently $\mathfrak{r}_0 = \mathfrak{r}'$.

(7.13). Again let $\mathfrak{r}_0$ be a right- and left-invariant sub-space. Then either $\mathfrak{r}_0 \varepsilon' = \mathfrak{r}'$ or it is empty ; in the former case ε' is in $\mathfrak{r}_0$. It follows from

$$\mathfrak{r}_0 = \mathfrak{r}_0 \varepsilon' + \mathfrak{r}_0 \varepsilon'' + \cdots$$

that $\mathfrak{r}_0$ is necessarily the sum of certain of the spaces $\mathfrak{r}', \mathfrak{r}'', \cdots$; when in particular $\mathfrak{r}_0$ is irreducible in the sense of right- and left-invariance it must coincide with one of the $\mathfrak{r}', \mathfrak{r}'', \cdots$. Hence the reduction (7.6) is unique. This further shows that every right- and left-invariant sub-space $\mathfrak{r}_0$ possesses a generating unit i which belongs to the central of the algebra, and that $\mathfrak{r}$ can be completely reduced into $\mathfrak{r}_0$ and a supplementary right- and left-invariant sub-space.

(7.14). If $\mathfrak{p}$ is an irreducible (left-) invariant sub-space with the generating unit e, then $\mathfrak{p}\varepsilon'$ is invariant, and since $\mathfrak{p}\varepsilon' = \varepsilon'\mathfrak{p}$ it is either 0 or $\mathfrak{p}$ itself. Since

$$\mathfrak{p} = \mathfrak{p}\varepsilon' + \mathfrak{p}\varepsilon'' + \cdots$$

the equation $\mathfrak{p}\varepsilon = \mathfrak{p}$ must hold for some *one* of the $\varepsilon', \varepsilon'', \cdots$, while for all others $\mathfrak{p}\varepsilon = 0$. We then say that ε *belongs to* $\mathfrak{p}$

and that conversely e or $\mathfrak{p}$ *belongs to* $\mathfrak{e}$. $\mathfrak{p}$ is a sub-space of the right- and left-invariant $\mathfrak{r}\,\mathfrak{e}$.

An algebra $\rho = \mathfrak{r}$, concerning which we only assume that it is completely reducible into irreducible invariant sub-spaces $\mathfrak{p}_i$, is necessarily obtainable by successive application of the following processes:

(A) Construction of a field;

(B) Transition to matrices: we take as elements the matrices of a fixed order r whose components are arbitrary elements of the field;

(C) Direct summation.

The processes (B) and (C) are formally completely determined and are therefore of an elementary character. Hence the construction of algebras is reduced to the construction of fields, i.e. of special algebras in which division is possible ("division algebras").

The converse is naturally also true: any algebra constructed by the three steps (A), (B) and (C) is completely reducible, for:

(A) If the algebra $\mathfrak{r}$ is itself a field, $\mathfrak{r}$ is itself an irreducible sub-space of $\mathfrak{r}$. For if α is any non-null element of the field then $\xi\alpha$ runs through the entire field with ξ; this is merely the content of the division axiom.

(B) The matrices (7.9) in which all components of every column except the ith vanish constitute the irreducible sub-space $\mathfrak{p}_i$, and the space $\mathfrak{r}$ of all matrices is the sum of these $\mathfrak{p}_i$. $\mathfrak{p}_i$ is irreducible; to show this we must prove that if a is any element in $\mathfrak{p}_i$ then any element of $\mathfrak{p}_i$ can be expressed in the form xa. a as well as $a' = xa$ has as its only non-vanishing column the ith; dropping the last index i, we denote these two columns by

$$(\alpha_1, \alpha_2, \cdots, \alpha_r), \quad (\alpha_1', \alpha_2', \cdots, \alpha_r'),$$

respectively. The equation $a' = xa$ is then

$$\alpha_i' = \sum_{k=1}^{r} \xi_{ik}\alpha_k;$$

we are therefore concerned with proving the theorem that any non-vanishing "vector" $(\alpha_1\alpha_2 \cdots \alpha_r)$ can be transformed into any given "vector" $(\alpha_1'\alpha_2' \cdots \alpha_r')$ by an appropriate linear correspondence. Since not all the α_k vanish take one of them, say α_2, which does not vanish and let all ξ_{ik} for which $k \neq 2$ be 0; ξ_{i2} is then to be determined by the equation

$$\alpha_i' = \xi_{i2}\alpha_2;$$

that this is possible is guaranteed by the division axiom.

(C) The assertion is self-evident for this step.

In general only the first step, (A), does not lend itself to an exhaustive formal treatment. *However, if the field over which the field ("division algebra") referred to in* (A) *is defined is algebraically closed* this step becomes extremely simple:

The only division algebra of finite order over an algebraically closed field is this field itself.

Proof. Consider an algebra of order v defined over an algebraically closed field. If a is an element of the algebra there must exist a linear dependence between the $v + 1$ powers $a^v, a^{v-1}, \cdots, a, 1$, i.e. a linear relation whose coefficients are numbers of the field. Hence a satisfies an algebraic equation of degree $m \leqq v$:

$$f(\lambda) = \lambda^m + \gamma_1\lambda^{m-1} + \cdots + \gamma_m$$
$$f(a) = a^m + \gamma_1 a^{m-1} + \cdots + \gamma_m 1 = 0.$$

Since the field is algebraically closed $f(\lambda)$ can be expressed as the product of linear factors:

$$f(\lambda) = (\lambda - \alpha_1)(\lambda - \alpha_2) \cdots (\lambda - \alpha_m).$$

Correspondingly

$$(a - \alpha_1 1)(a - \alpha_2 1) \cdots (a - \alpha_m 1) = 0. \tag{7.15}$$

We now introduce the assumption that the algebra of order v is a division algebra; then the product of two or more elements can vanish only if one of the factors is 0. Hence we may conclude from (7.15) that $a = \alpha_i 1$ for some i; the algebra then consists of the products of the modulus 1 with any number of the fundamental field, and therefore the algebra itself is simply isomorphic with this field.

If we are dealing in the field of all complex numbers the auxiliary theorem (7.3) can be replaced, in accordance with the above, by the more definite:

(7.3′). *All elements of the form $ex'e$ are zero if the primitive idempotent elements e, e' are inequivalent. If they are equivalent all such elements are multiples of one of them (which is different from* 0).

Further: *The number of times an irreducible representation appears in the regular representation is not merely a factor of the dimensionality of the representation; it is actually equal to it.* Our analysis has thus revealed the true source of this remarkable fact.

Under these circumstances the given ("semi-simple") algebra is the direct sum of simple matric algebras over the original field.

We obtain a complete set of basal units $e'_{ik}, e''_{\iota\kappa}, \cdots$:

$$a = \sum_{ik} \alpha'_{ik} e'_{ik} + \sum_{\iota\kappa} \alpha''_{\iota\kappa} e''_{\iota\kappa} \cdots, \tag{7.16}$$

for the algebra; these basal units satisfy the multiplication law of "matrix units," i.e. products of the type

$$e'_{il} e'_{lk} = e'_{ik} \tag{7.17}$$

and all others vanish. The correspondences

$$a \to \|\alpha'_{ik}\|, \quad a \to \|\alpha''_{\iota\kappa}\|, \quad \cdots$$

are the inequivalent irreducible representations $\mathfrak{h}', \mathfrak{h}'', \cdots$. The basal units $e'_{ii}, e''_{\iota\iota}, \cdots$ are the generating units $e'_i, e''_\iota, \cdots$ of the irreducible sub-spaces $\mathfrak{p}_i$ with which we began our construction. e'_{ik} is the element of character (ik) generated by the correspondence $\Gamma_i\Gamma_k^{-1}$ of $\mathfrak{p}_i$ on $\mathfrak{p}_k$, i.e. that element which this correspondence associates with e'_i.

After having obtained the irreducible representations in this constructive way we derive their *orthogonality properties* again from our present standpoint. For the moment let the ***trace*** *of* a denote the trace of the correspondence

$$x \to y = ax \tag{7.18}$$

of $\mathfrak{r}$ on itself which is associated with a in the regular representation. In terms of the co-ordinate system defined by the basal units above this correspondence becomes

$$\eta'_{ik} = \sum_{j=1}^{g'} \alpha'_{ij} \xi'_{jk}, \quad \cdots.$$

Each of the g' columns of variables

$$\xi'_{1k}, \xi'_{2k}, \cdots, \xi'_{g'k} \qquad (k = 1, 2, \cdots, g')$$

undergoes the transformation with matrix $\|\alpha'_{ij}\|$; the trace of a is accordingly

$$g' \cdot \sum_{i=1}^{g'} \alpha'_{ii} + \cdots.$$

By (7.16) this is equivalent to the equations

$$\operatorname{tr}(e'_{ik}) = \begin{cases} 0 & (i \neq k) \\ g' & (i = k) \end{cases}, \quad \cdots$$

for the basal units. Hence by (7.17)

$$\operatorname{tr}(e'_{ik} e'_{ki}) = g', \quad \cdots \tag{7.19}$$

and all other types of products of basal matric units have a vanishing trace.

If the algebra is the algebra of a group of order h the correspondence (7.18) is expressed in the original co-ordinate system, consisting of the elements $\mathbf{s}$ associated with the elements s of the group, by the equation

$$y(s) = \sum_t a(st^{-1})x(t).$$

From this it follows that the trace, as defined above, of a is equal to $h \cdot a(\mathbf{1})$; but in the case of a group algebra we have previously called $a(\mathbf{1})$ itself, without the factor h, the *trace* of a. On returning to this original definition of the trace we need merely to replace the right-hand side g' of the orthogonality relations (7.19) by g'/h. Equation (7.16) may now be solved explicitly for the coefficients:

$$\alpha'_{ik} = \frac{h}{g'}\mathrm{tr}\,(ae'_{ki}) = \frac{h}{g'} \cdot \sum_s a(s) \cdot e'_{ki}(s^{-1}). \tag{7.20}$$

The connection with the development in Chapter III, § 13, is obtained by noting that the

$$u'_{ik}(s) = \frac{h}{g'} \cdot e'_{ki}(s^{-1}) \tag{7.21}$$

are the components of the matrix $U'(s)$ associated with the element s of the group in the irreducible representation $\mathfrak{h}'$. The character of $\mathfrak{h}'$ is therefore

$$\chi'(s) = \frac{h}{g'} \cdot \varepsilon'(s^{-1}) \tag{7.22}$$

and (7.19) yields the orthogonality relations for the representations.

We have thus arrived at a constructive formulation of the theory, in which the fundamental concepts involved in and the range of validity of each step are clearly apparent. It supplies us with a constructive method for obtaining a complete set of irreducible representations, as well as establishing the orthogonality relations.

Additional remark. In dealing with the continuum of all complex numbers and a group algebra defined over this field we can, in accordance with the remark at the end of § 3, completely reduce the modulus 1 into *real* primitive e_i and the space $\mathfrak{r}$ into the corresponding *unitary-orthogonal* irreducible $\mathfrak{r}_i$. Further, the projections Γ_i can be normalized in such a way that e'_{ki} is conjugate to e'_{ik}. To show this we note that the conjugate of e'_{ik} is under these conditions an element of character (ki) and must therefore be the product of e'_{ki} by a number γ_{ik}:

$$\tilde{e}'_{ik} = \gamma_{ik} \cdot e'_{ki}. \tag{7.23}$$

The rules

$$e'_{ik}e'_{kl} = e'_{il}, \quad \tilde{e}'_{il} = \tilde{e}'_{kl}\tilde{e}'_{ik}$$

yield the conditions

$$\gamma_{ik}\gamma_{kl} = \gamma_{il}, \quad \gamma_{ii} = 1$$

on the coefficients. Further, γ_{ik} is real and positive, for from (7.23) and (7.19) we find

$$\sum_s |e'_{ik}(s)|^2 = \mathrm{tr}(e'_{ik}\tilde{e}'_{ik}) = \frac{g'}{h} \cdot \gamma_{ik}.$$

We then find that the γ_{ik} can be brought into the form $\gamma_{ik} = \beta_k^2/\beta_i^2$, where the β_i are positive real numbers (take, for example, $\beta_i^2 = \gamma_{1i}$). On replacing the original correspondences Γ_i by $\beta_i\Gamma_i$ we find that the new e'_{ki} is actually conjugate to the new e'_{ik}. Our representations $\mathfrak{h}'$, $\mathfrak{h}''$, $\cdots$ are accordingly thrown into unitary form.

B. Extension of the Theory and Physical Applications

§ 8. The Characters of the Symmetric Group and Equivalence Degeneracy in Quantum Mechanics

The notation employed in this section is as follows: $\pi = \pi_f$ is the symmetric permutation group of f things, $\mathfrak{r} = \rho = (\pi)$ the corresponding algebra, e a (primitive) idempotent element of ρ, $\mathfrak{p} = \mathfrak{r}e$ the (irreducible) invariant sub-space of $\mathfrak{r}$ generated by e, $\mathfrak{h}$ the representation induced in $\mathfrak{p}$ by the regular representation, g the dimensionality of $\mathfrak{p}$ and $\mathfrak{h}$, χ the character of $\mathfrak{h}$, ε that element of the set ε', ε'', $\cdots$ (7.14) to which the irreducible $\mathfrak{p}$ belongs; $\mathfrak{P}$ the corresponding symmetry class of tensors of order f, consisting of all tensors of the form $\hat{e}F$, $\mathfrak{H}$ the representation of the algebra Σ of symmetric transformations (and therefore of the linear group $\mathfrak{c}$) which is induced in $\mathfrak{P}$ by Σ itself. When further differentiation is necessary, we also denote this $\mathfrak{H}$ by $\mathfrak{H}(\chi)$ or $\mathfrak{H}_n(\chi)$. In case the considerations are valid for an arbitrary finite group π, h denotes the order of π ($= f!$ for π_f).

Determination of the Group Characters.

We begin by calculating the character of the representation $\mathfrak{h}$. To this end we construct the trace of the linear correspondence

$$x \to y = ax \tag{8.1}$$

of $\mathfrak{p}$ on itself; the considerations of the previous section show it to be

$$\sum_s a(s)\chi(s).$$

Now consider instead of (8.1) the projection

$$x \to y = axe \tag{8.2}$$

of the total space $\mathfrak{r}$ on $\mathfrak{p}$; it coincides with (8.1) within $\mathfrak{p}$ and sends *any* element x of $\mathfrak{r}$ into an element y of $\mathfrak{p}$. On choosing the co-ordinate system in $\mathfrak{r}$ in such a way that the first g fundamental vectors span the sub-space $\mathfrak{p}$, the last $h - g$ rows of the matrix of (8.2) consist only of zeros; hence the trace of the projection (8.2) of the total group space is equal to the trace of the correspondence (8.1) in $\mathfrak{p}$. In terms of components equation (8.2) is

$$y(s) = \sum a(t)x(s')e(t'), \qquad (ts't' = s)$$

and the trace is therefore

$$\sum_s \sum a(t)e(t')$$

where the inner sum is extended over the pairs t, t' of elements of the group which satisfy the equation $tst' = s$, or explicitly, the trace is

$$\sum_t \{a(t)\sum_s e(s^{-1}t^{-1}s)\}.$$

Hence the character χ of $\mathfrak{h}$ is given by

$$\chi(t) = \sum_s e(s^{-1}t^{-1}s)$$

or

$$\boxed{\chi(s) = \sum_r e(rs^{-1}r^{-1}).} \tag{8.3}$$

In particular, the dimensionality g of the representation $\mathfrak{h}$ (and the space $\mathfrak{p}$) is

$$\chi(\mathbf{1}) = h \cdot e(\mathbf{1}).$$

Resonance or Equivalence Degeneracy.

The significance of our results for quantum mechanics, as first recognized by *Wigner*, is the following.[7] The complete reduction of the tensor space $\Re^f$ into invariant sub-spaces $\mathfrak{P}_i$ implies a separation of the terms of the physical system I^f, consisting of f equivalent individuals I (electrons), into sets of terms which no dynamical influence whatever can cause to enter into combination with each other. We have further seen that the reduction of $\mathfrak{R}^f$ into the $\mathfrak{P}_i$ parallels the complete reduction of the total group space $\mathfrak{r}$ of the symmetric permutation group π into invariant sub-spaces $\mathfrak{p}_i$. *Hence there is a system of terms associated with every irreducible representation* $\mathfrak{h}$ *of* π—which we denote simply as the term system χ, using the character χ of $\mathfrak{h}$ as a name for the system—and the multiplicity of this term system is the number $m(\chi)$ of times that $\mathfrak{h}$ occurs in the regular representation. This suffers a slight modification in case $n < f$, for we must then ignore all $\mathfrak{p}_i$ which are not contained in $\mathfrak{r}_0 = \natural\mathfrak{R}^f$. But since $\mathfrak{r}_0$ is both right- and left-invariant, all sub-spaces which are equivalent to an irreducible invariant $\mathfrak{p}$ lying in $\mathfrak{r}_0$ are also in $\mathfrak{r}_0$. Hence the multiplicity of the term system χ is $m(\chi)$ or 0 according as that $\mathfrak{e}$ with which the character χ is associated by (7.22) is in $\mathfrak{r}_0$ or not. From the physical standpoint, the only additional fact of interest obtained from the more extended theory built up on the assumption that the number field in which we are operating is algebraically closed is that then the multiplicity $m(\chi)$ is equal to the dimensionality g of the representation $\mathfrak{h}$. Furthermore, it is impossible to resolve this multiplicity by any physical means whatever, for corresponding terms in these various term systems remain in coincidence under all dynamical influences.

We consider the resolution of terms in the case in which the interaction between the f individuals is expressed by a small perturbation energy λW, neglecting higher powers of the small parameter λ. *Assume for the moment that the energy levels* $E_1, E_2, \cdots$ *of a single individual* I *are non-degenerate.* On neglecting the perturbation I^f possesses *energy terms of the type*

$$E = E_1 + E_2 + \cdots + E_f ; \tag{8.4}$$

we first concern ourselves with such a term. Its multiplicity is $f!$ and the corresponding co-ordinates in tensor space are the coefficients $F(i_1, i_2, \cdots i_f)$ whose *indices* are any permutation s of $1, 2, \cdots, f$. This coefficient $F(i_1 i_2 \cdots i_f)$ is the component $x(s)$ of the element

$$\boldsymbol{x} = \boldsymbol{F}(1, 2, \cdots, f)$$

of the algebra (π). The separation of the term (8.4) is to a first approximation determined by the reduction of the correspondence

$$F(i_1 i_2 \cdots i_f) = \sum_{(k)} a(i_1 i_2 \cdots i_f ;\ k_1 k_2 \cdots k_f) F(k_1 k_2 \cdots k_f)$$

to diagonal form ; here the matrix of the coefficients a represents the energy and $i_1, i_2, \cdots, i_f ;\ k_1, k_2, \cdots, k_f$ are permutations

s, t of $1, 2, \cdots, f$. This equation may therefore be written in the form

$$\dot{x}(s) = \sum_t a(s, t)\, x(t). \tag{8.5}$$

The equation

$$a(i_{1'} \cdots i_{f'};\ k_{1'} \cdots k_{f'}) = a(i_1 \cdots i_f;\ k_1 \cdots k_f)$$

describing the symmetry of a, in which

$$1 \to 1', \cdots, f \to f'$$

s any fixed permutation r, is expressed by

$$a(sr, tr) = a(s, t)$$

for the only coefficients in which we are here interested; r is here considered as applied to the *indices* $1, 2, \cdots, f$ themselves rather than the sub-indices. Hence $a(s, t)$ depends only on st^{-1}:

$$a(s, t) = a(st^{-1}),$$

and equation (8.5) may now be written in the abbreviated form

$$(a):\quad \dot{x} = ax \tag{8.6}$$

where a, x, $\dot{x}$ are the symmetry elements of the algebra (π) with components $a(s)$, $x(s)$, $\dot{x}(s)$.

On restricting ourselves to an invariant irreducible sub-space $\mathfrak{P}$ of the system space $\mathfrak{R}^f$ the element x of (π) lies in the corresponding $\mathfrak{p}$. The g terms $W_1, W_2, \cdots, W_g$ into which (8.4) is resolved by the perturbation and which belong to the term system χ under consideration are, to the approximation involved in the perturbation theory, the characteristic numbers of the correspondence (8.6) *of* $\mathfrak{p}$ *on itself*. The sum of these terms must therefore equal the trace of this correspondence, or

$$W_1 + W_2 + \cdots + W_g = \sum_s a(s)\chi(s). \tag{8.7}$$

The sum of the squares of these terms, of their third powers, etc., are obtained by reiterating the correspondence (a), i.e.

$$W_1^\tau + W_2^\tau + \cdots + W_g^\tau = \sum_s a_\tau(s)\, \chi(s), \tag{8.7'}$$

where the $a_\tau(s)$ are the components of the symmetry element a^τ:

$$\left.\begin{aligned} a_0(s) &= 1 \text{ or } 0, \text{ according as } s = \mathbf{I} \text{ or } \neq \mathbf{I}, \\ a_{\tau+1}(s) &= \sum_t a_\tau(st^{-1})a(t). \end{aligned}\right\} \tag{8.8}$$

As soon as the "*exchange energies*" $a(s)$ are known we can apply this formula to calculate those of the terms arising from

(8.4) which are contained in the term system χ; for this we need only to know the character χ—it is not necessary to have an explicit expression for the idempotent generator e or the representation $\mathfrak{h}$ of π.

These considerations are immediately applicable only if we ignore the spin phenomena. If we take into account the perturbation due to the interaction of the electrons before that due to the spin, as in the case of normal term order, the mere existence of spin implies that each of the energies E_i is at least two-fold. We shall later concern ourselves with the far-reaching modifications caused by the spin and by the Pauli exclusion principle, which enables us to discard the majority of possible terms.

The unperturbed I^f will have, in addition to terms of the type (8.4), terms in which groups of two or more summands appear with the same indices. The multiplicity of the term

$$f_1E_1 + f_2E_2 + \cdots + f_\nu E_\nu \quad (f_1 + f_2 + \cdots + f_\nu = f) \tag{8.9}$$

with integral non-negative weights f_i is but

$$\frac{f!}{f_1!f_2! \cdots f_\nu!}. \tag{8.10}$$

The corresponding tensor coefficients $x(s)$ are those obtained from

$$F(\underbrace{1\,1 \cdots}_{f_1};\ \underbrace{2\,2 \cdots}_{f_2};\ \cdots)$$

by the permutations s of the f arguments. But a permutation p is without effect if it only permutes the first f_1 indices among themselves, the next f_2 among themselves, etc.; we may no longer distinguish between the permutations s and ps—they must be considered as giving rise to but one component. Such permutations p constitute a group $\pi' = \pi(f_1, f_2, \cdots)$ of order $h' = f_1!f_2! \cdots$, and two permutations s, t are to be considered as the same if they are left-equivalent with respect to this subgroup π', i.e. if $s \equiv t$ ($ps = t$, where p is an element of π'). The only elements x of the algebra (π) in which we are now interested are those which satisfy the equation

$$x(t) = x(s) \quad \text{when} \quad t \equiv s \ (\text{mod. } \pi');$$

they constitute a linear sub-space $\mathfrak{r}' = \mathfrak{r}(\pi')$ of dimensionality (8.10). More precisely, $\mathfrak{r}'$ is a right-invariant sub-algebra, for if $s \equiv t$ then also $sr \equiv tr$. Again $a(s, t) = a(st^{-1})$; further

$$a(ps) = a(s), \quad a(sp) = a(s)$$

if p is in π'.

We are now concerned with the correspondence $x \to \dot{x}$ in $\mathfrak{r}'$:

$$\dot{x}(s) = \sum_t a(st^{-1})x(t) \quad (\text{mod. } \pi'), \tag{8.11}$$

where the " mod. π' " indicates that both s and t run through a complete set of elements of the group which are inequivalent mod. π'. As x runs through $\mathfrak{r}'$, xe generates a sub-space $\mathfrak{p}'$ of $\mathfrak{r}'$ which is transformed into itself by the correspondence (8.11), and the reduction of this correspondence of $\mathfrak{p}'$ into diagonal form yields those terms arising from (8.9) and lying in the term system χ. The trace of (8.11) in $\mathfrak{p}'$ is equal to the trace of the correspondence $\mathbf{A}_e$: $x \to \dot{x}$ in $\mathfrak{r}'$ which is obtained from (8.11) by replacing x by xe, i.e. $x(t)$ by

$$\sum_r x(tr^{-1})e(r) = \sum_r x(r)e(r^{-1}t).$$

Hence

$$\text{tr}(\mathbf{A}_e) = \sum_{s,t \text{ mod. } \pi'} \{a(st^{-1}) \sum_{r \equiv s} e(r^{-1}t)\}.$$

Since $a(st^{-1}) = a(rt^{-1})$ when $r \equiv s$ (mod. π'), this trace may be written

$$\sum_{t \text{ mod. } \pi'} \sum_r a(rt^{-1})e(r^{-1}t).$$

Naturally this sum does not depend on which particular element t we have happened to choose from the set of group elements which are equivalent mod. π' ; hence on dropping the restriction on the range of t the above sum is multiplied by the order h' of π' :

$$\text{tr}(\mathbf{A}_e) = \frac{1}{h'} \sum_{r,t} a(rt^{-1})e(r^{-1}t) = \frac{1}{h'} \sum_s a(s)\chi(s). \tag{8.12}$$

Here again $\chi(s)$ is the character of $\mathfrak{h}$ as determined by (8.3). In particular, the dimensionality of $\mathfrak{p}'$, i.e. the number of terms in the system χ arising from (8.9), is obtained by replacing the symmetry element a in (8.12) by the element a_0 defined by

$$a_0(s) = 1 \text{ or } 0, \text{ according as } s \equiv \mathbf{1} \text{ (mod. } \pi') \text{ or not};$$

this number is consequently

$$\boxed{\frac{1}{h'} \sum_{s \text{ in } \pi'} \chi(s)} \tag{8.13}$$

We express this result, the validity of which is not restricted to permutation groups, in the theorem :

Let π' be a sub-group of π of order h' and let $\mathfrak{p}$ be a left-invariant sub-space of the group space $\mathfrak{r}$ of π. Consider the elements x of *the algebra* (π) *which satisfy the condition* $x(s) = x(t)$, *where* s *and* t *are any two elements of the group* π *which are left-equivalent* mod. π'; *the elements of* (π) *which are of this type and which lie in* $\mathfrak{p}$ *constitute a linear sub-space whose dimensionality is given by* (8.13), *where* χ *is the character of the regular representation in* $\mathfrak{p}$.

The sum of the terms is equal to the trace (8.12), and the sums of their powers are given by

$$\sum_W W^\tau = \frac{\sum_s a_\tau(s)\chi(s)}{f_1!f_2!\cdots} \tag{8.14}$$

The only way this result differs from (8.7′) is by the introduction of the denominator $f_1!f_2!\cdots$ and the fact that $a_\tau(s)$ is now defined by

$$a_{\tau+1}(s) = \sum_t a_\tau(st^{-1})a(t) \quad (\text{mod. } \pi').$$

Degenerate Case. Denote the numerically different energy levels of the individual I by E', E'', $\cdots$, and the multiplicity of $E^{(\nu)}$ by n_ν. We now distinguish between the various variables having the same " principal quantum number " ν by an " auxiliary quantum number " k_ν which assumes n_ν values. An energy level of the type

$$E' + E'' + \cdots + E^{(f)} \tag{8.15}$$

of the unperturbed total system I^f has the multiplicity

$$f!\, n_1 n_2 \cdots n_f,$$

and the corresponding tensor coefficients are those obtained from those of type

$$F\begin{pmatrix}1 & 2 & \cdots & f\\ k_1 & k_2 & \cdots & k_f\end{pmatrix}$$

by any permutation s of the f pairs $(\nu|k)$ of arguments; we write instead

$$x(s|k_1k_2\cdots k_f) \text{ or briefly } x(s|k).$$

Similarly the coefficients of the energy matrix are denoted by

$$a(s|k_1k_2\cdots k_f;\ t|l_1l_2\cdots l_f) = a(st^{-1}|k;\ l).$$

The energy levels W arising from (8.15) by the perturbation and lying in the term system χ are, to a first approximation, determined by

$$\sum W^\tau = \sum_{(k)}\sum_s a_\tau(s|k;\ k)\chi(s), \tag{8.16}$$

where $a_0(s|k\,;\ l) = 1$ or 0 according as $s = \mathsf{I}$, $k = l$ or not, and the composition is defined by

$$a_{\tau+1}(s|k\,;\ l) = \sum_{t,(m)} a_\tau(st^{-1}|k\,;\ m)a(t|m\,;\ l). \tag{8.17}$$

If the unperturbed energy level is of the form

$$f'E' + f''E'' + \cdot\cdot\cdot \qquad\qquad (f' + f'' + \cdot\cdot\cdot = f)$$

the tensor coefficients in which we are interested are those obtained from

$$F\Big(\underbrace{\begin{matrix}1 & 1 & \cdots\\ k_{11} & k_{12} & \cdots\end{matrix}}_{f'}\,;\ \underbrace{\begin{matrix}2 & 2 & \cdots\\ k_{21} & k_{22} & \cdots\end{matrix}}_{f''}\,;\ \begin{matrix}\cdots\\ \cdots\end{matrix}\Big).$$

Let exactly f_1' of the auxiliary quantum numbers $k_{1\nu}(\nu = 1, \cdots, f')$ have a certain value k_1, f_2' a different value k_2, etc.; $f_1' + f_2' + \cdots = f'$, and let f_1'', f_2'', $\cdots$ have the analogous meaning for the quantum numbers $k_{2\nu}(\nu = 1, \cdots, f'')$ associated with the principal quantum number 2, etc. Then those permutations p which leave the above tensor coefficient unchanged constitute a certain sub-group π_k', depending on the distribution of auxiliary quantum numbers k, of the group π' introduced in the non-degenerate case above; the order of π_k' is $[k] = f_1'!f_2'! \cdots f_1''! \cdots$. $a(s|k\,;\ l)$ is unchanged when s is multiplied on the left by an element of π_k' and on the right by an element of π_l'. The formula (8.16) now becomes

$$\sum W^\tau = \sum_k \left\{\frac{1}{[k]}\sum_s a_\tau(s|k\,;\ k)\chi(s)\right\} \tag{8.18}$$

$a_0(s|k\,;\ l) = 1$ or 0 according as $k = l$ and $s \equiv \mathsf{I}$ (mod. π_k') or not, and in the composition rule (8.17) we first sum with respect to t mod. π_m' and then over the various possibilities

$$m = (m_{11}, m_{12}, \cdots;\ m_{21}, \cdots;\ \cdots).$$

In every case we obtain explicit expressions for the sums of the various powers of the perturbed energy levels in terms of the character χ of the term system under consideration and the exchange energies $a(s)$.

§ 9. Relation between the Characters of the Symmetric Permutation and Affine Groups

The thorough correspondence existing between the representations of the symmetric permutation group π_f and the representations of order f of the linear group $\mathfrak{c}$ must lead to a simple relation between the corresponding characters. In dealing with the linear group it suffices to consider only the " principal transformations "

$$x_i \to \varepsilon_i x_i \quad (i = 1, 2, \cdots, n) \tag{9.1}$$

of the vector space $\mathfrak{R} = \mathfrak{R}_n$, for any linear transformation is conjugate within $\mathfrak{c}$ to a principal transformation—except for those cases in which two or more of the characteristic numbers ε_i coincide. Furthermore, if we restrict ourselves *ab initio* to the unitary group $\mathfrak{u}$—the one in which we are interested in physics—the result is valid without exception and the ε_i are complex numbers of unit absolute value. The problem here proposed is identical with that of investigating the distribution of the terms of I^f among the various term systems χ in the absence of interaction between the various individuals and when the single system I is non-degenerate, for on choosing a Heisenberg co-ordinate system x_i in the system space of I (i.e. one in which the operator representing the energy of I is in diagonal form) the variable x_i assumes the multiplicative factor $e\left(-\frac{E_i t}{h}\right)$ in time t.

We denote the characteristic * of the representation $\mathfrak{H}$ of the linear group whose substratum consists of all tensors of the form $\hat{e}F$ by $\mathrm{X}(S)$ or $\mathrm{X}(\varepsilon_1, \varepsilon_2, \cdots, \varepsilon_n)$ where the element S of $\mathfrak{c}$ is the principal transformation (9.1). The ε_1 are to be considered as n independent variables. The transformation of tensor space associated with (9.1) consists in multiplying the coefficient $F(i_1, i_2, \cdots, i_f)$ of the tensor F by $\varepsilon_{i_1} \cdot \varepsilon_{i_2} \cdots \varepsilon_{i_f}$. The sum of all these multipliers, extended over all linearly independent coefficients of a general tensor of the form $F' = \hat{e}F$, is the desired characteristic. A component in which f_1 of the arguments i are equal to 1, f_2 are equal to 2, $\cdots$ is multiplied by $\varepsilon_1^{f_1} \cdot \varepsilon_2^{f_2} \cdots \varepsilon_n^{f_n}$. But the number of linearly independent components of F' of this type is, by equation (8.13),

$$\frac{\Sigma \chi(s)}{f_1! f_2! \cdots}; \tag{9.2}$$

here χ is the character of the representation $\mathfrak{h}$ of π_f, the sum being extended over all elements s of the group $\pi' = \pi(f_1, f_2, \cdots)$ which permutes the first f_1 numerals among themselves, the next f_2 among themselves, etc. That this number (9.2) depends only

* We prefer, for the sake of clarity, hereafter to employ the word " characteristic " for continuous and " character " for finite groups.

on the character χ is a fact of greatest importance for our present considerations. The result is [8]

$$\mathrm{X}(\varepsilon_1, \varepsilon_2, \cdots) = \sum_{f_1, f_2, \cdots} \left\{ \frac{\varepsilon_1^{f_1} \varepsilon_2^{f_2} \cdots}{f_1! \, f_2! \cdots} \sum_s \chi(s) \right\}, \tag{9.3}$$

where the inner sum is extended over all the elements s of $\pi(f_1, f_2, \cdots)$. We denote the value of the character χ for an element s belonging to the class $\mathfrak{k}$ of conjugate elements of π_f by $\chi(\mathfrak{k})$; our formula may then be written

$$\mathrm{X} = \sum_{\mathfrak{k}} \left\{ \chi(\mathfrak{k}) \sum_{f_1, f_2, \cdots} \frac{c_{f_1 f_2 \cdots}(k)}{f_1! f_2! \cdots} \varepsilon_1^{f_1} \varepsilon_2^{f_2} \cdots \right\}, \tag{9.4}$$

where $c_{f_1 f_2} \cdots (k)$ is the number of elements of $\pi(f_1, f_2, \cdots)$ belonging to the class $\mathfrak{k}$. This number can be evaluated in an elementary manner.

Distribution of Permutations in Classes.

Any permutation s is a product of cycles, no two of which contain a common numeral. The 5-term cycle (1 3 7 2 4) is a permutation which sends 1 into 3, 3 into 7, 7 into 2, 2 into 4, and 4 into 1 again; writing these 5 numerals at equidistant intervals on the rim of a wheel, this permutation may be considered as the rotation of the wheel about the angle 2π 5. Given any permutation, for example

$$\begin{matrix} 1 & 2 & 3 & 4 & 5 & 6 & 7 & 8 & 9 \\ \downarrow & \downarrow & \downarrow & \downarrow & \downarrow & \downarrow & \downarrow & \downarrow & \downarrow \\ 3 & 4 & 7 & 1 & 9 & 8 & 2 & 6 & 5, \end{matrix} \tag{9.5}$$

the cycles may be separated out by first determining the number (3) into which 1 is transformed, then the number (7) into which 3 is transformed, etc., until a number is obtained which has already appeared in the cycle; this number can, of course, only be 1. After separating out the first cycle the remaining numbers can be handled in the same way, and the process may be continued until the desired result is obtained. The permutation (9.5) is, in terms of its 3 cycles,

$$(1\ 3\ 7\ 2\ 4)\ (5\ 9)\ (6\ 8). \tag{9.6}$$

The reduction of an arbitrary permutation into its cycles is obviously unique. This way of writing the permutation enables us to tell at a glance whether two given permutations are conjugate in π_f or not, for an element conjugate to (9.6) is obtained by replacing the numbers 1, 2, 3, 4, $\cdots$ by the same numbers in any order. The class $\mathfrak{k}$ to which an element s belongs is thus determined entirely by the number of cycles and the number of integers they contain; in particular, any permutation s and its inverse s^{-1} belong to the same class. We denote the class $\mathfrak{k}$ whose elements s consist of i_1 cycles with one numeral, i_2 with two, i_3 with three, $\cdots$ by $(i_1\, i_2\, i_3 \cdots)$ and write $\chi(\mathfrak{k}) = \chi(i_1\, i_2 \cdots)$; naturally

$$1i_1 + 2i_2 + 3i_3 + \cdots = f. \tag{9.7}$$

The number K of classes is the number of solutions of (9.7) with non-negative integers $i_1, i_2, i_3, \cdots$.

The number of elements in the class $\mathfrak{k} = (i_1 i_2 i_3 \cdots)$ is

$$n(\mathfrak{k}) = \frac{f!}{1^{i_1} i_1!\, 2^{i_2} i_2!\, 3^{i_3} i_3! \cdots}. \tag{9.8}$$

To show this we write the f integers $1, 2, \cdots, f$ in any of the $f!$ possible orders and divide off each of the first i_1 integers by parentheses, then divide off the next $2i_2$ in groups of 2, the next $3i_3$ in groups of 3, $\cdots$. The symbol so obtained is to be interpreted as the expression of permutation in terms of its cycles. Each of the $f!$ possible arrangements so obtained leads to a definite element s of the class $\mathfrak{k}$, and all such elements must be included. We must now investigate how often the same s occurs among these $f!$. Now the 5-term cycle (1 3 7 2 4) can also be read as (3 7 2 4 1), (7 2 4 1 3), etc.: the particular integer with which we begin is immaterial; such a cycle will occur five times. Hence those $1^{i_1}\, 2^{i_2}\, 3^{i_3} \cdots$ arrangements which differ only by a cyclic permutation of the numerals in each cycle are all associated with the same element s. Furthermore, the i_1 1-term cycles may be written down in any order, the i_2 2-term ones in any order, etc., and these $i_1!i_2! \cdots$ arrangements all lead to the same element s. Hence each element occurs exactly $1^{i_1} i_1! 2^{i_2} i_2! \cdots$ times, and the total number of elements in the class is accordingly given by (9.8).

We must also determine the number of elements of $\mathfrak{k}$ which are contained in the sub-group $\pi(f_1, f_2, \cdots)$. For this purpose we divide the numbers from 1 to f in sections of lengths f_1, $f_2, \cdots$ and consider only those permutations s which permute the numbers of the first section among themselves, the numbers of the second among themselves, etc. On dividing s into cycles as in the above some of the cycles will be contained in the first section, i.e. will consist only of numerals belonging to the first section, some will be contained in the second section, etc., and no cycle will consist of numerals belonging to different sections. Denoting the number of 1-term cycles contained in the first section by i_{11}, the number of 2-term cycles in this section by i_{12}, etc., whence necessarily

$$1i_{11} + 2i_{12} + 3i_{13} + \cdots = f_1,$$

the number of permutations of 1, 2, $\cdots$, f_1 satisfying this requirement is, by (9.8),

$$\frac{f_1!}{i_{11}!\, i_{12}! \cdots} \cdot \frac{1}{1^{i_{11}}\, 2^{i_{12}} \cdots}. \tag{9.9}$$

Proceeding analogously for the 2nd, 3rd, etc., sections, the number of permutations in $\pi(f_1 f_2 \cdots)$ satisfying all our requirements is given by the product of all numbers of the form (9.9) for the various sections. But such an element is a member of the class $\mathfrak{k} = (i_1 i_2 \cdots)$ if and only if

$$\sum_\alpha i_{\alpha 1} = i_1, \ \sum_\alpha i_{\alpha 2} = i_2, \ \cdots; \tag{9.10}$$

hence

$$c_{f_1 f_2 \cdots}(\mathfrak{k}) = \frac{1}{1^{i_1} 2^{i_2} \cdots} \sum_{(i)} \left\{ \prod_\alpha \frac{f_\alpha!}{i_{\alpha 1}!\, i_{\alpha 2}! \cdots} \right\},$$

where the sum is extended over the various solutions of equations (9.10) and

$$\sum_\nu \nu\, i_{1\nu} = f_1, \quad \sum_\nu \nu\, i_{2\nu} = f_2, \ \cdots.$$

The inner sum in (9.4) is accordingly

$$\frac{1}{1^{i_1} 2^{i_2} \cdots} \sum_{(i)} \left\{ \prod_\alpha \frac{\varepsilon_\alpha^{1 i_{\alpha 1}}}{i_{\alpha 1}!} \cdot \frac{\varepsilon_\alpha^{2 i_{\alpha 2}}}{i_{\alpha 2}!} \cdots \right\},$$

the only restriction on the sum being the conditions (9.10). Let

$$\begin{aligned} \sigma_1 &= \varepsilon_1 + \varepsilon_2 + \cdots + \varepsilon_n, \\ \sigma_2 &= \varepsilon_1^2 + \varepsilon_2^2 + \cdots + \varepsilon_n^2, \\ &\cdots\cdots\cdots \end{aligned}$$

Our results can be expressed entirely in terms of these sums of powers, for by the multinomial theorem

$$\begin{aligned} \sum_{(i)} \left\{ \prod_\alpha \frac{\varepsilon_\alpha^{1 i_{\alpha 1}}}{i_{\alpha 1}!} \right\} &= \frac{1}{i_1!} \sigma_1^{i_1}, \\ \sum_{(i)} \left\{ \prod_\alpha \frac{\varepsilon_\alpha^{2 i_{\alpha 2}}}{i_{\alpha 2}!} \right\} &= \frac{1}{i_2!} \sigma_2^{i_2}, \\ &\cdots\cdots\cdots \end{aligned}$$

where the variables $i_{\alpha 1}, i_{\alpha 2}, \cdots$, over which the sum is extended, are subject to the restrictions (9.10). We thus finally obtain the simple formula

$$\boxed{\mathrm{X}(\varepsilon_1, \varepsilon_2, \cdots, \varepsilon_n) = \sum_{\mathfrak{k}} \frac{\chi(i_1 i_2 \cdots)\sigma_1^{i_1}\sigma_2^{i_2}\cdots}{1^{i_1}2^{i_2}\cdots i_1!\, i_2!\cdots}} \tag{9.11}$$

We have so far made use only of the elementary connection between the groups π and $\mathfrak{c}$. If we now introduce the assumption that the number field over which our algebras are defined is algebraically closed, and is in particular the continuum of all complex numbers, the primitive characters of the finite group π have the orthogonality properties

$$\sum_{\mathfrak{k}} n(\mathfrak{k})\chi(\mathfrak{k})\chi(\mathfrak{k}^{-1}) = h,$$

$$\sum_{\mathfrak{k}} n(\mathfrak{k})\chi(\mathfrak{k})\chi'(\mathfrak{k}^{-1}) = 0 \quad (\chi \neq \chi').$$

Furthermore, the number of primitive characters is equal to the number K of classes. The above relations assert that the matrix of the $\chi(\mathfrak{k})$, where χ runs through the entire set of primitive characters and $\mathfrak{k}$ all classes, has as its reciprocal the matrix

$$\frac{1}{h} \cdot n(\mathfrak{k})\chi(\mathfrak{k}^{-1}).$$

Hence we also have

$$\sum_{\chi} \chi(\mathfrak{k})\chi(\mathfrak{k}^{-1}) = \frac{h}{n(\mathfrak{k})},$$

$$\sum_{\chi} \chi(\mathfrak{k}')\chi(\mathfrak{k}^{-1}) = 0 \text{ for } \mathfrak{k}' \neq \mathfrak{k}.$$

This is, in fact, merely an alternative form of the completeness theorem. In dealing with the symmetric permutation group π_f $\mathfrak{k}^{-1} = \mathfrak{k}$ and the order is $h = f!$.

On multiplying the expression (9.11) for the *primitive* character X by $\chi(i_1 i_2 \cdots)$ and summing over all the primitive characters χ of π_f, we obtain, with the aid of the relations derived above, the important formula

$$\boxed{\sigma_1^{i_1}\sigma_2^{i_2}\cdots = \sum_{\chi} \chi(i_1 i_2 \cdots)\mathrm{X}(\varepsilon_1, \varepsilon_2, \cdots, \varepsilon_n)} \tag{9.12}$$

where χ and X are the characters of corresponding irreducible representations of π_f and $\mathfrak{c}_n$.

§ 10. Direct Product. Sub-groups

Programme.

If two atoms or ions with f_1, f_2 electrons, respectively, come together to form a molecule we may to a first approximation neglect the interaction between the two atoms so long as the distance between them is relatively large. In this approximation the two kinds of electrons are dynamically different, for the electrons of each atom are influenced only by the nucleus and the remaining electrons of the same atom. The symmetry is therefore described by the sub-group π' of the symmetric group $\pi = \pi_f$ of $f = f_1 + f_2$ things in which the first f_1 and the last f_2 things are permuted among themselves. A similar situation arises when three or more atoms come together to form a molecule. These considerations immediately suggest the following problems.

I. The theory developed in §§ 2-4 is to be extended to the case in which the symmetric permutation group is replaced by any permutation group π'. Naturally the definition of a symmetric transformation in tensor space is to be adapted to the new situation: we require only that the coefficients $a(i_1 \cdots i_f;\ k_1 \cdots k_f)$ of (1.2) remain unchanged under an *arbitrary permutation belonging to the group* π' of the sub-indices $1, 2, \cdots, f$. We say that these transformations are ***symmetric with respect to*** π'; they constitute an algebra Σ' which is obviously more extensive than Σ.—This question is immediately settled by the remark that all our previous deductions are valid for an arbitrary permutation group π'. Here π' is considered as an independent group rather than as a sub-group of the symmetric group.

II. Let the set of integers from 1 to f be divided into two or more sub-sets. We consider, as an example, the case of two sub-sets: the " red " numerals from 1 to f_1 and the " green " ones from 1 to f_2; $f_1 + f_2 = f$. Let π' consist of all permutations of the red among themselves and the green among themselves. Hence a permutation $s' = (s_1, s_2)$ of π' consists of a permutation s_1 of the f_1 red numerals and a permutation s_2 of the green ones; π' is the *direct product* $\pi_1 \times \pi_2$ of the symmetric group π_1 of f_1 and π_2 of f_2 things. Or conversely, this direct product—the abstract definition of which has nothing to do with the group of permutations of f things—*may* be considered as a sub-group π' of the symmetric group of $f = f_1 + f_2$ things on arranging the sets of numerals, on which permutations of π_1, π_2 act, one after the other to form a single set. But here we are interested in the following problem (which can be proposed for arbitrary finite groups): to discuss the properties of a group $\pi_1 \times \pi_2$ which is the direct product of two finite groups π_1, π_2.

III. In order to discuss the structure of molecules we must eventually take into account the interaction between the various atoms or ions contained in the molecule. This means that we must finally return from the sub-group π' to the full symmetric group π, so we must examine the relations existing between the group π and its sub-group π'. Here again the problem is not restricted to permutation groups.

Direct Product.

Let π_1, π_2 be two finite groups of orders f_1, f_2 respectively. The elements of the direct product $\pi = \pi_1 \times \pi_2$ are the pairs (s_1, s_2) consisting of an element s_1 of π_1 and an element s_2 of π_2. An element of the ***algebra*** of π is accordingly a function $x(s_1, s_2)$, and it follows from this that the algebra of π is the product of the algebras (π_1) and (π_2):

$$(\pi) = (\pi_1) \times (\pi_2)$$

in the sense of the $\times$-multiplication of vector spaces introduced in II, § 10. An element $x_1 : x_1(s_1)$ of (π_1) and an element $x_2 : x_2(s_2)$ of (π_2) yield the element $x = x_1 \times x_2$ of (π), whose components are given by

$$x(s_1, s_2) = x_1(s_1) \cdot x_2(s_2).$$

Indeed, given any two algebras ρ_1, ρ_2, their direct product $\rho = \rho_1 \times \rho_2$ can be constructed and multiplication in ρ defined by

$$(a_1 \times a_2)(b_1 \times b_2) = (a_1 b_1 \times a_2 b_2)$$

whether they are group algebras or not.

If $\mathfrak{p}_\alpha$ is a linear sub-space of $\mathfrak{r}_\alpha = \rho_\alpha$ $(\alpha = 1, 2)$, an element $x : x(s_1, s_2)$ of (π) is in $\mathfrak{p} = \mathfrak{p}_1 \times \mathfrak{p}_2$ if and only if it belongs to $\mathfrak{p}_1$ when considered as a function of s_1, holding s_2 fixed, and to $\mathfrak{p}_2$ when s_1 is held fixed; indeed, any element of this kind can be expressed as a linear combination of products of the form $a_1 \times a_2$, where a_1 is in $\mathfrak{p}_1$ and a_2 in $\mathfrak{p}_2$. If $\mathfrak{p}_\alpha (\alpha = 1, 2)$ is an ***invariant*** sub-space of $\mathfrak{r}_\alpha$, generated by the idempotent element e_α and the representation space of the representation $\mathfrak{h}_\alpha$ of ρ_α induced in $\mathfrak{p}_\alpha$ by the regular representation, then $\mathfrak{p}$ is also invariant, has as generating idempotent element $e = e_1 \times e_2$ and is the substratum of the representation $\mathfrak{h}_1 \times \mathfrak{h}_2$ of ρ. It is evident that the equivalences $\mathfrak{p}_1 \sim \mathfrak{p}_1'$, $\mathfrak{p}_2 \sim \mathfrak{p}_2'$ imply the equivalence $\mathfrak{p}_1 \times \mathfrak{p}_2 \sim \mathfrak{p}_1' \times \mathfrak{p}_2'$.

Suppose the two $\mathfrak{p}_\alpha$ considered above are also irreducible with respect to their algebras ρ_α; the question then arises as to whether $\mathfrak{p}_1 \times \mathfrak{p}_2$ is irreducible (with respect to ρ) and whether $\mathfrak{p} = \mathfrak{p}_1 \times \mathfrak{p}_2$ is equivalent to $\mathfrak{p}' = \mathfrak{p}_1' \times \mathfrak{p}_2'$ ($\mathfrak{p}_\alpha'$ irreducible) *only if* $\mathfrak{p}_1 \sim \mathfrak{p}_1'$ $\mathfrak{p}_2 \sim \mathfrak{p}_2'$. $\mathfrak{p}$ and $\mathfrak{p}'$ are inequivalent if $exe' = 0$ identically in x, i.e. if the sub-space consisting of elements of character (e, e') contains only the element 0; here $e = e_1 \times e_2$, $e' = e_1' \times e_2'$. Now the formula

$$(e_1 \times e_2)(x_1 \times x_2)(e_1' \times e_2') = e_1x_1e_1' \times e_2x_2e_2'$$

shows immediately that the sub-space (e, e') is the direct product of the two sub-spaces (e_1, e_1') and (e_2, e_2'), and can consist merely of 0 only if one of these two sub-spaces consists merely of 0, i.e. only if $\mathfrak{p}_1$ is inequivalent to $\mathfrak{p}_1'$ or $\mathfrak{p}_2$ is inequivalent to $\mathfrak{p}_2'$. Our second question is thus answered in the affirmative—regardless of the nature of the field over which the algebras are defined.

The first question is answered in the affirmative in III, § 9, for the only case of physical interest, i.e. that in which the field is algebraically closed. If we are more interested in the reduction of the algebra than in the representations we can argue as follows. The algebra of elements of character (e, e) is the direct product of the *field* (*division algebra*) Φ_1 of elements of character (e_1, e_1) in ρ_1 and the field Φ_2 of character (e_2, e_2) in ρ_2. Assuming the original field is algebraically closed, all elements of Φ_α are multiples of e_α and consequently all elements of ρ with character (e, e) are multiples of e. This proves the irreducibility of $\mathfrak{p}_1 \times \mathfrak{p}_2$. If, however, the original field over which the algebras are defined is not algebraically closed our assertion is correct only if the direct product $\Phi_1 \times \Phi_2$ of the two fields is again a field, and this is by no means always the case. But in any case the question concerning the nature of the direct product of algebras is, as in the question concerning the structure of an algebra in § 7, reduced to the analogous problem for fields (division algebras).

Again taking the fundamental field to be the continuum of all complex numbers, the complete reduction

$$\mathfrak{r}_1 = \sum_i \mathfrak{p}_1^{(i)}, \quad \mathfrak{r}_2 = \sum_k \mathfrak{p}_2^{(k)}$$

into irreducible invariant sub-spaces $\mathfrak{p}_\alpha$ has as a consequence, in accordance with the above, the reduction of $\mathfrak{r} = \mathfrak{r}_1 \times \mathfrak{r}_2$ into invariant irreducible sub spaces $\mathfrak{p}_1^{(i)} \times \mathfrak{p}_2^{(k)}$.

Sub-groups.

Let π' be a sub-group of the given finite group π. An element x' of the algebra $\mathfrak{r}' = \rho' = (\pi')$ of π' consists of components $x'(s')$

associated with the various elements s' of π'. However, such an element can, and in the following will, at the same time be considered as an element of the algebra $\rho = (\pi)$; we need only to define the components $x'(s)$ associated with elements s of π which are not contained in π' as zero. This disturbs in no way the addition and multiplication of elements of (π') with each other or with arbitrary numbers of the field. An element x of (π) "belongs" to π' or "lies" in (π') if and only if all components $x(s)$ associated with elements s of the group that are not in π' vanish.

An irreducible invariant sub-space $\mathfrak{p}'$ of $\mathfrak{r}'$ is generated by a primitive idempotent element e' and is the substratum of a representation $\mathfrak{h}'$ of π' induced in $\mathfrak{p}'$ by the regular representation. On reducing the modulus 1 of π' into independent primitive idempotent elements

$$1 = \sum_{i=1}^{g'} e_i' + \cdots \tag{10.1}$$

a certain number, say g', of elements e_i' will appear which are equivalent to e'; the sub-spaces $\mathfrak{p}_i'$ which they generate are all equivalent to $\mathfrak{p}'$ and the regular representation of π' contains $\mathfrak{h}'$ g' times. Equivalent summands are added together into such partial sums. Considered as an element of the total algebra $\rho = (\pi)$ e' is, however, in general reducible into independent primitive idempotent elements:

$$e' = \sum_{\alpha=1}^{b} e_\alpha + \cdots \tag{10.2}$$

Here again equivalent summands on the right are collected together into partial sums; let the e_α in the first such partial sum generate the representation $\mathfrak{h}$ of π—we shall in the following be interested only in these. Let the sub-space $\mathfrak{p}$ with the generating unit e be a representative of the sub-spaces $\mathfrak{p}_\alpha$ generated by the e_α. The elements of (π) of the form xe' constitute an invariant sub-space $\langle\mathfrak{p}'\rangle$ which is the substratum of a representation $\langle\mathfrak{h}'\rangle$ of π induced in $\mathfrak{p}'$ by the regular representation of π. Our formula asserts that on reducing $\langle\mathfrak{h}'\rangle$ into its irreducible constituents $\mathfrak{h}$ occurs exactly b times.

In order to obtain a simple characterization of the elements of $\langle\mathfrak{p}'\rangle$ we divide the elements of the group π into sets of group elements which are equivalent mod. π'; the uth such class consists of the group elements $\sigma_u s'$, where s' runs through the sub-group π'. An element x of the algebra (π) has as components $x(\sigma_u s')$; the numbers $x(\sigma_u s')$ may, for fixed u, be considered as the components of an element x_u' of the algebra (π'), so that x

may be considered as the set of elements x'_u belonging to the algebra (π'). The formula $y = xe'$ then becomes $y'_u = x'_u e'$ in (π'): hence x belongs to $\langle \mathfrak{p}' \rangle$ if and only if all the partial elements x'_u lie in $\mathfrak{p}'$. The correspondence

$$x \to y = ax$$

may then be written

$$y(\sigma_u s') = \sum_v \sum_{t' \text{ in } \pi'} a(\sigma_u s' t'^{-1} \sigma_v^{-1}) x(\sigma_v t')$$

or

$$y'_u = \sum_v a'_{uv} x'_v$$

where a'_{uv} is the element of the algebra (π') defined by

$$a'_{uv}(s') = a(\sigma_u s' \sigma_v^{-1}).$$

The representation $\langle \mathfrak{h}' \rangle$ may therefore be constructed as follows: first associate with the element a of (π) the matrix $\|a'_{uv}\|$, the coefficients of which are elements of the algebra (π') instead of numbers, and then replace each a'_{uv} by the matrix A'_{uv} associated with it in the representation $\mathfrak{h}'$ of π'.

As we have seen in the earlier part of the present chapter, the representations are obtained with the aid of a double Peirce decomposition; we therefore consider the elements $x = e'xe'$ of character (e', e'). The idempotent elements $e_\alpha, \cdots$ appearing in (10.2) are of this character, and such an element x may be expressed in terms of its components

$$x = \sum_{\alpha, \beta = 1}^{b} e_\alpha x e_\beta + \cdots. \tag{10.3}$$

We now repeat the analysis of § 7 for our more restricted set of elements: let Γ_α be a one-to-one similarity correspondence of $\mathfrak{p}_\alpha$ on $\mathfrak{p}$ and let the element into which e_α is sent by the correspondence $\Gamma_\alpha \Gamma_\beta^{-1}$ be denoted by $e_{\alpha\beta}$*. If, as we now assume, the field over which the algebras are defined is algebraically closed $e_\alpha x e_\beta$ is necessarily a multiple $x_{\alpha\beta}$ of $e_{\alpha\beta}$. We then obtain instead of (10.3) the reduction

$$x = \sum x_{\alpha\beta}\, e_{\alpha\beta} + \cdots, \tag{10.4}$$

(where the $x_{\alpha\beta}$ are numbers) and the representations

$$x \to \|x_{\alpha\beta}\|, \cdots \tag{10.4'}$$

* Here, as in § 7, but in contrast with our usual notation, the product of two or more correspondences Γ is to be read from left to right.

Now if in particular x is in (π') then $x = e'xe'$ is a numerical multiple of (10.2), and the matrix $\|x_{\alpha\beta}\|$ associated with such an element is a multiple of the unit matrix.—The degree of the secular equation, the solutions of which determine the characteristic numbers, is thus decreased from g to b for an element x of character (e', e'). We now proceed to examine the cause of this.

Let $\Gamma_i'^{-1}$ be a one-to-one similar correspondence of $\mathfrak{p}'$ on $\mathfrak{p}_i'$ $(i = 1, 2, \cdots, g')$, and let the element into which it sends e' be b_i'. On considering an arbitrary element x of the algebra of π as the set x_u', we see that the correspondence

$$xe' \to xb_i'$$

is a one-to-one reciprocal and similar mapping of $\langle\mathfrak{p}'\rangle$ on $\langle\mathfrak{p}_i'\rangle$: the projection Γ_i' of $\mathfrak{p}_i'$ on $\mathfrak{p}$ gives rise to such a projection of $\langle\mathfrak{p}_i'\rangle$ on $\langle\mathfrak{p}'\rangle$. This projection associates with the reduction of $\langle\mathfrak{p}'\rangle$ into irreducible invariant sub-spaces a reduction of the same kind of the sub-space $\langle\mathfrak{p}_i'\rangle$; corresponding to equation (10.2) we obtain the equations

$$e_i' = \sum_{\alpha=1}^{b} e_{\alpha i} + \cdots. \qquad (10{\cdot}5)$$

On combining (10.1) and (10.5) we obtain a reduction of the modulus 1 into independent primitive idempotent elements of (π). Now consider the partial sums $\sum_i e_i'$ of 1 and their reductions (10.5) as written one above the other. Each row is then associated with a definite representation $\mathfrak{h}'$ of π' and each column on the right-hand side, the terms of which are sums of the form $\sum_i \sum_\alpha e_{\alpha i}$, is associated with a definite representation $\mathfrak{h}$ of π. We now collect together all the summands e_J occurring in the first column on the right, i.e. all those elements e_J which are equivalent to e. The set of indices J is then broken up into sub-sets, each of which is associated with one of the inequivalent irreducible representations $\mathfrak{h}', \cdots$ of π'; the first of these sub-sets, which is associated with $\mathfrak{h}'$, consists of the bg' double indices αi.

Let the similarity projection $\Gamma_i'\Gamma_k'^{-1}$ of $\mathfrak{p}_i'$ on $\mathfrak{p}_k'$ send e_i' into $e_{i;k}'$. If x' is an element of (π') the equation

$$x' = \sum_{i,k} e_i' x' e_k' + \cdots$$

yields the reduction

$$x' = \sum_{i,k} x_{ik}' e_{i;k}' + \cdots \qquad (10.6)$$

with numerical coefficients x'_{ik}, and $x' \to \|x'_{ik}\|$ is the representation $\mathfrak{h}'$. (The partial sums should preferably be written one above the other rather than horizontally.) Γ'_i may be considered as a similarity transformation of $\langle\mathfrak{p}'_i\rangle$ on $\langle\mathfrak{p}'\rangle$ and therefore contains a transformation of the same type of $\mathfrak{p}_{\alpha i}$ on $\mathfrak{p}_\alpha$; $\Gamma'_i\Gamma_\alpha$ then provides us with a similarity correspondence of $\mathfrak{p}_{\alpha i}$ on $\mathfrak{p}$. Let Γ_J be a fixed one-to-one similarity correspondence of $\mathfrak{p}_J$ on $\mathfrak{p}$ and let the similarity correspondence $\Gamma_J\Gamma_K^{-1}$ of $\mathfrak{p}_J$ on $\mathfrak{p}_K$ send e_J into $e_{J;\,K}$. We may take the correspondence $\Gamma'_i\Gamma_\alpha$ as Γ_J for the index $J = \alpha i$, and similarly for the remaining sub-sets. On applying the correspondence $\Gamma'_i\Gamma'^{-1}_k = \Gamma'_i\Gamma_\alpha(\Gamma'_k\Gamma_\alpha)^{-1}$ to equation (10.5) we find

$$e'_{i;\,k} = \sum_{\alpha=1}^{b} e_{\alpha i;\,\alpha k} + \cdot\cdot\cdot. \tag{10.7}$$

The equation

$$x = \sum_{J,K} e_J\, x\, e_K + \cdot\cdot\cdot = \sum_{J,\,K} x_{JK}\, e_{J;\,K} + \cdot\cdot\cdot \tag{10.8}$$

then determines the representations

$$\mathfrak{h} \;:\; x \to \|x_{JK}\|\,;\;\cdot\cdot\cdot.$$

By (10.6) and (10.7) the matrix associated with an element x of (π') is

$$x_{\alpha i;\,\beta k} = \delta_{\alpha\beta}\, x'_{ik},\;\; x_{JK} = 0$$

where the two indices J and K belong to different sub-sets. But this means that on restricting π to π' the representation $\mathfrak{h}$ is reducible into the irreducible representations $\mathfrak{h}'$, $\cdot\cdot\cdot$ of π', $\mathfrak{h}'$ appearing exactly b times. We have thus obtained a constructive proof of the theorem [6]:

First Reciprocity Theorem (for arbitrary groups). If $\langle\mathfrak{h}'\rangle$ contains the representation $\mathfrak{h}$ of π exactly b times, then on restricting the group π to π', $\mathfrak{h}$ contains the representation $\mathfrak{h}'$ of π' exactly b times.

If the sub-group π' consists merely of the unit element 1 this theorem reduces to our previous result: the number of times an irreducible representation appears in the regular representation is equal to its dimensionality. Both the complete theorem and this special case depend on the assumption that the field over which the algebra is defined is algebraically closed.

Connection with Symmetry Classes of Tensors.

We apply the results of our investigation III to the symmetric group π and make use of the correlation described in I above for

π as well as for its sub-group π'. An irreducible sub-space $\mathfrak{p}$ of (π) determines a symmetry class $\mathfrak{P} = \sharp\mathfrak{p}$ of tensors; let the corresponding representations of π and the linear group $\mathfrak{c}$ be $\mathfrak{h}$ and $\mathfrak{H}$, respectively. An irreducible invariant sub-space $\mathfrak{p}'$ of (π') determines a symmetry class $\mathfrak{P}'$ of tensors which is invariant with respect to the more extensive algebra Σ' of all transformations which are symmetric with respect to π'; as such $\mathfrak{P}'$ is irreducible. If e' is the generating unit of $\mathfrak{p}'$, $\mathfrak{P}'$ consists of all tensors of the form $\hat{e}'F$; but this is equivalent to saying that the symmetry element $\boldsymbol{F}$ of (π) belongs to $\langle\mathfrak{p}'\rangle$. Hence the reduction of $\mathfrak{P}'$ into irreducible invariant sub-spaces with respect to the more restricted algebra Σ parallels the reduction of $\langle\mathfrak{p}'\rangle$. Let $\mathfrak{h}'$ be that representation of π' induced in $\mathfrak{p}'$ by the regular representation of π' and $\mathfrak{H}'$ that representation of $\mathfrak{c}$ whose substratum consists of all tensors in the symmetry class $\mathfrak{P}'$. Hence our general theorem—or rather its converse, the truth of which follows immediately from the theorem itself—allows us to state the

Second Reciprocity Theorem (applicable only to permutation groups). If the irreducible representation $\mathfrak{h}$ of π contains the irreducible representation $\mathfrak{h}'$ of π' exactly b times when considered as a representation of the sub-group π', then conversely the representation $\mathfrak{H}'$ of $\mathfrak{c}$ contains the representation $\mathfrak{H}$ exactly b times.

Finally we take π' as $\pi_1 \times \pi_2$ as in step II above. $\mathfrak{p}'$ can then always be taken in the form $\mathfrak{p}_1 \times \mathfrak{p}_2$, and the irreducible invariant sub-space $\mathfrak{p}_\alpha$ of (π_α) determines a symmetry class $\mathfrak{P}_\alpha$ of tensors of order f_α $(\alpha = 1, 2)$. Denote the corresponding representations of π_α and $\mathfrak{c}$ by $\mathfrak{h}_\alpha$ and $\mathfrak{H}_\alpha$. The $\mathfrak{P}'$ associated with $\mathfrak{p}' = \mathfrak{p}_1 \times \mathfrak{p}_2$ consists of all tensors of order $f = f_1 + f_2$ which satisfy the symmetry conditions of $\mathfrak{P}_1$ with respect to their first f_1 indices and the symmetry conditions of $\mathfrak{P}_2$ with respect to the last f_2; i.e. $\mathfrak{P}' = \mathfrak{P}_1 \times \mathfrak{P}_2$. Our theorem now becomes:

Third Reciprocity Theorem (for permutation groups). If the irreducible representation $\mathfrak{h}$ of π contains, on restricting π to the sub-group $\pi' = \pi_1 \times \pi_2$, the representation $\mathfrak{h}_1 \times \mathfrak{h}_2$ of π' exactly b times ($\mathfrak{h}_\alpha$ an irreducible representation of π_α), then conversely the representation $\mathfrak{H}_1 \times \mathfrak{H}_2$ of $\mathfrak{c}$ contains the representation $\mathfrak{H}$ exactly b times.

§ 11. Perturbation Theory for the Construction of Molecules

We return to the investigation of the physical system II consisting of f electrons or equivalent individuals I. As long as we disregard the interaction between the individuals we obtain, among others, $f!$-fold energy levels E of the type (8.4). We consider in particular the case in which the E_i are different simple levels of the individual I. In order to follow the resolution of E, due to the mutual interactions of the electrons, to the approximation which characterizes the perturbation theory, we must first determine the elements a of the algebra of π, the components $a(s)$ of which are the exchange energies, and transform the matrices corresponding to a in the various irreducible representations of π into diagonal form by an appropriate change of co-ordinates (§ 8). We now assume that the most important of the exchange energies $a(s)$ are those belonging to the permutations s of a certain sub-group π' of π; all others shall be small in comparison with them ("quantities of 2nd order"). Our procedure is divided into two steps, corresponding to the investigation of sub-groups carried out in the preceding section. Let a' denote that element of the algebra (π') which is defined by

$$a'(s) = a(s) \text{ or } 0$$

according as s is an element of the sub-group π' or not, and let the matrices associated with a' in the irreducible representations $\mathfrak{h}'$ of π' be referred to principal axes; then

$$e'_i a' e'_k = 0 \quad (i \neq k), \quad e'_i a' e'_i = W_i \cdot e'_i.$$

The characteristic numbers W_i are the energy levels on neglecting perturbations of 2nd order; we assume they are all different. In order to examine the further resolution of such a term $W = W_i$ under the influence of the 2nd order perturbation we need, in accordance with the perturbation theory, to consider only that part

$$a^* = e'ae'$$

of a which is of character (e', e'), where we have written e' in place of e'_i. This term yields b terms W_α belonging to the symmetry class χ associated with the irreducible representation $\mathfrak{h}$ of π, the values of which are the characteristic numbers of the matrix $\|a^*_{\alpha\beta}\|$ associated with the element $a^* = e'ae'$ as in (10.4′). All the algebraic elements appearing in these considerations are real and the corresponding matrices are consequently Hermitian.

We apply the procedure to the process by which molecules are constructed from their constituent atoms.[10] We consider as an example two atoms joining to form a molecule, the one containing f_1 and the other f_2 electrons; $f = f_1 + f_2$. We consider the two nuclei as held fixed at a distance d apart, which is large compared with the linear dimensions of the atoms, and attempt to determine their interaction energy as a function of d. The sub-group $\pi' = \pi_1 \times \pi_2$ consists of all permutations which send no electron of one atom over into the other; we have seen in § 10 that we may then take the primitive idempotent elements $e'_i = e'$ of the algebra (π') in the form $e_1 \times e_2$, where e_1, e_2 are in (π_1), (π_2) respectively. On neglecting the interaction between the electrons of the one and the electrons of the other atom we obtain an energy term W which belongs to definite symmetry states of both atoms. e' generates a sub-space $\mathfrak{P}' = \mathfrak{P}_1 \times \mathfrak{P}_2$ (of the tensor space $\mathfrak{R}^f$) which is invariant under all symmetric transformations; that the state of the molecule is described by a tensor of this sub-space $\mathfrak{P}$ means that the state of the first atom is in $\mathfrak{P}_1$ and that of the second in $\mathfrak{P}_2$. Hence on reducing $\mathfrak{P}'$ in parallel with the reduction of $\langle\mathfrak{p}'\rangle$ into irreducible invariant sub-spaces:

$$e' = \sum_\alpha e_\alpha + \cdots, \ \langle\mathfrak{p}'\rangle = \sum_\alpha \mathfrak{p}^{(\alpha)} + \cdots, \ \mathfrak{P}' = \sum_\alpha \mathfrak{P}^{(\alpha)} + \cdots,$$

there occur b sub-spaces $\mathfrak{P}^{(\alpha)}$ which are equivalent to one another and which belong to a certain representation of π or to a certain symmetry class of terms of the total system. The procedure sketched in the preceding paragraph thus leads to b terms which (1) arise, due to the perturbation, from the given unperturbed term (8.4) and (2) which belong to certain given symmetry states χ_1, χ_2 and χ of the two atoms and the molecule. This reduction of the total system space $\mathfrak{R}^f$ into sub-spaces, each of which corresponds to a definite symmetry state of each of the atoms taken separately and of the molecule, naturally is not bound up with the approximate calculation of levels with the aid of perturbation theory; the connection between the two appears only on taking the above condition (1) into account—the very essence of which implies the assumption of small perturbations. This somewhat sketchy account of the situation arising from an unperturbed term of the type (8.4), in which the energies E_i of the individual I are non-degenerate, can readily be extended to cover other more complicated types of unperturbed terms. These other cases are of course of much greater physical interest, for we have seen in Chapter IV that all atomic energy levels, except S-terms, are necessarily degenerate.[11]

The fact that the total system may be in any one of several symmetry states $\mathfrak{P}$, corresponding to different energy levels (i.e. binding energies), when the symmetry states of the component atoms are given is of greatest importance. We shall later show that these possibilities, finite in number, coincide with those predicted by the empirical theory of the valence bond, and that consequently the symmetry state of an atom is that which chemists call its valence state. The situation thus arising cannot be described adequately in terms of classical models—e.g. the fact that the two H atoms constituting an H_2 molecule can behave in such a way that the state of the molecule may lie in either the space of symmetric or anti-symmetric tensors of order 2 ; only the first case can lead to an attraction which will bind the atoms together—the second always results in a repulsion.[12] The binding energy between two ions of total residual charges e_1, e_2 is naturally due mainly to the Coulomb potential e_1e_2/d (" ionic binding " or " polar bond "), but the corresponding energy for two neutral atoms is due for the most part to the interaction of the " exchange energies " $a(s)$ of the electrons of the two atoms (" atomic binding " or " non-polar bond "). This quantum-mechanical solution of the puzzle offered by the non-polar valence bond was first given by *F. London* and *W. Heitler*.

The following points are to be taken into consideration in applying the theory of perturbations to the actual evaluations. On neglecting the interaction between the various electrons each is subject only to the attraction of the two nuclei ; we should therefore perhaps begin with the characteristic numbers E_i and the corresponding characteristic functions $\psi_i(xyz)$ of *this* one-electron problem. The first approximation should then be obtained by taking into account the repulsions between the electrons of each of the atoms separately, thus introducing a dynamical difference between the two kinds of electrons. This procedure is naturally significant only so long as the distance d between the atoms is large in comparison with their linear dimensions a. But then it is also reasonable to take as our 0th approximation that in which each of the electrons is subject only to the attraction of its own nucleus (plus the closed shell of electrons which are not to be taken into explicit account in the calculations). Let this one-electron problem for the first atom have the characteristic values E_i and characteristic functions ψ_i, and let the corresponding quantities for the second atom be $E_{i'}$, $\psi_{i'}$. The fact that the ψ_i and the $\psi_{i'}$ together cannot constitute an orthogonal system—indeed, they are not even linearly independent, for the ψ_i alone constitute a complete orthogonal system—causes some difficulty. But if we break off the series of quantum states at a finite n—which can be chosen higher the larger the value of d/a under consideration—the finite set

$$\psi: \quad \psi_1, \psi_2, \cdot\cdot\cdot, \psi_n ; \; \psi_{1'}, \psi_{2'}, \cdot\cdot\cdot, \psi_{n'}$$

of functions ψ constitute an *almost* orthogonal system; the fundamental metric form G_0, the coefficients of which are the scalar products

$$g_{ik} = (\psi_i, \psi_k) = \int \bar{\psi}_i \psi_k \, dV$$

(where i and k run through the primed as well as the un-primed indices), differs but little from the unit form. Indeed, an integral of the form $(\psi_1, \psi_{1'})$ is of order of magnitude $e^{-d/a}$. To show this we note that if the two centres of force are nuclei or closed cores with "unit" residual charge, the normal states of the atoms are given by

$$\psi_1 = \frac{1}{\sqrt{\pi a^3}} \cdot e^{-r/a}, \quad \psi_{1'} = \frac{1}{\sqrt{\pi a^3}} \cdot e^{-r'/a},$$

where r and r' are the distances to the two cores. The integrand in

$$(\psi_1, \psi_{1'}) = \frac{1}{\pi a^3} \int e^{-(r+r')/a} \, dV$$

is everywhere $\leqq e^{-d/a}$. This integral can readily be exactly evaluated on introducing bi-polar co-ordinates (r, r', ϕ); the volume element is then

$$dV = \frac{2\pi}{d} r \, r' \, dr \, dr'$$

and the range of integration is defined by

$$r + r' \geqq d, \quad -d \leqq r - r' \leqq d.$$

On introducing

$$\frac{r + r'}{d} = \rho, \quad \frac{r - r'}{d} = \rho', \quad \frac{d}{a} = \lambda$$

we obtain

$$(\psi_1, \psi_{1'}) = \frac{\lambda^3}{4} \int_1^\infty \int_{-1}^{+1} (\rho^2 - \rho'^2) e^{-\lambda\rho} d\rho' d\rho$$

$$= \frac{\lambda^3}{2} \int_1^\infty \left(\rho^2 - \frac{1}{3}\right) e^{-\lambda\rho} d\rho = e^{-\lambda}\left(1 + \lambda + \frac{\lambda^2}{3}\right).$$

For the f-electron problem we therefore start with the functions

$$\psi(i_1, \cdots, i_f) = \prod_i \psi_i(xyz)$$

as approximations to the characteristic functions; in this product the co-ordinates are those of the f electrons and i runs through the values $i_1, i_2, \cdots, i_f$, each of which is one of the primed or un-primed indices between $1'$ and n' or 1 and n. The fundamental metric form $G = G_0 \times G_0 \times \cdots \times G_0$ has as components the scalar products of $\psi(i_1, i_2 \cdots i_f)$ with $\psi(k_1, k_2 \cdots k_f)$ and the components of the energy H, the potential part of which is obtained by adding together the potential energies resulting from the attractions and repulsions of the various electrons and the two cores, are the scalar products of $\psi(i_1 \cdots i_f)$ with the vector $H\psi(k_1 \cdots k_f)$ into which $\psi(k_1 \cdots k_f)$ is sent by the operator H. We consider the resolution of the unperturbed term

$$E = (E_1 + \cdots + E_{f_1}) + (E_{1'} + \cdots + E_{f_2'}).$$

The components

$$G(i_1 \cdots i_f\,;\; k_1 \cdots k_f) \text{ and } H(i_1 \cdots i_f\,;\; k_1 \cdots k_f), \tag{11.1}$$

in which the indices i, k are permutations s, t, respectively, of $1, \cdots, f_1, 1', \cdots, f_2'$, are of the form $G(st^{-1})$ and $H(st^{-1})$. We introduce the (real) elements $\boldsymbol{G}$ and $\boldsymbol{H}$ with components $G(s)$ and $H(s)$. $\boldsymbol{G}$ and $\boldsymbol{H}$ are next replaced by $\boldsymbol{G}'$ and $\boldsymbol{H}'$ with components $G(s)$ and $H(s)$ if s is in $\pi' = \pi_1 \times \pi_2$, and 0 otherwise; the justification for this lies in the fact that the components associated with an s which is not in π' are very small—they are of relative order $e^{-2d/a}$. $\boldsymbol{G}'$ is in fact the modulus, whereas $\boldsymbol{G}$ is not; the procedure employed previously must therefore be modified in the following purely formal respect. On repeating the reasoning, keeping in mind the fact that $\boldsymbol{G}$ is no longer the modulus, we find as the secular equation for the determination of the b terms $\lambda = W_\alpha$

$$| \lambda G_{\alpha\beta} - H_{\alpha\beta} | = 0, \tag{11.2}$$

in which

$$\boldsymbol{e}'\boldsymbol{G}\boldsymbol{e}' = \sum_{\alpha\beta} G_{\alpha\beta}\, \boldsymbol{e}_{\alpha\beta} + \cdots,$$

$$\boldsymbol{e}'\boldsymbol{H}\boldsymbol{e}' = \sum_{\alpha\beta} H_{\alpha\beta}\, \boldsymbol{e}_{\alpha\beta} + \cdots$$

in terms of the notation employed in the preceding section.

This procedure is open to the criticism that whereas the second order perturbations between the electrons of the same atom are neglected, the interaction between the two atoms, which is considered to be of second order, is taken into account. The results are therefore inapplicable to the limit $d/a \to \infty$ and can at most be applied successfully in cases in which d/a is considerably larger than 1 but not too large. On the other hand, we could begin by assuming that the solution of the quantum problem for the individual atoms is already known. Let the function ψ_1 of the co-ordinates of the first f_1 electrons be a characteristic function of the first atom corresponding to the energy term E_1 (so normalized that the integral of $\bar{\psi}_1\psi_1$ is unity); it will belong to a certain simple symmetry state of the first atom, i.e. there exists a certain real primitive idempotent element $\boldsymbol{e}_1$ of (π_1) such that $\hat{\boldsymbol{e}}_1\psi_1 = \psi_1$. Similarly, let ψ_2 be a characteristic function of the second atom for the term E_2, having a corresponding property $\hat{\boldsymbol{e}}_2\psi_2 = \psi_2$. Neglecting the interaction between the atoms, $\psi = \psi_1 \cdot \psi_2$ is a characteristic function of the molecule consisting of the two atoms and having the energy $E = E_1 + E_2$. $\boldsymbol{e}' = \boldsymbol{e}_1 \times \boldsymbol{e}_2$ is a primitive idempotent element of the algebra of $\pi' = \pi_1 \times \pi_2$ and ψ has the property

$$\hat{\boldsymbol{e}}'\psi = \psi.$$

The functions $\boldsymbol{s}\psi$, which are obtained from ψ by the totality of $f!$ permutations s of its arguments, span a linear function space $(\mathfrak{R})$ of a finite number of dimensions—in which the $\boldsymbol{s}\psi$ are naturally neither linearly independent nor mutually orthogonal. The theory of perturbations requires us to find those functions ϕ of $(\mathfrak{R})$ which are such that the orthogonal projection of $H\phi$ on $(\mathfrak{R})$ is proportional to ϕ itself; the factors of proportionality are then the values of the displaced terms, to a first approximation. We must therefore evaluate the integrals $G(s, t)$, $H(s, t)$ of

$$\boldsymbol{t}\bar{\psi} \cdot \boldsymbol{s}\psi \quad \text{and} \quad \boldsymbol{t}\bar{\psi} \cdot H(\boldsymbol{s}\psi)$$

and solve the secular equation

$$|\lambda G(s, t) - H(s, t)| = 0.$$

G and H depend only on $t^{-1}s$:*

$$G(s, t) = G(t^{-1}s), \quad H(s, t) = H(t^{-1}s).$$

This is proved by the fact that the integral of $\bar{\psi} \cdot \phi$ is unchanged on replacing ψ, ϕ by $\boldsymbol{r}\psi$, $\boldsymbol{r}\phi$ (r an arbitrary permutation); $H(\boldsymbol{s}\psi)$ is equal to $\boldsymbol{s}H\psi$ because of the symmetry of the operator H. Let $\boldsymbol{G}$ and $\boldsymbol{H}$ again be the elements of (π) with components $G(s)$, $H(s)$. They satisfy the equations

$$\boldsymbol{e}'\boldsymbol{G}\boldsymbol{e}' = \boldsymbol{G}, \quad \boldsymbol{e}'\boldsymbol{H}\boldsymbol{e}' = \boldsymbol{H}$$

* On comparing this with (11.1) it is to be remembered that there the permutations s and t operate on the indices and not on the arguments; hence the elements (11.1) are, in our present notation,

$$G(t^{-1}, s^{-1}) \quad \text{and} \quad H(t^{-1}, s^{-1}).$$

and are therefore of character (e', e'). Indeed, we have, for example,

$$\psi = \sum_r e'(r^{-1}) \cdot \boldsymbol{r}\psi, \quad \text{whence} \quad H(\boldsymbol{s}\psi) = \sum_r e'(r^{-1}) \cdot H(\boldsymbol{sr}\psi),$$

and on multiplying this latter by $\bar{\psi}$ and integrating we find

$$H(s) = \sum_r e'(r^{-1}) H(sr) \quad \text{or} \quad \boldsymbol{H} = \boldsymbol{He'}.$$

It then follows that also $\boldsymbol{H} = \tilde{\boldsymbol{e}}'\boldsymbol{H}$ whence, since e' is real, $\boldsymbol{H} = \boldsymbol{e'H}$ and consequently $\boldsymbol{H} = \boldsymbol{e'He'}$ as asserted.

The only non-vanishing elements of the matrix $\|H_{JK}\|$, which corresponds to the element $\boldsymbol{H}$ in the representation $\mathfrak{h}$, are (in the notation of § 10 with $e'_1 = e'$) those contained in the square sub-matrix of length b in which the row and column indices J and K are of the form $\alpha 1$. We are thus led directly to the secular equation

$$|\lambda G_{\alpha\beta} - H_{\alpha\beta}| = 0$$

of bth degree. (The most natural method of solving this equation consists in finding that linear transformation which sends the Hermitian form with coefficients $G_{\alpha\beta}$ into the unit form and at the same time reduces $\|H_{\alpha\beta}\|$ to diagonal form.) $\sum_\alpha H_{\alpha\alpha}$ is then the trace of the matrix belonging to $\boldsymbol{H}$ in the representation $\mathfrak{h}$, or

$$\sum_\alpha H_{\alpha\alpha} = \sum_s H(s)\chi(s).$$

If in particular $b = 1$ the above symmetry system of the molecule contains but a single term arising from the unperturbed term E; its value is, in accordance with the equation derived above, given by

$$\frac{\sum H(s)\chi(s)}{\sum G(s)\chi(s)} = \frac{E + \sum' H(s)\chi(s)}{1 + \sum' G(s)\chi(s)}. \tag{11.3}$$

The accent on the right-hand side indicates that these sums are to be extended over only those permutations s which do not belong to π'. This formula (11.3) is due to *F. London.*[13] It will be shown later that in the case of diatomic molecules b is always 1; we must expect, however, to find higher values of b in dealing with more complex molecules. The real difficulty from the physical standpoint naturally consists in getting information concerning the exchange energies $H(s)$. It is to be noted, however, that we need only to concern ourselves with the sums

$$\sum_r H(rsr^{-1}), \quad \sum_r G(rsr^{-1})$$

over the various classes, for since $\chi(s)$ is a class function all summands in (11.3) for elements in the same class $\mathfrak{k}$ may be added together to give the above coefficients multiplied by $\chi(\mathfrak{k})$.

Without doubt these investigations, which are as yet in their infancy, are of fundamental importance for theoretical chemistry; the non-polar bond is due to the exchange energies. *Heisenberg* has given an explanation of ferro-magnetism with the aid of these same principles.[14]

§ 12. The Symmetry Problem of Quantum Theory

On taking the spin into account the components of a vector $x(\iota i)$, which represents the state of a single electron, has two indices ι and i; the first of these refers to the spin and runs from 1 to ν, while the second refers to the translation and runs from 1 to n. Actually $\nu = 2$ and $n = \infty$ (as long as we do not restrict ourselves to the consideration of quantum states with fixed energy). Our vector space $\mathfrak{R}$ is accordingly $\mathfrak{R}_{\nu n} = \mathfrak{R}_\nu \times \mathfrak{R}_n$. The state of a system consisting of f electrons is now to be represented by a tensor of order f in this space: $F(\iota_1 i_1, \iota_2 i_2, \cdots, \iota_f i_f)$—a "double tensor" which stands, so to speak, with one foot (the Greek indices) in the space $\mathfrak{R}_\nu$ and the other (the Latin indices) in $\mathfrak{R}_n$. This tensor space is completely reducible, with respect to the algebra $\Sigma_{\nu n}$ of all symmetric transformations of the index pairs (ιi), into irreducible invariant sub-spaces, each of which is generated by an idempotent symmetry operator. The Pauli exclusion principle states that only one of these sub-spaces $\mathfrak{P}_{\nu n}$ is physically realized; it automatically abolishes the physically absurd existence of multiplicities which cannot be resolved and at the same time denies the existence of absolutely non-combining systems of terms. Furthermore, according to Pauli this $\mathfrak{P}_{\nu n}$ is the space $\{\mathfrak{R}^f_{\nu n}\}$ of all anti-symmetric double tensors.

On ignoring the spin perturbation, $\mathfrak{P}_{\nu n}$ is to be reduced as far as possible into sub-spaces $\mathfrak{P}$ which are invariant with respect to the special symmetric transformations of the form

$$F'(\iota_1 i_1 \cdots \iota_f i_f) = \sum_{(k)} c(i_1 \cdots i_f;\ k_1 \cdots k_f) \cdot F(\iota_1 k_1 \cdots \iota_f k_f) \quad (12.1)$$

which do not depend on the Greek indices at all; these constitute our old algebra $\Sigma = \Sigma_n$. This transition from $\Sigma_{\nu n}$ to Σ_n is to be accomplished in two steps. We first ignore the interaction between spin and translation, but allow the translations to interact among themselves in an arbitrary manner and similarly the spins among themselves; we must then consider only the symmetric transformations of the form

$$\gamma(\iota_1 \cdots \iota_f;\ \kappa_1 \cdots \kappa_f) \cdot c(i_1 \cdots i_f;\ k_1 \cdots k_f). \tag{12.2}$$

These transformations do not constitute an algebra themselves, but they belong to their "enveloping" algebra $\Sigma_\nu \times \Sigma_n$ which consists of all transformations whose coefficients

$$c(\iota_1 i_1 \cdots \iota_f i_f;\ \kappa_1 k_1 \cdots \kappa_f k_f)$$

are unaltered on subjecting the two rows $\iota_1 \cdots \iota_f;\ \kappa_1 \cdots \kappa_f$ of Greek indices to the same arbitrary permutation σ and the two rows of Latin indices to the same arbitrary permutation s. The second step then consists in letting γ in (12.2) be the identity. The first step thus consists merely in making the permutation of the Greek indices independent of the permutation of the Latin indices, and the second in restricting the first of these permutations to the identity.

In the first place, then, we introduce the elementary symmetry operator $\sigma \times s$ which, on applying it to the double tensor $F(\iota_1 i_1 \cdots \iota_f i_f)$, subjects the Greek indices to the permutation σ and the Latin to the permutation s. The general symmetry operator is then an arbitrary linear combination

$$\boldsymbol{a} = \sum_{\sigma, s} a(\sigma, s)(\boldsymbol{\sigma} \times \boldsymbol{s})$$

of these elementary ones; we have thus to deal with the algebra $\rho \times \rho$ of elements $\boldsymbol{x}$, the components $x(\sigma, s)$ of which are functions both of whose arguments run through the elements of the group π. We denote the element with components $F(\sigma, s) = (\boldsymbol{\sigma} \times \boldsymbol{s})F$ by $\boldsymbol{F}$; the equation $F' = \boldsymbol{a}F$ (F' the double tensor obtained from F by the operator $\boldsymbol{a}$) is equivalent to $\boldsymbol{F}' = \boldsymbol{F} \cdot \hat{\boldsymbol{a}}$. The group $\pi \times \pi$ of elements $\sigma \times s$ contains π itself as the sub-group consisting of elements $s \times s$. So far as the first step is concerned, our problem amounts to the following: Let $l(s)$ be the components of a primitive idempotent element of the algebra $\mathfrak{r} = \rho = (\pi)$; we set

$$\boldsymbol{l} = \sum_{s} l(s)(\boldsymbol{s} \times \boldsymbol{s})$$

and study the elements of the form xl in $\rho \times \rho$. They constitute an invariant sub-space $(\mathfrak{r} \times \mathfrak{r})_l$ which is to be reduced into its irreducible invariant constituents ; in Pauli's case we have in particular

$$l = \frac{1}{f!}\Sigma\,\delta_s(s \times s).$$

The procedure which it seems natural to follow is first of all to express the modulus 1 of ρ in any two ways as the sum of primitive independent idempotent elements :

$$1 = \sum_i e_i', \qquad 1 = \sum_j e_j. \tag{12.3}$$

An arbitrary element x of the algebra of $\rho \times \rho$ is reduced into independent constituents in accordance with the equation

$$x = \sum_{i,j} x(e_i' \times e_j) = \sum_{i,j} x_{ij}. \tag{12.4}$$

Now we know from § 10, II, that the elements of the form x_{ij} constitute an irreducible invariant sub-space $\mathfrak{p}_{ij}$; consider

$$xl = \sum_{i,j} x_{ij}\, l$$

in this light. The projection $x \rightarrow y = xl$ sends $\mathfrak{p}_{ij}$ over into a certain invariant sub-space $(\mathfrak{p}_{ij})$ of $(\mathfrak{r} \times \mathfrak{r})_l$. Since those x of $\mathfrak{p}_{ij}$ for which $xl = 0$ constitute an invariant sub-space of $\mathfrak{p}_{ij}$ we have only the two typical possibilities : either $(\mathfrak{p}_{ij}) = 0$ or this projection $x \rightarrow xl$ maps $\mathfrak{p}_{ij}$ in a one-to-one and similar manner on $(\mathfrak{p}_{ij})$. The sum

$$(\mathfrak{r} \times \mathfrak{r})_l = \Sigma(\mathfrak{p}_{ij}), \tag{12.5}$$

arranged in some particular order, is such that each term can, in virtue of its irreducibility, only either be contained in the sum of the preceding terms or be independent of this sum. On retaining only those terms arising from this second possibility, $(\mathfrak{r} \times \mathfrak{r})_l$ is completely reduced into the sum of certain of the $(\mathfrak{p}_{ij})$; the representation induced in $(\mathfrak{r} \times \mathfrak{r})_l$ by the regular representation of the group $\pi \times \pi$ is correspondingly reduced into its irreducible constituents of the form $\mathfrak{h}' \times \mathfrak{h}$. It will be remembered that this symbol stands for the correspondence

$$(\sigma, s) \rightarrow U'(\sigma) \times U(s), \tag{12.6}$$

where $\mathfrak{h}'$, $\mathfrak{h}$ are the irreducible representations $\sigma \rightarrow U'(\sigma)$, $s \rightarrow U(s)$ of π. This representation $\mathfrak{h}' \times \mathfrak{h}$ appears with a certain multiplicity $b(\chi', \chi)$ which is determined by the number of pairs ij in (12.5) whose e_i' generate the representation $\mathfrak{h}'$ and whose e_j generate $\mathfrak{h}$. These considerations are of course merely a repetition for the case at hand of the proof of theorem (6.1).

We now return to the space of double tensors and consider the sub-space $\mathfrak{L}$ defined by those of the form $\hat{\boldsymbol{I}}F$. It is the substratum of a certain representation $\mathfrak{L}(\Sigma_\nu \times \Sigma_n)$ of $\Sigma_\nu \times \Sigma_n$, and its complete reduction is given by the formula

$$\mathfrak{L}(\Sigma_\nu \times \Sigma_n) = \sum_{\chi', \chi} b(\chi', \chi)(\mathfrak{H}'_\nu \times \mathfrak{H}_n). \tag{12.7}$$

This remains correct even if ν or n is less than f. Earlier in this chapter we introduced the right- and left-invariant sub-space $\mathfrak{r}_0$ of $\mathfrak{r}$ as that sub-space consisting of all elements $\boldsymbol{F}$ which correspond to tensors F in the n-dimensional vector space $\mathfrak{R}_n$. On denoting this $\mathfrak{r}_0$, which depends on n (and only for $n \geqq f$ coincides with the entire $\mathfrak{r}$), by $\overset{n}{\mathfrak{r}}$ we should consider the algebra $\overset{\nu}{\mathfrak{r}} \times \overset{n}{\mathfrak{r}}$ instead of $\mathfrak{r} \times \mathfrak{r}$. But if e'_i is in $\overset{\nu}{\mathfrak{r}}$ and e_j in $\overset{n}{\mathfrak{r}}$, the manifold of elements $x(e'_i \times e_j)$ is not decreased on restricting x to $\overset{\nu}{\mathfrak{r}} \times \overset{n}{\mathfrak{r}}$, and every e'_k (e_l) which is equivalent to such an e'_i (e_j) also belongs to $\overset{\nu}{\mathfrak{r}}$ $(\overset{n}{\mathfrak{r}})$. This shows that (12.7) remains correct under this restriction to $\overset{\nu}{\mathfrak{r}} \times \overset{n}{\mathfrak{r}}$; the only effect is that those terms for which $\mathfrak{H}'_\nu \times \mathfrak{H}_n$ is the 0-dimensional representation are illusory We are now ready to take the second step: to perform the transition from the algebra $\Sigma_\nu \times \Sigma_n$ to $\Sigma = \Sigma_n$ by taking γ in (12.2) as the identity. We then see immediately that the representation $\mathfrak{L}(\Sigma)$ of Σ, whose substratum consists of the double tensors of $\mathfrak{L}$ in the sense of equation (12.1), is completely reduced into its irreducible constituents $\mathfrak{H}$, corresponding to the various primitive characters χ of π, in accordance with the equation

$$\mathfrak{L}(\Sigma) = \sum_{\chi} m(\chi) \cdot \mathfrak{H}.$$

The multiplicity $m(\chi)$ with which this representation $\mathfrak{H}$ occurs is given by

$$m(\chi) = \sum_{\chi'} b(\chi', \chi) N_\nu(\chi'), \tag{12.8}$$

where $N_n(\chi)$ is the dimensionality of the representation $\mathfrak{H}_n$, and the sum is extended over all the primitive characters χ' of π. Hence on disregarding the spin perturbation we obtain the same type of reduction into non-combining systems of terms as before, except that the multiplicity, which was previously equal to the dimensionality g of χ, is now given by (12.8).

(The spin perturbation causes weak inter-system combinations to take place and, in addition, resolves each term of the system χ into its $m(\chi)$ components. $m(\chi)$ is the multiplicity of the multiplet structure. Term systems χ for which $m(\chi) = 0$ do not appear at all.)

Our reciprocity theorem enables us to determine the constants b. As mentioned before, π is contained in $\pi \times \pi$ as the sub-group of elements of the form $s \times s$; the algebra $\rho = (\pi)$ appears in $\rho \times \rho$ as the totality of algebraic elements of the form $\sum_s a(s)(s \times \mathbf{s})$. The elements xl of the algebra ρ constitute an irreducible invariant sub-space $\mathfrak{p}_l$; let the irreducible representation of π which is induced in this sub-space by the regular representation be denoted by $\mathfrak{h}_l$ and its character by $\lambda(s)$. The space of all elements of the form xl in $\rho \times \rho$ is then $\langle \mathfrak{p}_l \rangle$ in the notation of § 10; it is the substratum of the representation $\langle \mathfrak{h}_l \rangle$ of $\rho \times \rho$. $\langle \mathfrak{h}_l \rangle$ contains the representation $\mathfrak{h}' \boldsymbol{\times} \mathfrak{h}$ exactly b times; the reciprocity theorem then tells us that the number of times the representation $\mathfrak{h}' \boldsymbol{\times} \mathfrak{h}$ contains the representation $\mathfrak{h}_l$ on restricting $\pi \times \pi$ to its sub-group π is also b. Now this restriction to π sends the representation (12.6) of $\pi \times \pi$ into the representation

$$(s, s) \rightarrow U'(s) \times U(s)$$

of π. This means, however, that $b(\chi', \chi)$ *is the number of times the representation* $\mathfrak{h}_l$ *of* π *is contained in the representation* $\mathfrak{h}' \times \mathfrak{h}$ *of* π (no longer with boldface multiplication sign!). Hence b is expressed by

$$b(\chi', \chi) = \mathfrak{M}\{\chi'(s)\chi(s)\lambda(s^{-1})\}. \tag{12.9}$$

With this we have carried our solution of the problem of determining the multiplicities $m(\chi)$ as far as is possible in the general case.

Consider in particular the special cases (1) complete symmetry, $\mathfrak{L} = [\mathfrak{R}^f]$, and (2) complete anti-symmetry, $\mathfrak{L} = \{\mathfrak{R}^f\}$—the Pauli case. For the first $\lambda(s) = 1$. With each irreducible representation χ is associated the contragredient representation with character $\hat{\chi}(s) = \chi(s^{-1})$; if the substratum of the first is generated by the idempotent element e the substratum of the latter is generated by $\hat{e}$. Or we may describe this situation by saying that χ and $\hat{\chi}$ are the characters of mutually contragredient representations. (Accidentally $\chi(s^{-1}) = \chi(s)$ for the complete symmetric group π; this does not hold for a general permutation group, however, whereas our entire theory does.) Equation (12.9) now becomes

$$b(\chi', \chi) = \mathfrak{M}\{\chi'(s)\hat{\chi}(s^{-1})\}.$$

But in virtue of the orthogonality property of characters this mean value is 1 or 0 according as the representation $\hat{\chi}$ is equivalent to χ' or not. The expression (12.8) for the multiplicity then assumes the simple form

$$\boxed{m(\chi) = N_\nu(\hat{\chi}).}$$

The theorem that the representation $\mathfrak{h} \times \mathfrak{h}$ contains the identical representation $s \to 1$ once or not at all according as $\mathfrak{h}'$ is equivalent to the contragredient of $\mathfrak{h}$ or not is nothing other than the fundamental theorem [III, (10.5)] on which the entire theory of representations was based.

In the second (anti-symmetric) case $\lambda(s) = \delta_s$. Now

$$\chi^*(s) = \delta_s \cdot \chi(s^{-1})$$

is the character of the "*dual*" representation $\mathfrak{h}^*$ associated with $\mathfrak{h}$; if $\mathfrak{h}$ is generated by the idempotent element e then $\mathfrak{h}^*$ is generated by the idempotent $e^*(s) = \delta_s \cdot e(s^{-1})$. Or if $\mathfrak{h} : s \to U(s)$ then $\mathfrak{h}^* : s \to \delta_s \cdot \breve{U}(s)$. The expression for the multiplicity is in this case

$$\boxed{m(\chi) = N_\nu(\chi^*)} \qquad (12.10)$$

If we denote the 1-dimensional representation $s \to \delta_s$ by $\{1\}$, the fundamental theorem mentioned above tells us immediately that $\mathfrak{h}' \times \mathfrak{h}$ contains the representation $\{1\}$ once or not at all according as $\mathfrak{h}'$ is equivalent to $\mathfrak{h}^*$ or not. (12.10) is the actual multiplet formula, for this second case is the one which is of interest for atomic physics.

Additional Remarks.

The only cases of importance for physics, (1) that of symmetric and (2) that of anti-symmetric double tensors, can be handled by elementary methods. We again refrain as long as possible from making restrictive assumptions concerning the field over which the algebras are defined. The method will be illustrated by application to case (1).

(12.11) *If* e_1, e_2 *are equivalent idempotent elements, then* $\hat{e}_1$, $\hat{e}_2$ *are also.*

Proof. Let $\mathfrak{p}_1$ be mapped on $\mathfrak{p}_2$ by a one-to-one similarity correspondence $\Gamma : x_2 = x_1 b$; b is here the element, of character (e_1, e_2), into which e_1 is sent by Γ. Let the inverse correspondence carry e_2 over into a, which is then of character (e_2, e_1). Γ carries a over into e_2; since the element associated with a by

Γ is ab we have $e_2 = ab$. Similarly, we find with the aid of Γ^{-1} that $e_1 = ba$. We then have

$$e_2 = ab, \quad e_1 = ba\,; \quad e_2ae_1 = a, \quad e_1be_2 = b.$$

Conversely, the existence of these equations guarantees that

$$x_2 = x_1b, \quad x_1 = x_2a$$

are reciprocal similarity correspondences $\mathfrak{p}_1 \rightleftarrows \mathfrak{p}_2$. That is, the existence of these four equations means that e_1 and e_2 are equivalent. We need only to "roof" these equations in order to conclude that $\hat{e}_1$ and $\hat{e}_2$ are then also equivalent—i.e., go over to the quantities $\hat{x}$ associated with each of these x by the definition $\hat{x}(s) = x(s^{-1})$. We have here neither assumed that the e are primitive nor that the field is algebraically closed.

(12.12). *The invariant sub-spaces* $\hat{\mathfrak{p}}$, $\mathfrak{p}$ *generated by* e, $\hat{e}$ *are the substrata of mutually contragredient representations.*

Proof. Let $\mathfrak{p}$ consist of all elements xe; we introduce in addition to this left-invariant sub-space the right-invariant sub-space $\mathfrak{q}$ consisting of all elements of the form ex. Let $\operatorname{tr}(xy)$ be the trace of the elements x and y, which may vary freely in $\mathfrak{p}$, $\mathfrak{q}$, respectively; we assert that it is a non-degenerate bilinear form. That is: if $\operatorname{tr}(ay) = 0$ identically in $\mathfrak{q}$ then the element a of $\mathfrak{p}$ must be 0, and if $\operatorname{tr}(xb) = 0$ identically in $\mathfrak{p}$ the element b of $\mathfrak{q}$ must be 0. Indeed, if z is any arbitrary element whatever and a is in $\mathfrak{p}$, then

$$az = ae \cdot z = a \cdot ez = ay,$$

where $y = ez$ is in $\mathfrak{q}$. Hence the assumption that $\operatorname{tr}(ay) = 0$ in $\mathfrak{q}$ implies that $\operatorname{tr}(az) = 0$ for *arbitrary* z, whence $a = 0$ [cf. § 4]. Similarly for the remaining case $\operatorname{tr}(xb) = 0$.

Now let $\mathfrak{p}$ and $\mathfrak{q}$ be referred to arbitrary co-ordinate systems and let the co-ordinates of x, y be $\xi_1, \xi_2, \cdots, \xi_g$; $\eta_1, \eta_2, \cdots, \eta_h$ respectively. Then $\operatorname{tr}(xy)$ is of the form

$$\operatorname{tr}(xy) = \sum_{(i,k)} s_{ik}\,\xi_i\,\eta_k.$$

The theorem above shows that $g \leqq h$ and $h \leqq g$, whence $h = g$, and that the coefficients s_{ik} may be considered as the coefficients of a non-singular linear transformation. Hence on choosing the co-ordinate system in $\mathfrak{q}$ in an appropriate manner $\operatorname{tr}(xy)$ may be reduced to the canonical form

$$\operatorname{tr}(xy) = \sum_{i=1}^{g} \xi_i\,\eta_i.$$

But then

$$\operatorname{tr}(xy) = \operatorname{tr}(yx) = \operatorname{tr}(yr^{-1} \cdot rx).$$

Hence the simultaneous substitution

$$x' = rx, \quad y' = yr^{-1},$$

which does not lead out of $\mathfrak{p}$, $\mathfrak{q}$ respectively, leaves the trace invariant. These two transformations are therefore contragredient in the new co-ordinate systems; our assertion (12.12) then follows immediately on writing the second of these equations in the "roofed" form $\hat{y}' = r\hat{y}$ and noting that $\hat{y}$ runs through the left-invariant sub-space $\hat{\mathfrak{p}}$ generated by $\hat{e}$ as y runs through $\mathfrak{q}$.

After this preliminary skirmish we apply the method employed before, somewhat modified, to the case (1) in which

$$l = \frac{1}{f!}\sum_s (s \times s).$$

We are now interested in the reduction (12.4) only for symmetric elements x, i.e. elements which satisfy the equations

$$x(\sigma r, sr) = x(\sigma, s) \tag{12.13}$$

for all r. This amounts to replacing x by xl; we subsequently note that $xl(e' \times e)$ is not symmetric and accordingly multiply again on the right by l. We thus replace $e' \times e$ by $l(e' \times e)l$ rather than $(e' \times e)l$ and proceed to obtain an explicit expression for the reduction, rather than calling on the aid of the reciprocity theorem. First, the components of $l(e' \times e)$ are (on ignoring the factor $1/f!$) given by

$$\sum_r e'(r\sigma)e(rs) = \sum_r \hat{e}(s^{-1}r^{-1})e'(r\sigma) = \hat{e}e'(s^{-1}\sigma).$$

This expression vanishes if $\hat{e}e' = 0$; for $e' = \hat{e}$ we find it is equal to $\hat{e}(s^{-1}\sigma) = e(\sigma^{-1}s)$. This suggests that we choose

$$1 = \sum_i \hat{e}_i, \quad 1 = \sum_i e_i$$

as the two complete reductions (12.3) of the modulus 1. The only terms in the sum (12.4) which then remain for symmetric $x = xl$ are those of the form $x(\hat{e}_i \times e_i)$, and the factor $l(\hat{e}_i \times e_i)$ is the element with components $e_i(\sigma^{-1}s)$. Since $x(\hat{e}_i \times e_i)$ has not been reduced identically to 0 on restricting x to the domain of symmetric elements, the sub-space which it generates is here, as before, equivalent to the irreducible $\hat{\mathfrak{p}}_i \times \mathfrak{p}_i$. The next step consists in multiplying on the right with l, whereby $e(\sigma^{-1}s)$ becomes, in accordance with (8.3) and (7.22),

$$\frac{1}{f!}\sum_r e(r^{-1}\sigma^{-1}sr) = \frac{1}{f!}\chi(s^{-1}\sigma) = \frac{1}{g} \cdot \varepsilon(\sigma^{-1}s).$$

Our final result is that any symmetric x can be reduced in accordance with

$$x = x\varepsilon' + x\varepsilon'' + \cdots, \quad \text{where} \quad \varepsilon(\sigma, s) = \frac{1}{f!}\varepsilon(\sigma^{-1}s)\,; \tag{12.14}$$

in deriving this result it is to be remembered that the number of times any irreducible representation appears in the regular one is given by its dimensionality.

It follows from the fact that $\varepsilon(s)$ is a class function that these elements $\varepsilon', \varepsilon'', \cdots$ constitute a set of independent idempotent elements in $\rho \times \rho$. This result is in fact obtainable by direct methods and is valid, regardless of whether the field in which we are operating is algebraically closed or not. To show this we note that any "symmetric" element $x(\sigma, s)$ is a function only of $s\sigma^{-1}$ in virtue of (12.13) : $x(\sigma, s) = x(s\sigma^{-1})$. Thus there exists a one-to-one correspondence between the symmetric elements of $\rho \times \rho$—the space of which we denote by $[\mathfrak{r} \times \mathfrak{r}]$—and the elements of $\mathfrak{r}$. Direct computation shows that this correspondence associates with each left-invariant sub-space of $[\mathfrak{r} \times \mathfrak{r}]$ a left- and right-invariant sub-space of $\mathfrak{r}$, and conversely; the reduction of $[\mathfrak{r} \times \mathfrak{r}]$ into left-invariant sub-spaces thus parallels the reduction of $\mathfrak{r}$ into sub-spaces which are both left- and right-invariant. The whole problem is thus much simpler for $[\mathfrak{r} \times \mathfrak{r}]$ than for $\mathfrak{r}$ itself; its solution is obtained by carrying over the equation

$$x = x\varepsilon' + x\varepsilon'' + \cdots \tag{7.5}$$

for the algebra ρ to $[\mathfrak{r} \times \mathfrak{r}]$, the result of which is (12.14). Nevertheless we must return to the previous less elementary analysis in order to see—and this result presupposes that the field is algebraically closed—that each of the irreducible invariant sub-spaces of $[\mathfrak{r} \times \mathfrak{r}]$ obtained in this way is equivalent to a sub-space of the algebra $\mathfrak{r} \times \mathfrak{r}$ of the form $\hat{\mathfrak{p}} \times \mathfrak{p}$ (where $\mathfrak{p}$ and $\hat{\mathfrak{p}}$ are irreducible invariant sub-spaces of $\mathfrak{r}$ with generating units e and $\hat{e}$).

The completely anti-symmetric case can be dealt with in a corresponding elementary way.

The complete reduction of the manifold $\mathfrak{R}_2^f$ of tensors in the 2-dimensional spin space $\mathfrak{R}_\nu$, $\nu = 2$, is accomplished with the aid ot the Clebsch-Gordan formula [III, (5.9)]. $(\mathfrak{c})^f$ is $\mathfrak{C}_1 \times \mathfrak{C}_1 \times \cdots \times \mathfrak{C}_1$ (f factors), where $\mathfrak{C}_1$ is the representation of the linear group $\mathfrak{c} = \mathfrak{c}_2$ by itself, and by the formula mentioned above this representation is completely reducible into the irreducible $\mathfrak{C}_v$, where v can assume only the values $f, f-2, f-4, \cdots$. The dimensionality of $\mathfrak{C}_v$ is $v + 1$, and to each of these possible dimensionalities there corresponds here but one irreducible representation. Formula (12.10) then tells us that *there exists only one term system having the multiplicity* $v + 1 (= f + 1, f - 1, f - 3, \cdots)$; compare the beginning of § 15 on this point.

The preceding analysis seems to me to be necessary in order to obtain a complete understanding of the relations implied by the permutation group without recourse to the approximation characteristic of the theory of perturbations. So far as the latter is concerned we proceed as follows. Again consider a term of the form (8.4) of the unperturbed system, the only degeneracy of which is that necessitated by the equality of the f electrons. The perturbation equation is then

$$\dot{F}(\iota_1 i_1, \cdots, \iota_f i_f) = \sum_t a(st^{-1}) \cdot F(\iota_1 k_1, \cdots, \iota_f k_f), \quad (12.15)$$

where the $a(s)$ are the exchange energies and $i_1 \cdots i_f$, $k_1 \cdots k_f$ are obtained from $1 \cdots f$ by the permutations s, t respectively. Let ϕ be the tensor in spin space defined by

$$F(\iota_1 1, \iota_2 2, \cdots, \iota_f f) = \phi(\iota_1 \iota_2 \cdots \iota_f);$$

the anti-symmetry of the double tensor F then tells us that

$$F(\iota_1 i_1, \cdots, \iota_f i_f) = \delta_s \cdot \mathbf{s}^{-1} \phi(\iota_1 \cdots \iota_f),$$

and on letting $a'(s) = \delta_s \cdot a(s)$, (12.15) becomes

$$\dot{\phi} = \mathbf{a}' \phi. \quad (12.16)$$

The problem is thus reduced to that of finding the characteristic numbers of this linear correspondence in the 2^f-dimensional space $\mathfrak{R}_2^f$.

Let $\psi_i(P)$ be the characteristic functions of the single electron. If the perturbation is due solely to the Coulomb forces between the various electrons, that part of the energy matrix $a(i_1 \cdots i_f; k_1 \cdots k_f)$ which is due to the perturbation is obtained additively from terms of the form

$$\int \cdots \int \frac{\bar{\psi}_{i_1}(P_1) \cdots \bar{\psi}_{i_f}(P_f) \cdot \psi_{k_1}(P_1) \cdots \psi_{k_f}(P_f)}{\overline{P_\alpha P_\beta}} dV_1 \cdots dV_f$$

where $\alpha \neq \beta$ and the denominator is the distance between the two points P_α and P_β. The orthogonality of the ψ tells us that this integral can be non-vanishing only if the permutation s, which sends the set of indices k into the set i (both of which are permutations of $1, 2, \cdots, f$), is either the identity or the transposition $(\alpha\beta)$. In this latter case we find

$$a(s) = E_{\alpha\beta} = \iint \frac{\bar{\psi}_\alpha(P)\psi_\beta(P) \cdot \bar{\psi}_\beta(P')\psi_\alpha(P')}{\overline{PP'}} dV\, dV'.$$

On the right-hand side of (12.16) we then have only the terms arising from $s = \mathbf{I}$ and the transpositions $s = (\alpha\beta)$:

$$\dot{\phi} = \{a(\mathbf{I}) - \sum_{(\alpha\beta)} E_{\alpha\beta}(\alpha\beta)\}\phi. \tag{12.17}$$

Dirac has given a remarkable formula for the transposition acting on a spin tensor. Let $\mathfrak{S}^\alpha$ be the spin of the αth electron ; S_x^α, S_y^α, S_z^α are then the operators

$$\begin{Vmatrix} 0 & 1 \\ 1 & 0 \end{Vmatrix}, \quad \begin{Vmatrix} 0 & -i \\ i & 0 \end{Vmatrix}, \quad \begin{Vmatrix} 1 & 0 \\ 0 & -1 \end{Vmatrix}$$

acting on the αth index of the tensor $\phi(\iota_1\iota_2 \cdots \iota_f)$. On calculating in particular

$$(\mathfrak{S}^1\mathfrak{S}^2) = S_x^1S_x^2 + S_y^1S_y^2 + S_z^1S_z^2$$

(which should perhaps be written $(\mathfrak{S}^1 \times \mathfrak{S}^2)$ instead, since $\mathfrak{S}^1$ affects only the first index and $\mathfrak{S}^2$ only the second), we find that it is the operator

ι_1 ι_2				
0 0	1			
1 0		-1	2	
0 1		2	-1	
1 1				1

acting on the first two indices, all other places being 0. Hence $\frac{1}{2}\{1 + \mathfrak{S}^1\mathfrak{S}^2)\}$ is the substitution

$$\phi(00) \rightarrow \phi(00), \ \phi(11) \rightarrow \phi(11) ; \quad \phi(10) \rightarrow \phi(01), \ \phi(01) \rightarrow \phi(10)$$

or the *transposition* of the first two indices. The energy (12.17) may then be written in the form

$$H = E_0 - \frac{1}{2} \sum_{\alpha < \beta} E_{\alpha\beta}(\mathfrak{S}^\alpha\mathfrak{S}^\beta). \tag{12.18}$$

This may be interpreted as saying that the coupling between the electrons α and β is responsible for the term $-\frac{1}{2}E_{\alpha\beta}(\mathfrak{S}^\alpha\mathfrak{S}^\beta)$ in the energy operator. However, the constant E_0 does not represent the energy of the unperturbed system.[15]

C. Explicit Algebraic Construction

§ 13. Young's Symmetry Operators

We now supplement the general theory developed above by an explicit algebraic construction of the irreducible representations of the symmetric permutation group $\pi = \pi_f$. This problem is, as we know, equivalent to that of constructing the primitive symmetry classes of tensors of order f by means of idempotent symmetry operators $\boldsymbol{e}$; here a " primitive " symmetry class is one such that the symmetry of the tensors belonging to it cannot be further increased by the addition of further symmetry conditions—such an additional condition either reproduces all the tensors of the class or reduces them all to 0. This construction is due to *A. Young* and *G. Frobenius*[16]; with its help we are able to verify step by step the entire theory of representations of the symmetry group in an explicit and elementary manner.

We are already acquainted with two very simple processes which yield tensors of maximum symmetry : " symmetrization," by means of which the tensor F yields the completely symmetric tensor $\sum_s \boldsymbol{s}F$, and " alternation," which sends F into $\sum_s \delta_s \cdot \boldsymbol{s}F$. The first of these processes can be readily generalized as follows : We divide the range from 1 to n of the " variables " $i_1 i_2 \cdots i_f$, on which the general tensor component $F(i_1 i_2 \cdots i_f)$ depends (or, what amounts to the same, the sub-indices $1, 2, \cdots, f$), into sub-sets of lengths $f_1, f_2, \cdots$; $f_1 + f_2 + \cdots = f$. We then symmetrize with respect to the indices of each of these sub-sets.

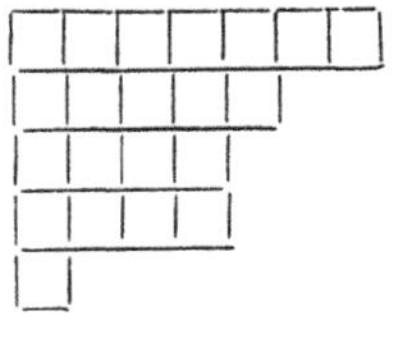

Pattern 7, 5, 4, 4, 1.

This distribution into sub-sets may be readily visualized with the aid of a " pattern " $P = P(f_1, f_2, \cdots)$ as illustrated in the accompanying figure [for the pattern $P(7, 5, 4, 4, 1)$] ; each of the f squares in the pattern is occupied by a different one of the f integers $1, 2, \cdots, f$. Each of the sub-sets mentioned above constitutes a horizontal row of the pattern, and the various rows are arranged one under another. The individual sub-sets may be arranged in order of decreasing length: $f_1 \geqq f_2 \geqq \cdots$; the pattern then consists of non-interrupted vertical columns as well as non-interrupted horizontal rows. Those permutations p which permute the members of each row among themselves constitute a sub-group (p) of π of order $f_1!f_2!\cdots$ [denoted in § 8 by $\pi(f_1, f_2, \cdots)$]. The symmetry operator described above, and which is to be applied to an arbitrary tensor, is

$$\boldsymbol{a} = \sum_p \boldsymbol{p};$$

henceforth p will always denote an arbitrary permutation which sends no numeral of one row into another row.

So far we have made no use of the process of alternation. If after having symmetrized with the aid of the operator $\boldsymbol{a}$ we alternate with respect to certain of the variables or sub-indices $1, 2, \cdots, f$, we certainly obtain 0 if any two of these numerals are in the same row, for the tensor obtained by the symmetrization is symmetric with respect to any two such numerals and the result of subsequently alternating with respect to them must be 0. To avoid this situation we choose *one* variable in each of the rows and alternate with respect to them; since the order of the variables in each row is so far immaterial we may place these chosen variables in the first column. We then disregard the first column and proceed to alternate with respect to a set of variables obtained by selecting one from each row of the remainder of the pattern; these variables may now be shifted into the second column. This process is continued until we have covered the entire pattern; the result is that *we have symmetrized with respect to the rows and have followed this symmetrization by alternation with respect to the columns.* Let q denote an arbitrary permutation which permutes the variables in each column among themselves; these q constitute a certain sub-group (q) of π. The alternation described above consists in applying the symmetry operator

$$\boldsymbol{b} = \sum_q \delta_q \cdot \boldsymbol{q},$$

and the entire process consists in applying the resultant operator

$$\boldsymbol{c} = \boldsymbol{ba} = \sum_{p,q} \delta_q \cdot \boldsymbol{qp}.$$

We call $\boldsymbol{c}$ the ***Young symmetry operator*** belonging to the pattern P.

In order to obtain a unique symmetry operator $\boldsymbol{c}$ associated with a given pattern P we must specify the way in which the numerals from 1 to n are to be distributed in P: they shall be introduced in such a way that on reading the pattern, as one would read a page of a book, they appear in their natural order $1, 2, \cdots, f$. If we write them in any other order, say that obtained from the standard form with the aid of the permutation r, we obtain a "conjugate" element $\boldsymbol{c}_r$ which, as is readily seen on considering the relation between the tensors generated by these two operators, is related to $\boldsymbol{c}$ by

$$\boldsymbol{c}_r\boldsymbol{r} = \boldsymbol{rc} \quad \text{or} \quad c_r(s) = c(r^{-1}sr).$$

Hence the introduction of r results merely in a new name.

From now on we operate with symmetry quantities, i.e. elements of the algebra (π), instead of tensors; we consider the invariant sub-space $\mathfrak{p}_c$ of $\mathfrak{r}$ consisting of all elements of the form $y = x\hat{c}$ and the representation $\mathfrak{h}_c$ of π induced in it by the regular representation. With $\mathfrak{p}_c$ is associated the symmetry class $\mathfrak{P}_c$ of all tensors of the form $\boldsymbol{c}F$. If we replace c by one of its conjugates c_r we obtain instead of $\mathfrak{p}_c$ an equivalent invariant sub-space; in this sense the order in which the variables are written in the pattern is quite immaterial. We hope that $\mathfrak{p}_c$ is irreducible and that the totality of representations $\mathfrak{h}_c$ associated with all possible patterns constitutes a complete set of inequivalent irreducible representations of π. This hope is strengthened by the fact that the total number of patterns is just equal to the number of inequivalent irreducible representations. To show this we note that the number of patterns is equal to the number of partitions of f into integral non-negative summands $f = f_1 + f_2 + \cdots$ which satisfy the condition $f_1 \geqq f_2 \geqq \cdots$. On writing

$$f_1 - f_2 = r_1, \quad f_2 - f_3 = r_2, \quad \cdots$$

we see that this number is equal to the number of solutions of the equation

$$1r_1 + 2r_2 + 3r_3 + \cdots = f$$

for non-negative integral r. But we have already seen that this is the number of classes of conjugate elements in π and, by the general theory, is therefore equal to the number of inequivalent irreducible representations of π.

If the dimensionality n of the vector space is less than f the only non-vanishing symmetry classes are those arising from patterns containing at most n rows, for if the first column is longer than n alternation with respect to the variables standing in it alone causes an arbitrary tensor to go over into 0. The only patterns which we need in this case are consequently those obtainable from the algebra $\mathfrak{r}_0$, instead of $\mathfrak{r}$, where $\mathfrak{r}_0 = \natural\mathfrak{R}^f$ as defined in § 2 above. The number of inequivalent irreducible invariant sub-spaces into which the tensor space $\mathfrak{R}^f$ can be reduced is accordingly decreased to the number of partitions of f into n integral summands $f = f_1 + f_2 + \cdots + f_n$ for which $f_1 \geqq f_2 \geqq \cdots \geqq f_n \geqq 0$.

A permutation $s = qp$ which is obtained by composition from a permutation p of (p) and a permutation q of (q) can be so obtained in only one way. This is an immediate consequence of the remark that the equation $qp = \mathsf{I}$ can be fulfilled only by $p = \mathsf{I}$, $q = \mathsf{I}$, for it asserts that $p = q^{-1}$ belongs to (p) as well as to (q). The components of the symmetry operator c can therefore be described as follows: *$c(s) = 0$ unless s belongs to the set $(q)(p)$; when s belongs to this set $c(s) = \pm 1$ according as the unique decomposition $s = qp$ yields an even or an odd permutation q.*

We must now prove the following three assertions concerning c:

(1) c is essentially idempotent; or, more precisely, c satisfies an equation $cc = \gamma \cdot c$, where γ is a non-vanishing numerical factor. Furthermore, γ is an integral positive number which is a factor of $f!$. Then replacing e by $e = c/\gamma$, e is idempotent.

(2) The sub-space $\mathfrak{p}_c$ is irreducible, the e introduced in (1) is primitive.

(3) Different patterns lead to inequivalent sub-spaces $\mathfrak{p}_c$.

The execution of this programme depends upon a simple combinatorial auxiliary theorem, which we now proceed to develop. Denote the lengths of the columns in the pattern P with rows of lengths $f_1, f_2, \cdots$ by $f_1^*, f_2^*, \cdots$:

$$f_1 \geqq f_2 \geqq \cdots, f_1^* \geqq f_2^* \geqq \cdots;$$
$$f_1 + f_2 + \cdots = f_1^* + f_2^* + \cdots = f.$$

We think of the pattern P as cut out of a rectangular chessboard consisting of f_1 horizontal rows and f_1^* vertical columns, and the permutation s as operating on f chess-men occupying the f fields. On interchanging rows and columns in P we obtain the *dual or transposed pattern P^**.

Auxiliary Theorem. A permutation s belongs to (qp) if and only if any two pieces originally in the same row are not sent into the same column by s.

Proof. It is evident that this condition is necessary in order that s belong to (qp). The change of position which one of the pieces suffers as a result of s can be accomplished in two moves, a horizontal and a vertical move (in this order). It is at first conceivable that the horizontal move could send the piece into a field of the original board which is not contained in the pattern P. If the decomposition $s = qp$ is possible p must represent the horizontal move and q the subsequent vertical one; it is clear that q and p are thus uniquely determined. Now if s satisfies the conditions enunciated in the above theorem the horizontal move can never throw them into the same column, i.e. the same field. It only remains to show that the horizontal move can never send any piece out of the pattern proper, or: *those pieces which s sends into a column of length f^* come from the first f^* rows of the pattern.* We divide the chess-board horizontally into an upper and a lower part, the upper consisting of the first f^* rows. The pieces which are sent into the *first* column by s are, by assumption, from f_1^* different rows; hence there are at least (and therefore exactly) $f_1^* - f^*$ of them which come from the lower part of the board and not from the first f^* rows. Note that $f_1^* - f^*$ is exactly the number of fields in the first column which lie in the lower part of the board. On applying this argument to each column in succession we find that the number of pieces which s sends into those columns which protrude into the lower part of the board is exactly equal to the number of fields in this part of the board. Hence all the pieces in the lower part of the pattern are sent into columns whose lengths are greater than f^*, and the only pieces s sends into a column of length f^* come from the upper part of the board.

This auxiliary theorem allows us to assert that if s does not belong to (qp) then there exist two pieces in a single row which are sent into the same column by s. If u denotes the transposition of the two pieces in their initial positions and v their transposition in the final then $su = vs$; here u belongs to (p) and v to (q).

§ 14. Irreducibility, Linear Independence, Inequivalence, and Completeness

We now examine the Young symmetry operators c associated with the various patterns. Obviously

$$c(sp) = c(s), \quad c(qs) = \delta_q \cdot c(s), \tag{14.1}$$

where p, q are, as usual, elements of (p), (q), respectively.[17]

Theorem (14.2). *Any element a of (π) which satisfies equations* (14.1):

$$a(sp) = a(s), \quad a(qs) = \delta_q \cdot a(s), \tag{14.3}$$

is a multiple of c.

To prove this theorem we first note that (14.3) implies

$$a(qp) = \delta_q \cdot a(\mathbf{I}) ;$$

on setting $a(\mathbf{I}) = \lambda$ the equation

$$a(s) = \lambda \cdot c(s),$$

which is to be proved, is certainly correct for all group elements s of the form qp. We must next show that $a(s) = 0$ if s does not belong to the set (qp). Such an s implies that there exist transpositions u and v, lying in (p) and (q) respectively, for which $su = vs$. But then by (14.3)

$$a(su) = a(s), \quad a(vs) = \delta_v \cdot a(s) = - a(s),$$

whence $a(s) = - a(s)$ or $a(s) = 0$.

Theorem (14.4). *Every element of* (π) *of the form* cxc *is a multiple of* c.

It was shown in the general theory that this theorem is valid if c is a primitive idempotent element of (π) and if the field in which we operate is algebraically closed ; here we approach it from the opposite direction, as we wish to show directly that it holds for c in order to prove that c is primitive. Now obviously any element of the form xc satisfies the first of equations (14.3) and any element cx the second ; hence any element of the form cxc has both properties and is consequently a multiple of c.

Theorem (14.5). $cc = \gamma c$ *and* γ *is a positive integer which is contained in* $f!$.

That cc is a multiple of c follows immediately from the previous theorem ; γ is therefore the number

$$\gamma = \sum_{tt' = \mathbf{I}} c(t)c(t') = \sum_s c(s) \cdot c(s^{-1}).$$

Let the sub-space $\mathfrak{p}_c$ of elements of the form $x\hat{c}$ be of dimensionality g. The projection

$$x \rightarrow y = x\hat{c} \tag{14.6}$$

projects any element x into an element lying in this sub-space and is, within $\mathfrak{p}_c$ itself, merely the multiplication $y = \gamma x$. Its trace is therefore γg ; to see this we need merely to adapt the co-ordinate system in group space to the sub-space $\mathfrak{p}_c$. On the other hand its trace is immediately obtainable from (14.6) or

$$y(s) = \sum_t x(t)c(s^{-1}t) ;$$

it is $f!c(\mathbf{I}) = f!$, hence

$$\gamma g = f!.$$

Consider the meaning of this fact that γ is positive, i.e. that $c(s)c(s^{-1})$ is oftener positive than negative!

$e = c/\gamma$ is idempotent; hence the character of the representation $\mathfrak{h}_c$ induced in $\mathfrak{p}_c$ by the regular representation is by (8.3)

$$\chi(s) = \frac{1}{\gamma}\sum_r c(r^{-1}sr). \tag{14.7}$$

We obtain as a by-product the fact that *the dimensionality g of the representation* $\mathfrak{h}_c$ *is a factor of* $f!$.

Theorem (14.8). $\mathfrak{p}_c$ *is irreducible.*

We know already that this theorem is a consequence of (14.4), but it may be instructive to prove it directly as follows. Let $e = c/\gamma$ be reduced into two independent idempotent elements $e_1 + e_2$; then

$$ee_1 = e_1e = e_1, \quad \text{whence} \quad ee_1e = e_1.$$

Now by theorem (14.4) any element of the form ee_1e is a multiple of e; hence $e_1 = \lambda e$. $e_1e_1 = e_1$ then yields the equation $\lambda^2 = \lambda$ for the number λ. Consequently either $\lambda = 1$ or $\lambda = 0$, i.e. either $e_1 = e$ or $e_1 = 0$.

We shall say that the pattern P' with rows of lengths $f'_1, f'_2, \cdots$ is *higher* than P if the first non-vanishing difference $f'_1 - f_1, f'_2 - f_2, \cdots$ is positive.

Theorem (14.9). *If the pattern P' is higher than P then* $c'c = 0$.

We do not here assume that the variables are written in the patterns P, P' in the normal form agreed upon in the previous section—i.e. in which the numerals appear in their natural order on reading the pattern as one would a page of a book. The proof is based on the fact (F) that there exist two numerals which are in the same row in the pattern P' and in the same column in the pattern P. If v is their transposition it belongs to the group (p') associated with the rows of P' and at the same time to the group (q) associated with the columns of P; hence

$$c'(sv) = c'(s), \quad c(vs) = -c(s).$$

On replacing vt in

$$c'c(s) = \sum_t c'(st^{-1})c(t) = -\sum_t c'(st^{-1})c(vt) \tag{14.10}$$

by t alone we find

$$c'c(s) = -\sum_t c'(st^{-1}v)c(t) = -\sum_t c'(st^{-1})c(t) = -c'c(s). \tag{14.11}$$

(F) is evident if the first row of P' is already longer than the first row of P, for it is impossible to distribute the f_1' numerals in the first row of P' over different columns of P if $f_1 < f_1'$. If $f_1' = f_1$ and the numerals of the first row of P' are actually distributed over different columns of P, we discard the first row of P' and the f_1 fields of P containing the same numerals as this row. On shifting the fields of P upward to fill in the gaps P is transformed into a pattern which has exactly the same appearance as if we discarded the first row of S; we are only interested in the fact that this process leaves all pieces in their original column. The proof can then be completed by mathematical induction—by assuming that it holds for the abbreviated patterns obtained by omitting the first rows of P and P'.

Theorem (14.12). *Let c, c', $\cdots$ be the Young symmetry operators associated with different patterns P, P'. $\cdots$; the corresponding sub-spaces $\mathfrak{p}_c$, $\mathfrak{p}_{c'}$, $\cdots$ are then linearly independent.*

Let the P, P', P'', $\cdots$ be arranged in such an order that P is higher than P', P' higher than P'', $\cdots$. An element x of $\mathfrak{p} = \mathfrak{p}_c$ is reproduced by right-multiplication with $\hat{c}/\gamma$ but, by the previous theorem, this process transforms all elements x' of $\mathfrak{p}'$, x'' of $\mathfrak{p}''$, $\cdots$ into 0. Assume there exists such a linear dependence

$$x + x' + x'' + \cdots = 0;$$

on right-multiplication with $\hat{c}$ we find $x = 0$ and consequently $x' + x'' + \cdots = 0$. The theorem is thus reduced to the same theorem for the smaller set P', P'', $\cdots$, and the proof follows by mathematical induction.

Theorem (14.13). *Different patterns P, P' give rise to inequivalent sub-spaces $\mathfrak{p}_c$, $\mathfrak{p}_{c'}$.*

The proof is accomplished by a direct derivation of the orthogonality relations. Let P' be higher than P. Since we did not assume in proving theorem (14.9) that the numerals were distributed in the same order in the two patterns P and P', we may replace the element c with components $c(s)$ by the "conjugate" element $c_{r^{-1}}$ with components $c(rsr^{-1})$:

$$\sum_t c'(st^{-1})c(rtr^{-1}) = 0.$$

Summation with respect to r yields

$$\sum_t c'(st^{-1}) \cdot \chi_c(t) = 0.$$

On writing $\chi = \chi_c$, $\chi' = \chi_{c'}$ this formula is equivalent to

$$\sum_t \chi'(st^{-1})\chi(t) = 0.$$

In particular

$$\sum_t \chi'(t^{-1})\chi(t) = 0.$$

If the two sub-spaces were equivalent we would have $\chi'(t) = \chi(t)$, and since $\chi(t^{-1}) = \chi(t)$ for the symmetric group the above equation would yield

$$\sum_s \chi^2(s) = 0.$$

But this is impossible, for by (14.7) the character $\chi(s)$ has rational components, and in particular $\chi(\mathbf{1}) = g \neq 0$.

This last conclusion is valid only if the number field in which we operate is non-modular; naturally this restriction is irrelevant for physics. Nevertheless it constitutes a blemish which should be removed, for the remainder of our deductions only introduce the minimum assumption that $f!$ is not 0 in the field under consideration. Now from the general theory we know that

Theorem (14.14). $\quad \sum_s \chi(s)\chi(s^{-1}) = f!.$

The blemish mentioned above is removed by proving this theorem directly. We must show that

$$\sum_s \chi(s^{-1}) \cdot e(s) = 1$$

or

$$\sum_{r,s} e(rs^{-1}r^{-1})e(s) = 1.$$

On replacing the summation variable s by sr, where r is fixed, this becomes

$$\sum_{r,s} e(sr)e(s^{-1}r^{-1}) = 1. \tag{14·15}$$

Consider next the function

$$a(s, s') = \sum_r e(sr)e(s'r^{-1});$$

as a function of s it satisfies the second condition in (14.3). But the first of these conditions is also satisfied, as can be seen immediately by replacing r in

$$a(sp, s') = \sum_r e(spr)e(s'r^{-1})$$

by the summation variable $p^{-1}r$. Hence by (14.2)

$$a(s, s') = c(s) \cdot \sum_r e(r)e(s'r^{-1}) = c(s) \cdot e(s') = \frac{1}{\gamma}c(s)c(s')$$

and therefore the left-hand side of (14·15) or

$$\sum_s a(s, s^{-1}) = \frac{1}{\gamma}\sum_s c(s)c(s^{-1})$$

is actually equal to 1.

The relations

$$\sum_s \chi(s)\chi'(s^{-1}) = 0 \text{ or } f! \tag{14.16}$$

show that the primitive characters obtained by our construction from the various symmetry patterns are linearly independent, and since their number is equal to the number of classes of conjugates in the group π, any class function can be represented as a linear combination of the $\chi(s)$. In particular, the function $1(s)$, which is 1 for $s = \mathbf{I}$ and otherwise 0, must possess such an expansion:

$$f! \cdot 1(s) = m\chi(s) + m'\chi'(s) + \cdots. \tag{14.17}$$

Multiplying by $\chi(s^{-1})$ and summing over s we obtain, with the aid of the orthogonality relations (14.16), the equation

$$f!\,\chi(\mathbf{I}) = f!\,m$$

or

$$\boxed{m = g} \tag{14.18}$$

for m. Since

$$\chi(s) = \sum_r e(rsr^{-1}) = \sum_r e_r(s),$$

equation (14.17) gives the reduction of the modulus **1** into primitive idempotent elements e_r. Hence the regular representation is reduced into the irreducible representations $\mathfrak{h}_c$ associated with the various symmetry patterns. Since $f!\,1(s)$ is the character of the regular representation, eq. (14.18) is a direct verification of the fact—proved in the general theory—that the number of times each irreducible representation appears in the regular representation is equal to its dimensionality. This completes our direct and elementary development of the theory of the representations of the symmetric group.

The method of proof employed in establishing theorem (14.9), i.e. that $cc' = 0$ if P' is *lower* than P, will now be used to answer another question. Let a be the operator, introduced in the previous section, which symmetrizes with respect to the ciphers occupying the rows of P:

$$a(s) = 1 \text{ or } 0 \text{ according as } s \text{ belongs to } (p) \text{ or not,}$$

and let the numerals be written in the pattern P', which is lower than P, in an arbitrary order. I assert that $ac' = 0$. There exist two numerals which occupy the same row in P and the same column in P'. If v is the transposition of these two numerals then

$$a(sv) = a(s), \quad c'(vs) = - c'(s),$$

and the assertion is proved with the aid of (14.10), (14.11) on replacing c', c there by a, c'. Hence also

$$\sum_t a(st^{-1})c'(rtr^{-1}) = 0,$$

i.e.

$$\sum_t a(st^{-1})\chi'(t) = 0 \quad \text{or} \quad \sum_r a(r^{-1})\chi'(rs) = 0.$$

That is, the sum of the $\chi'(t)$ extended over all elements $t = rs$ which are left-equivalent to s mod. (p) [i.e. r in (p)], is zero. In particular, $\sum_s \chi'(s) = 0$, where the sum is extended over all elements s of (p); χ' is the character associated with a pattern P' which is lower than P. On applying this result to the considerations of § 8 (in particular, to (8·13) ff.) we find:

If the individual I has the simple energy levels E_1, E_2, $\cdots$ the term

$$f_1E_1 + f_2E_2 + \cdots \quad (f_1 \geqq f_2 \geqq \cdots, \; f_1 + f_2 + \cdots = f)$$

of the unperturbed system I^f appears only in those symmetry classes of tensors whose pattern $P' = P(f_1', f_2' \cdots)$ is not lower than $P = P(f_1 f_2 \cdots)$.

Thus we saw in discussing the two-electron problem that terms of the form $E_1 + E_2$ appeared in the " anti-symmetric " as well as the " symmetric " term systems, whereas terms such as $2E_1$ appeared only in the latter.

Finally, we consider the relations existing between two dual patterns P and P^* with generators c, c^* and characters χ, χ^*. The group (p) which permutes the members of each row of P among themselves coincides with the group (q^*) which permutes the members in each column of P^* among themselves; similarly $(q) = (p^*)$. If $s = qp$ is in (qp), then $s^{-1} = p^{-1}q^{-1} = q^*p^*$ is in (q^*p^*), and conversely; for such an element

$$c(s) = \delta_q, \quad c^*(s^{-1}) = \delta_{q^*} = \delta_p.$$

Hence in general—even when s is not in (qp) and, consequently, s^{-1} is not in (q^*p^*)—we have

$$c^*(s^{-1}) = \delta_s \cdot c(s).$$

"Dual" elements c, c^* are therefore related to each other in exactly the same way as the "duals" introduced in § 12. Further

$$\gamma^* = \gamma\,; \quad \chi^*(s^{-1}) = \chi^*(s) = \delta_s \cdot \chi(s)\,; \quad g^* = g.$$

If P is higher than Q, then conversely P^* is lower than Q^*. For if we lower P by taking away the last field of one of the rows of P and adding it to the end of a later (shorter) row, one of the columns of P is increased at the expense of a later (shorter) column; by such a process of shifting individual fields, in which no gap is to occur in a row or a column, P can be transformed into the lower pattern Q.

§ 15. Spin and Valence. Group-theoretic Classification of Atomic Spectra

If the vector space $\mathfrak{R} = \mathfrak{R}_2$ is only 2-dimensional, the only symmetry patterns P which give rise to primitive symmetry classes of tensors of order f are those which consist of at most two rows. Let the first row contain $l + v$ fields and the second l; then

$$v = f - 2l.$$

The symmetry pattern P is thus uniquely characterized by the number v, which we call its *valence*, and v may assume any of the values $f, f - 2, f - 4, \cdots$. Let $\mathfrak{P}_v$ be the totality of tensors of the form $\boldsymbol{c}F$ obtained by applying the Young symmetry operator $\boldsymbol{c}$ associated with the pattern P to the totality of tensors F, and let $\mathfrak{H}_v$ be the representation of the linear group, the substratum of which is the tensor manifold $\mathfrak{P}_v$. A sufficiently general tensor of order f which is symmetric in the first as well as the second rows of indices is given by

$$\begin{array}{ll} \mathfrak{x} \times \mathfrak{x} \times \cdots \times \mathfrak{x} \times \mathfrak{x} & (l + v \text{ terms}) \\ \times\, \mathfrak{y} \times \mathfrak{y} \times \cdots \times \mathfrak{y} & (l \text{ terms}), \end{array}$$

where

$$\mathfrak{x} = (x_1, x_2), \quad \mathfrak{y} = (y_1, y_2)$$

are two arbitrary vectors. On alternating with respect to the columns we find that the representation $\mathfrak{H}_v$ of the linear group $\mathfrak{c} = \mathfrak{c}_2$ is that one which is induced on the quantities

$$(x_1 y_2 - x_2 y_1)^l \cdot x_1^{r_1} x_2^{r_2} \quad (r_1 + r_2 = v).$$

Hence $\mathfrak{H}_v$ is the representation of the linear group which was denoted in III, § 5, by $\mathfrak{C}_v$.

This remark supplies the connection with the symmetry problem of quantum mechanics as dealt with in § 12—on applying the Pauli exclusion principle when the existence of the spin, but not its dynamical effect, is taken into account.[18] Since the spin space is 2-dimensional, formula (12.10) tells us that the only patterns P which give rise to a term system are those whose duals P^* consist of at most two rows, i.e. those P which themselves have but two columns. If v is now the number of fields by which the first column of P exceeds the second we call v the *valence* of the term system or of the corresponding state of the atom. The multiplicity of the term system with valence v is $v + 1$, and to each of these possible multiplicities corresponds but one term system as we have already seen in § 12 (in particular p. 356). We previously (Chap. IV) called $s = v/2$ the "spin quantum number."

The fact that the longest column of P cannot exceed the dimensionality N of the vector space $\mathfrak{R}_t$ associated with the electron translation may result in a further restriction on the possible symmetry patterns P. This situation cannot arise as long as we deal with the total ∞-dimensional system space. On the other hand if we restrict ourselves, for example, to those states of the electron which are characterized by a fixed principal quantum number n and a fixed azimuthal quantum number l—and which therefore constitute a $(2l + 1)$-dimensional sub-space $\mathfrak{R}(nl)$ within $\mathfrak{R}_t$—i.e. if we consider only those states of the atom in which all the f electrons outside a closed core are in $\mathfrak{R}(nl)$, the dimensionality N is reduced to $2l + 1$. Then f cannot exceed $2(2l + 1)$ and the possible valences of the states under consideration are given by the following table:

$f =$	1,	2,	3,	4,	......,	$4l$,	$4l + 1$,	$4l + 2$
	1	0	1	0	....	0	1	0
v		2	3	2	....	2		
				4				

This table again gives us the alternation law, but shows that in addition the number of possibilities decreases from the middle of the table on. The possible multiplet numbers $2s + 1$ of terms in these states is one greater than v.

This "valence" v, which describes the symmetry state of the system, is actually the chemical valence, as was shown by *F. London.*[19] We allow two atoms, consisting of f_1, f_2 electrons respectively, to come together to form a molecule with $f = f_1 + f_2$ electrons. Let $\mathfrak{P}_1$, $\mathfrak{P}_2$ be irreducible invariant sub-spaces of the system spaces $\mathfrak{R}_t^{f_1}$, $\mathfrak{R}_t^{f_2}$ respectively. In order to find which symmetry states the molecule is capable of assuming when the first atom is in the state $\mathfrak{P}_1$ and the second in $\mathfrak{P}_2$ we must completely reduce the space $\mathfrak{P}_1 \times \mathfrak{P}_2$ into its irreducible constituents. If we consider this decomposition as taking place in the vector space of electron spin rather than in that of electron translation (the justification for which will be given below), the problem is solved by the Clebsch-Gordan series (III, 5.9); it tells us that if the valences of the symmetry states of the two atoms are v_1, v_2 the resulting symmetry states of the molecule are those with valences

$$v = v_1 + v_2,\ v_1 + v_2 - 2,\ v_1 + v_2 - 4,\ \cdots,\ |v_1 - v_2|. \tag{15.1}$$

This situation can be readily visualized in terms of the symmetry patterns as follows. Bring the two symmetry patterns P_1, P_2 of the two atoms into the positions shown in the accompanying diagram and then shove 2 vertically upwards, one field at a time, until one of the two columns of the combined pattern is closed; each of these steps represents a possible symmetry pattern for the molecule, in which v is the number of fields which are not paired horizontally. The saturation of the valence bonds here appears as the pairing of fields or, more physically, as the saturation of the spin of an electron in one of the atoms with that of an electron in the other. The empirical theory of the valence bond has therefore a rather profound significance.

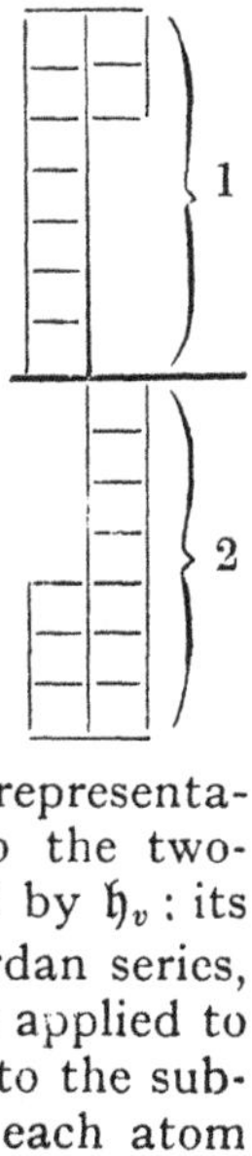

We have yet to justify our use of spin space rather than translation space in the above. Let the representation of the permutation group π_f corresponding to the two-columned symmetry pattern of valence v be denoted by $\mathfrak{h}_v$: its dual $\mathfrak{h}_v^*$ consists of but two rows. The Clebsch-Gordan series, together with the third reciprocity theorem of § 10 as applied to the linear group $\mathfrak{c} = \mathfrak{c}_2$, tells us that on restricting π to the subgroup $\pi' = \pi_1 \times \pi_2$ which permutes the electrons of each atom separately the representation $\mathfrak{h}_v^*$ of π contains the irreducible representation $\mathfrak{h}_{v_1}^* \times \mathfrak{h}_{v_2}^*$ of π' once or not at all, according as v is one of the values (15.1) or not. From this it follows immediately that the same result holds for the duals on reducing $\mathfrak{h}_v$ after restricting π to π'. Applying the same reciprocity theorem in the opposite direction for the case in which $\mathfrak{c} = \mathfrak{c}_n$ is the linear group in n dimensions, we find that the representation $\mathfrak{H}_{v_1} \times \mathfrak{H}_{v_2}$ of $\mathfrak{c}$ (or the algebra Σ) contains the representation

$\mathfrak{H}_v$ once or not at all according as v is one of the values (15.1) or not. On reducing $\mathfrak{H}_{v_1} \times \mathfrak{H}_{v_2}$ into its irreducible constituents we may expect to find other representations—which may even occur more than once—in addition to these simple $\mathfrak{H}_v$, but these additional representations will correspond to symmetry patterns with more than two columns and are, in virtue of the Pauli exclusion principle, of no importance for physics. The number b introduced in § 11 is accordingly at most equal to 1 in the case of diatomic molecules.

Molecules which consist of a larger number of atoms can be studied by the same method. If in particular we are interested in the case of three atoms and their valences are v_1, v_2, v_3, we can determine with the aid of the Clebsch-Gordan series the number b_v of times the representation $\mathfrak{C}_v$ occurs in the reduction of $\mathfrak{C}_{v_1} \times \mathfrak{C}_{v_2} \times \mathfrak{C}_{v_3}$. Those v for which $b_v \neq 0$ are the valences of the possible symmetry states of the molecule and $b = b_v$ (which may here be greater than 1) are the corresponding multiplicities. The characterization of the quantum and symmetry states of a molecule which is formed by the union of three atoms in given quantum and symmetry states requires, in addition to the valence v, a further index which distinguishes between the various b_v possible energy levels. But this description of the various possibilities differs from the empirical theory of the valence bond—the manifold of possible bindings is smaller.[20]

Classification of Spectral Terms.

Let the unitary or the complete linear group $\mathfrak{c}_{\nu n}$ in the system space $\mathfrak{R}$ of the single electron be restricted to the group $\mathfrak{c}_\nu \times \mathfrak{c}_n$ of transformations $S_\nu \times S_n$, the two factors of which are transformations of the spin and translation spaces $\mathfrak{R}_\nu$, $\mathfrak{R}_n$ respectively: $\mathfrak{R} = \mathfrak{R}_\nu \times \mathfrak{R}_n$. The space $\{\mathfrak{R}^f\}$ of anti-symmetric tensors of order f is then reducible into irreducible invariant sub-spaces with respect to the algebra of symmetric transformations of the form (12.2). We thus obtain a distribution (I) of spectral terms among the various symmetry classes; this step is of universal validity and is applicable to molecules as well as atoms.

The further classification of terms, as discussed in Chapter IV, A, refers to " simple " rather than " quantum " states, i.e. to those states which are related to spatial rotation and moment of momentum in the same way that the quantum states are related to displacement in time and energy. Naturally this application of the rotation group $\mathfrak{d} = \mathfrak{d}_3$ (the elements of which we now denote by σ, τ, $\cdots$) is significant only for atoms (or ions), the nuclei of which are considered as fixed centres of force. So long as we concern ourselves only with the electron translation and neglect the mutual perturbations of the electrons, which are characterized by principal and azimuthal quantum numbers n and l, each individual term of the system is characterized by the quantum numbers $(n_1, l_1;\ n_2, l_2;\ \cdots;\ n_f, l_f)$. The number of times such a term appears in a given symmetry system is equal to the dimensionality of the linear sub-space in which the atomic states under consideration lie. The resolution caused by the mutual perturbations parallels the reduction of this sub-space into its irreducible constituents $\mathfrak{R}_L$ with respect to the group $\mathfrak{d}$ of rotations; the resulting components of the term have the natural multiplicities $2L + 1$. The spin space is similarly to be reduced. Let $\mathfrak{d}$ induce the representations $\mathfrak{h}_\nu : \sigma \rightarrow U(\sigma)$ and $\mathfrak{E} : \sigma \rightarrow V(\sigma)$ in $\mathfrak{R}_\nu$ and $\mathfrak{R}_n$ respectively. This second step (II), in which the spin and translation spaces are considered separately, is interpreted from the stand-point of group theory as meaning that we associate with the element (σ, τ) of $\mathfrak{d} \times \mathfrak{d}$ the transformation $U(\sigma) \times V(\tau)$; we thus obtain a 6-parameter sub-group of $\mathfrak{c}_\nu \times \mathfrak{c}_n$, and on restricting $\mathfrak{c}_\nu \times \mathfrak{c}_n$ to this sub-group our original irreducible sub-space is further completely reducible into irreducible constituents. The irreducible representation of $\mathfrak{d} \times \mathfrak{d}$ induced in such a sub-space is of the type $\mathfrak{D}_s \times \mathfrak{D}_l$. The final step, (III), consists in introducing the coupling $\sigma = \tau$: the 6-parameter sub-group is thereby restricted to a 3-parameter sub-group, i.e. that sub-group induced in the total system space by the rotations $\mathfrak{d}$. The spin perturbation then resolves each such term multiplet into its (at most $2s + 1$) components:

$$\mathfrak{D}_s \times \mathfrak{D}_l = \sum_j \mathfrak{D}_j \quad (j = l + s,\ l + s - 1,\ \cdots,\ |l - s|);$$

naturally $\mathfrak{D}_s \times \mathfrak{D}_l$ is here a representation of $\mathfrak{d}$ instead of $\mathfrak{d} \times \mathfrak{d}$.

Actually $\nu = 2$, and the transformations induced in the spin space $\mathfrak{R}_2$ by the rotation group constitute the unitary group in two dimensions. Consequently the transition from $\mathfrak{c}_\nu$ to $\mathfrak{h}_\nu$ in step (II) involves no reduction in spin space—this is the essential simplification caused by the fact that $\mathfrak{R}_\nu$ has so small a dimensionality.

To the symmetry system of terms corresponds a certain irreducible representation of the unitary group $\mathfrak{u}$ in the space $\mathfrak{R}_t$ of the electron translation and with it a certain irreducible characteristic (§ 9)

$$\mathsf{X} = \mathsf{X}(\varepsilon_1, \varepsilon_2, \cdots).$$

The co-ordinates x_i in the space $\mathfrak{R}_t$ are broken up into classes in the manner described in Chapter IV, § 1 :

$$x(m) \quad [m = l, l-1, \cdot \cdot \cdot, -l] ;$$
$$x'(m') \quad [m' = l' \cdot l' - 1, \cdot \cdot \cdot, -l'] ; \quad \cdot \cdot \cdot .$$

Each of these classes describes a $(2l + 1)$-dimensional sub-space $\mathfrak{R}(nl)$ of $\mathfrak{R}_t$ in which the group $\mathfrak{d}_3$ of spatial rotations induces the irreducible representation $\mathfrak{D}_l$ and is characterized by the principal quantum number n and the azimuthal quantum number l. The arguments ε_i of **X** are correspondingly broken up into classes. To give the principal and azimuthal quantum numbers of the individual electrons—without stating how these numbers are distributed among the f electrons—we need only to state how many (f') electrons are represented by states in each of the various sub-spaces $\mathfrak{R}' = \mathfrak{R}(nl)$. If, for example, 3 of the electrons are in $\mathfrak{R}'$ and the remaining 5 in $\mathfrak{R}''$ $(f = 8)$ we must separate out that part of **X** which is of degree 3 in the variables ε_i belonging to $\mathfrak{R}'$ and of degree 5 in those belonging to $\mathfrak{R}''$. The multiplicity M of the corresponding term

$$E(n_1 l_1) + E(n_2 l_2) + \cdot \cdot \cdot + E(n_f l_f)$$

of the " unperturbed " atom in the symmetry system under consideration is then obtained from the part of **X** described above by setting all ε contained in it equal to unity. In order to determine how this M-fold term is broken up on taking the mutual influence of the electrons into account we replace the variables $\varepsilon(m)$ of the class $\mathfrak{R}(nl)$ by $\varepsilon(m) = \varepsilon^m$, the variables $\varepsilon'(m')$ of the class $\mathfrak{R}(n'\varepsilon')$ by $\varepsilon'(m') = \varepsilon^{m'}$ (with the same ε), etc. The resulting expression must be a linear combination of the sums

$$\sum_{m=-L}^{+L} \varepsilon^m = \frac{\varepsilon^{L+1} - \varepsilon^{-L}}{\varepsilon - 1}$$

with non-negative integral coefficients. This enables us to tell which of the various total azimuthal quantum numbers L appear, and how often, in the resolution of the above term ; each such L-term has still the multiplicity $2L + 1$.

Example. We consider, as an example, the case in which $f = 3$ and all three electrons are in the same sub-space $\mathfrak{R}(nl)$. The possible symmetry patterns are

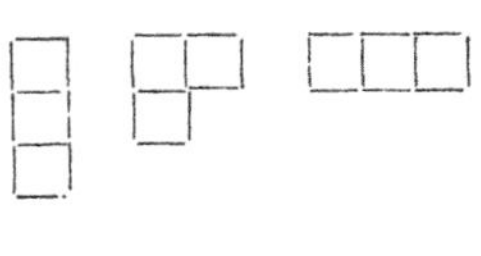

The Pauli exclusion principle allows only the first two; their valences are $v = 3$ and $v = 1$, and the corresponding terms are therefore quadruplets and doublets, respectively. The first pattern defines the anti-symmetric tensors of order 3 and the third the symmetric tensors. The corresponding characteristics are therefore

$$\mathrm{X}_1 = s_1 = \sum_{i<j<k} \varepsilon_i \varepsilon_j \varepsilon_k, \quad \mathrm{X}_3 = \sum_{i \leq j \leq k} \varepsilon_i \varepsilon_j \varepsilon_k.$$

On introducing

$$s_2 = \sum_{i \neq j} \varepsilon_i^2 \varepsilon_j, \quad s_3 = \sum \varepsilon_i^3$$

we have $\mathrm{X}_3 = s_1 + s_2 + s_3$. The dimensionalities of the representations of π_3 corresponding to these three patterns, and therefore the numbers of times the representations X_1, X_2, X_3 of $\mathfrak{c}$ appear in $(\mathfrak{c})^3$, are easily shown to be 1, 2, 1, (in accordance with the equation $3! = 1^2 + 2^2 + 1^2$). Now the characteristic of the representation $(\mathfrak{c})^3$ of $\mathfrak{c}$ is

$$t_1 = (\sum_i \varepsilon_i)^3 = s_3 + 3s_2 + 6s_1; \tag{15.2}$$

the equation

$$t_1 = \mathrm{X}_1 + 2\mathrm{X}_2 + \mathrm{X}_3 = (2s_1 + s_2 + s_3) + 2\mathrm{X}_2$$

then allows us to conclude that

$$\mathrm{X}_2 = s_2 + 2s_1.$$

We prefer to carry out the evaluation with the aid of the sums of powers

$$t_1, \quad t_2 = \sum_i \varepsilon_i^2 \cdot \sum_i \varepsilon_i, \quad t_3 = \sum_i \varepsilon_i^3;$$

we then have

$$t_2 = s_3 + s_2, \quad t_3 = s_3$$

in addition to (15.2). Consequently the characteristics in which we are interested are:

$$\text{Doublets:} \qquad \mathrm{X}_2 = \frac{1}{3}(t_1 - t_3), \tag{15.3}$$

$$\text{Quadruplets:} \quad \mathrm{X}_1 = \frac{1}{2}\left[\frac{1}{3}(t_1 - t_3) - (t_2 - t_3)\right]. \tag{15.4}$$

The solution of the problem discussed above is now obtained by replacing the $2l + 1$ variables ε_i by the set

$$\varepsilon^l, \varepsilon^{l-1}, \cdots, \varepsilon^{-l}$$

and then expressing t_1, t_2, t_3 as a sum $\Sigma a_L(L)$ of expressions of the form

$$(L) : \varepsilon^L + \varepsilon^{L-1} + \cdots + \varepsilon^{-L}$$

with integral multiplicities a_L. The computation is considerably simplified by multiplying both sides of the equation by $\varepsilon - 1$, as (L) then becomes $\varepsilon^{L+1} - \varepsilon^{-L}$. The multiplicities so obtained are given in the following tables :

$\boxed{t_1}$

	$L = 3l, 3l-1, 3l-2, \cdots, l$	$L = 0, 1, 2, \cdots, l$
Multiplicity :	1, 2, 3, $\cdots$	1, 3, 5, $\cdots$
	(increasing by 1 each step)	(increasing by 2 each step).

$\boxed{t_2}$

	$L = 3l, 3l-1, 3l-2, 3l-3, \cdots, l$
Multiplicity :	1, 0, 1, 0, $\cdots$, 1 ;
	(alternately 1 and 0)

$L = l, l-1, l-2, l-3, \cdots, 0$
1, -1, 1, -1, $\cdots$
(alternately 1 and -1)

$\boxed{t_3}$

	$L = 3l, 3l-1, 3l-2, 3l-3, 3l-4, 3l-5, \cdots$
Multiplicity .	1, -1, 0, 1, -1, 0, $\cdots$
	(repetition with period 3)

On applying these results to the computation of $\mathbf{X}_2$, $\mathbf{X}_1$ with the aid of (15.3) and (15.4) we find that the number of terms with total azimuthal quantum number L is as given in the following tables :

Doublet System

(1)

$L =$ 0,	1,	2,	3,	4,	5,	$\cdots$
0	1	2	2	3	4	$\cdots$

up to $L = l$. The period is here 3 ; the multiplicities in the second period are those of the first increased by 2, those in the third are obtained from those in the second by adding 2, etc.

(2)

$L = 3l$,	$3l-1$,	$3l-2$	$3l-3$,	$3l-4$,	$3l-5$,	$\cdots$
0	1	1	1	2	2	

down to $L = l$. The periodicity is again 3, but the multiplicities in each period are obtained from those in the previous one by adding 1 instead of 2.

Quadruplet System. The periodicity is here 6 instead of 3.

(1) For the values of L from 0 to l the first period of multiplicities ($L = 0, 1, 2, 3, 4, 5$) is for even l: 0 1 0 2 1 2 and for odd l: 1 0 1 1 2 1. The multiplicities increase by 2 from period to period.

(2) For values of L from $3l$ down to l the first period is 0 0 0 1 0 1 regardless of whether L is odd or even, and the multiplicities are increased by 1 from period to period.

§ 16. Determination of the Primitive Characters of $\mathfrak{u}$ and π

The guiding principle in the whole of the present chapter is the reciprocity between the symmetric permutation group π_f and the algebra Σ of symmetric transformations. But this latter can, as was shown in § 1, be replaced by the special symmetric transformations induced in tensor space by the linear transformations of vector space and which constitute a group $(\mathfrak{c})^f$ isomorphic with the linear group $\mathfrak{c}$. Indeed, we may even restrict $\mathfrak{c}$ to the unitary group $\mathfrak{u}$. The algebra Σ is thereby referred to a *group*—not to a finite group, it is true, but to a *closed* continuous group. Now we have seen in Chapter III that we may expect such groups to behave in a manner entirely analogous to that met in dealing with finite groups, at least if we concern ourselves only with unitary representations. As a rule we find in mathematics that the continuum is more easily handled than a discrete manifold; the formula (9.11), which expresses the fundamental reciprocity mentioned above, will therefore better serve to compute χ from X than the converse.

We therefore next evaluate the characteristics X of the continuous irreducible unitary representations of the n-dimensional unitary group $\mathfrak{u}$ by a direct method which is independent of our previous development. The case $n = 1$ has already been solved in III, § 8; the procedure there developed serves as a model for the present case. With this in mind we first prove the following auxiliary theorem:

A continuous function $f(\omega_1, \omega_2, \cdots, \omega_n)$ of absolute value 1 which possesses the period 2π in each of the n real arguments and which satisfies the functional equation

$$f((\omega + \omega')) = f((\omega))f((\omega'))$$

is necessarily of the form

$$f((\omega)) = e(h_1\omega_1 + h_2\omega_2 + \cdots + h_n\omega_n),$$

where the constants h are integers.

On introducing the n functions

$$f_1(\omega) = f(\omega, 0, 0, \cdots, 0), f_2(\omega) = f(0, \omega, 0, \cdots, 0), \cdots$$

of *one* variable, we are able to conclude from the functional equation above that

$$f(\omega_1, \omega_2, \cdots) = f_1(\omega_1)f_2(\omega_2) \cdots$$

It therefore suffices to prove the theorem for functions $f(\omega)$ of *one* variable, and this we have already done [III, § 8].

Every element S of the group $\mathfrak{u}$ is conjugate to a " principal " element E, i.e. to a transformation of the form

$$x_\nu \to \varepsilon_\nu x_\nu \quad (\nu = 1, 2, \cdots, n). \tag{16.1}$$

The numbers ε_ν are of unit modulus and may therefore be expressed as

$$\varepsilon_\nu = e^{i\omega_\nu} = e(\omega_\nu)$$

in terms of the " angles of rotation " $\omega_1, \omega_2, \cdots, \omega_n$ (which are only determined mod. 2π) of the unitary transformation S. In order to employ the orthogonality relations it is necessary to determine the volume dS of that portion of the group manifold $\mathfrak{u}$ whose elements have angles between ω_ν and $\omega_\nu + d\omega_\nu$. $a_1, a_2, \cdots, a_n$ being any n numbers, let $D(a_1, a_2, \cdots, a_n)$ denote the product

$$\prod_{i<k} (a_i - a_k) = | a^{n-1}, \cdots, a, 1 |$$

of differences ; the n rows of the determinant on the right are obtained by replacing a successively by $a_1, a_2, \cdots, a_n$. The evaluation of the volume element dS will be carried out in the following section ; we here anticipate the result

$$dS = \Delta\bar{\Delta}\, d\omega_1\, d\omega_2 \cdots d\omega_n, \quad \Delta = D(\varepsilon_1, \varepsilon_2, \cdots, \varepsilon_n). \tag{16.2}$$

The determination of the primitive characteristics of $\mathfrak{u}$ is accomplished by combining the following important facts.[21]

1. *Symmetry.*—Each element S of $\mathfrak{u}$ is conjugate to a principal element E, (16.1). Hence it suffices to determine the characteristic X of a continuous representation of $\mathfrak{u}$ for such a principal element. E goes over into a conjugate transformation within $\mathfrak{u}$ on permuting the ε_ν : *hence* X *is a continuous symmetric function of the angles* ω_ν *and is of period* 2π *in each of them.*

2. *Arithmetic Properties.*—The principal elements constitute an Abelian sub-group of $\mathfrak{u}$; on compounding two such elements E, E' the angles ω_ν, ω_ν' are added. The normal co-ordinates y_k in representation space $\mathfrak{R}$ can therefore be chosen in such a way that the principal elements correspond to principal transformations

$$E: \; y_k \to \rho_k y_k ;$$

indeed, we have shown in I, § 5, that any commutative system of unitary correspondences can be brought simultaneously into diagonal form. On compounding two principal elements the condition that E be a representation is expressed by the functional equation

$$\rho(\omega_1, \omega_2, \cdots)\rho(\omega_1', \omega_2', \cdots) = \rho(\omega_1 + \omega_1', \omega_2 + \omega_2', \cdots)$$

for each of the multipliers $\rho = \rho_k$. The auxiliary theorem then tells us that each ρ_k is of the form

$$e(h_1\omega_1 + \cdots + h_n\omega_n),$$

where the constants h are integers. The characteristic of the representation is the sum of these ρ_k; *hence* X *is a finite Fourier series in the arguments* ω *with integral non-negative coefficients.* The "*weights*" of a representation are the sets of exponents $(h_1, h_2, \cdots, h_n)$ of each term

$$e(h_1\omega_1 + h_2\omega_2 + \cdots + h_n\omega_n) = \varepsilon_1^{h_1}\varepsilon_2^{h_2} \cdots \varepsilon_n^{h_n}$$

which actually appears in X. The term $(h_1, h_2, \cdots, h_n)$ is said to be "higher" than $(h_1', h_2', \cdots, h_n')$ if the first non-vanishing difference $h_1 - h_1', h_2 - h_2', \cdots$ is positive.

3. *Orthogonality.*—For all primitive characteristics X the integral

$$\int_0^{2\pi} \cdots \int_0^{2\pi} \mathrm{X}\bar{\mathrm{X}}\Delta\bar{\Delta}\, d\omega_1 \cdots d\omega_n$$

must have the value

$$V = \int_0^{2\pi} \cdots \int_0^{2\pi} \Delta\bar{\Delta}\, d\omega_1 \cdots d\omega_n. \tag{16.3}$$

These orthogonality relations suggest that we introduce the quantities $\xi = \Delta \cdot \mathrm{X}$ in place of the characteristics X; they are also finite Fourier series, but they are *anti-symmetric* functions of the angles ω instead of symmetric ones. $h_1, h_2, \cdots, h_n$ being integers arranged in decreasing order

$$h_1 > h_2 > \cdots > h_n, \tag{16.4}$$

we construct the "elemental sum"

$$\xi(h_1, h_2, \cdots, h_n) = \Sigma \pm e(h_1\omega_1 + h_2\omega_2 + \cdots + h_n\omega_n), \tag{16.5}$$

.e. the alternating sum over the permutations of the arguments ω; the term which we have written down is the highest one in the sum. Every alternating Fourier series is a linear aggregate of such elemental sums; since the coefficients of these sums are integers, and in particular that of the "highest" term is 1, every alternating Fourier series, such as ξ, with integral coefficients can be expressed as a linear aggregate of the form

$$\xi = c \cdot \xi(h_1, h_2, \cdots) + c' \cdot \xi(h'_1, h'_2, \cdots) + \cdots \qquad (16.6)$$

with integral coefficients $c, c', \cdots$. Let this expansion be arranged in decreasing order, i.e. in such a way that the set $(h_1, h_2, \cdots)$ of exponents is higher than $(h'_1, h'_2, \cdots)$, etc.; $(h_1, h_2, \cdots)$ is then the highest term in ξ. Δ is itself an elemental sum, namely

$$\Delta = \xi(n-1, n-2, \cdots, 1, 0).$$

Hence if the highest term in $\mathbf{X}$ has exponents $f_1, f_2, \cdots$, we have

$$h_1 = f_1 + (n-1), \cdots, h_{n-1} = f_{n-1} + 1, h_n = f_n; \qquad (16.7)$$

in the following the numbers f_i and h_i are always in the relation (16.7) *with one another.*

We denote integration with respect to all the angles of rotation from 0 to 2π by a single integral sign and write $d\omega$ for $d\omega_1 d\omega_2 \cdots d\omega_n$. We now calculate

$$\int \bar{\xi}(h_1, h_2, \cdots)\xi(h'_1, h'_2, \cdots)d\omega;$$

the h and the h' are arranged in decreasing order in accordance with (16.4). Consequently no permutation of the h can coincide with a permutation of the h' unless

$$h_1 = h'_1, h_2 = h'_2, \cdots, h_n = h'_n; \qquad (16.8)$$

the integral of each of the $(n!)^2$ terms in the product

$$\bar{\xi}(h_1, h_2, \cdots)\ \xi(h'_1, h'_2, \cdots)$$

is therefore 0 unless (16.8) holds. In this latter case those $n!$ terms, for which the permutation of the h is the same as that of the h', each contribute $(2\pi)^n$ to the integral and all others contribute 0; hence

$$\int \bar{\xi}(h_1, h_2, \cdots)\xi(h'_1, h'_2, \cdots)d\omega = \begin{cases} n!\,(2\pi)^n \\ 0 \end{cases}$$

according as (16.8) holds or not. Applying this in particular to the elemental sum Δ, we find

$$\int \Delta \bar{\Delta} d\omega = V = n!\,(2\pi)^n.$$

On setting the expansion (16.6) in the equation

$$\int \xi \bar{\xi} d\omega = V$$

we find $|c|^2 + |c'|^2 + \cdots = 1$. Since the c, c', $\cdots$ are non-vanishing integers only the first term can appear in (16.6), and we must have $c = 1$ or -1, and since the coefficient of the highest term of ξ (as of X) must be positive we are restricted to the first alternative $c = 1$. We have thus shown that *every primitive characteristic is of the form*

$$\boxed{\mathsf{X} = \frac{\xi(h_1, h_2, \cdots)}{\Delta} = \frac{|\varepsilon^{h_1}, \varepsilon^{h_2}, \cdots, \varepsilon^{h_n}|}{|\varepsilon^{n-1}, \cdots, \varepsilon, 1|}}, \tag{16.9}$$

where the h_i are integers arranged in decreasing order : $h_1 > h_2 > \cdots$. The function defined by (16.9) is a finite Fourier series with the highest term $(f_1, f_2, \cdots, f_n)$; the coefficient of this term, its multiplicity, is 1.

4. *Completeness.*—The last question to be answered asks whether every function of the form (16.9) is conversely the characteristic of some irreducible representation of $\mathfrak{u}$ or not. Our explicit algebraic construction allows us to answer this question in the affirmative. To show this we first note that the representation of order f arising from the symmetry pattern with (at most n) rows of lengths $f_1, f_2, \cdots, f_n$ has as highest weight $(f_1, f_2, \cdots, f_n)$; this can be seen immediately by considering the representation as generated by alternation from the product of n vectors, the first of which occurs f_1 times as a factor, the second f_2, etc. (as in the simple case at the beginning of § 15). The f are here any integers satisfying the conditions

$$f_1 \geqq f_2 \geqq \cdots \geqq f_n \geqq 0.$$

On dividing the transformation corresponding to the arbitrary element S of $\mathfrak{u}$ in this representation by the lth power of the determinant of S (l being any fixed non-negative integer) the highest weight of the resulting transformation is $(f_1 - l, f_2 - l, \cdots, f_n - l)$; this simple device thus enables us to dispense with the restriction $f_n \geqq 0$. We have thus proved that *all irreducible unitary representations of the unitary group $\mathfrak{u}_n$ are obtainable by completely reducing the representations $(\mathfrak{u})^f$ for $f = 0, 1, 2, \cdots$ into their irreducible constituents and multiplying by the 1-dimensional representations*

$$S \rightarrow (\text{det. } S)^l \quad [l = 0, \pm 1, \pm 2, \cdots].$$

We have further shown that the characteristic of the irreducible representation $\mathfrak{H} = \mathfrak{H}(f_1, f_2, \cdots, f_n)$ of order f of $\mathfrak{u}$, which is generated by the symmetry pattern $P(f_1, f_2, \cdots, f_n)$, is given by equation (16.9).

We could also have obtained this last result with the more transcendental method of proof employed in steps 1 to 3. If we are operating in the continuum of all complex numbers rather than an arbitrary field the proof of the completeness of the irreducible representations of a finite group can be formulated in such a way that it can be taken over immediately for the case of a closed continuous group with the aid of the theory of integral equations. The particular application of this general group-theoretic completeness theorem to the group $\mathfrak{d}_2$ of rotations of a circle into itself yields the completeness of the Fourier orthogonal system $e^{im\phi}$ $(m = 0, \pm 1, \pm 2, \cdots)$. Its application to the closed group $\mathfrak{u}_n$ yields the following two facts: (1) Every expression of the form (16.9) is in fact a primitive characteristic. For if it were not it would be a non-vanishing function of position on the group manifold—in fact, a class function—whose Fourier coefficient with respect to each irreducible representation vanishes; it is indeed orthogonal to all other functions of the form (16.9). (2) We further find that the functions (16.9) constitute a complete set of orthogonal functions for symmetric periodic functions of $\omega_1, \omega_2, \cdots, \omega_n$; this result is of no particular interest, as it is a consequence of the completeness of Fourier's orthogonal system in one dimension. Our general considerations (1) to (4) yielded so many properties of primitive characteristics that we were able to obtain an explicit expression for them from these properties alone.

Consequences.—The assumption that $h_n = f_n \geqq 0$ constitutes no actual restriction; the characteristic is then a symmetric rational integral function of the ε of order f. The ε are in fact roots of the characteristic polynomial $f(\tau) = \det(\tau\mathbf{1} - S)$ of the unitary transformation S; it is therefore possible to express $\mathbf{X}$ rationally and integrally in terms of the coefficients of this polynomial, and therefore in terms of the coefficients of the matrix S. The restriction to the unitary group can then readily be removed, but we shall not go further into these considerations here.[22]

The dimensionality of the representation $\mathbf{X}$ is found by calculating $\mathbf{X}$ for the unit element, all of whose characteristic numbers ε_ν are 1. On substituting directly in (16.9) we obtain the indeterminate form $0/0$, so we proceed as follows. Take

$$\omega_1 = (n-1)\omega, \ \omega_2 = (n-2)\omega, \cdots, \omega_n = 0\omega$$

in terms of the single angle ω. The determinant in the numerator of (16.9) is then the alternating sum of the terms obtained from the product

$$e(h_1(n-1)\omega) \cdot e(h_2(n-2)\omega) \cdots e(h_n 0\omega)$$

by permutations of the numbers $n-1, n-2, \cdots, 0$; it is therefore equal to

$$\left|\{e(h\omega)\}^{n-1}, \cdots, \{e(h\omega)\}^1, 1\right|$$

or to the product of the differences of the expressions $e(h_1\omega)$, $e(h_2\omega), \cdots$ obtained by subtracting any member of the set from any of the earlier members. On allowing $\omega \to 0$ we have

$$e(h_1\omega) - e(h_2\omega) \sim i\omega(h_1 - h_2).$$

The dimensionality N of the representation denoted by $\mathfrak{H}(f_1, f_2, \cdots, f_n)$ in the above is consequently

$$\boxed{N = \frac{D(h_1, h_2, \cdots, h_n)}{D(n-1, \cdots, 1, 0)}}. \tag{16.10}$$

Evaluation of the Characters of π_f.—Having obtained explicit expressions for the characteristics of the representations of $\mathfrak{u}_n$ we now employ the connection between the representations of π_f and $\mathfrak{u}_n$ developed in § 9 to evaluate the primitive characters of π_f. In equation (9.12) χ is the character and $\mathbf{X}$ the characteristic of the irreducible representations of π_f and $\mathfrak{u}_n$, respectively, generated by the symmetry pattern $P(f_1, f_2, \cdots)$; in particular we must put $\mathbf{X} = 0$ if the pattern has more than n rows. The sum is extended over all possible symmetry patterns P with f fields. The expression (16.9) for $\mathbf{X}$ then allows us to enunciate the following rule for the evaluation of χ: Let

$$\chi_{f_1 f_2} \cdots (i_1 i_2 \cdots) \tag{16.11}$$

denote the value of the character of the irreducible representation $\mathfrak{h}(f_1, f_2, \cdots)$ of π_f, which is generated by the symmetry pattern $P(f_1, f_2, \cdots)$, for an element s belonging to the class $\mathfrak{k} = (i_1 i_2 \cdots)$. *Choose an arbitrary positive integer n and construct the sums $\sigma_1, \sigma_2, \cdots$ of powers of n independent variables $\varepsilon_1, \varepsilon_2, \cdots, \varepsilon_n$ and the product $D(\varepsilon_1, \varepsilon_2, \cdots, \varepsilon_n)$ of their differences. The term* (16.11) *is then the coefficient of the term $\varepsilon_1^{h_1} \varepsilon_2^{h_2} \cdots \varepsilon_n^{h_n}$ $[h_i = f_i + (n - i)]$ in the expansion of*

$$D(\varepsilon_1, \varepsilon_2, \cdots, \varepsilon_n) \cdot \sigma_1^{i_1} \sigma_2^{i_2} \cdots. \tag{16.12}$$

We here assume that the pattern P has at most n rows; hence if we wish to obtain *all* primitive characters of π_f we must choose $n \geqq f$. The rule shows that *the components of the characters are integers.*

This result was obtained by *Frobenius* in a purely algebraic manner, without introducing the continuous group $\mathfrak{u}$.[23] But

I believe that the real reason for the rule comes to light only when we consider this connection between the groups π_f and $\mathfrak{u}_n$—in particular, it enables us to understand why a second integer n in addition to f is involved.

The *dimensionality* g of $\mathfrak{h}(f_1, f_2, \cdots)$ is obtained by substituting the argument $s = \mathbf{I}$: $i_1 = f$, $i_2 = i_3 = \cdots = 0$ in the character χ. Formula (9.12) is then

$$\sigma_1^f = \Sigma g \mathrm{X},$$

where the sum is extended over all patterns $P(f_1, f_2, \cdots)$. Since σ_1 is the characteristic of the n-dimensional representation $\mathfrak{c}: S \to S$ of the group $\mathfrak{u}$ by itself, this merely means that in the complete reduction of $(\mathfrak{c})^f$ the irreducible representation $\mathfrak{H} = \mathfrak{H}(f_1, f_2, \cdots)$ appears exactly g times, as we already know. On substituting the explicit expression (16.9) for X we obtain

$$\sigma_1^f \cdot |\varepsilon^{n-1}, \cdots, \varepsilon, 1| = \Sigma g \cdot |\varepsilon^{h_1}, \varepsilon^{h_2}, \cdots, \varepsilon^{h_n}|.$$

g is accordingly equal to the coefficient of $\varepsilon_1^{h_1} \varepsilon_2^{h_2} \cdots \varepsilon_n^{h_n}$ in the expansion of the product on the left-hand side. The term $\pm \varepsilon_1^{k_1} \varepsilon_2^{k_2} \cdots \varepsilon_n^{k_n}$ in the expansion of the determinant must be multiplied by the term

$$\frac{f!}{(h_1 - k_1)!\,(h_2 - k_2)! \cdots} \varepsilon_1^{h_1 - k_1} \varepsilon_2^{h_2 - k_2} \cdots$$

of σ_1^f in order to obtain a contribution to the term $\varepsilon_1^{h_1} \varepsilon_2^{h_2} \cdots$ of the product. $(k_1, k_2, \cdots, k_n)$ here run through the permutations of $n - 1, \cdots, 1, 0$ and g is accordingly equal to the alternating sum

$$f! \sum_{(k)} \pm \frac{1}{(h_1 - k_1)!\,(h_2 - k_2)! \cdots}$$

over these permutations, or equal to the determinant

$$f! \left| \frac{1}{(h - n + 1)!}, \cdots, \frac{1}{(h - 1)!}, \frac{1}{h!} \right|$$
$$= \frac{f!}{h_1!\, h_2! \cdots} \left| h(h-1) \cdots (h - n + 2), \cdots, h, 1 \right|.$$

The rows of this determinant consist, on reading from right to left, of polynomials in h of degrees $0, 1, \cdots, (n - 1)$ with highest coefficient 1. The determinant is therefore

$$|h^{n-1}, \cdots, h, 1|$$

and we finally obtain the simple formula

$$\boxed{g = \frac{f!\, D(h_1, h_2, \cdots, h_n)}{h_1!\, h_2! \cdots h_n!}}. \tag{16.13}$$

n is to be taken at least as large as the number of rows in the pattern $P(f_1, f_2, \cdots)$; the reader should convince himself by direct calculation that the value of (16.13) remains unchanged on replacing n by $n + 1$.

Frobenius' rule for the character and this formula for the dimensionality are vastly superior to (14.7) for purposes of practical evaluation.

As an example, we carry through the computations for the case of four electrons; the results are given in the table below. The group π_4 contains twenty-four elements which are divided into five classes of conjugates; each of these classes is designated in the second column of the table by the values $(i_1 i_2 \cdots)$ associated with it. The first column contains the number of elements in each of these classes, and the sign + or — indicates whether the class consists of even or odd permutations. Each of the five remaining columns contains the values of a primitive character for the classes in whose row they stand. The symmetry pattern to which each of these characters belongs is indicated at the head of the column by the numbers $f_1, f_2, \cdots$ of elements in its rows. The first and the last of these columns may be filled in immediately, and the second and third with the aid of Frobenius' rule. The fourth is then obtained from the second on noting that its symmetry pattern is the dual of that of 2; we need then merely to replace the values in the second column by their negative for the (-)-classes. Since patterns 2 and 3 contain but two rows we may take $n = 2$. Hence on writing x, y in place of ε_1, ε_2 we have merely to find the coefficients of x^4y (for the column 31) and x^3y^2 (for the column 22) in the following polynomials:

$$\begin{aligned}
&(x - y)(x + y)^4,\\
&(x - y)(x + y)^2(x^2 + y^2) = (x + y)(x^2 - y^2)(x^2 + y^2)\\
&\qquad\qquad\qquad\qquad\qquad\quad = (x + y)(x^4 - y^4),\\
&(x - y)(x^2 + y^2)^2,\\
&(x - y)(x + y)(x^3 + y^3) = (x^2 - y^2)(x^3 + y^3),\\
&(x - y)(x^4 + y^4).
\end{aligned}$$

The dimensionalities of the five irreducible representations are contained in the first row; they are 1, 3, 2, 3, 1. The verification of the orthogonality relations is left to the reader.

No. Elements.	Pattern. / Class.	4	31	22	211	1111
1+	4	1	3	2	3	1
6−	21	1	1	0	−1	−1
3+	02	1	−1	2	−1	1
8+	101	1	0	−1	0	1
6−	0001	1	−1	0	1	−1

§ 17. Calculation of Volume on $\mathfrak{u}$

Consider the line elements going out from the unit point $\mathbf{I}$ on the group manifold $\mathfrak{u}$, i.e. the infinitesimal unitary transformations $\delta S = \|\delta s_{\alpha\beta}\|$. We may take as the real components of this " vector " the n quantities $\frac{1}{i} \cdot \delta s_{\alpha\alpha}$ and the real and imaginary parts of the $n(n-1)/2$ quantities $\delta s_{\alpha\beta}\,(\alpha < \beta)$; the total number of components is thus n^2, which is therefore the dimensionality of the group manifold $\mathfrak{u}$. Now in a linear algebra of this kind we may replace any two real quantities a, b by the complex quantities $a + ib$, $-a + ib$ obtained from them by a simple linear substitution ; we may therefore replace the real and imaginary parts of $\delta s_{\alpha\beta}\,(\alpha < \beta)$ by $\delta s_{\alpha\beta}$ itself and $-\overline{\delta s_{\alpha\beta}} = \delta s_{\beta\alpha}$.

On transporting such an infinitesimal vector to the point S on the group manifold by a left-translation its terminus goes into the point $S + dS = S(\mathbf{1} + \delta S)$, $dS = S \cdot \delta S$; we must therefore consider the infinitesimal element $\delta S = S^{-1}dS$ as the " vector " which leads from S to $S + dS$. Our definition of volume on the group manifold [III, § 12] consisted in the following : the parallelepiped defined by n^2 vectors δS leading from the fixed point S to the neighbouring points $S + dS$ has as volume the absolute value of the determinant formed from the components of the n^2 vectors δS. In accordance with the above remarks we may take as components of the vector $\delta S = \|\delta s_{\alpha\beta}\|$ the totality of coefficients $\delta s_{\alpha\beta}$ themselves.

Any S can be expressed in the form

$$S = UEU^{-1} \tag{17.1}$$

where E is a principal (diagonal) element of $\mathfrak{u}$ and U is unitary. S is unchanged on multiplying U on the right by any principal element. We employ a geometrical terminology which will allow us to visualize our procedure by means of an analogy. Two elements U, U' of $\mathfrak{u}$ which are right-equivalent with respect to the group of principal elements: $U' = UE$, will be said to "lie on the same *vertical* $[U]$." From the n^2-dimensional manifold $\mathfrak{u}$ we obtain by projection the $(n^2 - n)$-dimensional manifold $[\mathfrak{u}]$ of verticals $[U]$ on considering all points of $\mathfrak{u}$ which belong to the same vertical to be coincident. This process of identifying equivalent elements was described in general in the beginning of Chapter III—we had, in fact, already met it in I, § 1, in the special case of projection in affine space. We may now consider U in (17.1) merely as a representative element of the vertical $[U]$; on allowing $[U]$ to run through the entire manifold $[\mathfrak{u}]$ and the angles ω_ν of E:

$$E = \left\| \begin{matrix} e(\omega_1) & & & \\ & e(\omega_2) & & \\ & & \ddots & \\ & & & e(\omega_n) \end{matrix} \right\|$$

to vary independently over the complete range $0 \leqq \omega < 2\pi$ the element S defined by (17.1) describes the manifold $\mathfrak{u}$ exactly $n!$ times.

The vector $\delta U = U^{-1} dU$ leads from the point U of the vertical $[U]$ to the neighbouring point $U + dU$ of the vertical $[U + dU]$. The totality of all points on $[U + dU]$ which are in the neighbourhood of U is given by expressions of the form

$$(U + dU)(1 + \delta E) = U + (dU + U\,\delta E)$$

where δE is an arbitrary infinitesimal principal element with coefficients $i\,\delta\omega_\nu$ on the principal diagonal; the corresponding vectors are $\mathring{\delta} U = \delta U + \delta E$. Since the terms in the principal diagonal of δU are pure imaginary, E may be uniquely determined in such a way that all terms in the principal diagonal of $\mathring{\delta} U$ vanish; we call this transition from $[U]$ to $[U + dU]$ the "*horizontal* transition from U."—The transition from some other point UE of the vertical $[U]$ to the point $(U + dU)E$ of $[U + dU]$ is accomplished by means of the vector

$$\delta' U = E^{-1} \cdot \delta U \cdot E. \tag{17.2}$$

That this linear transformation (17.2) determined by E, which sends δU into $\delta' U$, is unimodular follows from our general remarks concerning closed continuous groups—and can in this case be readily verified by direct computation. Naturally this same equation holds for the horizontal transitions $\mathring{\delta}U$, $\mathring{\delta}'U$ from U, UE respectively:

$$\mathring{\delta}'U = E^{-1} \cdot \mathring{\delta}U \cdot E. \tag{17.3}$$

$n^2 - n$ horizontal vectors δU leading out from U determine an infinitesimal "parallelogram" whose content is measured by the absolute value of the determinant of the $n^2 - n$ components $\delta u_{\alpha\beta}$ $(\alpha \neq \beta)$ of the various vectors δU. On allowing each point U on the periphery of the parallelogram to describe the vertical $[U]$ we obtain a tube whose horizontal sections are parallelograms; its projection on $[\mathfrak{u}]$ is the original element of volume, the "parallelogram" defined by the δU. Since the linear transformation (17.3), $\mathring{\delta}U \to \mathring{\delta}'U$, is unimodular, the content of each horizontal section is the same, and may therefore be considered as the content of the volume element on $[\mathfrak{u}]$.

We now examine the variations in $[U]$ and E in (17.1) when S goes over into $S + dS$. We have

$$SU = UE$$

and therefore

$$dS \cdot U + S \cdot dU = dU \cdot E + U \cdot dE.$$

On multiplying both sides of this equation by $U^{-1}S^{-1} = E^{-1}U^{-1}$ we find

$$U^{-1} \cdot \delta S \cdot U + \delta U = E^{-1} \cdot \delta U \cdot E + \delta E$$

or

$$\delta'S \equiv U^{-1} \cdot \delta S \cdot U = \{E^{-1} \cdot \mathring{\delta}U \cdot E - \mathring{\delta}U\} + \delta E. \tag{17.4}$$

The components of the matrix contained in parentheses are

$$\delta u_{pq}\left(\frac{\varepsilon_q}{\varepsilon_p} - 1\right).$$

We now define a parallelepiped at S which shall serve as a volume element in the following manner: $n^2 - n$ of the n^2 sides δS are obtained from (17.4) on allowing the angles of rotation to remain fixed, i.e. $\delta E = 0$, and drawing $n^2 - n$ horizontal vectors $\mathring{\delta}U$ from the point U to form a volume element of magnitude $d[U]$ on $[\mathfrak{u}]$; the remaining n vectors δS are then chosen in such a way that for each of them one and only one of the angles ω_r changes by $d\omega_r$ and $[U]$ remains unchanged. The corresponding n^2 vectors $\delta'S$ define, in accordance with (17.4), an element of volume of magnitude

$$\prod_{p \neq q}\left(\frac{\varepsilon_q}{\varepsilon_p} - 1\right) \cdot d[U] \cdot d\omega_1\, d\omega_2 \cdots d\omega_n. \tag{17.5}$$

Since the linear transformation $\delta S \to \delta' S = U^{-1} \cdot \delta S \cdot U$ is unimodular this volume is equal to that of the element defined by the δS themselves. Since $\bar{\varepsilon} = 1/\varepsilon$ the product Π in (17.5) can be written

$$\prod_{q<p}\left(\frac{\varepsilon_q}{\varepsilon_p} - 1\right)\left(\frac{\bar{\varepsilon}_q}{\bar{\varepsilon}_p} - 1\right) = \prod_{q<p}\frac{(\varepsilon_q - \varepsilon_p)(\bar{\varepsilon}_q - \bar{\varepsilon}_p)}{\varepsilon_p\bar{\varepsilon}_p}$$

$$= \prod_{q<p}(\varepsilon_q - \varepsilon_p)(\bar{\varepsilon}_q - \bar{\varepsilon}_p) = \Delta \cdot \bar{\Delta}.$$

The final result is : *The volume element described by S on allowing* $[U]$ *in* (17.1) *to describe an infinitesimal volume element of magnitude* $d[U]$ *on* $[\mathfrak{u}]$ *and on allowing the angles of rotation* ω_ν *to vary by* $d\omega_\nu$ *has the magnitude*

$$\Delta\bar{\Delta}\, d\omega_1\, d\omega_2 \cdots d\omega_n \cdot d[U]. \tag{17.6}$$

On integrating with respect to $d[U]$ over $[\mathfrak{u}]$ we obtain the theorem, already applied in the preceding section, concerning the magnitude of that portion of $\mathfrak{u}$ in which the angles of rotation have values lying between ω_ν and $\omega_\nu + d\omega_\nu$.

These considerations remain valid on restricting ourselves to the *group* $\mathfrak{u}$ *of unitary transformations with determinant* 1. The angles of rotation are then subjected to the restriction

$$\omega_1 + \omega_2 + \cdots + \omega_n = 0, \tag{17.7}$$

and the only difference in the result is that the factor $d\omega_n$ in (17.6) is to be omitted. Condition (17.7) allows us to normalize the linear form $h_1\omega_1 + \cdots + h_n\omega_n$ in the angles of rotation in such a way that $h_n = 0$; the exponents $(h_1, h_2, \cdots, h_n)$ in the weights of the representations of $\mathfrak{u}$ are then non-negative integers. It is desirable, however, not to impose this normalization $h_n = 0$; we need then only to remark that only the *differences* between the h_i are of significance: the irreducible representations $\mathfrak{H}(f_1, f_2, \cdots, f_n)$ of $\mathfrak{u}$ are unchanged on increasing each of the f_i by the same integer. In particular, these considerations justify the expression used in Chapter III for the volume on the group manifold of the unimodular unitary group $\mathfrak{u}_2$, and the results of the preceding section constitute a direct proof, which is independent of the completeness theorem, of the fact that the representations of $\mathfrak{u}_2$ denoted by $\mathfrak{C}_v$ constitute a complete set of inequivalent irreducible representations of $\mathfrak{u}_2$.

§ 18. Branching Laws

Finally, we show the usefulness of our formula for the characters by deriving two simple " branching laws " from them.

1. *Branching law for the Permutation Group.*

The irreducible representation of π_f with the symmetry pattern $P(f_1, f_2, \cdots)$ reduces, on restricting π_f to the sub-group π_{f-1} of permutations of $f-1$ things, into the sum of those irreducible representations of π_{f-1} associated with the patterns

$$\begin{array}{l} P(f_1 - 1, f_2, f_3, \cdots); \\ P(f_1, f_2 - 1, f_3, \cdots); \\ \cdots\cdots\cdots\cdots\cdots\cdots ; \end{array}$$

those patterns in which the rows are not arranged in decreasing length are to be omitted. Each such constituent appears exactly once. (In words, these patterns are obtained from the original one by removing a field in turn from the end of each row which is actually longer than the following one.)

Proof. Let s be a permutation of the numbers $1, 2, \cdots, f-1$ belonging to the class $(i_1 - 1, i_2, i_3, \cdots)$. Considered as a permutation of the f numbers $1, 2, \cdots, f$, s leaves the last number fixed; the number of one-term cycles is thus increased by 1, and s, considered as an element of π_f, belongs to the class $(i_1, i_2, i_3, \cdots)$. In the expansion

$$\Delta \cdot \sigma_1^{i_1 - 1} \sigma_2^{i_2} \cdots = \Sigma a_{h'_1 h'_2 \cdots} \varepsilon_1^{h'_1} \varepsilon_2^{h'_2} \cdots \tag{18.1}$$

we have as the coefficients of those terms for which $h'_1 \geqq h'_2 \geqq \cdots$

$$a_{h'_1 h'_2 \cdots} = 0 \quad \text{or} \quad \chi_{f'_1 f'_2 \cdots}(s) \tag{18.2}$$

according as any of the signs $\leqq$ in the above inequalities is actually $=$ or not. $\chi_{f'}$ is the primitive character of π_{f-1} belonging to the symmetry pattern $P(f'_1, f'_2, \cdots)$. On the other hand, the coefficient of $\varepsilon_1^{h_1} \varepsilon_2^{h_2} \cdots$ $[h_1 > h_2 > \cdots]$ in $\Delta \cdot \sigma_1^{i_1} \sigma_2^{i_2} \cdots$ is equal to the character $\chi_{f_1 f_2 \cdots}(s)$ of the representation of π_f with pattern $P(f_1, f_2, \cdots)$. Hence on multiplying (18.1) with $\sigma_1 = \varepsilon_1 + \varepsilon_2 + \cdots + \varepsilon_n$ we find

$$\chi_{f_1 f_2 \cdots}(s) = a_{h_1 - 1, h_2, h_3, \cdots} + a_{h_1, h_2 - 1, h_3, \cdots} + \cdots.$$

Our branching law follows from this result and (18.2). The branching law leads to a recurrence formula for the dimensionalities $g(f_1, f_2, \cdots)$.

2. *Branching law for* $\mathfrak{c}_n$.

On restricting $\mathfrak{c}_n$ *to the sub-group of linear transformations of an* $(n-1)$*-dimensional sub-space the irreducible representation* $(f_1, f_2, \cdots)$ *of* $\mathfrak{c}_n$ *reduces into the sum of all those representations* $(f_1', f_2', \cdots)$ *of* $\mathfrak{c}_{n-1}$ *for which*

$$f_1 \geqq f_1' \geqq f_2 \geqq f_2' \geqq \cdots \geqq f_{n-1}' \geqq f_n; \qquad (18.3)$$

each of these constituents appears exactly once.

Proof. The linear transformations S of the sub-space $\mathfrak{c}_{n-1}$: $x_n = 0$ are simply isomorphic to those linear transformations S of the variables $x_1, x_2, \cdots, x_n$ in which $x_n \to x_n$. Hence ε_n is to be replaced by 1 in the characteristic (16.9). The denominator is then

$$D(\varepsilon_1, \varepsilon_2, \cdots, \varepsilon_{n-1}) \cdot (\varepsilon_1 - 1)(\varepsilon_2 - 1) \cdots (\varepsilon_{n-1} - 1),$$

as can be seen by subtracting the last column of $D(\varepsilon_1, \varepsilon_2, \cdots, \varepsilon_{n-1}, 1)$ from each of the previous ones and factoring the resulting $(n-1)$-row determinant. In order to divide the determinant in the numerator by the factor $(\varepsilon_1 - 1)(\varepsilon_2 - 1) \cdots (\varepsilon_{n-1} - 1)$ we subtract the second column from the first, the third from the second, $\cdots$, and finally the nth from the $(n-1)^{\text{st}}$. The last row then is $0, 0, \cdots, 0, 1$; the determinant is thus reduced to a determinant of order $(n-1)$. Now divide each element in the νth row by $\varepsilon_\nu - 1$ in accordance with

$$\frac{\varepsilon^{h_1} - \varepsilon^{h_2}}{\varepsilon - 1} = \varepsilon^{h_1 - 1} + \cdots + \varepsilon^{h_2}.$$

The result is that we then have in the numerator the determinant

$$|\varepsilon^{h_1-1} + \cdots + \varepsilon^{h_2}, \; \varepsilon^{h_2-1} + \cdots + \varepsilon^{h_3}, \cdots| \qquad (\varepsilon = \varepsilon_1, \varepsilon_2, \cdots, \varepsilon_{n-1}).$$

But this is the sum of all $(n-1)$-rowed determinants of the form

$$|\varepsilon^{h'_1}, \varepsilon^{h'_2}, \cdots, \varepsilon^{h'_{n-1}}|,$$

$$h_1 > h_1' \geqq h_2 > h_2' \geqq h_3 > \cdots > h_{n-1}' \geqq h_n \qquad (18.4)$$

On subtracting $n-1$ from h_1, $n-2$ from h_1' and $h_2, \cdots, 0$ from h_{n-1}' and h_n, in order to obtain the numbers f [(16.7)], the inequalities (18.4) become the inequalities (18.3) and our theorem is proved.

APPENDIX 1

Proof of an Inequality

(*Page* 77.)

In order to prove the inequality stated on page 77 we must show that any continuous and differentiable function ψ, which is defined for all values of the real variable x, satisfies the condition

$$\frac{1}{4}\left(\int_{-\infty}^{+\infty}\bar{\psi}\psi dx\right)^2 \leqq \int_{-\infty}^{+\infty}x^2\bar{\psi}\psi dx \cdot \int_{-\infty}^{+\infty}\frac{d\bar{\psi}}{dx}\frac{d\psi}{dx}dx, \qquad (*)$$

provided, of course, that the integrals involved actually exist.

The Schwarz inequality

$$|a_1b_1 + \cdots + a_nb_n|^2 \leqq (a_1\bar{a}_1 + \cdots + a_n\bar{a}_n)(b_1\bar{b}_1 + \cdots + b_n\bar{b}_n)$$

employed in Chapter I becomes, on replacing the sums by integrals—or rather each sum by two integrals—

$$\left|\int f_1g_1dx + \int f_2g_2dx\right|^2 \leqq \left(\int f_1\bar{f}_1dx + \int f_2\bar{f}_2dx\right)\left(\int g_1\bar{g}_1dx + \int g_2\bar{g}_2dx\right).$$

Applying this inequality to

$$x\frac{d}{dx}(\psi\bar{\psi}) = x\psi \cdot \frac{d\bar{\psi}}{dx} + x\bar{\psi} \cdot \frac{d\psi}{dx}$$

by taking

$$f_1 = x\psi,\ f_2 = x\bar{\psi},\ g_1 = \frac{d\bar{\psi}}{dx},\ g_2 = \frac{d\psi}{dx},$$

and transforming the integral

$$\int x\frac{d}{dx}(\psi\bar{\psi})dx \quad \text{into} \quad -\int \psi\bar{\psi}dx$$

by partial integration over the range $-\infty$, $+\infty$, we obtain the desired relation (*) provided the term $x\psi\bar{\psi}$, which is integrated out, approaches 0 as $x \to \pm\infty$. That this is actually the case if the two integrals on the right of (*) converge can be seen by the following indirect proof. Let ε be any pre-assigned positive constant and consider a positive value of x for which $x\,|\psi(x)|^2 > \varepsilon$ and which is so large that $\int_x^\infty \left|\frac{d\psi}{dx}\right|^2 dx \leqq \frac{1}{4}$. The Schwarz inequality

$$\left|\int_x^{x'} \frac{d\psi}{dx}dx\right|^2 \leqq (x' - x) \cdot \int_x^{x'} \left|\frac{d\psi}{dx}\right|^2 dx$$

then tells us that for $x \leqq x' \leqq x + \frac{\varepsilon}{x}$

$|\psi(x') - \psi(x)|^2 \leqq \frac{1}{4} \cdot \frac{\varepsilon}{x}$, whence $|\psi(x')| \geqq |\psi(x)| - \frac{1}{2}\sqrt{\frac{\varepsilon}{x}} > \frac{1}{2}\sqrt{\frac{\varepsilon}{x}}$

The integral of $x^2\,|\psi|^2$ over the range from x to $x + \frac{\varepsilon}{x}$ is then

$$> x^2 \cdot \frac{1}{4}\frac{\varepsilon}{x} \cdot \frac{\varepsilon}{x} = \frac{\varepsilon^2}{4}.$$

Hence it follows that conversely

$$\int_x^\infty \left|\frac{d\psi}{dx}\right|^2 \leqq \frac{1}{4}, \qquad \int_x^\infty x^2\,|\psi|^2 dx \leqq \frac{\varepsilon^2}{4}$$

imply the inequality

$$x\,|\psi(x)|^2 \leqq \varepsilon.$$

APPENDIX 2

A Composition Property of Group Characters

(*Page* 169.)

THE fundamental property of the irreducible representation $\mathfrak{H} : s \rightarrow U(s)$ which is expressed in the equation

$$U(st) = U(s)U(t)$$

is paralleled by the relation

$$\chi(s)\chi(t) = \frac{g}{h}\sum_r \chi(sr^{-1}tr). \qquad (*)$$

Proof. If $\boldsymbol{x}$, $\boldsymbol{y}$ are two elements of the algebra of the group, the second of which belongs to the central, and if

$$\boldsymbol{x} \rightarrow X, \quad \boldsymbol{y} \rightarrow Y \quad \text{in} \quad \mathfrak{H},$$

then $Y = \frac{\eta}{g}\mathbf{1}$. The matrix associated with $\boldsymbol{z} = \boldsymbol{xy}$ in $\mathfrak{H}$ is $\frac{\eta}{g}X$ and its trace is $\frac{\xi\eta}{g}$:

$$\sum_r z(r)\chi(r) = \frac{1}{g}\sum_s x(s)\chi(s) \cdot \sum_t y(t)\chi(t).$$

On setting

$$z(r) = \sum x(s)y(t) \quad (st = r)$$

we find

$$\sum_{s,t} x(s)\, y(t)\, \chi(st) = \frac{1}{g}\sum_{s,t} x(s)\, y(t)\, \chi(s)\, \chi(t).$$

Since $y(t)$ depends only on the class of conjugate elements to which t belongs we may replace

$$\chi(st) \quad \text{by} \quad \frac{1}{h}\sum_r \chi(sr^{-1}tr)$$

on the left-hand side of the previous equation. Then the coefficient of $x(s)y(t)$ on either side of the equation depends only on the class to which the element t belongs, and since $x(s)$ is an arbitrary function, $y(t)$ an arbitrary class function, the assertion (*) follows from the fact that the two coefficients must agree.

We have omitted mention of this equation (*) in the text in order not to interrupt the systematic development of the theory of representations, which is completely described by the orthogonality relations and the completeness theorem.

APPENDIX 3

A Theorem Concerning Non-degenerate Anti-symmetric Bilinear Forms

(*Page* 274.)

We consider the given non-degenerate anti-symmetric bi-linear form

$$\sum_{i,k=1}^{f} c_{ik} x_i y_k \qquad (c_{ki} = -c_{ik})$$

as the "anti-symmetric product" $[\mathfrak{x}\mathfrak{y}]$ of the two vectors $\mathfrak{x} = (x_1, x_2, \cdots, x_f)$ and $\mathfrak{y} = (y_k)$. Let $\mathfrak{e}_1$ be any non-vanishing vector; then by assumption $[\mathfrak{e}_1\mathfrak{x}]$ cannot vanish identically in $\mathfrak{x}$, and consequently a second vector $\mathfrak{e}_2$ can be found such that $[\mathfrak{e}_1\mathfrak{e}_2] = 1$. The simultaneous equations

$$[\mathfrak{e}_1\mathfrak{x}] = 0, \quad [\mathfrak{e}_2\mathfrak{x}] = 0$$

then have $f - 2$ linearly independent solutions $\mathfrak{e}_3, \cdots, \mathfrak{e}_f$. These vectors are furthermore such that no linear dependence can exist between them and $\mathfrak{e}_1$, $\mathfrak{e}_2$, for if

$$\mathfrak{x} = \xi_1\mathfrak{e}_1 + \xi_2\mathfrak{e}_2 + \xi_3\mathfrak{e}_3 + \cdots + \xi_f\mathfrak{e}_f = 0,$$

it follows on building the anti-symmetric products $[\mathfrak{e}_1\mathfrak{x}] = \xi_2$, $[\mathfrak{e}_2\mathfrak{x}] = -\xi_1$ that $\xi_1 = \xi_2 = 0$. We may therefore choose $\mathfrak{e}_1, \mathfrak{e}_2, \cdots, \mathfrak{e}_f$ as a co-ordinate system, i.e. as a basis from which all vectors may be constructed. Let the anti-symmetric product be expressed in terms of the components ξ_i, η_k of $\mathfrak{x}$, $\mathfrak{y}$ in this new co-ordinate system by

$$[\mathfrak{x}\mathfrak{y}] = \sum_{i,k=1}^{f} \gamma_{ik}\xi_i\eta_k.$$

The manner in which the new fundamental vectors were determined requires that of the coefficients $\gamma_{ik} = [\mathfrak{e}_i\mathfrak{e}_k]$

$$\gamma_{11} = 0, \quad \gamma_{12} = 1; \quad \gamma_{13} = 0, \cdots, \gamma_{1f} = 0,$$
$$\gamma_{21} = -1, \; \gamma_{22} = 0; \quad \gamma_{23} = 0, \cdots, \gamma_{2f} = 0.$$

In consequence of the anti-symmetry all γ_{i1}, γ_{i2} with $i = 3, \cdots, f$ vanish, and the matrix of the γ_{ik} is completely reduced into the 2-rowed square sub-matrix

$$\begin{Vmatrix} 0 & 1 \\ -1 & 0 \end{Vmatrix}$$

and an $(f-2)$-dimensional anti-symmetric matrix. Mathematical induction with respect to the dimensionality f yields the desired theorem that f is necessarily even and that the original form can be transformed into

$$(\xi_1\eta_2 - \xi_2\eta_1) + (\xi_3\eta_4 - \xi_4\eta_3) + \cdots \quad (f/2 \text{ terms})$$

by an appropriate linear transformation.

BIBLIOGRAPHY

CHAPTER I

PAGE

(1) **39.** Proved in complete generality by M. PLANCHEREL, Circ. Mat. Palermo, **30,** 330 (1910), and E. C. TITCHMARCH, Proc. Lond. Math. Soc. (2), **23,** 279 (1924).

(2) **39.** Acta Mathematica, **45,** 29; **46,** 101; **47,** 237 (1924-26). H. WEYL, Math. Ann. **97,** 338 (1926), gives a brief development of the subject, which is most intimately related to the completeness theorem of the theory of representations of continuous groups to be considered in Chaps. III and V—particularly in V, § 16.

(3) **39.** D. HILBERT, Grundzüge einer allgemeinen Theorie der linearen Integralgleichungen, Leipsic 1912, Part 4 (collected papers appearing in the Gött. Nachr. 1904-1910). E. HELLINGER, Journ. f. d. reine und angew. Math. **136,** 210 (1909).

(4) **39.** A. WINTNER, Spektraltheorie der unendlichen Matrizen (Leipsic 1929).

(5) **40.** Math. Ann. **102,** 49 and 370 (1929). In addition to J. v. NEUMANN and A. WINTNER see M. H. STONE, Proc. Nat. Acad. Sci. **15,** 198 and 423 (1929).

(6) **40.** Gött. Nachr. 1927, 1, Part V.

CHAPTER II

(1) **41.** In addition to the books by SOMMERFELD, RUARK-UREY and GERLACH cited in the introduction, see: M. PLANCK, The Theory of Heat Radiation (English translation by M. MASIUS, Philadelphia 1914); the article "Quantentheorie," by W. PAULI, in GEIGER and SCHEEL's Handbuch der Physik, Vol. 23, p. 1 (Berlin 1926). For a careful treatment of the theory of quantized orbits as developed during the years 1913-25 see M. BORN, The Mechanics of the Atom (English by J. W. FISHER and D. R. HARTREE, London 1927); the second volume, "Elementare Quantenmechanik" (Berlin), which deals with the new quantum mechanics, was written jointly with P. JORDAN and has not been translated into English. For treatises on quantum mechanics see: L. DE BROGLIE, An Introduction to the Study of Wave Mechanics (English by H. T. FLINT, London 1930); E. U. CONDON and P. M. MORSE, Quantum Mechanics (New York 1929); P. A. M. DIRAC, The Principles of Quantum Mechanics (Oxford 1930); J. FRENKEL, Einführung in die Wellenmechanik (Berlin 1929); A. HAAS, Wave Mechanics and the New Quantum Theory (English by L. W. CODD, London 1928); W.

PAGE
HEISENBERG, The Physical Principles of the Quantum Theory (English by C. ECKART and F. C. HOYT, Chicago 1930); O. HALPERN and H. THIRRING, Die Grundzüge der neueren Quantentheorie (Berlin 1929). For a connected report on the subject see: H. HÖNL and C. ECKART, Grundzüge und Ergebnisse der Wellenmechanik, Phys. Zeits. **31,** 89 and 145 (1930); E. C. KEMBLE and E. L. HILL, General Principles of Quantum Mechanics, Rev. Mod. Phys. **1,** 157; **2,** 1 (1929-30).

(2) **42.** Ann. d. Phys. **17,** 132 (1905); Ann. d. Phys. **20,** 199 (1906).

(3) **46.** J. FRANCK and P. JORDAN, Anregung von Quantensprüngen durch Stösse (Berlin 1926), or article with same title in GEIGER and SCHEEL's Handbuch der Physik, Vol. 23, p. 641.

(4) **48.** In addition to the fundamental paper of W. HEISENBERG, Zeits. f. Phys. **33,** 879 (1925), see: M. BORN and P. JORDAN, Zeits. f. Phys. **34,** 858 (1925); M. BORN, W. HEISENBERG, and P. JORDAN, Zeits. f. Phys. **35,** 557 (1926); P. A. M. DIRAC, Proc. Roy. Soc. **109** (A), 642 (1926). L. DE BROGLIE, Ann. de phys. (10) **2,** 22 (1925); Ondes et mouvements (Paris 1926); E. SCHRÖDINGER, Ann. d. Phys. **79,** 361, 489 and 734; **80,** 437; **81,** 109 (1926). For English translations see: Selected Papers on Wave Mechanics by L. BRILLOUIN and L. DE BROGLIE (trans. by J. F. SHEARER and W. M. DEANS, London 1928); Collected Papers on Wave Mechanics by E. SCHRÖDINGER (trans. by W. M. DEANS, London 1928).

(5) **49.** A. EINSTEIN, Phys. Zeits. **18,** 121 (1917).

(6) **50.** A. EINSTEIN, Sitzungsber. Preuss. Akad. 1924, 261; 1925, 3 and 18. S. N. BOSE, Zeits. f. Phys. **26,** 178 (1924). For relation to wave mechanics see E. SCHRÖDINGER, Physikal. Zeits. **27,** 95 (1926).

(7) **50.** Phys. Rev. **30,** 705 (1927); see also W. ELSASSER, Naturwiss. **16,** 720 (1928). For diffraction of electrons passing through thin metal foils see G. P. THOMSON, Proc. Roy. Soc. **117** (A), 600 (1928); also Wave Mechanics of Free Electrons (New York 1930). Further: S. KIKUCHI, Jap. Journ. Phys. **5,** 83 (1928); E. RUPP, Ann. d. Phys. **85,** 981 (1928) and Naturwiss. **17,** 174 (1929); H. MARK and R. WIERL, Naturwiss. **18,** 205 (1930) and Zeits. f. Phys. **60,** 741 (1930).

(8) **58, 69.** The completeness proof for the Hermitian polynomials [**58**] can be obtained from the development in COURANT-HILBERT, Methoden der mathematischen Physik I. The same methods suffice to guarantee that the one-electron ion has no discrete terms other than those obtained in § 5 [**69**]. For their connection with the LAGUERRE polynomials see E. SCHRÖDINGER, Collected Papers.

(9) **67.** W. GERLACH and O. STERN, Ann. d. Phys. **74,** 673 (1924). A. EINSTEIN and P. EHRENFEST, Zeits. f. Phys. **11,** 31 (1922).

(10) **68.** For the theory of solutions of differential equations in the neighbourhood of a regular singular point see L. BIEBERBACH, Theorie der Differentialgleichungen, 3rd ed., Part II, Chap. IV (Berlin 1930); E. L. INCE, Ordinary Differential Equations, 160 and 365 (London 1927); J. PIERPONT, Theory of Functions of a Complex Variable, Chap. XIII, § 209 (Boston 1914).

(11) **70.** For the general theory of continuous spectra of differential equations see H. WEYL, Math. Ann. **68,** 220 (1910); applications to quantum theory E. SCHRÖDINGER, l.c. (4); E. FUES, Ann. d. Phys. **81,** 281 (1926). For transition probabilities between

PAGE
positive and negative energy levels in the field of a central charge *Ze* see Y. SUGIURA, Inst. Chem. and Phys. Research, Tokyo, Sci. Papers, No. 193 (1929).
(12) **74.** The collision problem discussed in the text gave rise to M. BORN's statistical interpretation of quantum mechanics; Zeits. f. Phys. **37,** 863; **38,** 803 (1926). G. WENTZEL, Zeits. f. Phys. **40,** 590 (1926); J. R. OPPENHEIMER, Zeits. f. Phys. **43,** 413 (1927). Exact calculation, W. GORDON, Zeits. f. Phys. **48,** 180 (1928).
(13) **74.** M. BORN, Gött. Nachr. 1927, 146. P. A. M. DIRAC, Proc. Roy. Soc. **114** (A), 243 (1927). W. ELSASSER, Zeits. f. Phys. **45,** 522 (1927). For a report on the entire field see G. WENTZEL, Die unperiodischen Vorgänge in der Wellenmechanik, Phys. Zeits. **29,** 321 (1928); E. U. CONDON, Quantum Mechanics of Collision Processes, Rev. Mod. Phys. **3,** 43 (1931).
(14) **74.** G. GAMOW, Zeits. f. Phys. **51,** 204; **52,** 496 and 510; **53,** 601 (1928-1929). R. W. GURNEY and E. U. CONDON, Nature, **122,** 439 (1928); Phys. Rev. **33,** 127 (1929). These are followed by a series of papers in the Zeits. f. Phys. **52-60** by v. LAUE, KUDAR, SEXL, MÖLLER, BORN, ATKINSON, HOUTERMANS. Further: R. H. FOWLER and A. H. WILSON, Proc. Roy. Soc. **124** (A), 493 (1929); O. K. RICE, Phys. Rev. **35,** 1538 (1930). Another application by R. H. FOWLER and L. NORDHEIM, Proc. Roy. Soc. **119** (A), 173 (1928). General investigation by E. SCHRÖDINGER, Sitzungsber. Preuss. Akad. 1929, 2.
(15) **76.** See H. WEYL, Philosophie der Mathematik und Naturwissenschaft, Part II, Naturwissenschaft, II Methodologie (Munich 1926).
(16) **76.** For the fundamental interpretation of the quantum theory see: W. HEISENBERG, Zeits. f. Phys. **40,** 501 (1926); N. BOHR, Naturwiss. **16,** 245 (1928), or Nature, **121,** 580 (1928). Also J. V. NEUMANN, Gött. Nachr. 1927, 1; H. WEYL, Zeits. f. Phys. **46,** 1927, 1; A. SOMMERFELD, Phys. Zeits. **30,** 866 (1929); thoroughly treated in HEISENBERG's book cited in (1).
(17) **78.** Gött. Nachr. 1927, 245.
(18) **79.** Gesammelte Abhandlungen, Vol. 2, XXV: Theorie der konvexen Körper, § 12, 157 (Leipsic 1911).
(19) **80.** J. V. NEUMANN, Gött. Nachr. 1927, 273; Zeits. f. Phys. **57,** 30 (1929). E. SCHRÖDINGER, Ann. d. Phys. **83,** 15 (1927). L. NORDHEIM, Proc. Roy. Soc. **119** (A), 689 (1928). W. PAULI, SOMMERFELD-Festschrift: Probleme der modernen Physik, 30 (Leipsic 1928).
(20) **80.** P. A. M. DIRAC, Proc. Cambr. Phil. Soc. I, **25,** 62 (1928).
(21) **86.** For a detailed account see the books by HUND and PAULING-GOUDSMIT cited in the Introduction.
(22) **87.** This formulation of the fundamental problem of perturbation theory refers to the more general case in which the perturbation function *W* also depends explicitly on the time *t*; equation (8.1) is valid in any case. See M. BORN's investigation of the adiabatic principle in quantum mechanics, Zeits. f. Phys. **40, 167 (1927),** and H. WEYL, l.c. (16). E. FERMI and F. PERSICO, Rend. Acc. d. Lincei (6) **4,** 452 (1926); M. BORN and V. FOCK, Zeits. f. Phys. **51,** 165 (1928).
(23) **95.** The recognition of the non-commutativity of multiplication and the discovery of these commutation rules was a most important step in HEISENBERG's first paper and in the further

PAGE
development of the new quantum mechanics in the papers by BORN, HEISENBERG and JORDAN cited in (4).

(24) **96.** An excellent account of the HAMILTON-JACOBI theory of dynamics and of the perturbation theory of classical mechanics is to be found in the chapters of GEIGER and SCHEEL's Handbuch der Physik on these subjects by L. NORDHEIM and E. FUES : Vol. V, Chaps. III and IV. The English reader may refer to the book by M. BORN cited in (2). For canonical transformations in quantum mechanics see P. JORDAN, Zeits. f. Phys. **37,** 383 ; **38,** 513 (1926) ; F. LONDON, Zeits. f. Phys. **40,** 193 (1926) ; P. A. M. DIRAC, Proc. Roy. Soc. **113** (A), 621 (1927).

(25) **100.** H. WEYL, Raum-Zeit-Materie, 5th ed., §§ 40, 41 (Berlin 1923) ; or the English translation by H. L. BROSE, Space, Time and Matter, § 35 (London 1922). E. SCHRÖDINGER, Zeits. f. Phys. **12,** 13 (1922). F. LONDON, Zeits. f. Phys. **42,** 375 (1927).

(26) **102.** Collected Papers, p. 76. New data by J. S. FOSTER and L. CHALK, Proc. Roy. Soc. **123** (A), 108 (1929).

(27) **104.** This result is easily obtained by elementary methods for a rectangular parallelepiped. For the general proof see H. WEYL, Journ. f. d. reine u. angew. Math. **141,** 163 ; **143,** 177 (1912-13) ; Rend. d. Circ. Mat. Palermo, **39,** 1 (1915). R. COURANT has carried over the method from integral to differential equations : see Chap. VI in COURANT-HILBERT, Methoden der mathematischen Physik I.

(28) **104.** P. A. M. DIRAC, Proc. Roy. Soc. **114** (A), 243 (1927). In addition to this paper on emission and absorption see also the one on dispersion to be found on p. 710 of the same volume. For JEANS' treatment of black body radiation, which led to the RAYLEIGH-JEANS radiation law, see J. H. JEANS, Phil. Mag. **10** (6), 91 (1905). P. DEBYE, Ann. d. Phys. (4), **33,** 1427 (1910), introduced the quantum of action into this theory.

(29) **109.** Led by arguments of a general statistical nature, EINSTEIN had recognized the necessity for introducing stimulated emission long before the development of the new quantum mechanics and had derived equations (13.9), (13.10) : Phys. Zeits. **18,** 121 (1917). The new quantum mechanics completes the derivation by obtaining the probability coefficient A, eq. (13.8), from the structure of the atom.

(30) **109.** V. WEISSKOPF and E. WIGNER, Zeits. f. Phys. **63,** 54 (1930).

CHAPTER III

(1) **110.** For the general foundations of the theory of groups and the development of the theory of finite groups see : W. BURNSIDE, Theory of Groups of Finite Order, 2nd ed. (Cambridge 1911) ; G. A. MILLER, H. F. BLICHFELDT and L. E. DICKSON, Theory and Applications of Finite Groups (New York 1916) ; A. SPEISER, Theorie der Gruppen von endlicher Ordnung, 2nd ed. (Berlin 1927).

(2) **112.** Vergleichende Betrachtungen über neuere geometrische Forschungen (Erlangen 1872) ; Math. Ann. **43,** 63 (1893) ; or F. KLEIN, Gesammelte mathematische Abhandlungen, Vol. I, 460 (Berlin 1921).

PAGE

(3) **120.** Following the fundamental results of T. MOLIEN on the theory of hyper-complex numbers (Math. Ann. **41** and **42,** 1893), the theory of representations of finite groups was developed principally by G. FROBENIUS (Sitzungsber. Preuss. Akad. 1896-99). The most important general results were re-discovered by BURNSIDE—cf. his book cited in (1) above. The method developed by I. SCHUR, Neue Begründung der Theorie der Gruppencharaktere, Sitzungsber. Preuss. Akad. 1905, 406, is particularly recommended for its clarity.

(4) **134.** The development of § 6 follows E. NOETHER, Math. Zeits. **30,** 641 (1929), in particular §§ 3 and 16. The uniqueness of complete reduction rather than reduction follows in general W. KRULL, Math. Zeits. **23,** 161 (1925) ; O. SCHMIDT, Math. Zeits. **29,** 34 (1928) ; R. BRAUER and I. SCHUR, Sitzungsber. Preuss. Akad. 1930, 209.

(5) **152.** SCHUR's treatment of the theory of representations, cited in (3) is based on this lemma.

(6) **153.** W. BURNSIDE, Proc. Lond. Math. Soc. (2), **3,** 430 (1905).

(7) **156.** G. FROBENIUS and I. SCHUR, Sitzungsber. Preuss. Akad. 1906, 209.

(8) **161.** The method of integration over the group manifold is due to A. HURWITZ, Gött. Nachr. 1897, 71, although it was applied by him to the theory of invariants rather than to the theory of groups. I. SCHUR first obtained the orthogonality properties of the characteristics of the continuous rotation group in this way and used them to prove the completeness of the system of known representations : Sitzungsber. Preuss. Akad. 1924, 189, 297, and 346.

(9) **166.** For a modern book on algebra see L. E. DICKSON, Algebras and their Arithmetics (Chicago 1923) ; the German edition, Algebren und ihre Zahlentheorie (trans. by J. J. BURCKHARDT and E. SCHUBARTH, Zurich 1927), follows an author's revision which has not appeared in English. Also B. L. VAN DER WAERDEN, Moderne Algebra II (Berlin 1931). An algebra was previously called a " system of hyper-complex numbers," and is at present to some extent in the German literature ; the algebra of a group is there referred to as a " Gruppenring." The usual procedure in modern algebra consists in reducing the algebra into simple matric algebras, in which case the theorems on realization by linear transformations appear as corollaries ; this development will be followed in Chap. V.

(10) **173.** See R. WEITZENBÖCK, Invariantentheorie (Groningen 1923). The foundation for the proof of the fundamental theorem of the theory of invariants is the HILBERT basis theorem : D. HILBERT, Math. Ann. **36,** 473 (1890). The author has shown (Math. Zeits. **24,** 392, 1926) that the fundamental theorem is valid for any closed and for any semi-simple continuous group. The older theory of invariants was almost exclusively concerned with the group $\mathfrak{c}_n$ of all linear transformations with unit determinant. A really modern book on the theory of invariants is lacking.

(11) **175.** The theory has been presented by S. LIE himself, with the assistance of F. ENGEL, in a huge three-volume work : Theorie der Transformationsgruppen (Leipsic 1893, 1930). See also S. LIE, Vorlesungen über kontinuierliche Gruppen, ed. by G. SCHEFFERS

PAGE
(Leipsic 1893) and the brief presentation in H. WEYL, Mathematische Analyse des Raumproblem's, 5th lecture and Appendix 8 (Berlin 1923). The exclusively English reader may be referred to J. E. CAMPBELL, Introductory Treatise on LIE's Theory of Finite Continuous Transformation Groups (Oxford 1903).

(12) **180.** E. CARTAN, Bull. Soc. Math. d. France **41,** 53 (1913). See also H. WEYL, Math. Zeits. **23,** 275 (1925); M. BORN and P. JORDAN, Elementare Quantenmechanik, Chap. IV.

(13) **181.** For the more profound theory of ray representations see I. SCHUR, Journ, f. d. reine u. angew. Math. **127,** 20; **132,** 85; **138,** 155 (1904-11).

(14) **184.** This theorem is contained in my investigations on the representations of semi-simple groups: Math. Zeits. **23, 271**; **24,** 328, 377 and 789 (1925-26). To this type of group belong: the groups $\mathfrak{c}_n$ of all linear transformations with unit determinant, the rotation groups $\mathfrak{d}_n$ and the "complex group" of all linear transformations which leave a non-degenerate ant-symmetric bi-linear form in two arbitrary vectors in a $(2n)$-dimensional space invariant. The first and second of the above papers are concerned with these most important cases. The topological investigation of the rotation group is to be found in Chap. II, § 5 (**24, 346**).

CHAPTER IV

(1) **191.** The theory of atomic spectra, which is developed in this and the following chapter, is to be compared constantly with the empirical data; in particular see the books by HUND, PAULING-GOUDSMIT and GROTRIAN cited in the Introduction. The application of the theory of representations of the 3-dimensional rotation group to atomic spectra is treated by E. WIGNER, Zeits. f. Phys. **43,** 624 (1927); J. v. NEUMANN and E. WIGNER, Zeits. f. Phys. **47,** 203; **49,** 73 (1928). The subject has also been treated systematically recently by E. WIGNER: Gruppentheorie und ihre Anwendung auf der Quantenmechanik der Atomspektren (Braunschweig 1931); for a report on the subject see C. ECKART, Application of Group Theory to the Quantum Dynamics of Monatomic Systems, Rev. Mod. Phys. **2,** 305 (1930). The inner quantum number was introduced, on basis of the empirical data, by A. SOMMERFELD, Ann. d. Phys. **63,** 221 (1920); **70,** 32 (1923).

(2) **191.** The theory of the terms of diatomic molecules is treated in the following fundamental papers by F. HUND: Zeits. f. Phys. **36,** 657 (1926); **40,** 742 (1927); **42,** 93 (1927); **43,** 805 (1927); **51,** 759 (1928); **63,** 719 (1930). Further see: R. S. MULLIKAN, Phys. Rev. **32,** 186 and 761 (1928); **36,** 699 and 1440 (1930). M. BORN and J. R. OPPENHEIMER, Ann. d. Phys. (4) **84,** 457 (1927). E. U. CONDON, Phys. Rev. **28,** 1182 (1926); **32,** 858 (1928). A series of reports and discussions on this subject is to be found in Trans. Faraday Soc. **25,** 611-949 (1929); for a detailed report on the entire field of molecular spectra see R. S. MULLIKAN, Rev. Mod. Phys. **2,** 60 (1930); **3,** 89 (1931); see also the text by RUARK and UREY cited in the Introduction.

(3) **191.** W. ELERT, Zeits. f. Phys. **51,** 8 (1928). Cf. H. BETHE, Ann. d. Phys. (5), **3,** 133 (1929); E. HÜCKEL, Zeits. f. Phys. **60,** 423 (1930).

PAGE
(4) **194.** Appropriate methods for carrying through the perturbation calculations (method of the "self-consistent field") have been developed by: D. R. HARTREE, Proc. Cambr. Phil. Soc. **24**, 89 (1928). J. A. GAUNT, Proc. Cambr. Phil. Soc. **24**, 328 (1928). Also J. C. SLATER, Phys. Rev. **32**, 339 (1928); **34**, 1293 (1929); **36**, 57 (1930). E. U. CONDON, Phys. Rev. **36**, 1121 (1930); E. U. CONDON and G. H. SHORTLEY, Phys. Rev. **37**, 1025 (1931). V. FOCK, Zeits. f. Phys. **61**, 126; **62**, 795 (1930). G. BREIT, Phys. Rev. **35**, 569; **36**, 383 (1930). W. HEITLER and G. RUMER, Zeits. f. Phys. **68**, 12 (1931).

(5) **201.** See the report by H. HÖNL, Ann. d. Phys. (4) **79**, 273 (1926). For a derivation of the formulæ on quantum mechanics, although not from the group-theoretic standpoint, see M. BORN, W. HEISENBERG and P. JORDAN, Zeits. f. Phys. **35**, 557 (1926). Also in Chap. IV of BORN and JORDAN, Elementare Quantenmechanik.

(6) **203.** W. PAULI, Zeits. f. Phys. **43**, 601 (1927).

(7) **203.** G. E. UHLENBECK and S. GOUDSMIT, Naturwiss. **13**, 953 (1925); Nature **117**, 264 (1926).

(8) **205.** O. RICHARDSON, Phys. Rev. **26**, 248 (1908). A. EINSTEIN and W. J. DE HAAS, Verhandl. d. Deutsch. Phys. Ges. **17**, 152 (1915); **18**, 173 (1916). E. BECK, Ann. d. Phys. (4), **60**, 109 (1919). S. J. and L. J. H. BARNETT, Phys. Rev. **17**, 404 (1921). A. P. CHATTOCK and L. F. BATER, Phil. Trans. Roy. Soc. **223**, 287 (1922).

(9) **207.** A report on a unified notation for the designation of terms of atomic spectra in terms of quantum numbers has been presented by H. N. RUSSELL, A. G. SHENSTONE and L. A. TURNER, Phys. Rev. **33**, 900 (1929). It has also been found necessary to ascribe a spin to the atomic nucleus in order to account for the hyper-fine structure: E. BACK and S. GOUDSMIT, Zeits. f. Phys. **43**, 321 (1927); **47**, 174 (1928); S. GOUDSMIT and R. F. BACHER, Phys. Rev. **34**, 1501 (1929); S. GOUDSMIT, Phys. Rev. **37**, 663 (1931). J. HARGREAVES, Proc. Roy. Soc. **124** (A), 568 (1929). E. FERMI, Zeits. f. Phys. **60**, 320 (1930). G. BREIT, Phys. Rev. **37**, 51 (1931).

(10) **209.** E. BACK and A. LANDÉ, Zeemaneffekt und Multiplettstruktur (Berlin 1925) A. LANDÉ, Zeits. f. Phys. **15**, 189 (1923). W. PAULI, Zeits. f. Phys. **16**, 155; **20**, 371 (1923). A. LANDÉ, Zeits. f. Phys. **25**, 46 (1924). W. HEISENBERG and P. JORDAN, Zeits. f. Phys. **37**, 263 (1926). K. DARWIN, Proc. Roy. Soc. **118** (A), 264 (1928). For (*jj*) and (*sl*) coupling see J. H. BARTLETT, Phys. Rev. **35**, 229 (1930).

(11) **210.** H. WEYL, Math. Zeits. **23**, 292 (1925). J. v. NEUMANN and E. WIGNER, Phys. Zeits. **30**, 467 (1929).

(12) **210.** Proc. Roy. Soc. **117** (A), 610; **118**, 351 (1928). C. G. DARWIN, Proc. Roy. Soc. **118** (A), 654 (1928). A. LANDÉ, Zeits. f. Phys. **48**, 601 (1928); in the same volume F. MÖGLICH, 852, and J. v. NEUMANN, 868. V. FOCK, Zeits. f. Phys. **55**, 127 (1929). For the older work concerning the interaction of spin and orbital moment of momentum see L. H. THOMAS, Nature, **117**, 514 (1926); J. FRENKEL, Zeits. f. Phys. **37**, 243 (1926); W. HEISENBERG and P. JORDAN in the same volume, 863.

(13) **217.** P. A. M. DIRAC in Quantentheorie und Chemie, Leipziger Vorträge, 1928, 83 (Leipsic 1928).

(14) **220.** H. WEYL, Proc. Nat. Acad. Sci. **15**, 323 (1929); Zeits. f. Phys. **56**, 330 (1929). V. FOCK, Zeits. f. Phys. **57**, 261 (1929). V. AMBARCUMIAN and D. IVANENKO, C. R. Acad. sc. USSR. 1930, 45.

PAGE

(15) **224.** See WENTZEL's report cited in II (13); A. SOMMERFELD, Wave Mechanics; BORN and JORDAN, Elementare Quantenmechanik; O. KLEIN and Y. NISHINA, Zeits. f. Phys. **52,** 853 (1929). Y. NISHINA, same volume, 869.

(16) **237.** A. SOMMERFELD, Ann. d. Phys. (4) **51,** 1 (1916). For the significance of these results for the theory of X-ray spectra see SOMMERFELD'S book cited in the introduction. Perturbation calculation in the new quantum mechanics, W. HEISENBERG and P. JORDAN, l.c. (10); exact derivation by means of the Dirac theory of the electron; W. GORDAN, Zeits. f. Phys. **48,** 11 (1928); C. G. DARWIN, l.c. (12); A. SOMMERFELD, Wave Mechanics, p. 257 ff.

(17) **241.** W. HEISENBERG, Zeits. f. Phys. **38,** 411 (1926). Corresponding energy calculation for He atom; W. HEISENBERG, Zeits. f. Phys. **39,** 499 (1926). P. A. M. DIRAC, Proc. Roy. Soc. **112** (A), 661 (1926). J. A. GAUNT, Proc. Roy. Soc. **122** (A), 513 (1929); Phil. Trans. Roy. Soc. **228** (A), 151. Y. SUGIURA, Zeits. f. Phys. **44,** 190 (1927). W. V. HOUSTON, Phys. Rev. **33,** 297 (1929). J. C. SLATER, Phys. Rev. **32,** 349 (1928). G. BREIT, Phys. Rev. **34,** 553 (1929); **36,** 383 (1930). The "symmetric" sub-space leads to the Einstein-Bose statistics, which is discussed in the references cited in II (6) above. The statistics arising from the "anti-symmetric" sub-space was developed by E. FERMI, Zeits. f. Phys. **36,** 902 (1926) and applied by W. PAULI, Zeits. f. Phys. **41,** 81 (1927), to the explanation of paramagnetism and by A. SOMMERFELD to the electron theory of metals: A. SOMMERFELD, W. V. HOUSTON and C. ECKART, Zeits. f. Phys. **47,** 1 (1928).

(18) **244.** E. C. STONER, Phil. Mag. (6) **48,** 719 (1924). W. PAULI, Zeits. f. Phys. **31,** 765 (1925). It is to be remembered that this development antedates the new quantum theory and the theory of the spinning electron, and that Pauli's introduction of the four quantum numbers n, l, j, m demanded a complete re-classification of all spectroscopic material.

(19) **248.** P. A. M. DIRAC, Proc. Roy. Soc. **114** (A), 243 (1927). On taking the interaction of the particles into account: P. JORDAN and O. KLEIN, Zeits. f. Phys. **45,** 751 (1927).

(20) **250, 280.** P. JORDAN and E. WIGNER, Zeits. f. Phys. **47,** 631 (1928).

(21) **253.** P. JORDAN and W. PAULI, Zeits. f. Phys. **47,** 151 (1928). G. MIE, Ann. d. Phys. **85,** 711 (1928). W. HEISENBERG and W. PAULI, Zeits. f. Phys. **56,** 1 (1929); **59,** 168 (1930); W. HEISENBERG, Zeits. f. Phys. **65,** 4 (1930); Ann. d. Phys. **9,** 338 (1931). L. ROSENFELD, Zeits. f. Phys. **63,** 574 (1930). J. R. OPPENHEIMER, Phys. Rev. **35,** 461 (1930). G. BREIT, l.c. (17). E. FERMI, Rend. Acc. d. Lincei (6) **9,** 181 (1929). L. LANDAU and R. PEIERLS, Zeits. f. Phys. **62,** 188 (1930). L. ROSENFELD, Ann. d. Phys. (5) **5,** 113 (1930).

(22) **257.** H. WEYL, Journ. f. d. reine u. angew. Math. **141,** 163 (1912).

(23) **261.** See P. JORDAN, Die Lichtquantenhypothese, in: Ergebnisse der exacten Wissenschaften, **7,** 158 (1928).

(24) **262.** P. A. M. DIRAC, Proc. Roy. Soc. **126** (A), 360 (1930); Proc. Cambr. Phil. Soc., **26,** 361 (1930). J. R. OPPENHEIMER, Phys. Rev. **35,** 939 (1930). For a report on this theory see P. A. M. DIRAC, Nature, **126,** 605 (1930). For an attempt to avoid the negative energy levels by a reduction of all operators see E. SCHRÖDINGER, Sitzungsber. Preuss. Akad. 1931, 63.

PAGE
(25) **264.** See articles by HEISENBERG-PAULI and ROSENFELD cited in [21].
(26) **276.** H. WEYL, Zeits. f. Phys. **46,** 1 (1927).
(27) **280.** A rigorous proof of these theorems for ∞-dimensional space has been announced by M. H. STONE, Proc. Nat. Acad. Sci. **16,** 172 (1930) ; J. v. NEUMANN informs me in a recent letter that he has also obtained a proof of this theorem.

CHAPTER V

(1) **284.** The transition from the group Σ_0 to the algebra Σ, which is suggested by quantum mechanics, has also improved the theory from the purely mathematical standpoint ; see H. WEYL, Ann. of Math. (2) **30,** 499 (1929). The connection between the representations of $\mathfrak{u}_n$ or $\mathfrak{c}_n$ and π_f was first clearly seen by I. SCHUR in his Dissertation (Berlin 1901). Further see : H. WEYL, Math. Zeits. **23,** 271 (1925) ; I. SCHUR, Sitzungsber. Preuss. Akad. **1927,** 58 ; 1928, 100. On the symmetry classes of tensors see : A. YOUNG, Proc. Lond. Math. Soc. **33,** 97 (1900) ; **34,** 361 (1901). H. WEYL, Rend. Circ. Mat. Palermo, **48,** 29 (1924).
(2) **287.** This has been emphasized by P. A. M. DIRAC, Proc. Roy. Soc. **123** (A), 714 (1929).
(3) **291.** G. FROBENIUS used the term " characteristic unit " for this concept (see Sitzungsber. Preuss. Akad. **1903,** 328), and this name has been taken over into the physical literature. But in the meantime the term " idempotent " has been used in systematic investigations on algebras. The notions of " right- and left-invariant sub-algebra " and " left-invariant sub-algebra " correspond with those of " ideal " and " left-ideal " in arithmetic when all the elements of the algebra are considered as " integers."
(4) **303.** E. STEINITZ, Journ. f. d. reine u. angew. Math. **137,** 167 (1910).
(5) **307.** Our proof of this theorem follows E. NOETHER, Math. Zeits. **30,** 641 (1929).
(6) **313.** In the older investigations T. MOLIEN (Math. Ann. **41** and **42,** 1893) and G. FROBENIUS operate in the field of all complex numbers. The extension to arbitrary fields is due to J. H. M. WEDDERBURN, and is also valid for algebras which are not completely reducible—a branch of the subject into which we have not entered : J. H. M. WEDDERBURN, Proc. Lond. Math. Soc. (2) **6,** 99 (1907) ; Bull. Am. Math. Soc. **31,** 11 (1925). See also the book by DICKSON referred to in III [9]. Our proof follows E. NOETHER, l.c. [5]. See further E. ARTIN, Abh. Math. Semin. Hamburg, **5,** 251 (1927) ; G. KÖTHE, Math. Zeits. **32,** 161 (1930).
(7) **320.** E. WIGNER, Zeits. f. Phys. **40,** 492 and 883 (1926-27). W. HEITLER, Zeits. f. Phys. **46,** 49 (1927). Only the simplest case, that in which the unperturbed term of I^f consists of f different, non-degenerate terms of the individual I, is considered in detail in these papers.
(8) **328.** This direct derivation follows H. WEYL, Math. Zeits. **23,** 271 (1925).
(9) **338.** See G. FROBENIUS, Sitzungsber. Preuss. Akad. 1898, 501.
(10) **340.** W. HEITLER and F. LONDON, Zeits. f. Phys. **44,** 455 (1927). W. HEITLER, Zeits. f. Phys. **46,** 47 (1927) ; F. LONDON, in the same

PAGE
volume, 455. W. HEITLER, Gött. Nachr. 1927, 368 ; Zeits. f. Phys. **47,** 835 (1928). F. LONDON, Zeits. f. Phys. **50,** 24 (1928). W. HEITLER, Zeits. f. Phys. **51,** 805 (1928). M. DELBRÜCK, Zeits. f. Phys. **51,** 181 (1928). F. LONDON, in : Quantentheorie und Chemie, Leipziger Vorträge 1928, 59 (Leipsic 1928) ; Zeits. f. Phys. **63,** 245 (1930). M. BORN, Zeits. f. Phys. **64,** 729 (1930). J. C. SLATER, Phys. Rev. **37,** 481 ; **38,** 1109 (1931). L. PAULING, Journ. Ann. Chem. Soc. **53,** 1367 (1931).

(12) **342.** The calculation is carried through in the first paper by HEITLER and LONDON cited in (10). Further see : Y. SUGIURA, Zeits. f. Phys. **45,** 484 (1927). S. C. WANG, Phys. Rev. **31,** 579 (1928) ; **28,** 663 (1927). E. C. KEMBLE and C. ZENER, Phys. Rev. **33,** 512 (1929). P. M. MORSE and E. C. G. STÜCKELBERG, Phys. Rev. **33,** 932 (1929).

(13) **346.** Zeits. f. Phys. **50,** 24 (1928).

(14) **347.** Zeits. f. Phys. **49,** 619 (1928) ; SOMMERFELD-Festschrift : Probleme der modernen Physik (Leipsic 1929).

(15) **357.** P. A. M. DIRAC, l.c. (2). For a detailed term calculation following this scheme and examples see papers by SLATER, CONDON, CONDON-SHORTLEY, BORN-RUMER cited in (4) above.

(16) **358.** The introduction of the symmetry operators c into the theory of invariants is due to A. YOUNG, l.c. (1). But he proved the irreducibility of neither $\mathfrak{h}_c$ nor $\mathfrak{H}_c$; that of the first was proved by G. FROBENIUS, Sitzungsber. Preuss. Akad. 1903, 328, and that of the latter by E. CARTAN, Bull. Soc. Math. d. France, **41,** 53 (1913) and H. WEYL, l.c. (8). The symmetry classes were re-discovered in quantum mechanics by F. HUND, Zeits. f. Phys. **43,** 788 (1927).

(17) **362.** The development from theorem (14.2) to (14.8) follows a train of thought communicated to the author in a letter from J. v. NEUMANN.

(18) **370.** See F. HUND, l.c. (16) ; J. v. NEUMANN and E. WIGNER, Zeits. f. Phys. **47,** 203 ; **49,** 73 (1928).

(19) **370.** F. LONDON, Zeits. f. Phys. **46,** 455 (1928).

(20) **372.** W. HEITLER, Zeits. f. Phys. **51,** 805 (1928).

(21) **378.** Follows H. WEYL, l.c. (8). In the same way the characteristics of the rotation group in n-dimensional space, the " complex group " and all semi-simple groups can be calculated : Math. Zeits. **24,** 328, 377 and 789 (1926).

(22) **382.** L.c. (8). On removing the unitary restriction, the proof that we here obtain all irreducible representations of $\mathfrak{c}_n$ requires the use of the infinitesimal elements of the group. The knowledge won for $\mathfrak{u}_n$ has been carried over to $\mathfrak{c}_n$ under the broadest assumptions by J. v. NEUMANN, Sitzungsber. der Preuss. Akad. 1927, 26 ; Math. Zeits. **30,** 3 (1929) ; and I. SCHUR, Sitzungsber. Preuss. Akad 1928, 100.

(23) **383.** Sitzungsber. Preuss. Akad. 1900, 516.

OPERATIONAL SYMBOLS

The number refers to the page on which the symbol is defined

$\rightarrow$	with . . . is associated . . . 110, 114.
$\prec$	is contained in 290.
$\bar{x}$	conjugate complex of x 15.
$*$	transposition: for operators 13, symmetry quantities 352, symmetry patterns 361.
$\sim$	Hermitian conjugate: for operators 17, elements of an algebra 167.
$\wedge$	$\hat{a}(s) = a(s^{-1})$ 296.
$\smile$	contragredient matrix 123, representation 123.
$\simeq$	equivalent as correspondences of the ray field 21.
$\sim$	transforms as 145.
( )	scalar product 16, 32.
[ ]	vector product (in 3-dimensional space) 27; commutator $[HA] = \frac{1}{i}(HA - AH)$ 264.
$\langle\ \rangle$	temporal mean value 88.
$\times$	for vectors 90, vector spaces 90, correspondences 91, representations 126, groups 127, algebras and their elements 333.
$\boldsymbol{\times}$	multiplication of representations of two groups 127.
$+$	addition of representations 113.
$\sharp$	transition from $\mathfrak{p}$ to $\mathfrak{P}$ 287.
$\natural$	transition from $\mathfrak{P}$ to $\mathfrak{p}$ 290.

LETTERS HAVING A FIXED SIGNIFICANCE

The number refers to the page on which the quantity is defined

LATIN

c velocity of light; a Young symmetry operator 359.
e primitive idempotent element (generating unit 291); $-e$ charge of the electron.
$e(x) = e^{ix}$.
$(E_x, E_y, E_z) = \mathfrak{E}$ electric field strength 99.
E_i energy level 44.
f number of electrons, order of a tensor 139, 281.
f_α 4-vector potential multiplied by e/ch 214.
$f_{\alpha\beta}$ curl of f_α $\left(= \frac{\partial f_\beta}{\partial x^\alpha} - \frac{\partial f_\alpha}{\partial x^\beta}\right)$ 216.
F action of the electro-magnetic field 215.
$F(i_1, i_2, \ldots, i_f)$ tensor 139, 281.
g dimensionality of a group representation 120, Landé g-factor 204, 207.
h Planck's quantum of action divided by 2π 51, order of a finite group 118.
H energy 51.
$(H_x, H_y, H_z) = \mathfrak{H}$ magnetic field strength 99.
I signature 188.
j, $\mathcal{J}$ inner quantum number 189, 190.
$\mathcal{J}_\alpha$ total energy-momentum vector 220.
k auxiliary quantum number 228.
l, L azimuthal quantum number 64, 185, 194—for $s, p, d, f, g, \ldots$ terms $l = 0, 1, 2, 3, 4, \ldots$.
$(L_x, L_y, L_z) = \mathfrak{L}$ orbital moment of momentum 63.
m magnetic quantum number 64, 193, multiplicity of a representation 321, 350; $(= \mu)$ mass of the electron.
$m_0 = mc/h$.
M, M' action of the material field 211.
$(M_x, M_y, M_z) = \mathfrak{M}$ total moment of momentum 179, 187.
n dimensionality of a vector space 1; principal quantum number 69.

LATIN

p, q canonically conjugate variables 94, a permutation in the rows, columns of a symmetry pattern 359.

$(p_x, p_y, p_z) = \mathfrak{p}$ linear momentum of a particle 51.

P symmetry pattern 358.

$(q_x, q_y, q_z) = \mathfrak{q}$ electric dipole moment 83.

r distance from centre.

s element of a group ; spin quantum number 206.

$(s_x, s_y, s_z) = \mathfrak{s}$ electric current density 218, s^α charge-current 4-vector 214.

$(S_x, S_y, S_z) = \mathfrak{S}$ spin 178, 203.

$$S_0 = \begin{Vmatrix} 1 & 0 \\ 0 & 1 \end{Vmatrix}, S_1 = \begin{Vmatrix} 0 & 1 \\ 1 & 0 \end{Vmatrix}, S_2 = \begin{Vmatrix} 0 & -i \\ i & 0 \end{Vmatrix}, S_3 = \begin{Vmatrix} 1 & 0 \\ 0 & -1 \end{Vmatrix} \text{ 148.}$$

t_α^β energy-momentum tensor 218.

T interchange of ψ_1, ψ_2 and ψ_1', ψ_2' 149.

v valence 369.

W perturbation energy 86, total action 216.

$x_0\, x_1\, x_2\, x_3$ or $t\, x\, y\, z$ co-ordinates of space time ($t = x_0$ 98, or $ct = x_0$ 211).

GERMAN. (For 3-dimensional vectors see their components under Latin letters.)

$\mathfrak{c} = \mathfrak{c}_n$ group of (unimodular) linear transformations in n dimensions 128.

$(\mathfrak{c})^f$ representation of $\mathfrak{c}$ whose substratum is the tensors of order f 125.

$\mathfrak{C}_v = \mathfrak{D}_j (v = 2j)$ representation of vth degree of $\mathfrak{c}_2$ or $\mathfrak{u}_2 \sim \mathfrak{d}_3$ 128, 142.

$\mathfrak{d}_n$ orthogonal group in n dimensions 142 ; $\mathfrak{d}_n'$ same but including improper rotations 143.

$\mathfrak{D}^{(m)}$ 1-dimensional representation of rotation group $\mathfrak{d}_2$ 141.

$\mathfrak{e}_1, \mathfrak{e}_2, \ldots, \mathfrak{e}_n$ co-ordinate system in vector space 2.

$\mathfrak{E}$ unitary representation of the rotation group induced in the function space of $\psi(x\, y\, z)$ 143.

$\mathfrak{g}$ abstract group 114.

$\mathfrak{k}_a$ conjugation 118.

$\mathfrak{M}$ mean value 158.

$\mathfrak{R}$ representation of the rotation group induced in system space 187.

$\mathfrak{p}, \mathfrak{P}$ invariant sub-space of $\mathfrak{r}$, $\mathfrak{R}^f$ respectively 287, 282.

$\mathfrak{r}$ an algebra considered as a vector space 286, $\mathfrak{r}_0 = \overset{n}{\mathfrak{r}} = ǂ\, \mathfrak{R}^f$ 290, 350.

GERMAN

$\mathfrak{R}$ vector space |, $\mathfrak{R}^f$ corresponding space of tensors of order f, $[\mathfrak{R}^f]$ space of the symmetric tensors, $\{\mathfrak{R}^f\}$ space of the anti-symmetric tensors, 239, 242.

$\mathfrak{R}_t$, $\mathfrak{R}_a$ system space of electron translation, spin 196.

$\mathfrak{t}_a$ left-translation 116.

$\mathfrak{u} = \mathfrak{u}_n$ (unimodular) unitary group in n dimensions 139.

$\mathfrak{B}$ ray representation giving rise to algebra of complex quaternions 182.

$\mathfrak{x}$ vector in n dimensional vector space 1.

GREEK

$\alpha = e^2/ch$ fine structure constant 216.

δ_{ik} Kronecker symbol = 1 or 0 according as $i = k$ or $i \neq k$ 17.

$\delta(x)$ Dirac δ-function (= 0 except for $x = 0$ and $\int_{-\epsilon}^{+\epsilon}\delta(x)dx = 1$) 255.

$\delta_s = \pm 1$ according as s is an even or an odd permutation 121.

δ signature 201.

$\Delta = \frac{\partial^2}{\partial x^2} + \frac{\partial^2}{\partial y^2} + \frac{\partial^2}{\partial z^2}$ Laplace's operator α 52.

$\nabla = \sum_\alpha S_\alpha \frac{\partial}{\partial x^\alpha}$ 212.

ε generating element of a right- and left-invariant sub-space 311.

θ, ϕ polar co-ordinates 60.

$\mu(= m)$ mass of the electron.

ν frequency 50.

$o = \frac{e|\mathfrak{H}|}{2\mu c}$, Larmor factor—unit of Zeeman separation.

$\pi = \pi_f$ symmetric group of permutations of f objects 121.

ρ electric charge density 218, an algebra 304.

ϕ_α electro-magnetic 4-vector potential 98.

ψ vector defining the state of the material field 49.

χ, X group characteristics, 150, 151.

ω angle of rotation 151.

INDEX

The numbers refer to pages of the text, those in boldface to the pages where the concepts introduced in boldface are defined

A CATALOG OF SELECTED

DOVER BOOKS

IN SCIENCE AND MATHEMATICS

Astronomy

BURNHAM'S CELESTIAL HANDBOOK, Robert Burnham, Jr. Thorough guide to the stars beyond our solar system. Exhaustive treatment. Alphabetical by constellation: Andromeda to Cetus in Vol. 1; Chamaeleon to Orion in Vol. 2; and Pavo to Vulpecula in Vol. 3. Hundreds of illustrations. Index in Vol. 3. 2,000pp. 6⅛ x 9¼.
Vol. I: 0-486-23567-X
Vol. II: 0-486-23568-8
Vol. III: 0-486-23673-0

EXPLORING THE MOON THROUGH BINOCULARS AND SMALL TELESCOPES, Ernest H. Cherrington, Jr. Informative, profusely illustrated guide to locating and identifying craters, rills, seas, mountains, other lunar features. Newly revised and updated with special section of new photos. Over 100 photos and diagrams. 240pp. 8¼ x 11. 0-486-24491-1

THE EXTRATERRESTRIAL LIFE DEBATE, 1750–1900, Michael J. Crowe. First detailed, scholarly study in English of the many ideas that developed from 1750 to 1900 regarding the existence of intelligent extraterrestrial life. Examines ideas of Kant, Herschel, Voltaire, Percival Lowell, many other scientists and thinkers. 16 illustrations. 704pp. 5⅜ x 8½. 0-486-40675-X

THEORIES OF THE WORLD FROM ANTIQUITY TO THE COPERNICAN REVOLUTION, Michael J. Crowe. Newly revised edition of an accessible, enlightening book re-creates the change from an earth-centered to a sun-centered conception of the solar system. 242pp. 5⅜ x 8½. 0-486-41444-2

ARISTARCHUS OF SAMOS: The Ancient Copernicus, Sir Thomas Heath. Heath's history of astronomy ranges from Homer and Hesiod to Aristarchus and includes quotes from numerous thinkers, compilers, and scholasticists from Thales and Anaximander through Pythagoras, Plato, Aristotle, and Heraclides. 34 figures. 448pp. 5⅜ x 8½. 0-486-43886-4

A COMPLETE MANUAL OF AMATEUR ASTRONOMY: TOOLS AND TECHNIQUES FOR ASTRONOMICAL OBSERVATIONS, P. Clay Sherrod with Thomas L. Koed. Concise, highly readable book discusses: selecting, setting up and maintaining a telescope; amateur studies of the sun; lunar topography and occultations; observations of Mars, Jupiter, Saturn, the minor planets and the stars; an introduction to photoelectric photometry; more. 1981 ed. 124 figures. 25 halftones. 37 tables. 335pp. 6½ x 9¼. 0-486-42820-8

AMATEUR ASTRONOMER'S HANDBOOK, J. B. Sidgwick. Timeless, comprehensive coverage of telescopes, mirrors, lenses, mountings, telescope drives, micrometers, spectroscopes, more. 189 illustrations. 576pp. 5⅝ x 8¼. (Available in U.S. only.)
0-486-24034-7

STAR LORE: Myths, Legends, and Facts, William Tyler Olcott. Captivating retellings of the origins and histories of ancient star groups include Pegasus, Ursa Major, Pleiades, signs of the zodiac, and other constellations. "Classic."–Sky & Telescope. 58 illustrations. 544pp. 5⅜ x 8½. 0-486-43581-4

Chemistry

THE SCEPTICAL CHYMIST: THE CLASSIC 1661 TEXT, Robert Boyle. Boyle defines the term "element," asserting that all natural phenomena can be explained by the motion and organization of primary particles. 1911 ed. viii+232pp. 5⅜ x 8½.
0-486-42825-7

RADIOACTIVE SUBSTANCES, Marie Curie. Here is the celebrated scientist's doctoral thesis, the prelude to her receipt of the 1903 Nobel Prize. Curie discusses establishing atomic character of radioactivity found in compounds of uranium and thorium; extraction from pitchblende of polonium and radium; isolation of pure radium chloride; determination of atomic weight of radium; plus electric, photographic, luminous, heat, color effects of radioactivity. ii+94pp. 5⅜ x 8½.
0-486-42550-9

CHEMICAL MAGIC, Leonard A. Ford. Second Edition, Revised by E. Winston Grundmeier. Over 100 unusual stunts demonstrating cold fire, dust explosions, much more. Text explains scientific principles and stresses safety precautions. 128pp. 5⅜ x 8½. 0-486-67628-5

MOLECULAR THEORY OF CAPILLARITY, J. S. Rowlinson and B. Widom. History of surface phenomena offers critical and detailed examination and assessment of modern theories, focusing on statistical mechanics and application of results in mean-field approximation to model systems. 1989 edition. 352pp. 5⅜ x 8½.
0-486-42544-4

CHEMICAL AND CATALYTIC REACTION ENGINEERING, James J. Carberry. Designed to offer background for managing chemical reactions, this text examines behavior of chemical reactions and reactors; fluid-fluid and fluid-solid reaction systems; heterogeneous catalysis and catalytic kinetics; more. 1976 edition. 672pp. 6⅛ x 9¼. 0-486-41736-0 $31.95

ELEMENTS OF CHEMISTRY, Antoine Lavoisier. Monumental classic by founder of modern chemistry in remarkable reprint of rare 1790 Kerr translation. A must for every student of chemistry or the history of science. 539pp. 5⅜ x 8½. 0-486-64624-6

MOLECULES AND RADIATION: An Introduction to Modern Molecular Spectroscopy. Second Edition, Jeffrey I. Steinfeld. This unified treatment introduces upper-level undergraduates and graduate students to the concepts and the methods of molecular spectroscopy and applications to quantum electronics, lasers, and related optical phenomena. 1985 edition. 512pp. 5⅜ x 8½. 0-486-44152-0

A SHORT HISTORY OF CHEMISTRY, J. R. Partington. Classic exposition explores origins of chemistry, alchemy, early medical chemistry, nature of atmosphere, theory of valency, laws and structure of atomic theory, much more. 428pp. 5⅜ x 8½. (Available in U.S. only.) 0-486-65977-1

GENERAL CHEMISTRY, Linus Pauling. Revised 3rd edition of classic first-year text by Nobel laureate. Atomic and molecular structure, quantum mechanics, statistical mechanics, thermodynamics correlated with descriptive chemistry. Problems. 992pp. 5⅜ x 8½. 0-486-65622-5

ELECTRON CORRELATION IN MOLECULES, S. Wilson. This text addresses one of theoretical chemistry's central problems. Topics include molecular electronic structure, independent electron models, electron correlation, the linked diagram theorem, and related topics. 1984 edition. 304pp. 5⅜ x 8½. 0-486-45879-2

Engineering

DE RE METALLICA, Georgius Agricola. The famous Hoover translation of greatest treatise on technological chemistry, engineering, geology, mining of early modern times (1556). All 289 original woodcuts. 638pp. 6¾ x 11. 0-486-60006-8

FUNDAMENTALS OF ASTRODYNAMICS, Roger Bate et al. Modern approach developed by U.S. Air Force Academy. Designed as a first course. Problems, exercises. Numerous illustrations. 455pp. 5⅜ x 8½. 0-486-60061-0

DYNAMICS OF FLUIDS IN POROUS MEDIA, Jacob Bear. For advanced students of ground water hydrology, soil mechanics and physics, drainage and irrigation engineering and more. 335 illustrations. Exercises, with answers. 784pp. 6⅛ x 9¼. 0-486-65675-6

THEORY OF VISCOELASTICITY (SECOND EDITION), Richard M. Christensen. Complete consistent description of the linear theory of the viscoelastic behavior of materials. Problem-solving techniques discussed. 1982 edition. 29 figures. xiv+364pp. 6⅛ x 9¼. 0-486-42880-X

MECHANICS, J. P. Den Hartog. A classic introductory text or refresher. Hundreds of applications and design problems illuminate fundamentals of trusses, loaded beams and cables, etc. 334 answered problems. 462pp. 5⅜ x 8½. 0-486-60754-2

MECHANICAL VIBRATIONS, J. P. Den Hartog. Classic textbook offers lucid explanations and illustrative models, applying theories of vibrations to a variety of practical industrial engineering problems. Numerous figures. 233 problems, solutions. Appendix. Index. Preface. 436pp. 5⅜ x 8½. 0-486-64785-4

STRENGTH OF MATERIALS, J. P. Den Hartog. Full, clear treatment of basic material (tension, torsion, bending, etc.) plus advanced material on engineering methods, applications. 350 answered problems. 323pp. 5⅜ x 8½. 0-486-60755-0

A HISTORY OF MECHANICS, René Dugas. Monumental study of mechanical principles from antiquity to quantum mechanics. Contributions of ancient Greeks, Galileo, Leonardo, Kepler, Lagrange, many others. 671pp. 5⅜ x 8½. 0-486-65632-2

STABILITY THEORY AND ITS APPLICATIONS TO STRUCTURAL MECHANICS, Clive L. Dym. Self-contained text focuses on Koiter postbuckling analyses, with mathematical notions of stability of motion. Basing minimum energy principles for static stability upon dynamic concepts of stability of motion, it develops asymptotic buckling and postbuckling analyses from potential energy considerations, with applications to columns, plates, and arches. 1974 ed. 208pp. 5⅜ x 8½. 0-486-42541-X

BASIC ELECTRICITY, U.S. Bureau of Naval Personnel. Originally a training course; best nontechnical coverage. Topics include batteries, circuits, conductors, AC and DC, inductance and capacitance, generators, motors, transformers, amplifiers, etc. Many questions with answers. 349 illustrations. 1969 edition. 448pp. 6½ x 9¼. 0-486-20973-3

Mathematics

FUNCTIONAL ANALYSIS (Second Corrected Edition), George Bachman and Lawrence Narici. Excellent treatment of subject geared toward students with background in linear algebra, advanced calculus, physics and engineering. Text covers introduction to inner-product spaces, normed, metric spaces, and topological spaces; complete orthonormal sets, the Hahn-Banach Theorem and its consequences, and many other related subjects. 1966 ed. 544pp. 6⅛ x 9¼. 0-486-40251-7

DIFFERENTIAL MANIFOLDS, Antoni A. Kosinski. Introductory text for advanced undergraduates and graduate students presents systematic study of the topological structure of smooth manifolds, starting with elements of theory and concluding with method of surgery. 1993 edition. 288pp. 5⅜ x 8½. 0-486-46244-7

VECTOR AND TENSOR ANALYSIS WITH APPLICATIONS, A. I. Borisenko and I. E. Tarapov. Concise introduction. Worked-out problems, solutions, exercises. 257pp. 5⅜ x 8¼. 0-486-63833-2

AN INTRODUCTION TO ORDINARY DIFFERENTIAL EQUATIONS, Earl A. Coddington. A thorough and systematic first course in elementary differential equations for undergraduates in mathematics and science, with many exercises and problems (with answers). Index. 304pp. 5⅜ x 8½. 0-486-65942-9

FOURIER SERIES AND ORTHOGONAL FUNCTIONS, Harry F. Davis. An incisive text combining theory and practical example to introduce Fourier series, orthogonal functions and applications of the Fourier method to boundary-value problems. 570 exercises. Answers and notes. 416pp. 5⅜ x 8½. 0-486-65973-9

COMPUTABILITY AND UNSOLVABILITY, Martin Davis. Classic graduate-level introduction to theory of computability, usually referred to as theory of recurrent functions. New preface and appendix. 288pp. 5⅜ x 8½. 0-486-61471-9

AN INTRODUCTION TO MATHEMATICAL ANALYSIS, Robert A. Rankin. Dealing chiefly with functions of a single real variable, this text by a distinguished educator introduces limits, continuity, differentiability, integration, convergence of infinite series, double series, and infinite products. 1963 edition. 624pp. 5⅜ x 8½. 0-486-46251-X

METHODS OF NUMERICAL INTEGRATION (SECOND EDITION), Philip J. Davis and Philip Rabinowitz. Requiring only a background in calculus, this text covers approximate integration over finite and infinite intervals, error analysis, approximate integration in two or more dimensions, and automatic integration. 1984 edition. 624pp. 5⅜ x 8½. 0-486-45339-1

INTRODUCTION TO LINEAR ALGEBRA AND DIFFERENTIAL EQUATIONS, John W. Dettman. Excellent text covers complex numbers, determinants, orthonormal bases, Laplace transforms, much more. Exercises with solutions. Undergraduate level. 416pp. 5⅜ x 8½. 0-486-65191-6

RIEMANN'S ZETA FUNCTION, H. M. Edwards. Superb, high-level study of landmark 1859 publication entitled "On the Number of Primes Less Than a Given Magnitude" traces developments in mathematical theory that it inspired. xiv+315pp. 5⅜ x 8½. 0-486-41740-9

TENSOR CALCULUS, J.L. Synge and A. Schild. Widely used introductory text covers spaces and tensors, basic operations in Riemannian space, non-Riemannian spaces, etc. 324pp. 5⅜ x 8¼. 0-486-63612-7

ORDINARY DIFFERENTIAL EQUATIONS, Morris Tenenbaum and Harry Pollard. Exhaustive survey of ordinary differential equations for undergraduates in mathematics, engineering, science. Thorough analysis of theorems. Diagrams. Bibliography. Index. 818pp. 5⅜ x 8½. 0-486-64940-7

INTEGRAL EQUATIONS, F. G. Tricomi. Authoritative, well-written treatment of extremely useful mathematical tool with wide applications. Volterra Equations, Fredholm Equations, much more. Advanced undergraduate to graduate level. Exercises. Bibliography. 238pp. 5⅜ x 8½. 0-486-64828-1

FOURIER SERIES, Georgi P. Tolstov. Translated by Richard A. Silverman. A valuable addition to the literature on the subject, moving clearly from subject to subject and theorem to theorem. 107 problems, answers. 336pp. 5⅜ x 8½. 0-486-63317-9

INTRODUCTION TO MATHEMATICAL THINKING, Friedrich Waismann. Examinations of arithmetic, geometry, and theory of integers; rational and natural numbers; complete induction; limit and point of accumulation; remarkable curves; complex and hypercomplex numbers, more. 1959 ed. 27 figures. xii+260pp. 5⅜ x 8½. 0-486-42804-8

THE RADON TRANSFORM AND SOME OF ITS APPLICATIONS, Stanley R. Deans. Of value to mathematicians, physicists, and engineers, this excellent introduction covers both theory and applications, including a rich array of examples and literature. Revised and updated by the author. 1993 edition. 304pp. 6⅛ x 9¼. 0-486-46241-2

CALCULUS OF VARIATIONS, Robert Weinstock. Basic introduction covering isoperimetric problems, theory of elasticity, quantum mechanics, electrostatics, etc. Exercises throughout. 326pp. 5⅜ x 8½. 0-486-63069-2

THE CONTINUUM: A CRITICAL EXAMINATION OF THE FOUNDATION OF ANALYSIS, Hermann Weyl. Classic of 20th-century foundational research deals with the conceptual problem posed by the continuum. 156pp. 5⅜ x 8½. 0-486-67982-9

CHALLENGING MATHEMATICAL PROBLEMS WITH ELEMENTARY SOLUTIONS, A. M. Yaglom and I. M. Yaglom. Over 170 challenging problems on probability theory, combinatorial analysis, points and lines, topology, convex polygons, many other topics. Solutions. Total of 445pp. 5⅜ x 8½. Two-vol. set. Vol. I: 0-486-65536-9 Vol. II: 0-486-65537-7

INTRODUCTION TO PARTIAL DIFFERENTIAL EQUATIONS WITH APPLICATIONS, E. C. Zachmanoglou and Dale W. Thoe. Essentials of partial differential equations applied to common problems in engineering and the physical sciences. Problems and answers. 416pp. 5⅜ x 8½. 0-486-65251-3

STOCHASTIC PROCESSES AND FILTERING THEORY, Andrew H. Jazwinski. This unified treatment presents material previously available only in journals, and in terms accessible to engineering students. Although theory is emphasized, it discusses numerous practical applications as well. 1970 edition. 400pp. 5⅜ x 8½. 0-486-46274-9

Math–Decision Theory, Statistics, Probability

INTRODUCTION TO PROBABILITY, John E. Freund. Featured topics include permutations and factorials, probabilities and odds, frequency interpretation, mathematical expectation, decision-making, postulates of probability, rule of elimination, much more. Exercises with some solutions. Summary. 1973 edition. 247pp. 5⅜ x 8½.
0-486-67549-1

STATISTICAL AND INDUCTIVE PROBABILITIES, Hugues Leblanc. This treatment addresses a decades-old dispute among probability theorists, asserting that both statistical and inductive probabilities may be treated as sentence-theoretic measurements, and that the latter qualify as estimates of the former. 1962 edition. 160pp. 5⅜ x 8½. 0-486-44980-7

APPLIED MULTIVARIATE ANALYSIS: Using Bayesian and Frequentist Methods of Inference, Second Edition, S. James Press. This two-part treatment deals with foundations as well as models and applications. Topics include continuous multivariate distributions; regression and analysis of variance; factor analysis and latent structure analysis; and structuring multivariate populations. 1982 edition. 692pp. 5⅜ x 8½. 0-486-44236-5

LINEAR PROGRAMMING AND ECONOMIC ANALYSIS, Robert Dorfman, Paul A. Samuelson and Robert M. Solow. First comprehensive treatment of linear programming in standard economic analysis. Game theory, modern welfare economics, Leontief input-output, more. 525pp. 5⅜ x 8½. 0-486-65491-5

PROBABILITY: AN INTRODUCTION, Samuel Goldberg. Excellent basic text covers set theory, probability theory for finite sample spaces, binomial theorem, much more. 360 problems. Bibliographies. 322pp. 5⅜ x 8½. 0-486-65252-1

GAMES AND DECISIONS: INTRODUCTION AND CRITICAL SURVEY, R. Duncan Luce and Howard Raiffa. Superb nontechnical introduction to game theory, primarily applied to social sciences. Utility theory, zero-sum games, n-person games, decision-making, much more. Bibliography. 509pp. 5⅜ x 8½. 0-486-65943-7

INTRODUCTION TO THE THEORY OF GAMES, J. C. C. McKinsey. This comprehensive overview of the mathematical theory of games illustrates applications to situations involving conflicts of interest, including economic, social, political, and military contexts. Appropriate for advanced undergraduate and graduate courses; advanced calculus a prerequisite. 1952 ed. x+372pp. 5⅜ x 8½. 0-486-42811-7

FIFTY CHALLENGING PROBLEMS IN PROBABILITY WITH SOLUTIONS, Frederick Mosteller. Remarkable puzzlers, graded in difficulty, illustrate elementary and advanced aspects of probability. Detailed solutions. 88pp. 5⅜ x 8½.
0-486-65355-2

PROBABILITY THEORY: A CONCISE COURSE, Y. A. Rozanov. Highly readable, self-contained introduction covers combination of events, dependent events, Bernoulli trials, etc. 148pp. 5⅜ x 8¼. 0-486-63544-9

THE STATISTICAL ANALYSIS OF EXPERIMENTAL DATA, John Mandel. First half of book presents fundamental mathematical definitions, concepts and facts while remaining half deals with statistics primarily as an interpretive tool. Well-written text, numerous worked examples with step-by-step presentation. Includes 116 tables. 448pp. 5⅜ x 8½. 0-486-64666-1

Math–Geometry and Topology

ELEMENTARY CONCEPTS OF TOPOLOGY, Paul Alexandroff. Elegant, intuitive approach to topology from set-theoretic topology to Betti groups; how concepts of topology are useful in math and physics. 25 figures. 57pp. 5⅜ x 8½. 0-486-60747-X

A LONG WAY FROM EUCLID, Constance Reid. Lively guide by a prominent historian focuses on the role of Euclid's Elements in subsequent mathematical developments. Elementary algebra and plane geometry are sole prerequisites. 80 drawings. 1963 edition. 304pp. 5⅜ x 8½. 0-486-43613-6

EXPERIMENTS IN TOPOLOGY, Stephen Barr. Classic, lively explanation of one of the byways of mathematics. Klein bottles, Moebius strips, projective planes, map coloring, problem of the Koenigsberg bridges, much more, described with clarity and wit. 43 figures. 210pp. 5⅜ x 8½. 0-486-25933-1

THE GEOMETRY OF RENÉ DESCARTES, René Descartes. The great work founded analytical geometry. Original French text, Descartes's own diagrams, together with definitive Smith-Latham translation. 244pp. 5⅜ x 8½. 0-486-60068-8

EUCLIDEAN GEOMETRY AND TRANSFORMATIONS, Clayton W. Dodge. This introduction to Euclidean geometry emphasizes transformations, particularly isometries and similarities. Suitable for undergraduate courses, it includes numerous examples, many with detailed answers. 1972 ed. viii+296pp. 6⅛ x 9¼. 0-486-43476-1

EXCURSIONS IN GEOMETRY, C. Stanley Ogilvy. A straightedge, compass, and a little thought are all that's needed to discover the intellectual excitement of geometry. Harmonic division and Apollonian circles, inversive geometry, hexlet, Golden Section, more. 132 illustrations. 192pp. 5⅜ x 8½. 0-486-26530-7

THE THIRTEEN BOOKS OF EUCLID'S ELEMENTS, translated with introduction and commentary by Sir Thomas L. Heath. Definitive edition. Textual and linguistic notes, mathematical analysis. 2,500 years of critical commentary. Unabridged. 1,414pp. 5⅜ x 8½. Three-vol. set.
Vol. I: 0-486-60088-2 Vol. II: 0-486-60089-0 Vol. III: 0-486-60090-4

SPACE AND GEOMETRY: IN THE LIGHT OF PHYSIOLOGICAL, PSYCHOLOGICAL AND PHYSICAL INQUIRY, Ernst Mach. Three essays by an eminent philosopher and scientist explore the nature, origin, and development of our concepts of space, with a distinctness and precision suitable for undergraduate students and other readers. 1906 ed. vi+148pp. 5⅜ x 8½. 0-486-43909-7

GEOMETRY OF COMPLEX NUMBERS, Hans Schwerdtfeger. Illuminating, widely praised book on analytic geometry of circles, the Moebius transformation, and two-dimensional non-Euclidean geometries. 200pp. 5⅝ x 8¼. 0-486-63830-8

DIFFERENTIAL GEOMETRY, Heinrich W. Guggenheimer. Local differential geometry as an application of advanced calculus and linear algebra. Curvature, transformation groups, surfaces, more. Exercises. 62 figures. 378pp. 5⅜ x 8½. 0-486-63433-7

History of Math

THE WORKS OF ARCHIMEDES, Archimedes (T. L. Heath, ed.). Topics include the famous problems of the ratio of the areas of a cylinder and an inscribed sphere; the measurement of a circle; the properties of conoids, spheroids, and spirals; and the quadrature of the parabola. Informative introduction. clxxxvi+326pp. 5⅜ x 8½.
0-486-42084-1

A SHORT ACCOUNT OF THE HISTORY OF MATHEMATICS, W. W. Rouse Ball. One of clearest, most authoritative surveys from the Egyptians and Phoenicians through 19th-century figures such as Grassman, Galois, Riemann. Fourth edition. 522pp. 5⅜ x 8½. 0-486-20630-0

THE HISTORY OF THE CALCULUS AND ITS CONCEPTUAL DEVELOPMENT, Carl B. Boyer. Origins in antiquity, medieval contributions, work of Newton, Leibniz, rigorous formulation. Treatment is verbal. 346pp. 5⅜ x 8½. 0-486-60509-4

THE HISTORICAL ROOTS OF ELEMENTARY MATHEMATICS, Lucas N. H. Bunt, Phillip S. Jones, and Jack D. Bedient. Fundamental underpinnings of modern arithmetic, algebra, geometry and number systems derived from ancient civilizations. 320pp. 5⅜ x 8½. 0-486-25563-8

THE HISTORY OF THE CALCULUS AND ITS CONCEPTUAL DEVELOPMENT, Carl B. Boyer. Fluent description of the development of both the integral and differential calculus–its early beginnings in antiquity, medieval contributions, and a consideration of Newton and Leibniz. 368pp. 5⅜ x 8½. 0-486-60509-4

GAMES, GODS & GAMBLING: A HISTORY OF PROBABILITY AND STATISTICAL IDEAS, F. N. David. Episodes from the lives of Galileo, Fermat, Pascal, and others illustrate this fascinating account of the roots of mathematics. Features thought-provoking references to classics, archaeology, biography, poetry. 1962 edition. 304pp. 5⅜ x 8½. (Available in U.S. only.) 0-486-40023-9

OF MEN AND NUMBERS: THE STORY OF THE GREAT MATHEMATICIANS, Jane Muir. Fascinating accounts of the lives and accomplishments of history's greatest mathematical minds–Pythagoras, Descartes, Euler, Pascal, Cantor, many more. Anecdotal, illuminating. 30 diagrams. Bibliography. 256pp. 5⅜ x 8½.
0-486-28973-7

HISTORY OF MATHEMATICS, David E. Smith. Nontechnical survey from ancient Greece and Orient to late 19th century; evolution of arithmetic, geometry, trigonometry, calculating devices, algebra, the calculus. 362 illustrations. 1,355pp. 5⅜ x 8½. Two-vol. set. Vol. I: 0-486-20429-4 Vol. II: 0-486-20430-8

A CONCISE HISTORY OF MATHEMATICS, Dirk J. Struik. The best brief history of mathematics. Stresses origins and covers every major figure from ancient Near East to 19th century. 41 illustrations. 195pp. 5⅜ x 8½. 0-486-60255-9

Physics

OPTICAL RESONANCE AND TWO-LEVEL ATOMS, L. Allen and J. H. Eberly. Clear, comprehensive introduction to basic principles behind all quantum optical resonance phenomena. 53 illustrations. Preface. Index. 256pp. $5\frac{3}{8} \times 8\frac{1}{2}$. 0-486-65533-4

QUANTUM THEORY, David Bohm. This advanced undergraduate-level text presents the quantum theory in terms of qualitative and imaginative concepts, followed by specific applications worked out in mathematical detail. Preface. Index. 655pp. $5\frac{3}{8} \times 8\frac{1}{2}$. 0-486-65969-0

ATOMIC PHYSICS (8th EDITION), Max Born. Nobel laureate's lucid treatment of kinetic theory of gases, elementary particles, nuclear atom, wave-corpuscles, atomic structure and spectral lines, much more. Over 40 appendices, bibliography. 495pp. $5\frac{3}{8} \times 8\frac{1}{2}$. 0-486-65984-4

A SOPHISTICATE'S PRIMER OF RELATIVITY, P. W. Bridgman. Geared toward readers already acquainted with special relativity, this book transcends the view of theory as a working tool to answer natural questions: What is a frame of reference? What is a "law of nature"? What is the role of the "observer"? Extensive treatment, written in terms accessible to those without a scientific background. 1983 ed. xlviii+172pp. $5\frac{3}{8} \times 8\frac{1}{2}$. 0-486-42549-5

AN INTRODUCTION TO HAMILTONIAN OPTICS, H. A. Buchdahl. Detailed account of the Hamiltonian treatment of aberration theory in geometrical optics. Many classes of optical systems defined in terms of the symmetries they possess. Problems with detailed solutions. 1970 edition. xv + 360pp. $5\frac{3}{8} \times 8\frac{1}{2}$. 0-486-67597-1

PRIMER OF QUANTUM MECHANICS, Marvin Chester. Introductory text examines the classical quantum bead on a track: its state and representations; operator eigenvalues; harmonic oscillator and bound bead in a symmetric force field; and bead in a spherical shell. Other topics include spin, matrices, and the structure of quantum mechanics; the simplest atom; indistinguishable particles; and stationary-state perturbation theory. 1992 ed. xiv+314pp. $6\frac{1}{8} \times 9\frac{1}{4}$. 0-486-42878-8

LECTURES ON QUANTUM MECHANICS, Paul A. M. Dirac. Four concise, brilliant lectures on mathematical methods in quantum mechanics from Nobel Prize-winning quantum pioneer build on idea of visualizing quantum theory through the use of classical mechanics. 96pp. $5\frac{3}{8} \times 8\frac{1}{2}$. 0-486-41713-1

THIRTY YEARS THAT SHOOK PHYSICS: THE STORY OF QUANTUM THEORY, George Gamow. Lucid, accessible introduction to influential theory of energy and matter. Careful explanations of Dirac's anti-particles, Bohr's model of the atom, much more. 12 plates. Numerous drawings. 240pp. $5\frac{3}{8} \times 8\frac{1}{2}$. 0-486-24895-X

ELECTRONIC STRUCTURE AND THE PROPERTIES OF SOLIDS: THE PHYSICS OF THE CHEMICAL BOND, Walter A. Harrison. Innovative text offers basic understanding of the electronic structure of covalent and ionic solids, simple metals, transition metals and their compounds. Problems. 1980 edition. 582pp. $6\frac{1}{8} \times 9\frac{1}{4}$. 0-486-66021-4

CATALOG OF DOVER BOOKS